NATURE'S GHOSTS

CONFRONTING EXTINCTION

NATURE'S

FROM THE AGE OF JEFFERSON

GHOSTS

TO THE AGE OF ECOLOGY

MARK V. BARROW, JR.

The University of Chicago Press
CHICAGO AND LONDON

MARK V. BARROW, JR.,
is associate professor of history at Virginia Tech and the author of
A Passion for Birds: Ornithology after Audubon.

The University of Chicago Press, Chicago 60637
The University of Chicago Press, Ltd., London
© 2009 by The University of Chicago
All rights reserved. Published 2009
Printed in the United States of America

17 16 15 14 13 12 11 10 09   1 2 3 4 5

ISBN-13: 978-0-226-03814-8 (cloth)
ISBN-10: 0-226-03814-9 (cloth)

Library of Congress Cataloging-in-Publication Data

Barrow, Mark V., 1960–
Nature's ghosts : confronting extinction from the age of Jefferson
to the age of ecology / by Mark V. Barrow, Jr.
p. cm.
Includes bibliographical references and index.
ISBN-13: 978-0-226-03814-8 (cloth: alk. paper)
ISBN-10: 0-226-03814-9 (cloth: alk. paper) 1. Wildlife conservation—
United States—History. 2. Endangered species—Law and legislation—
United States—History. 3. Extinction (Biology) I. Title.
QL84.2.B377 2009
333.95'220973—dc22
2008049085

TO MY PARENTS, MARY B. BARROW AND MARK V. BARROW, SR.,
who not only ignited my curiosity about the natural world but also
taught me how endlessly fascinating the past can be.

The greatest tragedy in nature is the extinction of species.

ARTHUR A. ALLEN, 1942

CONTENTS

PREFACE

The idea for this book came to me almost two decades ago, while pursuing research at one of the world's great museums: the American Museum of Natural History in New York. One muggy August afternoon in 1990, I was slogging my way through the twenty-odd file drawers of historic correspondence in the museum's Department of Ornithology when I came upon a curious manila envelope with "Chapman on Parakeet" scrawled across its front. Intriguing, I thought to myself, for I already knew that Frank Michler Chapman had collected the now-extinct Carolina parakeet in the late 1880s, not long after abandoning a promising banking career to become a lowly museum assistant. I also knew that he would spend the next half-century at the American Museum, where he would earn a well-deserved reputation as one of the nation's leading ornithologists and most celebrated wildlife conservationists.

My curiosity piqued, I paused to investigate further. Inside the tattered envelope I found the field notebook Chapman had kept during his expedition to Florida in March 1889, a precious document that had long been lost. Across its yellowed pages, Chapman expressed not only his delight at securing no less than fifteen specimens of this exceedingly rare and magnificent creature, the only parrot native to the eastern United States, but also occasional pangs of guilt for what he had done. For all he knew at the time, he had shot every single remaining example of the beleaguered bird over the course of his brief visit. What had led Chapman to do such a thing? How did he reconcile his actions with his increasing involvement in the American bird protection movement that had been stirring at precisely that same moment in history? And what role did scientists like Chapman play in that movement?

I offered answers to those initial questions in my first book, A *Passion for Birds*. But inspired by the discovery I had made that sultry afternoon, I also determined to set out on a new project that would focus on how naturalists had grappled with the issue of wildlife extinction over a much wider swath of time. Over the next eighteen years, I would consult more than a dozen archives and read countless articles and books in at attempt to learn more about how and why naturalists had become haunted by the specter of extinction and how changing practices within the field of natural history shaped their deepening engagement with that issue.

I soon found that even though they may have been quite willing to collect endangered species in the name of science (just as Chapman had done in Florida), naturalists have also proven absolutely essential in identifying extinction as a problem, while simultaneously promoting policies that sought to reverse the wildlife decline they witnessed going on around them. I also found that the taxonomic tradition that had long characterized natural history—the ongoing effort to produce an authoritative inventory of the world's flora and fauna—proved central in efforts to create the first endangered species lists. Perhaps most surprisingly, I discovered that naturalists had been engaged with the issue of extinction for more than a century prior to Chapman's parakeet expedition. This book chronicles that extensive involvement, beginning in the late eighteenth century, when virtually all naturalists resisted the notion of extinction of any sort, until 1973, when concerned scientists, conservationists, and government officials rode the wave of the modern environmental movement to secure passage of the Endangered Species Act, landmark legislation that aimed to stem the tide of species loss that was sweeping the globe. That broad, prohibitive, and, as it would later turn out, controversial law remains with us to this day.

While the rediscovery of Chapman's field notebook provided the immediate impetus for this project, earlier life experiences also shaped it in important ways. In addition to being a talented physician, my father is something of an amateur naturalist himself. Over the years, he has taken a keen interest (sometimes bordering on obsession), in photographing wildflowers, collecting marine shells, cultivating native plants, and watching birds. When we were growing up, he and Mom would regularly load all five of us kids in the family station wagon to take us camping, hiking, and canoeing across the state of Florida, where there was still much natural beauty waiting to be discovered and enjoyed. We would complain, however, usually quite vocally, when in the middle of one of those trips Dad would suddenly veer off the highway to engage in what seemed like an endless photography session involving some obscure roadside weed and his latest piece of camera equipment. But even as we grumbled, slowly but surely we grew to cherish the native organisms to which we had been regularly exposed as children. Simple contact with the natural world alone is surely not sufficient to foster concern about the fate of the planet, but as Richard Louv has recently argued in *Last Child in the Woods,* it seems absolutely necessary. And given the proper cultural milieu, that contact can blossom not only into concern about the fate of species but also genuine love and respect, just as it did for the many naturalists chronicled in this book. In hindsight, given the arc of my own personal life journey, perhaps I should not have been surprised that naturalists have long cared so much about the future of the organisms they studied.

During the course of researching and writing this book, I have incurred numerous debts. A National Endowment for the Humanities Fellowship provided invaluable release time to pursue this project, as did two sabbatical semesters and a Dean's Faculty Research Grant from the College of Liberal Arts and Human Sciences at Virginia Tech. A pair of Humanities Summer Stipends, a Dean's Millennium Grant, and yearly departmental travel grants provided the funds for numerous research visits. I am grateful for the generous support, without which this book would not have been possible.

My experiences with the staff at the many libraries and archives I consulted have been uniformly positive. I appreciate the help, especially from those who went beyond the call of duty, including Mary LeCroy, of the Department of Ornithology at the American Museum of Natural History; Stephen Johnson, of the World Conservation Society; and Cheryl Oakes, of the Forest History Society. An exceedingly long list of individuals helped secure copies of and permissions to use the many illustrations contained in this book, for which I am grateful, but I especially need to thank Charlie Craighead, Pamela Wright Lloyd, and Nancy Tanner for sharing family photos. I also appreciate the conscientious work of the interlibrary loan staff at Virginia Tech's Newman Library.

Numerous friends, colleagues, and associates have provided all manner of help along the way. I would especially like to thank Peter Alagona, Ben Cohen, Gregory Cushman, Fritz Davis, Kurk Dorsey, Tom Dunlap, Paul Farber, Bill FitzPatrick, Kevin Francis, Betsy Hanson, Pamela Henson, Bruce Hull, Lloyd Kiff, Ed Larson, Keri Lewis, Ralph Lutts, Helen Macdonald, Mark Madison, Donna Mehos, Lisa Mighetto, Gregg Mitman, Amy Nelson, Megan Nelson, John Nielsen, Duncan Porter, Ron Rainger, (the late) Fritz Rehbock, Nigel Rothfels, Martin Rudwick, Mikko Saikku, Noel Snyder, David Tomblin, and Karen Wonders. I greatly appreciate the assistance and sincerely apologize to anyone I have missed. I should also admit that I have not always followed the sage advice my generous colleagues have provided, so I must assume responsibility for any errors that remain.

The staff at the University of Chicago Press have done a marvelous job of shepherding this book through the publication process. I especially need to thank Mark Reschke for his sharp eye and sound judgment on all matters concerning language and style, Levi Stahl for his outstanding publicity work, Maia Wright for her exquisite eye for design, and Abby Collier for her help on a variety of tasks, both large and small. Most of all, I am grateful to have had the opportunity to work with Christie Henry, a masterful editor whose thoughtful feedback, unflagging encouragement, and generous support have made all the difference in the world.

Finally, I want to thank my family, which has been incredibly supportive throughout the long process of creating this book. I feel truly blessed to have parents, siblings, and extended family members who care so deeply for each other. My wife, Marcia, has been an extraordinarily warm, generous, and loving soul mate for more than three decades now; I can't wait for the next three. My children, Mark, Alex, Hannah, and Lizzie, have been a constant source of pride and joy, and I regret that I have not been as diligent in introducing them to the wonders of nature as my parents were for me. I especially need to thank Mom and Dad for all the love and support they have so freely shared over the years. As a small token of my profound appreciation, I dedicate this book to them.

INTRODUCTION

*Empathy for living things comes from many years of observing them
in their natural environments, which is why field biologists have always been
among the most adamant defenders of wild Nature.*

REED F. NOSS, 1996

With surprisingly little debate and by unusually broad margins, in 1973 Congress approved one of the most sweeping environmental initiatives in American history: the Endangered Species Act.[1] The goal of this ambitious new law was to stem the rising tide of extinction that had swept away more than a thousand plants and animals since the beginning of the seventeenth century. According to the committee report accompanying the House version of the bill, countless organisms had "appeared, changed, and disappeared" during the more than four billion years since the earth had begun to support life. In recent decades, however, the pace of extinction had been "accelerating" at an alarming rate. The culprit in this unprecedented, tragic, and largely preventable biotic loss was humans: "As we homogenize the habitats in which these plants and animals evolved, and as we increase the pressure for products that they are in a position to supply . . . we threaten their—and our own—genetic heritage."[2]

Despite its comprehensive scope and strict prohibitions on harming species facing extinction, the Endangered Species Act attracted strong bipartisan support. The four versions of the bill introduced into Congress in early 1973 generated little discussion, even less controversy, and the Senate proposal breezed through in July by a unanimous 92–0 vote.[3] Two months later, after Rep. John Dingell reported that he had "yet to hear a whisper of opposition" to similar legislation he had introduced in the House, Dingell's bill passed by an equally impressive 390–12 margin.[4] In a statement issued after signing the new law on

December 28, 1973, President Nixon also declared unwavering support for the far-reaching measure.[5]

It is tempting to view the Endangered Species Act simply as a product of its times. The post–World War II period, particularly the late 1960s and 1970s, witnessed a remarkable environmental awakening in the United States and the West more broadly.[6] Higher levels of affluence, improved access to higher education, strong economic growth, continued technological development, growing media attention, and the diffusion of ecological concepts granted environmental issues unprecedented traction. Increasingly concerned not only about the myriad threats facing the natural world but also their potential impact on human health, Americans expressed their anxiety about the state of the environment in opinion polls, joined environmental groups in droves, and even took to the streets in protest. On April 22, 1970, more than 20 million participants celebrated the first Earth Day, a nationwide event billed as the "largest organized demonstration in human history."[7] Eager to appease their worried constituents, politicians responded with an outpouring of environmental legislation. Within the span of a decade, Congress passed the Wilderness Act (1964), the National Environmental Policy Act (1969), the Clean Air Act (1970), the Water Pollution Control Act (1972), the Marine Mammal Protection Act (1972), and no less than three versions of the Endangered Species Act (1966, 1969, and 1973). That expansive legal framework remains largely in place to this day.

While the Endangered Species Act undoubtedly rode the crest of the postwar environmental movement, it would be a mistake to see the measure solely as a consequence of that movement. History is characterized not only by change over time but also by continuity, and to properly understand the origins of this extraordinary legislation, it is important to recognize that the Endangered Species Act emerged from a long-standing dialogue about the issue of human-induced extinction. One logical starting point for exploring that extended discussion (and the one I have chosen for this book) is the eighteenth century, when a handful of commentators began to suspect that prehistoric humans might have wiped out some of the strange beasts whose fossil remains were being dug up at numerous locations around the world. The authors of these early speculations remained a distinct minority, however, for most educated Westerners at the time firmly rejected the idea that any kind of extinction—whether human-caused or otherwise—could take place. For them, the loss of any creature violated deeply held views about the overall stability and perfection of the natural world.

The meticulous, widely publicized studies of living and extinct elephants that the French naturalist Georges Cuvier completed at the end of the eighteenth and the beginning of nineteenth centuries erased all doubts about the possibility of extinction. Soon thereafter, a series of pioneering reports documenting the loss of more contemporary island fauna—like the dodo, the moa, and the auk—offered additional evidence supporting the reality of extinction. Later in the nineteenth century, the precipitous decline of the bison, the passenger pigeon, and other once abundant species proved beyond a shadow of doubt that humans could even wipe out organisms with vast continent-wide distributions. Yet, as the power of humanity to extirpate plants and animals became increasingly apparent, so did efforts to rescue declining species from the fate of oblivion.

Naturalists—individuals who pursued natural history either as hobby or a profession—not only proved central to the discovery of extinction but were also among the earliest groups to condemn the careless destruction of plants and wildlife going on around them. Although it has deep roots in Western society, the practice of natural history has experienced significant changes over the ages.[8] The thirty-seven books of Pliny the Elder's *Natural History* (ca. AD 77), which touched on a dizzying array of topics ranging from anthropology to zoology, illustrate the once expansive breadth of the enterprise. During the early modern period, Pliny's successors gradually narrowed the scope of natural history to encompass nonhuman natural objects—principally, animals, plants, and minerals—as well as natural phenomena such as the weather and tides. Accompanying this restriction in scope was a shift from highly stylized to more naturalistic illustrations and a corresponding elimination of the folklore, fables, astrological references, religious symbolism, and imaginary creatures that had once graced the pages of bestiaries, herbals, and other early natural history publications.

Between the sixteenth and the nineteenth centuries, a more popular interest in natural history also took root and blossomed. First in Europe and later in America, the well-to-do began cultivating the practice by subscribing to lavishly illustrated publications, amassing cabinets of curiosity, planting pleasure gardens, stocking menageries, and sponsoring collectors who journeyed to the far corners of the earth. The first natural history museums and botanical gardens, established in Renaissance Europe, expanded greatly in size and number following the discovery of the New World.[9] In a practice that would later be dubbed "bioprospecting," governments, corporations, and wealthy individuals scrambled to exploit the commercial potential of exotic biota discovered during overseas explorations.[10] And natural history enthusiasts of all stripes banded together to form clubs that organized field trips, established museums, and issued publications.[11] An aesthetic interest in nature, a conviction that the natural world offered a window onto the mind of God, a belief in the didactic value of nature study, and a desire to exploit the commercial potential of natural resources undergirded the continued expansion of popular interest in natural history.

During the eighteenth and nineteenth centuries, natural history not only became increasingly fashionable, but its structure and practice also transformed in two fundamental ways.[12] First, it gradually splintered into a series of more specialized disciplines, each with its own particular subject matter, agenda, and techniques. This process of discipline formation began with botany, zoology, and geology, which, in turn, spawned even more specialized fields. Individuals who focused on the study of birds, insects, and mosses became known as ornithologists, entomologists, and bryologists, to cite just a few of the many emerging specialties. By the second half of the nineteenth century, practitioners in these new fields began organizing their own scientific societies, publishing in narrowly focused technical journals, and issuing esoteric monographs.[13] A few of the luckier ones also secured professional positions related to their area of expertise, working for museums, governmental agencies, and universities at a time when such employment opportunities remained exceedingly rare.

Second, and closely related to this first transformation, the practitioners in these emerging specialities increasingly focused on taxonomy, the naming, description, and classification of living organisms. At one time, naturalists considered virtually any nugget of information about a species as not only valuable but also relevant to their studies. During the late eighteenth and the nineteenth centuries, however, they increasingly honed in on the external morphology—that is, the form, shape, and size—of organisms, while largely ignoring their life history, behavior, and relationship with their environment. This narrowing gaze coincided with a narrowing research agenda. The central organizing pursuit for most natural history fields became the creation of an exhaustive, authoritative, and rationally ordered inventory of the particular group on which they specialized. In *Systema naturae* (1735), one starting point for modern biological taxonomy, the Swedish botanist Carl Linnaeus declared that "classification and name-giving will be the foundation of our science."[14] But where Linnaeus remained a generalist who sought to construct systems of classification and nomenclature that were not only broadly applicable to all flora and fauna but also relatively easy for novices to use, his successors produced increasingly unwieldy classification schemes as well as publications that were narrow, technical, and generally inaccessible.

The twentieth century presented both opportunities and threats to the natural history tradition. Popular natural history continued to experience robust growth. Books, magazines, films, museums, zoos, (and later) television featured widely accessible examples of nature's diversity that could simultaneously entertain and educate the public while cultivating a sense of connection with the natural world.[15] At the same time, more and more Americans regularly pursued outdoor recreation opportunities that offered firsthand contact with nature. Birdwatching, for example, which had begun on a modest scale at the end of the nineteenth century, took flight in the middle decades of the twentieth century, following the appearance of the first edition of Roger Tory Peterson's field guide to birds in 1934.[16] As a result of the widespread popularity of natural history, the term came to connote a popular, superficial study of animals and plants.[17]

Academic natural history also experienced transformation and growth in the century leading up to the Endangered Species Act of 1973. With the advent of graduate education in the late nineteenth and early twentieth centuries, self-training and apprenticeship, once the typical routes to expertise in natural history, gave way to more formal, systematic, and broadly based instruction.[18] Even as formal educational opportunities in universities opened up, though, the rise of laboratory biology threatened to eclipse natural history, stealing away its funding, prestige, and most promising students.[19] Responding to this threat and the narrow taxonomic focus that had long gripped the enterprise, naturalists created new disciplines—like ecology and ethology—that broadened the scope of natural history once again while incorporating both observational and experimental research techniques.[20] In a move that signaled both a connection with and a distancing from tradition, in 1927 the British naturalist Charles Elton began referring to ecology, a field he was helping to create, as "scientific natural history."[21] Despite these changes, by the end of the twentieth century, some naturalists openly worried about what one set of prominent authors decried as the "Impending Extinction of Natural History." Traditional naturalists, whose work was characterized by "the close observation of organisms—their origins, their evolution, their behavior, and their relationships with other species," found themselves increasingly "pushed to the margins of academe," lacking in adequate funding and falling precipitously in the "academic pecking order."[22]

Whatever the status of natural history, the activities of collecting, cataloging, and classifying the earth's diversity placed naturalists in a unique position to appreciate the decline and extinction of species. In the process of locating, observing, and gathering specimens in the field, they not only discovered much about changes in the abundance of those species but also about transformations in their habitat, the appearance of diseases, the introduction of exotic species, increases in the rate of commercial exploitation, and other potential threats to their survival in the wild. Even if naturalists lacked firsthand experience with a given species in the field, they enjoyed access to several other sources of data about its population status. Vast far-flung networks of collectors had long provided the steady stream of specimens naturalists needed to build museum collections and complete taxonomic studies. When properly labeled with the date and site of capture, those same specimens could be used to track the contraction of a given species' range, a signal that it faced imminent danger of extinction. Methodical examination of published local lists of organisms, long a favorite genre in natural history, offered additional clues about changes in the population status and range of species.

As naturalists struggled to catalog the earth's biodiversity, they encountered numerous organisms that seemed to be either teetering on the brink of extinction or lost entirely. Until the middle of the nineteenth century, they tended to respond to this discovery with some degree of resignation. After all, the fossil record revealed myriad species that had been lost to extinction in the past, a fate that naturalists attributed to a variety of causes: climate change, natural disasters, disease, competition, and species senescence (the belief that species eventually died out, just as individuals did). The contemporary declines that they were witnessing were not really that much different from these earlier extinctions, they initially surmised, even if humans now seemed responsible. But as examples of human-caused extinction multiplied in the second half of the nineteenth century, naturalists grew increasingly haunted by the specter of extinction.

They acted on that deepening concern in a variety of ways. On the most basic level, they publicized examples of past extinctions while expressing alarm about impending losses. This is exactly what a young curator at Harvard's Museum of Comparative Zoology, the mammalogist and ornithologist Joel A. Allen, did in a pioneering set of articles published in 1876. While many Americans were celebrating the economic and material progress their nation had enjoyed in the century since its founding, Allen lamented a by-product of that progress: the imminent passing of the bison. He also decried the decline or loss of numerous other North American birds and mammals since Europeans first set foot in New World. In essence, what Allen did was extend the methods and practices of natural history to lost and vanishing species. But rather than fashion a catalog of existing species, as he and his colleagues had traditionally done, he documented their decline. Through fieldwork, the examination of museum specimens, and the scrutiny of published accounts of naturalists, explorers, and travelers, he crafted one of the first (albeit limited) inventories of endangered species. Over the next century, such inventories would not only become increasingly common but also more comprehensive before culminating in the official lists that International Union for the Conservation of Nature began maintaining in 1963 and the U.S. Fish and Wildlife Service's Committee on Rare and Endangered Wildlife Species initiated a year later.

But why should anyone care if the vanishing species recorded in these inventories were to be lost? To answer this most basic question, naturalists developed and deployed seven basic arguments for species preservation during the two centuries leading up to the Endangered Species Act of 1973. First was the aesthetic beauty of plants and animals. The gradual diffusion of romanticism in the eighteenth and nineteenth centuries promoted thinking about the natural world in deeply aesthetic terms.[23] Writing in his journal in 1856, for example, the American author, transcendentalist, and naturalist Henry David Thoreau lamented the fact that many of the "nobler animals" had been exterminated in the area surrounding Concord. He then compared the biologically impoverished countryside surrounding his home to a vandalized book of poetry: "I take infinite pains to know all the phenomenon of spring . . . thinking that I have here the entire poem, and then, to my chagrin, I hear that it is but an imperfect copy that I possess and have read, that my ancestors have torn out many of the finest leaves and grandest passages, and mutilated its page."[24] Fifty years later, William Beebe, the colorful deep-sea explorer and curator of birds at the New York Zoological Society, used a similar metaphor when he linked living organisms with great works of fine art. If an example of the latter were destroyed, it might be reconstructed, but "when the last individual of a race of living beings breathes no more, another heaven and another earth must pass before such a one can be again."[25]

The second argument for preserving endangered species involved more crassly utilitarian concerns: plants and animals provide a variety of useful and economically valuable resources, such as food, drugs, leather, fiber, and oils. On one level, this argument simply stated the obvious. Since first landing on American shores, Europeans had tended to view the New World as a vast storehouse of "merchantable commodities," a view that became both pervasive and deeply engrained.[26] Indeed, by the end of the nineteenth century, commercial exploitation of wildlife was so extensive that it posed increasingly obvious threats to numerous species. In *Our Vanishing Wildlife* (1913), for example, the outspoken naturalist, conservationist, and zoo administrator William Temple Hornaday cautioned that the fate of the once ubiquitous passenger pigeon offered a sobering lesson about the consequences of unrestrained wildlife commodification: "Any wild bird or mammal species can be exterminated by commercial interests in twenty years time, or less."[27] A decade earlier, the botanist and paleontologist Frank Hall Knowlton made a similar point we he warned that if the monetary value of rare plants were publicized as part of a campaign to save them, unscrupulous dealers and collectors would eagerly descend upon them and "their doom would be fixed."[28] So when early conservationists spoke of the value of plants and wildlife, they often framed their discussion in terms of the *services* rather than the *products* that those species provided. One example they often invoked involved songbirds, which supposedly boosted agricultural production by destroying pest insects, weeds, and rodents.[29]

After securing a series of state and federal laws that severely restricted commercial exploitation of most native game species, conservationists felt more comfortable touting the potentially marketable products that wild species provided. While writing in support of nature protection in Latin America on the eve of America's involvement in World War II, for example, the American mammalogist and international conservationist Harold J. Coolidge asserted unapologetically that wild flora and fauna represented "an immeasurable capital, a capital that ought not to be ruthlessly spent, but rather stewarded with care."[30] By the second half of the twentieth century, economic arguments loomed increasingly large in support of endangered species preservation.[31]

Yet, according to the twentieth-century naturalist and wildlife manager Aldo Leopold, relying on an economic justification for species preservation proved problematic because "most members of the land community have no economic value."[32] Each and every organism, however, does play a crucial role in maintaining the stability of the natural world, naturalists asserted in the third argument for species preservation. As early as the mid-eighteenth century, Linnaeus began developing protoecological arguments about the interrelationships between individual organisms within a given biological community. By the early twentieth century, the emerging science of ecology more deeply illuminated the complex, multifaceted connections within what would become known as "ecosystems" by the post–World War II period. In a series of publications issued in the middle of that century, Leopold warned that wildlife management not only needed to become more ecologically informed but also to make species preservation its top priority. After all, he argued, "To keep every cog and wheel is the first precaution of intelligent tinkering."[33]

Complementing the ecological argument for species preservation was the evolutionary argument, the fourth justification for attempting to prevent extinction. The plants and animals that surround us are of "great antiquity," the paleontologist, museum administrator, and Columbia University professor Henry Fairfield Osborn pointed out repeatedly at the dawn of the twentieth century.[34] Rapidly diminishing creatures like the bison, deer, wapiti, and pronghorn antelope "were not made in a day, nor in a thousand years, nor in a million years." Rather, they gradually emerged from the "ceaseless trials of nature," over vast eons of time. Naturalists understood and appropriately respected the fact that every extant organism serves as a "living monument of adaptation and beauty, which connects the past with the present," while the public already cherished ancient *cultural* monuments, like the Parthenon and Westminster Abbey. Now it was time for scientists to cultivate a broader "veneration of age sentiment" for living creatures too, before they were lost to extinction.

In addition to being the products of evolutionary change over vast spans of time, naturalists argued, native plants and animals also played a central role in shaping American identity. During the late eighteenth and early nineteenth centuries, refined North Americans embraced the notion that much of their distinctiveness as a people stemmed from the abundant natural world that surrounded them. The New World might lack the cultural achievements of the Old, but it did possess unique, wild features that were largely absent in long-settled Europe. Following independence, a fervent nationalism not only became a key motivation for studying natural history in the United States but also informed early calls for nature preservation.[35] As industrialization and urbanization increasingly took hold in the second half of the nineteenth century, commentators worried openly not only about the prospects of America's wild heritage but also about the plight of its citizens who would be denied access to that rapidly vanishing legacy.

The pursuit of science offered a sixth rationale for protecting vanishing species and their habitats. In the 1840s, the British naturalist Hugh Strickland warned that civilization's inexorable march forced the disappearance of numerous organisms across the globe. One result, he argued, was that "the Zoologist or Botanist of future ages will have a much narrower field for his researches than that which we enjoy at present."[36] Strickland urged his fellow naturalists to redouble their efforts to catalog the earth's flora and fauna before it fell victim to human-initiated destruction. Over the next century, however, his successors refashioned the looming threat of biological extinction into a call for nature preservation to ensure future opportunities for the scientific study of living organisms in the field. Writing in the early 1920s, for example, the American naturalist Francis B. Sumner called on his colleagues to rescue "a few fragments of vanishing nature" to guarantee the possibility of future field studies.[37] Here Sumner had in mind not only the perpetuation of the individual species essential to the pursuit of taxonomically focused natural history, but also of assemblages of organisms indispensable to the emerging field of ecology. One of Sumner's colleagues, Charles C. Adams, argued that it was vital for ecologists to have continued access to relatively undisturbed natural areas to provide what he called a "bionomic baseline" for ecological study, an argument that would be raised repeatedly over the ensuing decades.[38]

Naturalists also developed explicitly ethical arguments for species preservation. Near the end of his famous thousand-mile walk from Wisconsin to the Gulf of Mexico in 1867, the American nature writer and amateur geologist John Muir chided humanity for failing to show appropriate regard for fearsome beasts like the alligator: "How narrow we selfish, conceited creatures are in our sympathies! How blind to the rights of the rest of creation!"[39] This was not the first time anyone had suggested the possibility of extending the notion of rights to nonhuman beings; indeed, humanitarians had begun making similar arguments for domesticated species as early as the late eighteenth century. But according to the historian Roderick Nash, Muir's insight represented the first "association of rights" with the natural world more broadly and wildlife specifically.[40] Nearly a century later, Leopold drew extensively from the science of ecology to develop his notion of "the land ethic." Leopold's bold new vision called for an individual to begin viewing him- or herself as a "plain member and citizen" of the land community rather than "conqueror" of it, as modern humans had typically done. That ethical stance, in turn, implied "respect for his fellow-members, and also respect for the community as such."[41] It also supported the idea that all species have a "biotic right" to exist.[42] What Muir, Leopold, and their later colleagues were stumbling toward here was a notion of the intrinsic value of species, a value independent of human needs and concerns.

Naturalists also developed a second set of ethical arguments involving a notion of responsibility to future human generations. Heedlessly destroying wild species rendered them permanently unavailable to fulfill the many roles they had traditionally played both in nature and human society. Or as Hornaday put it, Americans had a solemn duty to ensure that future citizens would continue to enjoy the many values associated with wildlife: "The wild things of this earth are *not* ours, to do with as we please. They have been given to us *in trust,* and we must account for them to the generations which will come after us and audit our accounts."[43]

Beyond the aesthetic, economic, ecological, evolutionary, cultural, scientific, and ethical arguments they articulated to the public, naturalists were also motivated to act on behalf of endangered species from their emotional attachment to and identification with many of those species. Indeed, those links, typically forged during youthful forays into natural history, often played a decisive role in their decision to take up formal study of those species in the first place.[44] As the ornithologist Frank M. Chapman once claimed, birds "have not only a beauty which appeals to the eye, but often a voice whose message stirs emotions to be reached only through the ear. . . . They further possess humanlike attributes which go deeper still, arousing in us feelings which are akin to those we entertain toward our fellow-beings."[45] Chapman, who was more explicit than most of his colleagues in discussing the emotional bond he and other naturalists forged with wild creatures, even entitled his 1933 memoirs *Autobiography of a Bird-Lover.* Six decades later, the founder of conservation biology Michael Soulé declared that "most biologists *love* plants or animals—they love different ones. Some like lizards, some like grasses. But there's a certain affinity we have and even identification we have with the objects of our study. So it's hard for me to imagine why a person would not want to protect the diversity of those entities in the group he or she is interested in."[46]

Recognizing that generating publicity and cogent arguments alone would not save rapidly declining species, naturalists pursued a variety of other strategies as well. Beginning in the 1880s, they not only established conservation committees within newly emerging scientific societies but also wildlife protection organizations open to members of the public. Chapman, for example, played a key leadership role in the Audubon bird protection movement, which, from the end of the nineteenth century, relied on the tripartite strategy of education, legislation, and enforcement to snatch numerous threatened birds from the jaws of extinction. He also began publication of *Bird-Lore*, a popular journal of ornithology that served as the official organ of state Audubon Societies and (later) the National Association of Audubon Societies, established in 1905.

Other naturalists used their scientific know-how and political savvy to benefit endangered wildlife more directly. Working under the auspices of the New York Zoological Society and the American Bison Society, William Temple Hornaday undertook captive breeding experiments with the American bison, and he used the resulting progeny to restock newly established federal wildlife reserves in the West. Soon, captive breeding joined public education, protective legislation, habitat manipulation, predator control, and refuge creation as standard techniques in the wildlife conservationists' toolbox. Hornaday and later naturalists also successfully lobbied for a long series of wildlife protection laws and treaties that culminated in the Endangered Species Act of 1973, the first law of its kind anywhere in the world.

That is not to suggest that all naturalists firmly embraced the idea of political engagement with the issue of extinction. In 1902, when the American ornithologist Charles Cory was being recruited to join a local Audubon Society, as many of his colleagues had already done, he reportedly responded: "I don't protect birds. I kill them!"[47] Cory's reply may have been offered tongue-in-cheek, but many of the naturalists so busily engaged in cataloging the earth's biota openly challenged efforts to restrict their collecting activities. They resisted not only by vocally defending their right to collect but also by consistently (and probably justifiably) arguing that, relative to other threats, their activities resulted in little overall impact on most plant and animal populations. They also offered more active resistance, by continuing to seek out rare and vanishing species, even in defiance of local, state, and federal law. For well into the twentieth century, for example, long after the range of the once abundant heath hen had been reduced to a single island off the coast of Massachusetts and serious concerns about the species' future had begun to be broadly voiced, America's preeminent ornithologist, William Brewster, continued to pursue multiple specimens of the prized rare bird for his extensive personal collection. He did so while serving as president of the Massachusetts Audubon Society, an early, particularly active bird protection organization. Even the stalwart Frank Chapman collected multiple specimens of the critically endangered Carolina parakeet during expeditions to Florida in the 1890s and 1900s.[48] Taken as whole, though, the remarkable thing about naturalists' growing involvement with the issue of wildlife extinction is not that it provoked tensions, contradictions, and occasional opposition, but that the level of engagement was so prolonged and so deep.

While my focus is on extinction-related ideas, policies, and institutions promoted by American naturalists, I have also tried to show how they participated in a longstanding (though sometimes tenuous) transnational conversation.[49] European naturalists loom particularly large in the first two chapters, which explore the discovery of extinction and its wide acceptance as a recurring phenomenon by the first half of the nineteenth century, but they also make occasional appearances at other suitable points in the narrative. My decision to concentrate on the ideas and practices of American naturalists is based on several considerations: the fact that U.S. initiatives often played a formative role on the world conservation stage (e.g., the establishment of national parks and the passage of the Endangered Species Act of 1973), the broad chronological sweep of this study, a related need to reign in an already unwieldy narrative, and my own particular interests and expertise.[50]

Similarly, I have restricted this account almost exclusively to concern about wildlife extinction. Following on the heels of the successful Audubon movement that began at the end of the nineteenth century, a handful of American botanists began publicly expressing alarm about the fate of the nation's flora. Worried about the commercial exploitation of rare plants, the growing army of outdoor recreation enthusiasts who stripped the landscape of its wildflowers, and the continued habitat destruction that accompanied the nation's ongoing economic expansion, in 1900, flora enthusiasts in the Boston area organized the Society for the Protection of Native Plants. Two years later, the American botanist Elizabeth N. Britton helped found a competing institution, the Wild Flower Preservation Society, based out of the New York Botanical Garden. By the late 1920s, these and similar organizations successfully lobbied for legislation to protect at least some plants in more than twenty-two states, though that conservation legislation generally covered only a few species and was rarely enforced.

Compared to their colleagues struggling to rescue endangered wildlife, however, plant protectionists faced several distinct obstacles. Most fundamental was a common law tradition that considered wild animals as ferae naturae, a public good and the property of no one until captured or killed.[51] Related to this long-standing doctrine was the presumed right of the government to regulate the taking of game, a right first asserted in feudal Europe. In a landmark case handed down in 1896, *Geer v. Connecticut*, the U.S. Supreme Court affirmed that states were entitled to "control and regulate the common property in game," and they were to do so as "a trust for the benefit of the people." Plants, on the other hand, possessed an entirely different legal status. Rather than being considered public property, and hence subject to governmental regulation, vegetation was generally thought to be the property of whoever owned the land on which it grew.[52]

Compounding this basic legal difficulty, rare plants were also much more difficult both to bond and identify with. While many of them clearly possessed appealing aesthetic characteristics, they were less amenable to being conceptualized in human terms than animals, particularly birds and mammals that formed lasting pair bonds, reared their young, and openly communicated with one another.[53] For these and other reasons, by the time the Endangered Species Act gained passage in 1973, the scope and scale of plant conservation in the United States proved negligible compared to efforts to save endangered wildlife.[54]

THE BASIC ARGUMENT that runs through this book then is straightforward: naturalists' interest in collecting, describing, and classifying the earth's organisms alerted them to the growing problem of human-caused loss of plants and animals while fostering sympathy for their desperate plight. Increasingly haunted by specter of extinction, naturalists mobilized to act. They publicized the decline of wildlife in technical and popular publications, formulated arguments for their preservation, established conservation organizations, developed techniques for studying and saving rare species, and lobbied for protective reserves and legislation. This is by no means the first book to recognize the central role that naturalists have played in American wildlife conservation, but it is the first to provide a long view on their sustained and substantial efforts to discover, problematize, and respond to the issue of extinction over a roughly two-century period leading up to the Endangered Species Act of 1973. Without their ongoing campaign to rescue vanishing species, the natural world would undoubtedly be facing an even more bleak and biologically impoverished future.

CHAPTER ONE

BONES OF CONTENTION

THE AMERICAN INCOGNITUM
AND THE DISCOVERY OF EXTINCTION

*Such is the œconomy of nature, that no instance can be produced of her
having permitted any one race of her animals to become extinct;
or her having formed any link in her great work so weak as to be broken.*

THOMAS JEFFERSON, 1784

JEFFERSON'S DILEMMA

At the height of the American Revolution, while the outcome of the rebellion
against Great Britain remained uncertain, Thomas Jefferson grappled with the
problem of fossils. The specific context of his engagement with this thorny issue was a manuscript that he began sometime in the summer or early autumn of
1780. At the time the thirty-seven-year-old governor of Virginia and author of
the Declaration of Independence already enjoyed a considerable reputation for
accomplishment in the political sphere. Less well known to his contemporaries
was his keen interest in science. Jefferson was an inveterate reader of scientific
treatises, a zealous recorder of natural phenomena, and an eager correspondent
with others who shared his enthusiasm. He would later reflect on his longstanding curiosity about the natural world by declaring that "Science is my passion,
politics my duty."[1]

In 1780, when the secretary to the French minister in Philadelphia, François
Marbois, circulated a detailed questionnaire regarding the political and natural
history of Virginia, Jefferson seized the opportunity to organize his abundant
notes. Faced with a long series of personal and political crises—including the

FIGURE 1. Jefferson with Declaration of Independence and scientific instruments, 1801. Engraving by Cornelius Tiebout. In addition to his considerable accomplishment in the political sphere, Jefferson was also a skilled naturalist who made early, important contributions to the field of paleontology. Courtesy of the Prints and Photographs Division, Library of Congress, LC-USZ62-75384.

death of his daughter, the prolonged illness of his wife, a nasty injury sustained in a fall from his horse, forced retreat first from the state capital in Richmond and then from his estate in Monticello, and accusations that he had engaged in dishonorable conduct during the period of British occupation—Jefferson endured some of the darkest days of his entire life. Yet, at Popular Forest, his beloved rural retreat nestled in the Blue Ridge Mountains, he found solace in the hours devoted to fulfilling Marbois's request. By the time he returned to Monticello in August 1781, he was nearly finished drafting the manuscript. The resulting publication, issued three years later as *Notes on the State of Virginia*, proved the only book Jefferson would publish during his lifetime. It is now widely considered a classic, "one of America's first permanent literary and intellectual landmarks."[2]

Although he had long been fascinated with science, *Notes on the State of Virginia* signaled the beginning of Jefferson's active interest in fossil vertebrates.[3] Interspersed among his discussion of political philosophy, his ideas about religious freedom, and his famous condemnation of slavery, Jefferson meticulously cataloged the natural resources of his home state and the surrounding region. The first animal he described was also the one to which he devoted the most attention: "the Mammoth, or big buffalo," a creature he judged to be "six times the size of an elephant," which it seemed to resemble.[4] What little was known about this "incognitum" (as it was often referred to at the time) came from tales about the beast that Indians had passed down through the generations and from fossilized remains that had been uncovered in America beginning in the early eighteenth century. Since that initial discovery, the creature's teeth and bones had become highly sought after additions to institutional museums and private cabinets of curiosity on both sides of the Atlantic. In America, for example, not only Jefferson, but also George Washington and Benjamin Franklin owned prized specimens of the grinders from these mysterious beasts.[5]

For Jefferson, the incognitum proved not only an intrinsically fascinating creature, it also provided ammunition in an ongoing campaign to refute the theory of one of Europe's most renowned naturalist during the second half of the eighteenth century, Georges Louise Leclerc, Comte de Buffon, who was Intendant of the Jardin des Plantes and Keeper of the Royal Cabinet of Natural History in Paris. In his best-selling, thirty-six-volume *Histoire naturelle, générale et particulière* (1749–89), Buffon argued that the New World's generally cool and moist environment had forced its native inhabitants to degenerate over time, rendering them punier, less vigorous, and less fertile than their Old World counterparts.[6] Stung by the assault on his homeland and its people, Jefferson felt compelled to respond. In a discussion with explicitly nationalistic overtones, he asserted the morality, fecundity, and intelligence of America's aboriginal inhabitants. He also produced two tables showing that the mammals native to the land of his birth fared quite well when their weight or overall numbers were compared with those in Europe. The bones of the mysterious "incognitum," the "largest of all terrestrial beings," however, offered Jefferson with the strongest potential evidence in his bid to refute Buffon's troubling theory.[7] The problem was Jefferson could not prove that the prodigious beast still roamed the earth.

For the remainder of his life, Jefferson vigorously pursued the American incognitum and other quadrupeds whose fossilized remains were periodically uncovered across North America. He not only personally financed numerous expeditions to retrieve fossil remains but also encouraged others to follow his lead. For example, on December 19, 1781, the day before he sent his completed Virginia manuscript to the French consul in Philadelphia, Jefferson wrote to General George Rogers Clark, an old friend, Abermarle County native, and the commanding officer of the Army of the West. Jefferson's note, delivered by none other than Daniel Boone, asked Clark to venture to Big Bone Lick, Kentucky, on the banks of the Ohio River, to retrieve bones of the American incognitum. The threat of Indian attack kept Clark away from the area, so a year later Jefferson repeated his request, declaring: "A specimen of each of the several species of bones now to be found is to be the most desirable object in Natural History, and there is no expense of package or safe transportation which I will not gladly reimburse to procure them safely."[8] Despite his strict constructionist principles, as president of the United States Jefferson provided federal support so the Philadelphia artist, naturalist, and museum owner Charles Willson Peale could unearth a complete skeleton of the incognitum from a soggy marl pit along the Hudson River. With the bones he found, Peale mounted and displayed one of the earliest virtually intact fossil skeletons ever to be reconstructed, an outcome that thrilled Jefferson.

Jefferson enjoyed collecting fossil vertebrates—and proudly displayed them at Monticello—but he was also quite interested in furthering scientific understanding of these mysterious creatures. In keeping with that goal, he freely placed prized specimens in the hands of museums at home and abroad. The bones of the American incognitum that he sent to the prestigious National Museum of Natural History in Paris, for example, proved useful to French naturalists working in the nascent fields of paleontology and comparative anatomy.[9] He even published a paper of his own describing a new species of large fossil mammal, which he dubbed the megalonyx, based on bones recovered by workers digging saltpeter from a cave in Greenbrier County, Virginia (now West Virginia).[10] Although he thought the megalonyx was a fearsome gigantic clawed beast that dwarfed the African lion, thereby providing additional evidence against Buffon's theory of degeneracy, it later turned out to be a massive sloth that still bears his name, *Megalonyx jeffersoni*.

Although it strikes the modern reader as rather odd, Jefferson also firmly believed these creatures still survived somewhere in the unexplored regions of the continent. In his table of American and European mammals found in *Notes on the State of Virginia*, Jefferson listed the mammoth first. In defense of this decision he wrote: "It may be asked, why I insert the Mammoth as if it still existed? It may be asked in return, why I should omit it, as if it did not exist? Such is the œconomy of nature, that no instance can be produced of her having permitted any one race of her animals to become extinct; of her having formed any link in her great work so weak as to be broken."[11] Beyond this basic philosophical objection to extinction was the "traditionary testimony of the Indians, that this animal still existed in the northern and western parts of America," regions that remained "in their aboriginal state, unexplored and undisturbed."[12]

To Jefferson and many of his contemporaries, the very idea of species extinction seemed anathema. Intellectuals of his day recognized that settlement often resulted in the local extermination of wildlife. But the complete disappearance of a species was another matter altogether. The loss of any organism across its entire range implied an unacceptable imperfection in God's creation, while violating deep-seated assumptions about the balance of nature and the great chain of being that proved central to Western understandings of how that creation was ordered. In the hope that living examples of these beasts might still be found wandering somewhere in the unexplored regions of North America, Jefferson urged the explorers Meriwether Lewis and William Clark to keep a sharp lookout out for species animals "deemed to be rare or extinct," like the American incognitum, during their famous western exploring expedition. The Corps of Discovery found a host of new plant and animal species during their arduous two-year journey, but they encountered no lumbering elephants.[13]

While Jefferson's doubts about the possibility of extinction remained commonplace at the time he penned *Notes on the State of Virginia*, by the time of his death in 1826, most naturalists on both sides of the Atlantic had experienced a sea change in their ideas on the subject. Central to this transformation was the work of the brash young French naturalist, Georges Cuvier. With access to specimens provided by a transatlantic fossil network and training from prominent German anatomists, Cuvier deployed the principles of comparative anatomy to offer convincing evidence that extinction had been a regular part of the earth's history. Cuvier was the first naturalist to clearly distinguished between the two living species of elephant and two kinds of extinct fossil elephant, the mammoth and the mastodon, the latter of which he clearly differentiated and named in 1806. During the first several decades of the nineteenth century, he went on to describe a virtual zoo of lost creatures, thereby laying the foundations for modern paleontology. Within a surprisingly short period of time, the reality of extinction became central to most educated Westerners' understanding of the earth's history. Later in life, even Jefferson himself privately admitted that some species might have been lost.[14] Yet, as we shall see later, some of the ideas that had led him and most other naturalists to deny the reality of extinction—for example, the notion of plentitude that proved central to the great chain of being and the idea that nature was finely balanced—remained important to thinking about the natural world long after the possibility of extinction became widely accepted.

PROVIDENTIAL NATURAL HISTORY
AND THE ORDER OF NATURE

Jefferson and most of his contemporaries were certain that the natural world was orderly, static, and new. Most importantly, and one of the beliefs undergirding these convictions, they also firmly believed that it was the product of a divine mind.[15] One way to understand that mind, and at the same time to ensure that science and religion, reason and faith remained firmly reconciled, was a set of practices and beliefs known as natural theology.[16] As with so many foundational concepts in the Western world, the basic idea of natural theology—that careful scrutiny of the natural world revealed attributes of its creator—dates back to the ancient Greeks. During the Middle Ages, Aquinas and the scholastics labored to show how reason melded harmoniously with faith to demonstrate the existence and attributes of a Christian god. At end of the seventeenth and the beginning of the eighteenth centuries, these ideas found full expression when the famed British naturalist John Ray published *Wisdom of God Manifested in the Works of the Creation* (1691). Ray's influential, widely reprinted book combined observations and specific arguments made by previous authors with his own considerable knowledge of flora, fauna, and systematics.[17]

Ray leaned heavily on a line of reasoning central to the natural theology of his day: the argument from design.[18] This argument held that since the obvious order and complexity of the world could not have possibly emerged from nature itself, an intelligent designer must have imposed it. Just as human-made buildings and machines "do necessarily infer the being and operation of some intelligent Architect or Engineer," Ray argued, "why shall not also the Works of Nature, that Grandeur and Magnificence, that excellent contrivance for Beauty, Order, Use, &c., which is observable in them, wherein they do as much transcend the Effects of human Art as infinite Power and Wisdom exceeds finite, infer the existence of an Omnipotent and All-wise Creator?" Proponents of the argument from design applied it on multiple scales, ranging from individual organs (like the eye or the human hand) to particular species (like the honey bee or the beaver) to the larger patterns of relationship between species (e.g., the adaptations of prey to escape predation). In all cases, though, structure seemed to be perfectly adapted to function, while the order and complexity of nature was thought to reveal the wisdom, power, and beneficence of God.

When it came to conceptualizing the specific patterns of relationship between species, however, many possibilities presented themselves. One deeply entrenched way of thinking about that order was the chain of being or *scala naturae*, the idea that the diversity of the natural world could best be understood as a long chain containing every possible kind of organism in a linear, continuous series.[19] In its most expansive form, the great chain of being was thought to encompass not just living organisms, but all kinds of being from "nothing to the Deity."[20] The idea has its roots in the Platonic view that the world is full and all possible kinds of things exist (the notion of plentitude) and the Aristotelian belief that all creatures could be lined up in a hierarchical series, with no gaps between them (the notions of continuity and gradation).

Well into the eighteenth century, naturalists struggled to reconcile the expanding, increasingly detailed observations of known organisms into a single, hierarchical, continuous series. The idea of the chain of being proved central, for example, to the renowned Swedish naturalist Carl Linnaeus, who not only introduced the binomial system of scientific nomenclature but also developed widely adopted systems of botanical and zoological classification.[21] Linnaeus once wrote that "the closer we get to know the creatures around us, the clearer is the understanding we obtain of the chain of nature, and its harmony and system, according to which all things appear to have been created."[22] Similarly, in the preliminary discourse to his *Histoire naturelle,* Buffon argued that if man placed himself at the "head of all created beings, he would see with astonishment that one could descend by almost imperceptible degrees from the most perfect creature to the most shapeless matter, from the most organized animal to the crudest mineral; he would recognize that these imperceptible nuances were the greatest work of Nature."[23] Nor was the idea of the great chain of being confined to biological circles; rather, it remained the common cultural heritage of most educated Europeans and Americans until the end of the eighteenth century.[24]

The pervasive idea of the great chain of being had strong implications for how the notion of extinction was received. For if God had created every conceivable form and no discernable gaps existed between them, then the loss of any creature threatened to bring down the entire edifice. The British poet Alexander Pope simultaneously celebrated the chain of being while expressing concern about the implication of extinction in his *Essay on Man* (1733–34):

> Vast Chain of Being! which from God began,
> Natures aetherial, human, angel, man,
> Beast, bird, fish, insect, what no eye can see,
> No glass can reach; from Infinite to thee,
> From thee to nothing.—On superior pow'rs
> Were we to press, inferior might on ours:
> Or in the full creation leave a void,
> Where, one step broken, the great scale's destroyed:
> From Nature's chain whatever link you strike,
> Tenth or ten thousandth, breaks the chain alike.[25]

The great chain of being offered one widely adopted model for thinking about the apparent order of the world; the notion that nature was balanced provided a different (though complimentary) way of conceptualizing that order.[26] Not surprisingly, the idea that nature exists in some kind of overall balance also has deep roots in antiquity. Indeed, as the historian of ecology Frank Egerton has argued, "In one way or another a balance-of-nature concept is part of most cosmologies."[27] Early discussions of the idea tended to be vague and general, but by the end of the seventeenth and the beginning of the eighteenth centuries, naturalists like Ray began to marshal specific biological evidence—like the existence of finely tuned predator-prey relationships—to show how God ensured nature's balance.

Perhaps not surprisingly, the inveterate namer and classifier Linnaeus first provided a title for the balance-of-nature concept, while at the same time laying down the early foundations for the science of ecology. In 1749, he published an influential essay, "The Oeconomy of Nature," which declared that everything in the universe was "chained together" and that this interconnection demonstrated the "infinite wisdom of the Creator": "To perpetuate the established course of nature in a continued series, the divine wisdom has thought fit, that all living creatures should constantly be employed in producing individuals, that all natural things should contribute and lend a helping hand toward preserving every species, and lastly that the death and destruction of one thing should always be subservient to the restitution of another."[28] We see hints of the chain of being in this statement, which appeared in the introduction to the essay, but here Linnaeus is much more interested in the functional relationship between organisms in a given environment than a simple recounting of the series of creation. Central to this functional relationship is the observation that each animal not only feeds upon its own distinct prey but also is preyed upon by other species. Linnaeus offered several examples of this general phenomenon, including one that included multiple layers of predation: "Thus the *tree-louse* lives upon plants. The fly called *musca aphidivora* lives upon the *tree-louse*. The *hornet* and *wasp fly* upon *musca aphidivora*. The *dragon fly* upon the *hornet* and *wasp fly*. The *spider* on the *dragon fly*. The *small birds* upon the *spider*. And lastly, the *hawk* kind on the *small birds*."[29] In effect, what Linnaeus did here was delineate the ecological concept of the food chain, though naturalists would not adopt that precise name until nearly two centuries later.

Linnaeus also enumerated other patterns within the "economy of nature." Predators have fewer offspring, are less numerous in overall population size, and tend to live shorter lives than their prey. Each species exists within its own geographic range and eats a certain kind of food. Even particularly voracious "wild beasts and hawks" cannot "destroy a whole species." Rather, Providence has created an order that continually ensures a "just proportion among all the species," an order in which the complete loss of any species is inconceivable.[30] Linnaeus believed that ultimately this entire interconnected world had been made for the sake of humans—both to provide the things they need to survive and to offer a tangible reminder of God's power and glory.

Thus the wide acceptance of the chain of being and notions of a balanced nature both contributed to a generally static view of the world. Species could not go out of existence or come into being without fundamentally threatening that natural order. Ray, for example, remained unequivocal on this point: "The Number of true species in nature is fixed and limited, and as we may reasonably believe, constant and unchangeable from the first creation to the present day."[31] By the second half of the eighteenth century, some naturalists became more receptive to the possibility of limited change in organic nature. Linnaeus and his students, for example, toyed with the idea that hybridization might produce new species, while Buffon argued that the diversity of the natural world evident in his day was the result of the degeneration of a discrete number of basic forms.[32] But most naturalists refused to entertain the possibility of dramatic change in organic nature, and the complete loss of any species remained extremely problematic.

Scientists and other intellectuals in the West not only saw the world as static; they also believed it was relatively young.[33] Scriptural tradition played a central role in this assessment; the Bible recorded that the world had been created in six days, and it provided a written account of the early generations of patriarchs. In the seventeenth century, Archbishop James Ussher tried to nail down the exact date of creation by tracing these lineages while cross-referencing the people and events named in the Bible to the (still sketchy) historical record available at the time. Ussher's estimate—that the world began on the nightfall preceding Sunday, October 23, 4004, BC—gained further credence when church authorities incorporated it into annotated editions of the influential King James Bible translation.[34] While not everyone accepted Ussher's overly exact calculation, until the end of the eighteenth century, most educated Westerners seemed to embrace the notion that earth history and human history were roughly coeval and that the age of the earth was scarcely a few thousand years old.

THE PROBLEM OF FOSSILS

The discovery of fossils challenged firmly entrenched ideas about the order, stability, and age of the earth. Fossils—which we now define as the preserved remains of once living beings—had long been noticed, if not accurately identified. In a fascinating recent book, the folklorist Adrienne Mayor has documented the deep interest in fossils in the ancient world, where the "bones of gigantic beings were treasured as relics of the mythic past and displayed as natural wonders in temples and other public places."[35] She argues that many of the mythical beasts from this period—the griffin, the centaur, and others—had their origins in fossil skeletons that were widely collected, measured, and displayed throughout the lands known to the Greeks and Romans.

That knowledge about fossil beings seems to have been largely forgotten, though, until the Renaissance, when the science of paleontology first began to stir.[36] By then, the term "fossil" denoted any distinctive object found below the earth or lying on its surface. It thus referred not only to fossils in the modern sense, but also to mineral ores, crystals, and rocks of all sorts. During the sixteenth century, the systematic study of fossils first took off when Conrad Gesner and other scholars began amassing large collections of these curious stones, producing illustrated publications describing them, and corresponding with individuals who shared their interests. These fossil objects tended to be interpreted within either Neoplatonic frameworks—which saw a correspondence between the hidden and visible worlds while positing a pervasive molding force or "plastic virtue" responsible for that correspondence—and Aristotelian frameworks that saw simple fossils as spontaneously generating and more complex ones as growing from a specific "seed" within the earth. Whichever explanation sixteenth-century naturalists adopted for the origin of fossils, they failed to show a particular interest in discriminating between organic and inorganic objects. And neither explanation for their origin raised the touchy issue of extinction.[37]

During the seventeenth century, naturalists not only began to take more interest in the resemblance between fossil forms and living forms, but also began to call those with a significant likeness to living beings "organized fossils" or "extraneous fossils." As historian of science Martin Rudwick notes in his pioneering study on the origins of paleontology, interpreting the meaning of these apparently organic fossil forms posed difficult challenges for early modern naturalists. While it was relatively easy to make the connection between so-called tongue stones and the teeth of living sharks, for example, other fossils presented difficulties because of confusing modes of preservation or a lack of familiarity with the organisms from which they had originated. Even in the case of fossils that clearly resembled living species, explaining their position (e.g., shells embedded high on mountaintops) proved difficult without adequate models to explain geological change. And in cases of fossils that appeared to be organic but did not seem to have living analogs, like ammonites, the specter of extinction haunted naturalists because it challenged deeply held notions about plentitude, the balance of nature, and the age of the earth.[38]

One scholar who struggled mightily with the problem of fossils in the late seventeenth century was John Ray. Ray was too accomplished a naturalist not to appreciate the strong resemblance between many fossil and living forms; moreover, his commitment to natural theology suggested to him that nothing in nature had been done in vain. He also benefited from the earlier publications of Nicolas Steno and Robert Hooke, who had argued convincingly for the organic origins of certain fossils in the 1660s and 1670s by using examples of remains whose form, composition, and position were relatively easy to account for.[39] Following a tour on the Continent, Ray completed an essay on the problem of fossils in which he called the hypothesis of organic origin the most "probable opinion," but one which he nonetheless found troubling.[40] First, the morphological differences between living and fossil species were enough to suggest that the latter were the remains of species that were extinct, a conclusion that flew in the face of some of Ray's most cherished assumptions. And second, the appearance of fossils in highly elevated areas, like the Alps, was difficult to explain using ideas of mountain upheaval as generally understood at the time, or by invoking the Great Flood, which soon became a common means of addressing this particular issue. As Rudwick points out, "the only way out of the dilemma was to argue that fossil species might not really be extinct at all" a conclusion that he terms "perfectly justifiable" at the time.[41] Most of the organisms in dispute were marine animals, which were little studied, especially those forms inhabiting remote areas. For the specific case Ray mentioned, fossil stalked crinoids, Ray's sense that it was too early to write off the species as lost gained vindication fifty years after his death by the discovery of living examples in the West Indies.[42]

The problem of fossils continued to challenge naturalists throughout the eighteenth century. The general confusion they provoked is amply illustrated in an episode from the early part of that century, when purported fossil remains became the center of a notorious scientific hoax.[43] Johann Bartholomew Adam Beringer, senior professor and dean of the faculty of medicine at the University of Würzberg, was a well-known collector of and expert on fossils. In May 1725, he hired three young men to excavate a promising site, a hill about a mile outside the town where the university was located. Over the next six months, his assistants dug up hundreds of fossil mollusks and small figured fossils from this site. But there was something odd about these particular stones: they featured a dazzling variety of forms, including a radiant sun with a human face, stars and comets, lizards, fishes, bees, frogs, and even Jehovah's name in Hebrew! Beringer was suspicious, as he should have been. Nonetheless, a year after he began trying to make sense of the bizarre stones, he published an illustrated book featuring more than two hundred specimens from this site. After considering and reject-ing various explanations for the fossils, he ultimately concluded that they had been created by "the Author of Nature" but were not the remains of any living creature. Only later did he discover that he had been the victim of an elaborate hoax perpetrated by two jealous colleagues.

Nonorganic explanations for fossils died out slowly during this period, and, if Rudwick is correct, the humiliation that Beringer suffered may have played a small role in their decline.[44] Yet, until the end of the eighteenth century, relatively few naturalists seemed willing to entertain the idea that fossils represented the relics of extinct species. Linnaeus, for example, wrote of the fossil *Anomiae:* "the animals which inhabited these 'wild mussels,' as well as unaltered shells, are nowadays unknown to us . . . , nor do we know what in the world may have become of them. Still, we shall never believe that a species has entirely perished from the earth."[45]

Some naturalists at the time did begin arguing that the age of the earth was much greater than the generally accepted date of only several thousand years old. Based on estimates of the earth's cooling rate and his knowledge of fossils and sedimentation rates, in the late 1770s, Buffon argued that the world was more than seventy thousand years old, while privately speculating that the deposition of known geological strata would have required at least 10 million years.[46] Buffon thus became one of the first major Western thinkers to appreciate the concept of "deep time," a wonderfully evocative phrase coined by the twentieth-century American writer John McPhee.[47] The Scottish geologist James Hutton went even further in his landmark book *Theory of the Earth* (1795). There he argued that much of the earth's surface consisted of the relics of sea animals that had been deposited on the ocean floor, consolidated into strata, and then pushed upward by the heat of the earth. All this took so much time, Hutton argued, that when it came to estimating the age of the earth, there was "no vestige of a beginning,—no prospect of an end."[48] Even with these and other expansions in the earth's time scale, though, resistance to the idea of extinction remained solid. As Rudwick summarizes the situation at the end of the eighteenth century: "Whether any spe-cies had truly been 'lost' from the world thus remained a question as uncertain and debatable at the end of Buffon's life as it had been nearly a century earlier at the end of Ray's. Many groups of fossils, such as ammonites and belemnites, were now recognized beyond all doubt as organic remains differing radically from any known living animals; but it could still be asserted with good reason that they might be living in deep water or in some remote part of the world."[49] The discovery and careful examination of large fossil mammals would soon pres-ent insurmountable obstacles to that claim.

Given the profound challenges in interpreting fossil remains, it is easy to see why the discovery of large vertebrate bones in the New World would ignite such interest. Apparently, Native Americans began noticing, collecting, and trying to explain the meaning of fossils well before European contact.[50] In 1519, Cortes's conquistadors became the first Europeans to view fossil vertebrates in the Western Hemisphere when the defeated Tlascalan peoples of central Mexico surrendered the bones of what now appears to have been a mastodon. The Tlascalans believed the bones to be the remains of a giant race of humans that their ancestors had annihilated.[51]

Nearly two centuries later, in 1705, American colonists offered a similar explanation for a fist-sized tooth weighing about five pounds discovered near the Hudson River, not far from the frontier outpost of Albany.[52] While the root had badly decayed, its intact enamel vaguely resembled that of a human eyetooth. An assemblyman from Albany, who traded the farmer who found the tooth a half pint of rum for it, eventually presented the fossil to the governor of New York Province, who then dispatched it to the Royal Society of London, a prestigious scientific organization that had been founded nearly five decades earlier. Soon other large teeth and bones of the American incognitum began circulating throughout New York and New England. After viewing the curious remains, the Boston clergyman and scholar Cotton Mather argued that they provided scientific confirmation for the accounts of antediluvian human giants found in the Bible. Mather, a corresponding member of the Royal Society, took particular pride in the fact that the bones of this giant had been unearthed in America. The nationalistic sentiment associated with remains of the American incognitum long continued to shape discussions about the mysterious beast, and the Hudson River valley became one of two major North American sites that would yield large quantities of these bones.[53]

The other particularly productive site was along the banks of the Ohio, in an area that became known as Big Bone Lick, Kentucky.[54] In 1739, Charles LeMoyne, baron de Longueuil, led a French military expedition from Montreal to map the region. After a group of Indian scouts stumbled onto several gigantic bones, Longueuil and his men returned to the area to gather numerous specimens, which they shipped to the Jardin du Roi in Paris.[55] Nearly three decades later (1765), the Indian trader George Croghan ventured to the site and found a "vast quantity of bones," including two six-foot-long tusks that his men loaded onto one of their flatboats.[56] This first set of bones was lost when Indians attacked the party; though struck in the head, Croghan survived and, undaunted, returned to the

FIGURE 2. Mastodon molar from the collection of Thomas Jefferson. Following their initial discovery in North America during the early eighteenth century, fossil remains like this one not only ignited much interest but also much speculation about their origin. Courtesy of the Academy of Natural Sciences, Philadelphia.

area a year later. This time he managed to collect hundreds of pounds of bones, which he shipped to Benjamin Franklin and Lord Shelbourne in London.

As collections of American incognitum bones mounted, so did speculation about their meaning. From exactly what kind of beast did these fossils originate? In his *Natural History of Carolina, Florida and the Bahama Islands* (1731–43), the British artist and naturalist Mark Catesby made passing reference to the discovery of "three of four Teeth of a large Animal" that slaves had uncovered in Biggin Swamp, near Charleston, South Carolina. Ironically, these slaves recognized the resemblance between the molars they had dug up and those of the African elephant, with which they were already familiar, thus providing what the twentieth-century paleontologist George Gaylord Simpson has deemed "the first technical identification of an American fossil vertebrate."[57] In the 1750s and 1760s, Buffon and his anatomist, Louis Jean-Marie Daubenton, compared the bones of the American incognitum that Longueuil had excavated from the Kentucky salt lick with a number of seemingly related species.[58] Those included the living elephant and the mammoth, another large elephant-like creature whose remains had been discovered in Siberia beginning in the early eighteenth century and whose identity was often confused with the fossil elephants of North America.[59] Based on his examination, Daubenton argued that all fossil and living elephants belonged to the same species; to explain the decidedly un-elephant-like teeth found with the remains at Big Bone Lick, he suggested that somehow the teeth of a large hippopotamus had been inadvertently mixed with the femurs and tusks of the Ohio valley specimens. Initially, in his discussion of carnivores in an earlier volume of *Histoire naturelle*, Buffon had suggested that the natural world contained competing species that might drive one another to extinction, and he listed the mammoth as one of the species that could have been lost by this means. But several years later, at the end of his volume devoted to elephants, Buffon accepted Daubenton's conclusions while denying that the Siberian and American creatures were either larger than elephants or extinct. To address the issue of how such creatures—which now only lived in the tropics—had survived in a much colder climate, Buffon later postulated the idea that the earth had been created when a comet struck the sun. As the earth gradually cooled, these now tropical species had migrated from the poles.

Other naturalists reached different conclusions about the nature of this puzzling creature. In an important essay published by the Royal Society in 1769, the British physician and anatomist William Hunter declared that the "American *incognitum*" was "some carnivorous animal, larger than an elephant."[60] Hunter also argued that the Siberian mammoth and the American incognitum were probably the same species. He based his conclusions on an examination of fossil remains in the Tower of London, Kentucky specimens from the Shelbourne and Franklin collections, jaws of the living elephant belonging to his younger brother (John Hunter), and elephant tusks in the warehouses of London ivory dealers. Since grinders resembling those of the incognitum had been dug up in various other locations, Hunter suggested that the beast had at one time been "a general inhabitant of the globe." He concluded his essay with an expression of relief that the species was now apparently lost: "And if this animal was indeed carnivorous, which I believe cannot be doubted, though we may as philosophers regret it, as men we cannot but thank Heaven that its whole generation is *probably* extinct."[61] In an age that continued to venerate providential natural history, Hunter understood that his opinion would be controversial, and he seems to have been quite cautious here in his choice of words.

The British naturalist Thomas Pennant's opinion proved more acceptable at the time: "Providence maintains and continues every created species, and we have as much assurance, that no race of animals will any more cease while the Earth remaineth, than seed time and harvest, cold and heat, summer and winter, day and night." In his *Synopsis of Quadrupeds*, published only two years after Hunter's essay, Pennant argued that we he called the "American elephant" must still be alive "in some of the vast new continent, unpenetrated yet by Europeans."[62] Clearly, Jefferson was not alone in his belief that the American incognitum must still roam the earth.

Before the mystery of the American incognitum could be definitively solved, a second fossil incognitum came to light. In the spring of 1796, Jefferson received a letter from a friend in western Virginia, Colonel John Stuart. Laborers digging for saltpeter in a cave near his home in Greenbrier County had found a small collection of bones that Stuart thought might interest the man who was soon to be elected vice president.[63] Jefferson thanked Stuart for the specimens and requested any additional fossils from the site he might find. He was especially interested in obtaining a thighbone of the creature, which would help him estimate its overall size. During this exchange, Jefferson reiterated his views about the implausibility of extinction: "I cannot however help believing that this animal as well as the Mammoth are still existing. The annihilation of any species is so unexampled in any parts of the economy of nature which we see, that we have a right to conclude, as to the parts we do not see, that the probabilities against such annihilation are stronger than those for it."[64]

Extinct or not, Jefferson not only believed the creature was new to science but also that it would provide another opportunity to refute Buffon's "pretended degeneracy of animal nature in our continent."[65] He informed David Rittenhouse, the president of the American Philosophical Society, of his intent to write up a formal description of the beast, which he thought belonged to the "family of the lion, tyger, panther, &c. but as preeminent over the lion in size as the Mammoth is over the elephant."[66] After announcing his plans regarding what he began referring to as the "American lion," Jefferson gained election to the vice presidency of the United States and, following the death of Rittenhouse, presidency of the American Philosophical Society. He assumed the first office only reluctantly; the latter he called "the most flattering incident of my life."[67] Accompanying him on his trip from Monticello to Philadelphia were a box of fossils from Greenbrier County and a completed draft of his manuscript.

On March 10, 1797, the secretary of the American Philosophical Society read Jefferson's paper at a meeting over which he presided. Jefferson had written the first draft with the intent of announcing to the world that he had discovered a giant American lion or tiger. But sometime between the society's meeting and his arrival in Philadelphia a week earlier, he had stumbled upon a copy of a recent issue of *Monthly Magazine,* a British publication containing an abstracted description and a crude engraving of a previously unknown quadruped that had been dubbed the *Megatherium* or "huge beast."[68] The author of the original paper on which this abstract was based was Georges Cuvier, who named and described this clawed fossil creature based solely on an engraving of what was probably the first nearly intact fossil skeleton ever assembled.[69] The bones for the skeleton of the megatherium had been discovered in Buenos Aires, then part of Spanish South America, shipped back to Madrid, and reconstructed by Juan Bautista Bru, a conservator at the Royal Museum. A French official traveling through the area obtained a proof engraving of Bru's fossil skeleton and then passed it on to

Cuvier. Much to Jefferson's disappointment, Cuvier's megatherium seemed to strongly resemble his "American lion," and the self-assured French naturalist had placed it in the family of sloths and anteaters. Based on this newly gained knowledge, Jefferson scrambled to revise his paper. Where he had once unhesitatingly classified the creature with "the lion, tyger, panther, &c.," he now spoke of it more generically as an animal "of the clawed kind" and named it *Megalonyx* (or "great claw").[70] In a postscript, Jefferson noted similarities between his megalonyx and Cuvier's megatherium, while admitting that the former was probably not carnivorous. Yet, he continued to make repeated comparisons between the megalonyx and the lion throughout his paper, even writing that "if the bones of the megalonyx be ascribed to the lion, they must certainly have been of a lion of more than three times the volume of the African."[71] Ironically, filed away among his papers and forgotten, Jefferson owned a drawing of the megatherium skeleton, in Bru's own hand, which had been sent from Madrid in 1789.

But what had become of the megalonyx? As he had done earlier with the American incognitum in *Notes on the State of Virginia,* Jefferson pursued two (related) lines of argument to deny that it was extinct. First, he claimed the creature might still live in the unexplored portions of the continent: "In the present interior of our continent there is surely space and range enough for elephants and lions, if in that climate they could subsist; and for mammoths and megalonyxes who may subsist there. Our entire ignorance of the immense country to the West and North-West, and of its contents, does not authorise us to say what it does not contain."[72] Second, he resorted to the philosophical ideas of the great chain of being and the economy of nature to explain why it must still survive:

> In fine, the bones exist: therefore the animal has existed. The movements of nature are in a never ending circle. The animal species which has once been put into a train of motion, is still probably moving in that train. For if one link in nature's chain might be lost, another and another might be lost, till this whole system of things should evanish by piece-meal; a conclusion not warranted by the local disappearance of one or two species of animals, and opposed by thousands and thousands of instances of the renovating power constantly exercised by nature for the reproduction of all her subjects, animal, vegetable or mineral.[73]

Indeed, Jefferson continued, according to commonly accepted notions of the "economy of nature," it was quite reasonable to expect that a large beast like the megalonyx would be "the rarest of animals": "If lions and tygers multiplied as rabbits do, or eagles as pigeons, all other animal nature would have been long ago destroyed, and themselves would have ultimately extinguished after eating out their pasture."[74]

Several months after Jefferson presented his megalonyx research, the Phila-delphia judge George Turner offered a contentious paper on the American in-cognitum to the American Philosophical Society. Turner argued that there was not one incognitum, as had previously been assumed, but actually two incognita, both of which had been mistakenly lumped together under the name "mam-moth." Both were creatures about the size of an elephant, with wide and at least partially coextensive geographic ranges, but one had teeth suggesting it was her-bivorous and the other was apparently carnivorous or mixed. Moreover, their large size and once-extensive ranges convinced him that both were now extinct: "I have no hesitation in believing, that they belonged to some link in the chain of animal creation, which . . . has long been lost."[75] Noting the recent discovery of Jefferson's megalonyx, Cuvier's megatherium, and various skeletons of the mammoth found in Europe, Asia, and America, Turner argued that there were likely to be additional large creatures discovered in the future. It was difficult for him believe that "so many and such stupendous creatures could exist for centuries and be concealed from the prying eye of inquisitive man."[76] He discounted a be-lief in the chain of being or the oral traditions of Indians as sufficient grounds for insisting that these creatures must still survive somewhere. Turner supposed that the carnivorous American incognitum was a powerful beast that could spring a great distance to capture its prey: "With the agility and ferocity of the tiger; with a body of unequalled magnitude and strength, it is possible the Mammoth may have been at once the terror of the forest and of man!—And may not the human race have made the extirpation of this terrific disturber a common cause?"[77] As naturalists began to accept the idea of extinction in the early nineteenth century, the idea of predation by humans would be one of several theories that they used to explain how and why that species loss may have occurred.

NATIONALISM, NATURE, AND
THE EXHUMATION OF THE MASTODON

During the revolution and the early years of the young republic, nationalism became an increasingly important lens through which Americans saw the natu-ral world. When it came to thinking about the American landscape—its flora, fauna, and geography—there were various, and sometimes conflicting, ways in how this nationalistic sentiment played itself out.[78] On the one hand, European settlers in the New World had long viewed themselves as a chosen people with a calling to bring the light of civilization to the dark wilderness of the North American continent. That imperative was both moral and economic. Colonists felt they had a God-given mandate to make the wilderness economically productive, to transform an evil wasteland into a profitable enterprise. At the same time, as Jefferson's *Notes on the State of Virginia* suggests, by the end of the eighteenth century Americans also began to take pride in some of the wild features of that landscape, especially its sublime scenery and unique plants and animals. Soon they began arguing that America's wilderness was an important legacy that compensated for the young nation's lack of cultural heritage. In either case— whether the American landscape was something to be commodified, cherished, or both—the practice of natural history, the creation of a systematic inventory of the continent's natural riches, increasingly came to be viewed as a patriotic duty.[79] It was in this context that the American incognitum—believed to be the largest beast ever to roam the earth—became a widespread source of fascination and pride. As the historian Paul Semonin has argued, during the American Revolution "the bones began to take their place in the nation's public culture, celebrated in American literature and displayed in the nation's first national history museums."[80]

Charles Willson Peale's museum, in the capital city of Philadelphia, proved central to securing the incognitum's central place in American culture.[81] The museum has its origins in an episode that occurred late in the war, when George Washington authorized Christian Friedrich Michaelis, the physician-general to the Hessian mercenaries serving the British, to search for fossil bones on a farm in the Hudson River valley. Washington was quite familiar with the site in Orange County, seventy miles north of New York City, from a visit he had made in 1780 to view the large fossil grinders that workers had found there while draining a shallow swamp. Although Washington provided Michaelis with a dozen men, tools, and wagons, heavy rains thwarted efforts to excavate any more teeth or bones. Michaelis had to settle for a few specimens the site's owner gave him. When he returned to Philadelphia, Michaelis examined a collection of bones owned by Dr. John Morgan, whose brother had accompanied George Crohgan to Big Bone Lick nearly two decades earlier. Michaelis commissioned Peale, a well-known portrait painter, to make drawings of these Kentucky specimens. The enthusiasm the large bones generated among visitors to his studio set Peale to seriously thinking about creating a natural history museum to accompany his portrait gallery.

In 1786, three years after he completed the set of drawings for Michaelis, Peale began soliciting specimens for "a Repository for Natural Curiosities" that he hoped to append to his portrait gallery and home.[82] After several failed experiments, Peale perfected the technique of taxidermy, and he placed his stuffed animals in glass-fronted cases with painted landscape backgrounds, thus creating some of the first habitat groups that would later become standard in natural his-

FIGURE 3. Charles Willson Peale, *The Artist in His Museum*, 1822.
In this self-portrait, painted near then end of his career, Peale lifts
a curtain to reveal his famed natural history museum, which featured
a reconstruction of the mastodon as its centerpiece. Courtesy of the
Pennsylvania Academy of Fine Arts, Philadelphia, Gift of Mrs. Sarah
Harrison (The Joseph Harrison, Jr. Collection), accession no. 1878.1.2.

tory museums. Shells, minerals, fossils, live specimens, and portraits of famous
Americans rounded out the growing collection. Within a few years, Peale was
referring to the enterprise as the "American Museum," a name that was not only
befitting the nationalistic impulses that informed the institution but also his long-
standing (though ultimately unsuccessful) attempts to gain public funding to
support it. Cramped for space, in 1794 he moved to larger quarters in the American Philosophical Society, the first and most prestigious scientific organization
in the United States, which he had been invited to join eight years earlier. With
endorsements from Jefferson and other key political leaders, eventually Peale's

museum would become the repository for the specimens found on western exploring expeditions and an important source for a series of American natural history publications.

At about this same time, calls for an intact skeleton of the American incognitum grew more frequent. Near the end of his memoir, for example, Turner pointed out that whoever achieved this goal would "render,—not to his country alone, but to the world,—a most valuable present."[83] Turner was a member of an American Philosophical Society committee that sent out a printed circular requesting information on American antiquities in 1799. A special object of the committee's appeal was "one of more skeletons of the Mammoth, so called, and of such other such unknown animals as either have been, or hereafter may be discovered in America."[84] The committee, which also included Jefferson, Peale, and the Philadelphia anatomist Casper Wistar, suggested that Big Bone Lick was the place where fossil searchers were most likely to meet with success. That same year, however, workmen digging for marl to manure fields in Orange County, New York, uncovered a thighbone that was nearly four feet long and eighteen inches in diameter. When a minister and several physicians convinced the owner of the farm where the marl pit was located, John Masten, that the bones might be worth saving, he invited nearby residents to help him dig for more. A boisterous crowd of a hundred or so people managed to retrieve numerous additional bones, many of which were carelessly broken in the process, before being forced to stop by water oozing into the pit.

When Charles Willson Peale read about the discovery in 1801, he wasted little time in making preparations to visit the Masten farm.[85] If he could recover a complete skeleton of the American incognitum, not only would he help solve a compelling scientific mystery, but he would also bring fame and a steady stream of paying customers to his perennially cash-strapped museum. When he arrived at the site, he found the huge dark bones laid out in Masten's granary, where the farmer charged a small entrance fee for the privilege of viewing them. Anxious not to drive up the asking price, Peale pretended to be interested only in making life-sized drawings of the bones. During dinner later that day, Masten's son asked Peale if might be interested in purchasing them. The next morning Peale offered $200 for the specimens thus far recovered and $100 for the right to search for more on his property. Masten requested several additional items—a new double-barreled shotgun for his son and gowns from New York for his daughters—before agreeing to the deal. After packing up his prized specimens, Peale returned to Philadelphia to begin making arrangements for further excavations.

Peale recognized that retrieving additional bones from Masten's water-soaked marl pit would prove difficult, time consuming, and costly. But, if he managed to recover a complete skeleton, the scientific and financial payoffs would be great. He proudly displayed the Masten specimens at the American Philosophical Society and sought a $500 loan to finance an expedition to recover more. His letter requesting the loan appealed to the members' nationalism by pointing out that "foreigners . . . have hitherto deprived us of numerous collections of bones from every spot where they have been found."[86] He also requested federal aid from Jefferson, now president of the United States, in the form of a water pump from the navy and tents from the army. In his reply, Jefferson congratulated Peale on his find as well as his efforts to recover additional specimens of "these great animal monuments."[87]

Peale returned to Newburg, New York, in July 1801. After sizing up the land around Masten's farm, he contrived an ingenious device for removing the water that constantly streamed into the bone-encrusted marl pit. Rather than relying on inefficient hand pumps, he would erect a large human-powered mill wheel. A series of buckets attached to this wheel could lift the water to a wooden trough, which would then drain it off to the other side of a small hill. Peale hired a wheelwright and a carpenter to build the device, which he estimated could move more than a thousand gallons of water per hour. He then recruited volunteers from the large crowd that had gathered around the site to power the wheel and hired twenty-five local laborers to dig for bones. Despite nagging problems with water intrusion and collapses in the pit's muddy banks, Peale managed to retrieve most of the bones and parts of the tusk he was still missing but found no underjaw and only parts of the skull. He commemorated this triumph in one of his most famous paintings, *The Exhumation of the Mastodon*, completed five years later.[88]

Anxious to locate the bones still needed for a complete skeletal reconstruction, Peale also searched several nearby sites where bones of the American incognitum had previously been found. To aid in the hunt, his son Rembrandt designed a special device—a long slender steel rod with a wooden handle—that could locate hard underground objects. Using Rembrandt's invention, the expedition party recovered numerous additional bones, including the much-desired lower jaw. In all, Peale estimated they had recovered enough bones for two nearly complete skeletons.

With help from Wistar and others, Peale used a laborious process of trial and error to fit together the bones, many of which were badly broken, into a nearly complete skeleton. With an elephant's skull for a guide, he shaped the missing portion of the skull out of papier-mâché, marking the area with a horizontal red line to indicate that it was conjectural. When completed, the Masten skeleton, which took up most of the Peale family parlor, stood eleven feet high at the shoulder and measured seventeen feet, six inches long from its tusk to its tail,

FIGURE 4. Charles Willson Peale, *The Exhumation of the Mastodon,* 1806. In this, one of his most famous paintings, Peale celebrates his success in excavating a nearly complete mastodon skeleton from a marl pit in the Hudson River valley. Courtesy of the Maryland Historical Society.

quite a bit larger than the estimate of paleontologists today.[89] With nearly $2,000 in debts and his own money tied up in the skeleton, Peale was anxious to recoup his investment. On December 24, 1801, just over three months after he returned from New York, he invited members of the American Philosophical Society to a special viewing of the skeleton. Soon after this premier, he opened the special "Mammoth Room" located on the southwest corner of his museum, charging an admission fee of fifty cents for the privilege of gazing at the beast. A handbill promoted the reconstruction as the "ANTIQUE WONDER" and the "LARGEST of Terrestrial Beings!"[90]

The public flocked to the exhibit, the second fossil reconstruction ever completed. Beyond its continuing popularity, however, we have little sense of how viewers responded to Peale's skeleton. One wag, a British tourist, commented on seeing it: "Perhaps we ought to imagine Noah found it too large and troublesome to put in the ark, and therefore left the poor animal to perish."[91] We do know that the beast, which had long been of interest to naturalists and scholars, now seemed to capture the public's imagination as well, and the term "mammoth"

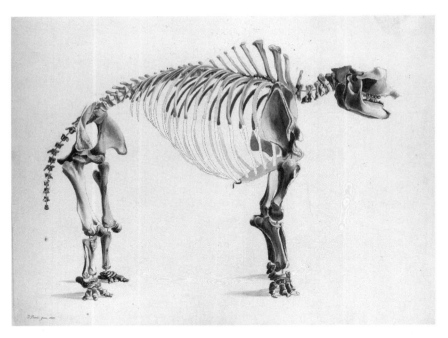

FIGURE 5. Titian Ramsay Peale sketch of Peale's mastodon, 1821. This reconstruction, which Charles Willson Peale completed with the help of the anatomist Casper Wistar, proved a huge hit with visitors to Peale's museum. Courtesy of the American Philosophical Society.

quickly entered into common usage to describe nearly any oversized item.[92] We also know that within a year, Peale was able to show a healthy $2,000 profit for his considerable efforts in locating and reconstructing the creature. As with other examples of wild America, Peale's cherished bones had turned out to be an exceedingly valuable commodity.

Hoping to capitalize on the interest in the beast abroad, Rembrandt and his brother Rubens took the second skeleton on a European tour. To promote the exhibit, Rembrandt authored a lengthy pamphlet entitled *Historical Disquisition on the Mammoth, or Great American Incognitum*. There he detailed how a "perfect skeleton of the "MAMMOTH—the first of American animals" had been unearthed, reconstructed, and displayed "in the first of the American Museums."[93] He also provided a history of various theories attempting to explain what kind of creature it was. Like most of his colleagues, the older Peale remained firmly committed to the idea of the chain of being and therefore reluctant to accept the idea of extinction. His son, however, suffered fewer qualms, declaring in a phrase that clearly echoed Jefferson's recent denial of extinction "the bones exist—the animals do not!"[94] Like Hunter had done earlier, Rembrandt Peale concluded that the American incognitum was a carnivorous beast that was now utterly lost. The elder Peale placed a copy of this pamphlet in a series of gilded frames that he hung in the Mammoth Room at the museum. Alongside the pamphlet, he also hung a copy of his painting commemorating the discovery of the beast, this glorious symbol of nationalism and one of his most enduring legacies.

CUVIER, COMPARATIVE ANATOMY, AND THE REALITY OF EXTINCTION

Prior to the departure of the Lewis and Clark expedition, Jefferson wrote to his friend, the French naturalist Bernard Lacépède, that he hoped the intrepid explorers might find a "living mammoth and megatherium" still roaming in the West.[95] Jefferson was disappointed that the expedition failed to locate either beast. Lewis did, however, view numerous fossil bones of the American incognitum in Cincinnati, on his way to meet Clark in 1803, including the upper portion of the skull that was missing from Peale's reconstruction. These remains belonged to Dr. William Goforth, who had obtained them from Big Bone Lick. The following year Goforth sent his collection to Pittsburgh, hoping that he could get it transported to Philadelphia and interest either Peale or the American Philosophical Society in buying it, but his agent sold it in London and absconded with the money. In 1807, one year after the return of the Corps of Discovery, Jefferson commissioned William Clark to venture to Big Bone Lick to recover additional fossil bones. With the help of ten hired laborers working for several weeks, Clark discovered a large collection of specimens that he dispatched to Washington. Jefferson spread the bones out in a large unused room in the east wing of the White House and invited Caspar Wistar to help him sort them. He sent one set to Peale, to complete his skeleton, kept a set of duplicates for himself, and sent the bulk to the National Institute in France, where Cuvier had been issuing a series of pioneering publications in comparative anatomy that authoritatively identified the American incognitum and established the reality of natural extinction.

Georges Cuvier had been born in Montbéliard, Würtenberg, an area that while linguistically French was also steeped in German cultural traditions.[96] As a child he stumbled upon a copy of Buffon's *Histoire naturelle* that belonged to his uncle. He read it avidly, spent many hours copying its figures, and developed a keen interest in collecting natural history specimens. At the Stuttgart Academy, Cuvier attended the lecturers of Friedrich Kielmeyer, who became his close friend and taught him the basics of comparative anatomy. During this period, German scholars were developing a new historicist perspective that portrayed the world neither as static nor progressing, but as constantly unfolding.[97]

According to this emerging view, geological strata might be conceived as the "archive" of the planet's unfolding. Rather than being hostile to the notion of extinction, this view tolerated, even embraced the idea of radical change that the permanent loss of species implied.

One of the principle proponents of these ideas was the comparative anatomist J. F. Blumenbach, who taught Kielmeyer and a raft of other influential students at the University of Göttingen. Blumenbach developed a philosophy of nature that portrayed fossils as monuments of past geological ages, and he became a towering figure in the development of paleontology, a role for which has only begun to be appreciated by modern historians. Unlike many of his contemporaries in France, England, and the United States, Blumenbach rejected the notion of the chain of being and the economy of nature. "Nature," he once wrote, "will not go to pieces if one species of creature dies out, or another is newly created,—and it is more than merely probable, that both cases have happened before now,—and all this without the slightest danger to order, either in the physical or the moral world, or for religion in general."[98]

Cuvier adopted Blumenbach's ideas and elaborated them into a highly productive research program. Confident to the point of being cocky, he arrived in Paris in 1795 and shortly afterward gained appointment as an assistant in animal anatomy at the newly reformed National Museum of Natural History, one of the world's premier scientific institutions. With interaction with brilliant colleagues, access to a constantly growing specimen collection, and regular dissections of deceased animals from the museum zoo, Cuvier quickly rose to prominence as a first-rate comparative anatomist. His deep, firsthand knowledge of the structure and function of living vertebrates would prove key to the success of his fossil reconstructions.[99] Based on a belief in the "correlation of parts"—the interdependence of an organism's organs and systems—Cuvier felt confident making educated guesses about the overall structure of a given animal even when he had precious little empirical evidence to guide him. He also developed the idea of "subordination of parts" to identify key structures that were most helpful in classifying organisms. He had just begun his anatomical studies in earnest when he received an unpublished engraving of Bru's reconstruction of a fossil beast from Paraguay. As we have already seen, he dubbed this animal the "megatherium," classified it as a kind of giant sloth, and declared that it was probably extinct since no living example was known.

Cuvier realized that he was now in the position to provide definitive proof of the reality of extinction. With each passing year, he thought it less and less likely that any large new mammal—especially one the size of an elephant, hippopotamus, or rhinoceros—would be discovered still roaming the earth.[100] If he used

FIGURE 6. Georges Cuvier seated at his desk, 1798. Cuvier's research on fossil and living elephants provided the first definitive proof that extinction had taken place. Based on an oil portrait by Mathieu-Ignace van Brée. From L. Bultingaire, "Iconographie de Georges Cuvier," *Archives du museum*, 6th ser., vol. 9 (1932), frontispiece.

the principles of comparative anatomy to show that the bones of large mammals contained in the collections of the National Museum of Natural History and elsewhere were distinct from any known living species, he would demonstrate that extinction had in fact occurred. The fossil elephants, whose bones had been recovered from many locations and been the subject of so much controversy, provided just such an opportunity. In his first paper on the subject in 1796, Cuvier marshaled evidence to show that the two living elephants, from India and Africa, were distinct species, and that the fossil elephant found in Siberia and in Europe was both distinct from these two species and undoubtedly extinct.[101] He felt at the time that the American incognitum, which he called the Ohio animal, was also a distinct species, but he lacked sufficient evidence to declare it as such. Cuvier was not the first to suggest these ideas, but in the words of Martin Rudwick, his fossil elephant paper provided "detailed and almost irrefutable evidence for the reality of extinction."[102]

Ten years after his initial publication on fossil elephants, Cuvier revisited the pachyderms.[103] This time he benefited from access to Rembrandt Peale's *Historical Disquisition*, various fossil casts of the American incognitum that the elder Peale had sent him, and other previously unavailable material. In his 1806 paper, he identified the incognitum as the mastodon (or "breast tooth"), based on the animal's characteristic grinders, which were quite distinct from elephant teeth and once the source of so much confusion for Buffon, Daubenton, and other naturalists. The American mastodon was only one of several species of this newly named beast that Cuvier was able to differentiate, and he dubbed the Ohio animal the "grand Mastodonte." At last, the American incognitum had a proper name.

Between his first fossil elephant paper and his return to the subject a decade later, Cuvier described a slew of extinct fossil species—including not only the Siberian mammoth and various mastodons but also lost species of the rhinoceros, the hippopotamus, the Irish elk, a purported bear from caves in Germany, a doglike carnivore from Paris, and others that proved more difficult to characterize.[104] Cuvier concluded that each of these remarkable beings had disappeared, destroyed in one of a series of sudden geological catastrophes or "revolutions" that periodically swept across portions of the globe. Cuvier also realized that fossils could be used as markers of geological strata, and the "older the beds in which these bones are found, the more they differ from those of animals we know today."[105] With each new fossil beast Cuvier described, it proved increasingly difficult to continue to deny the reality of extinction.[106]

EMBRACING EXTINCTION

In an address delivered in 1807, the physician, naturalist, and University of Pennsylvania professor Benjamin Smith Barton argued that one of the "principal desiderata" in natural history was more research on the mammoth, megatherium, and other animals that "no longer exist."[107] Barton not only unapologetically embraced the idea of extinction in his address but also belittled those who continued to argue otherwise: "I speak of these animals as *extinct*. In doing this, I adopt the language of the first naturalists of the age. No naturalist, no philosopher; no one tolerably acquainted with the history of nature's works and operations, will subscribe to the puerile opinion, that Nature does not permit any of her species of animals, or of vegetables, to perish." While certain of the reality of extinction, Barton also continued to champion a belief in the order of nature, which he referred to as "a harmony beautiful and divine." In addition, he accepted the idea of gradation of one species into the next. But he vehemently denied the essential connection between species that had been implied in the notions of plentitude and the economy of nature: "THERE IS NO SUCH THING AS A CHAIN OF NATURE: an absolutely necessary dependence (on this earth) of one species upon another."[108] Barton hoped to publish a volume on extinct quadrupeds, but, like many of his ambitious plans, he never got much beyond announcing his intentions. He did, however, issue *Archaeologiea Americanae* (1814), which consisted of a series of short memoirs and letters on the subject. In a prescient letter to Thomas Jefferson on the mammoth and the mastodon reproduced in that volume, Barton reiterated his views about extinction while hinting that the processes that had destroyed the mammoth were sure to continue operating into the future. Apparently the swift pace of settlement in North America led Barton to declare that "the steps of this vast and generally unlooked for change, are rapidly preparing, in different parts of the world; and in none, I think, more rapidly than in the portion of it which we inhabit."[109]

Barton was silent on how one might reconcile an acceptance of extinction with the more established notion of a divinely ordered creation. One of his contemporaries abroad, however, the London physician and fossil collector James Parkinson, approached this issue head-on in his three-volume *Organic Remains of a Former World* (1804–11). While many "good and learned men" had dismissed the possibility of the "loss of a single link, in the chain of creation" as impossible, Parkinson argued otherwise: "that plan, which prevents the failure of a genus, or species, from disturbing the general arrangement, and œconomy of the system, must manifest as great a display of the power and wisdom, as could any fancied chain of beings, in which the loss of a single link would prove the destruction of the whole."[110] Rather than entirely jettisoning the chain of being, however, Parkinson reconceptualized one of its major components—the notion of plentitude—in more dynamic terms. The successive creation and destruction of species represented phases in the progress of nature toward a more perfect form: the human species.[111]

Two major compilations of American mammals published in the mid-1820s both included lists of fossil animals, and neither author felt compelled to explicitly defend the increasingly common assumption that these species had gone extinct. In *Fauna Americana* (1825), the Philadelphia comparative anatomist and Cuvier-enthusiast Richard Harlan included eleven fossil species of mammals that (he stated matter-of-factly) "no longer exist in a living state, in this, or any other country."[112] A year later, Harlan's chief rival, the Philadelphia anatomist and physician John Godman, began offering a competing three-volume treatise on North American mammals, *American Natural History* (1826–28). While the two naturalists disagreed on many points, Godman was also unflinching in his acknowledgement of extinction. Writing of the mastodon he reflected: "Enormous as were these creatures during life, and endowed with faculties proportioned to the bulk of their frames, the whole race has been extinct for ages . . . leaving nothing but the 'mighty wreck' of their skeletons, to testify that they were once among the living occupants of this land."[113]

During this same decade, the Scottish zoologist and minister John Fleming began arguing for a connection between historically documented declines in animal populations and the prehistoric extinction of animals whose fossil remains were being discovered in peat bogs, marl pits, and caves throughout Great Britain.[114] In his view, that common link was humans. "Man," Fleming asserted, "whether we view him as a savage or a citizen, is induced, by various motives, to carry on a destructive warfare against many animals, which he finds to be his fellow residents on the globe."[115] The ability of "rude tribes" to destroy animals was "limited in its objects, and uncertain in its results." But as human society progressed, "the objects of the chace ceased to be limited, while the methods of capture, and engines of death, become more numerous, complicated, and effectual."[116] It is possible, Fleming continued, to show that contemporary humans have greatly altered the geographical range of many species and "may have succeeded in effecting the total destruction of a few."[117] In recent times, the expansion of agricultural production and increased firearm use had reduced the population of many once abundant British birds and quadrupeds. Others, like the horse, the oxen, and the boar, no longer existed in the wild. Extinct animals whose fossil remains were uncovered in Britain—fantastic beasts like the Irish elk, the rhinoceros, the hippopotamus, elephants, hyenas, and many others—might have also fallen victim to earlier humans, whose bones and implements were found in "similar situations with their remains."[118] In many cases, fossil remains of extinct species were interspersed with the bones of species still living, so no universal catastrophe, like a flood, could explain why only some animals had been lost. While disease, local floods, and severe weather might have contributed to population declines, Fleming concluded that the "weapons of the huntsman completed the extinction" of these animals.[119]

A decade later, the British geologist Charles Lyell went a step further by arguing that extinction was part of the "regular and constant order of Nature."[120] In the second volume of his influential book, *Principles of Geology* (1832), Lyell embraced the notion that the earth had been changing for vast eons of time.[121] Mountains had risen and fallen, land bridges between continents had emerged and subsided, and bodies of water had expanded and contracted. It would be absurd to assume, Lyell concluded, that through all these profound geological changes, existing associations of plants and animals remained completely unaltered. Rather, as the landscape changed, animals migrated, while seeds were widely dispersed. As they gained a foothold in new regions, these newcomers competed with native flora and fauna, thereby threatening to overturn the previously established balance of that region. Nature, he argued, was fundamentally violent, characterized by a constant "struggle for existence."[122]

To illustrate his point, Lyell suggested considering the impact of the polar bear's migration into a region it had not formerly inhabited. The breakup of ice around Greenland periodically landed polar bears on nearby Iceland, severely frightening the human inhabitants of that island, who banded together to destroy the fearsome beast. Now imagine what might have happened the first time this notorious predator had landed on the island, before humans had begun living there: the "havoc which they would make among the species previously settled in the island would be terrific. The deer, foxes, seals, and even birds, on which these animals sometimes prey, would soon be thinned down."[123] Cascade effects would ensue as the decline of these species led to changes in grass, insect, and fish populations, which in turn would promote the increase of other insects and birds. The "numerical proportion" of many inhabitants of the island might be permanently altered through the arrival of this single species, possibly even driving some of the former inhabitants over the brink of extinction. After some time, though, a new population "equilibrium" would again be established.[124]

Closely following Fleming, Lyell argued that humans were also quite capable of disrupting the balance of nature. Lyell discussed the decline of British indigenous animals using many of Fleming's examples and concluded that they involved only the "larger and more conspicuous animals inhabiting a small spot on the globe."[125] He asked his readers to imagine the "enormous revolutions which, in the course of several thousand years, the whole human species must have effected, and will continue to go on hereafter, in certain regions, in a still more rapid ratio, as the colonies of highly-civilized nations spread themselves over the occupied lands." We should not lament or feel guilty about the havoc we commit, Lyell argued, for extinction, rather human induced or not, was perfectly natural:

> We have only to reflect that in thus obtaining possession of the earth by conquest, and defending our acquisitions by force, we exercise no exclusive prerogative. Every species which has spread itself from a small point over a wide area, must in like manner, have marked its progress by the diminution, or the entire extirpation, of some other, and must maintain its ground by a successful struggle against the encroachments of other plants and animals. . . . The most insignificant and diminutive species, whether in the animal or vegetable kingdom, have each slaughtered their thousands, as they disseminated themselves over the globe, as well as the lion, when first it spread itself over tropical regions in Africa.[126]

Lyell's words reveal just how much had changed in the fifty years since Jefferson penned his first account of the American incognitum. A natural world that was once thought to be orderly, static, and new was now incomprehensibly old and in constant flux. Large fossil mammals had been reconstructed, named, and placed on public exhibition. Extinction, which had been considered anathema, increasingly came to be viewed as a recurring part of the earth's long history. And humans, both historic and prehistoric, were recognized as possible agents of extinction. For any doubters who might remain, naturalists would soon provide compelling evidence that humans possessed the ability to destroy other species.

CHAPTER TWO

PARADISE LOST

UNRAVELING THE MYSTERIES OF
INSULAR SPECIES

We need not marvel at extinction; if we must marvel, let it be at our
presumption in imagining for a moment that we understand
the complex contingencies on which the existence of species depends.

CHARLES DARWIN, 1859

Should civilized man ever reach these distant lands, and bring moral, intellectual,
and physical light into the recesses of these virgin forests, we may be sure that
he will so disturb the nicely balanced relations of organic and inorganic nature as
to cause the disappearance, and finally the extinction, of these very beings
whose wonderful structure and beauty he alone is fitted to appreciate and enjoy.

ALFRED RUSSEL WALLACE, 1869

ISLAND INSIGHTS

By the early decades of the nineteenth century, the idea that myriad species had
suffered extinction over the eons of time since the earth's origin presented a
formidable challenge to the hoary notion of a static universe. Naturalists who
remained firmly in the grips of providential natural history turned to the Noachian flood or other divinely driven catastrophes in an attempt to gloss over
the apparent anomaly of recurring extinction across multiple geological strata.
Efforts to reconcile fossil evidence with theologically grounded theory became
increasingly problematic, however, as the number and diversity of newly discovered fossil forms continued to proliferate. Though the fossil record contained

many gaps, it was already sufficiently complete to raise troubling questions about God's intentions for and role in the natural world. Why had so many species been lost? How had they been destroyed? And what was the significance of these losses?

While grappling with the meaning of fossils, naturalists also struggled to construct a detailed inventory of the world's living species.[1] Beginning in the late Middle Ages and the Renaissance—the dawn of the so-called Age of Discovery— the nations of Europe developed an interest in exploration that strengthened with each passing century.[2] Spurred on by religious zeal, mercantile ambitions, and imperial longings, vessels destined for the far corners of the earth began regularly departing from European ports. They returned home with tales of wondrous new landscapes, accounts of strange new peoples, and specimens of unique plants, animals, rocks, and minerals that soon overwhelmed existing systems of classification. Explorers and naturalists sought out new organisms for a diversity of potential practical uses—as food, fur, and fiber, drugs and dyes, leather and lumber. They also hoped to contribute to the enterprise of cataloging the globe's flora and fauna. Living examples of exotic species found homes in newly established botanical and zoological gardens, institutions that served as biological clearinghouses to promote the movement of potentially useful organisms between colonial powers and their overseas possessions.[3] Preserved specimens were gathered into cabinets of curiosity, museums, and herbaria.[4] The ongoing campaign to produce a complete inventory of the world's plants and animals received a decided boost in the mid-eighteenth century, when Linnaeus developed relatively easy-to-use and widely adopted systems of biological classification and nomenclature to replace the competing systems that had been in place up to that point.[5]

In addition to naming, describing, and classifying newly discovered species, by the end of the eighteenth century naturalists also began trying to discern patterns in how those species were distributed across the globe.[6] Initially, they tended to base their ideas about geographical distribution on a fairly literal reading of the biblical story of the Ark; in short, all the earth's species descended from an original pair that had dispersed from Mount Ararat following the Noachian deluge. Though firmly wedded to the notion that God was directly responsible for the order in the world, Linnaeus also created intellectual space for more secular understandings of geographical distribution. He did so by arguing that all life, including human life, had been created in pairs on a high mountain that included a wide variety of ecological conditions. This mountain paradise contained belts of tropical, temperate, and arctic zones as well as the appropriate assemblage of plants and animals suited for each of these zones. As the primeval waters subsided, Linnaeus speculated, species migrated to their appropriate places on the globe. By the early nineteenth century, the German naturalist Alexander von Humboldt and the French botanist Augustin Pyramus de Candolle established the study of geographical distribution on much firmer scientific grounds through research on plant communities.

As areas with relatively clear boundaries and unusual biota, islands proved a particularly appealing site for naturalists struggling to understand how and why species came to be distributed. By the 1770s, for example, the German naturalist Eberhardt Zimmermann recognized that the number of species common to continents and their neighboring islands decreased as those islands became more distant from the mainland. Writing at about the same time, Johann Forster, a German naturalist residing in Britain who had accompanied Captain Cook on his second tour of the southern hemisphere, suggested that the total number of species on any given island was generally proportional to its circumference.[7] In the second volume of his *Principles of Geology* (1832), Charles Lyell summarized several other well-established generalizations about the geographical distribution of insular flora and fauna: the number of plants declined as islands became more distant from continents, while the proportion of unique (what he and other naturalists were beginning to call "endemic") plants increased and the number of quadrupeds decreased.[8]

Because they are relatively small and isolated, islands also provided early demonstrations of the human power to radically transform the natural world. In a provocative book published just over a decade ago, the environmental historian Richard Groves argued that during the eighteenth century naturalists and colonial administrators on islands like St. Helena, Mauritius, and Cape Colony offered pioneering critiques of environmentally destructive practices that had accompanied European settlement, including wide-scale deforestation, the introduction of exotic species, and the overhunting of native species.[9] Fearful that their island Edens were being despoiled, these naturalist and government officials also formulated policies designed to mitigate large-scale, human-induced environmental transformations. Surely, Groves is on to something important here, but it is unclear whether these initial calls of environmental alarm reached a wider audience or remained buried in dusty archives and obscure, long-forgotten publications.

What is certain is that by the middle decades of the nineteenth century, naturalists on both sides of the Atlantic developed a growing awareness both of the high rates of endemism among island species and the high degree of vulnerability those species faced following encounters with humans and the domesticated animals that invariably accompanied them. At the same time that Charles Darwin was groping toward his theory of evolution by natural selection—aided by his extensive firsthand experience with insular flora and fauna during the voyage of the *H.M.S. Beagle* (1831–36)—several of his colleagues were dramatically bringing home the reality of historic extinction through a series of publications treating several lost or vanishing flightless island birds: the dodo, moa, and great auk. These publications not only demonstrated that humans had managed to exterminate other species, they also prepared their contemporaries to appreciate the larger-scale continent-wide process of wildlife destruction going on around them.

This chapter explores the growing interest in insular species and insular extinctions during the middle decades of the nineteenth century. The first part shows how, within a relatively short period in the 1830s and 1840s, naturalists reconstructed the story of the dodo, the moa, and the great auk. These pioneering studies provided early, well-documented, and widely accessible examples of human-induced extinction, and, in the case of the dodo, offered what would become the paradigmatic case of this phenomenon. The second part of the chapter examines the critical role that observations about insular species played in the development of the theory of evolution by natural selection. Although Darwin appears not to have been directly influenced by knowledge about the fate of the dodo, moa, or auk, the extensive experience he gained with insular flora and fauna during his global travels proved crucial in the formulation of his ideas about how new species came into being. As it turned out, by providing a convincing explanation for why island species often seemed so vulnerable to predation from humans and newly introduced species, Darwin's theory of evolution provided a coherence to the previously puzzling patterns of insular biogeography. The controversial theory also played a key role in later debates about the meaning of human-induced extinction.

DEAD AS A DODO

During the middle of the nineteenth century, the dodo became the first prototypical symbol of extinction, a veritable icon of oblivion. The dodo (*Raphus cucullatus*) was a swan-sized, flightless bird that once inhabited Mauritius, one of three Mascarene Islands that dotted the vast Indian Ocean, approximately five hundred miles east of Madagascar.[10] It possessed a blue-gray plumage, a large dark bill with a reddish hooked tip, small yellow wings, yellow legs with black feet, and a tuft of white feathers on its rear end. Overall, it had the appearance of an ungainly juvenile bird, almost as though it were a caricature of itself.[11] Little is known about the dodo's life history or behavior except that it seems to have

FIGURE 7. Engraving of Roelandt Savery's dodo portrait, 1848. Based on overstuffed specimens or animals kept in captivity, early depictions of the dodo (like this one completed ca. 1625) often portrayed it is a plump, sluggish, and ungainly bird. Scientists now believe the dodo was much slimmer and more fleet of foot than these early images suggest. From Hugh E. Strickland and Alexander G. Melville, *The Dodo and Its Kindred* (1848).

inhabited coastal areas, where it nested on the ground and apparently lived on fruits, berries, and seeds.

In addition to the dodo of Mauritius (an island about 700 square miles in extent), the neighboring islands of Réunion (a little larger than and 100 miles southwest of Mauritius) and Rodrigues (a tiny island 360 miles east of Mauritius) each harbored their own distinct species of related birds. Even less is known about these species than the dodo of Mauritius. The Réunion solitaire (*Raphus solitarius*), also called the white dodo, was yellowish-white in color, with black-tipped wings, and about the same size as its distant cousin on Mauritius. The brownish Rodrigues solitaire (*Pezophaps solitaria*), on the other hand, was taller, more slender, and had a shorter bill than the dodo.

Although Arab merchants visited the Mascarene Islands sometime before the end of the first millennium, they failed to settle there, and we know nothing about their long-term impact on the islands' flora or fauna.[12] The Portuguese began regularly stopping on Mauritius in the early sixteenth century, while plying the Indian Ocean to trade with Africa and the Indies. Only scanty accounts of their encounters with the region survive, but they did introduce the first monkeys and goats to the island, a common practice aimed at ensuring a ready food supply for future voyages. The Dutch became the first to stake a formal claim on Mauritius and the first to introduce the dodo to the Western world following Jacob Cornelius van Neck's expedition to the island in 1598. Seven years later, the French botanist Clusius offered the earliest scientific description of the species based on a foot preserved at the house of his friend, the anatomist Peter Pauw, together with accounts from ships' logs and sailors.[13]

The dodo soon gained a reputation for being gawky, dim-witted, and gastronomically suspect. The Dutch referred to it as the *walckvögel*, meaning "disgusting bird," apparently a reference to the toughness of its flesh.[14] Despite whatever qualms they might have had about how it tasted, the fearless bird offered an easy mark that proved irresistible to hungry sailors. The exact derivation of the term "dodo" remains uncertain. The *Oxford English Dictionary* relates it to the Portuguese word *duodo*, which means "simpleton" or "fool," while other scholars have associated it with the Dutch words *dod-aarsen*, which means "fat rump" or *doddor*, which means "sluggard." When Linnaeus provided the dodo with a scientific name—*Didus inpetus* or "clumsy dodo"—even he perpetuated the stereotypes that had developed around the species.[15] Buffon did the same, comparing the dodo to a "turtle that wrapped herself in the remains of a bird, and Nature, by giving it these useless frills, seems to have wanted to top them with the embarrassment of excessive weight, the ineptitude of movements, and the inertia of the mass, and to make its heavy thickness even more shocking because it belongs to a bird." At least most other flightless birds, like the ostrich, were capable of running fast; the dodo, on the other hand, "seems immobilized by its own weight, and to be able only of dragging itself; it seems to be composed of brute, inactive matter, where not enough living molecules were used."[16] As we shall see, the portrayal of the dodo as too slow, stout, and stupid to evade the new predators that arrived on the shores of Mauritius now seems a classic case of blaming the victim.[17]

Less than a century following its discovery the species had utterly vanished. The last definitive sighting of a dodo on Mauritius was in 1681, while the white dodo and the Rodrigues solitaire probably hung on for a few decades longer.[18] Overhunting and the proliferation of introduced animals—pigs, goats, rats, and monkeys—played havoc with the species that had evolved in an environment with few predators and thus showed no fear of the humans or the introduced species that either pursued it, its young, and its eggs or that competed for its habitat and food. The dodo was, in the words of the science writer David Quammen, an "ecologically naive" species that through eons of time had flourished by successfully adapting itself to an unusually benign insular landscape.[19] That mode of living proved fatal, however, when Europeans and their biota suddenly arrived on the islands.

The destruction of the dodo and the record of its life proved so complete that by the end of the eighteenth century, naturalists began questioning whether it had ever truly existed.[20] Only the most fleeting of evidence—a handful of published and sometimes-contradictory accounts in voyages of discovery, a half-dozen illustrations (several of which seemed to be copies of each other), and a few scattered bone fragments in museums—was all that remained of the species. Was the dodo another mythological creature—like the mermaid, unicorn, phoenix, or griffin—the product of overactive imaginations and undisciplined minds? Or was it a hoax, as the British Museum curator J. E. Gray suggested after noting that the painting of the bird under his care appeared to have artificially joined the head of a vulture-like bird with the feet of a gallinaceous species?[21]

In the 1840s, the British naturalist Hugh Strickland sought to definitively establish the dodo's reality as a once living, breathing creature that humans had driven to extinction. Strickland, who developed an interest in collecting birds and fossils from a young age, began making systematic geological studies following matriculation from Oriel College, Oxford, in the late 1820s.[22] By 1842, his scientific standing was sufficiently advanced that he played a leading role in a British Association for the Advancement of Science committee charged with framing a set of standard rules for zoological nomenclature. Charles Darwin and Richard Owen were among the other prominent members of this committee, whose findings soon gained wide acceptance. Not long afterward, Strickland developed an interest in the plight of the dodo. In addition to authoring several articles and offering reports at scientific meetings on the species, he and the anatomist and physician A. G. Melville published *The Dodo and Its Kindred* (1848), the first scientific monograph on the bird. There Strickland declared that the dodo and its cousins provided the "first clearly attested instances of extinction of organic species through human agency," although he did note that humans also probably killed off the Irish elk and the aurochs—a form of wild ox—in ancient times and the northern dugong more recently. The process of human-induced extinction seemed to be accelerating, he noted, and numerous animals and plants were now "undergoing this inevitable process of destruction before the ever-advancing tide of human population."[23] Strickland shared the widely held belief that species had definite life spans, not unlike individuals. But he also thought that they could be cut off prematurely by "violent or accidental causes," especially "the agency of Man."[24]

Although the specter of anthropogenic species loss clearly haunted Strickland, like his colleague Lyell, he faced the prospect with some degree of resignation:

> We cannot see without regret the extinction of the last individual of any race of organic beings, who progenitors colonized the pre-adamite Earth; but our consolation must be found in the reflection that Man is destined by his Creator to be "fruitful and multiply and replenish the Earth and subdue it." The progress of man in civilization, no less than his numerical increase, continually extends the geographical domain of Art by trenching on the territories of Nature, and hence the Zoologist or Botanist of future ages will have a much narrower field for his researches than that which we enjoy at present.

If human-caused extinction were an inevitable fact of modern life, as Strickland predicted, then naturalists had a duty to busy themselves with collecting, cataloging, and describing "these extinct and expiring organisms" before it was too late.[25] Strickland's response to the threat of extinction—his sense that his colleagues ought to actively gather vanishing organisms while it was still possible—would remain common within the scientific community for at least the next half century.

Strickland supported his claim about the reality of the dodo with historical, pictorial, and anatomical evidence. First, he reviewed fifteen "original and independent" accounts of the species, including a mid-seventeenth-century notice of a live specimen exhibited in London.[26] He then analyzed the five known oil paintings depicting the dodo, including three that were completed by the well-known Dutch artist Roelandt Savery and a fourth done by his nephew John Savery.[27] Finally, he turned to a discussion of the bird's meager surviving fragments: a foot in the British Museum, a cranium in the Gottorf Museum in Copenhagen, and a desiccated head and foot in the Ashmolean Museum in Oxford. The latter two fragments were all that remained of a once complete specimen that had decayed over the course of a century and barely escaped destruction during a bout of spring cleaning at the museum.[28] The episode prompted Lyell to remark: "The death of a *species* is so remarkable an event in natural history, that it deserves commemoration, and it is with no small interest that we learn, from the archives of the University of Oxford, the exact day and year when the remains of the last specimen of the dodo, which had rotted in the Ashmolean Museum, were cast away"—January 8, 1755.[29] Strickland noted that paleontologists often had much

FIGURE 8. Engraving of dodo fragment in Oxford's Ashmolean Museum, 1848. This specimen represented one of the few surviving remains of the dodo at the time of its publication. The initial paucity of physical evidence for the bird led many to question whether the dodo had actually ever existed. From Hugh E. Strickland and Alexander G. Melville, *The Dodo and Its Kindred* (1848).

more data with which to characterize species that perished "myriads of years ago" than the surviving physical evidence for the dodo and its kin.[30] To rectify this shortcoming, he urged naturalists to search diligently in caves and alluvial deposits throughout Mauritius to locate additional remains for study.

Based on his limited range of sources, Strickland concluded that the dodo was a "massive, clumsy bird, ungraceful in form, with a slow waddling motion."[31] Still, despite his generally disparaging description, he remained tied to a providential view of natural history; he firmly believed that the Creator had assigned each class of animals a definite type or structure and that the overall form of each species was perfectly adapted to the circumstances in which it lived. The doddering dodo was no exception, and prior to the arrival of humans, it had apparently thrived on the island. As to its affinities with other known birds, Strickland rejected the arguments of his naturalist colleagues, who until then had speculated that it was a form of ostrich, vulture, albatross, or fowl. The dodo and its cousins were a kind of oversized pigeon, he asserted, an opinion that gained confirmation in the mid-1860s, when George Clark, a school teacher on Mauritius, located a large number of dodo bones in a marsh known as the *Mare aux Songes* (Pond of Dreams).[32] Clark unearthed enough bones to make several more-or-less complete skeletons, and after shipping some to England, he sold the remainder to museums throughout the world.[33]

These and other newly discovered bones then became the object of a long series of scientific publications. Richard Owen, who had already published several articles on the flightless birds of New Zealand, for example, published his *Memoir on the Dodo* in 1866 from some of Clark's specimens. There he speculated that since the species had been "exempt from the attacks of any enemy" over the course of successive generations, it might have gradually lost its capacity for flight while gaining size and strength through the Lamarckian principle of use and disuse.[34] While the absence of predators had produced the species, their introduction proved fatal to it.[35] The Cambridge professor Alfred Newton and his brother Edward Newton, both founders of the British Ornithologists' Union, also showed a keen interest in the dodo.[36] Part of their curiosity about the lost bird was undoubtedly related to the fact that Edward served in the foreign service in Mauritius from 1859 to 1877, rising to the rank of colonial secretary. In addition to their researches on the Mauritius dodo, Edward visited Rodrigues island in 1864, where he unearthed bones of the Rodrigues solitaire in a cave.[37]

The fame of the dodo and its status as an archetype of extinction, however, ultimately rested not so much on the labors of Strickland, Owen, Newton, and other naturalists, or even the many bones placed on exhibit at leading museums around the world. Rather, public knowledge of the species probably owed much more to the labors of the eccentric mathematician and Oxford don Charles L. Dodgson (a.k.a. Lewis Carroll). In 1865, the same year that Clark reported on the cache of dodo bones he had discovered, Carroll incorporated the creature into the fantasy adventure *Alice in Wonderland*. Carroll's book, which became one of the most popular children's stories of all time, began as a series of tales he began recounting to the children of his colleagues and neighbors.[38] The illustrator of the first edition, the well-known *Punch* magazine cartoonist John Tenniel, based his depiction of the dodo on the ungainly image that Roelandt Savery had created two centuries earlier.[39]

Although the dodo remains the ultimate symbol of human-caused extinction, naturalists have recently begun to rehabilitate its reputation as a fat, clumsy bird. Following a series of calculations based on its bones, the Scottish naturalist Andrew Kitchener has concluded that the prevailing image of the dodo is false.[40] That stereotype, he argues, emerged from drawings and observations made of captive birds that had grown excessively large from overfeeding, improper diet, and lack of exercise. The species was swifter and leaner in the wild, Kitchener claims, a view that he confirmed after discovering several early drawings of the bird that had apparently been sketched in the field. While many questions remain about the dodo, one thing remains certain: history has not been kind to the species or its image.

A few years before Strickland began reconstructing the story of the lost dodo, the paleontologist and comparative anatomist Richard Owen was busy studying another enigmatic flightless bird and erstwhile island dweller: the moa. The Hunterian Professor of Comparative Anatomy at the Royal College of Surgeons, Owen had first earned a reputation as one of Britain's leading naturalists in part through his careful analysis of the mammalian fossils that Charles Darwin brought back from South America.[41] Later, following the publication of the *Origin of Species* (1859), he would become one of Darwin's staunchest critics. Standing at six feet tall, Owen was literally and figuratively a towering presence in British natural history circles, with a list of honorary distinctions that contained over a hundred items and a scientific output that numbered a staggering six hundred papers and twelve book-length studies. He also introduced countless creatures to the public, including the archaeopteryx and many dinosaurs. His family and close colleagues considered him to be warm, affable, and charming. Yet, he could also be a petty, autocratic, and jealous, the sort of fellow who was clearly not one to hide his light under a bushel.

In late 1839, one of Owen's colleagues at the Royal College of Surgeons, John Rule, approached him with an unusual bone fragment, a portion of a femur that he had received as a gift two years earlier.[42] The six-inch-long segment had come from Rule's nephew, a mercantile agent and native trader who had settled at Poverty Bay, on the east coast of North Island, New Zealand, and taken a Maori bride. When he gave the fragment to Rule, who was then practicing surgery in Australia, Harris indicated that local Maori tradition held that it belonged to a "bird of the Eagle kind, which has become extinct. They call it 'A Movie.' They are found buried on the banks of rivers."[43] After returning to London in 1839, Rule compared his femur to those of other birds found in several museum collections. To his delight, his specimen seemed larger, and at some point, it dawned on him that he might parlay the gift into a tidy profit. After failing to interest the British Museum in the fragment, he sent a letter to Owen offering it for ten guineas. There Rule speculated that the femur belonged to a large, extinct "bird of flight" from New Zealand.[44]

When Rule visited the Royal College of Surgeons with the bone in hand, Owen was busy preparing for a lecture and thus understandably distracted. Nonetheless, even from a quick glance, he immediately surmised that the bone before him appeared too massive to have come from a bird with the powers of flight. He initially dismissed Rule as a crank and his fragment as "a tavern delicacy" from an ox, a "marrow-bone like those brought to the table wrapped up in a napkin."[45] When Rule pointed out the unusually porous, cancellated structure of the bone's interior wall, and produced a greenstone *mere,* a distinctive type of Maori war club, to emphasize the bone's geographic origin, he finally gained Owen's attention. The famed comparative anatomist agreed to get back to Rule as soon as he had the chance to examine the specimen more carefully.

After comparing the bone in question to the skeleton of an ox and a number of other domesticated animals, Owen quickly realized that he had clearly erred in his initial hunch about its origin. Turning to the ostrich, he found Rule's bone was similar in size but different in shape from the femur of this well-known African bird. Following comparisons with several other large flightless birds—the cassowary, emu, and rhea—he became convinced that Rule's specimen belonged to some undescribed bird that was a more immense and sluggish species than its African relative. Later Owen and his supporters would tout the discovery as a demonstration of Owen's prodigious ability and a triumph of Cuvierian functionalism.[46] The British geologist and *Iguandon* discoverer Gideon Mantell, for example, praised Owen's deduction as "the most striking and sagacious application" of Cuvier's principles ever, "the felicitous prediction of genius enlightened by profound scientific knowledge."[47] Just as the famed Cuvier had long boasted that he could successfully predict the overall appearance of an animal from a single bone, so too had Owen—who once studied with Cuvier for several months—been able to predict the overall appearance of the moa from Rule's fragment. In actuality, the story of the discovery of the moa was much more complex than that, and Rule's initial information and his insistence that Owen take a closer look at the specimen deserve some credit.

At first his colleagues remained skeptical about Owen's characterization of the fragment.[48] The museum committee of the College of Royal Physicians, for example, refused to purchase Rule's specimen, while the publication committee of the Zoological Society inserted a disclaimer into Owen's brief account of the moa declaring that "the responsibility of the paper must rest with the author." Many naturalists questioned whether such a large bird could have inhabited such a modest-sized island.

In the meantime, several New Zealand residents with a keen interest in natural history had begun independently gathering information on the moa. The first mention of the bird in print actually came in 1838, four years after Joel Polack, the owner of a trading cutter damaged by storms off the east coast of New Zealand, had been forced ashore for repairs just north of Poverty Bay. The Maori there showed him several large "fossil ossifications" that they claimed belonged

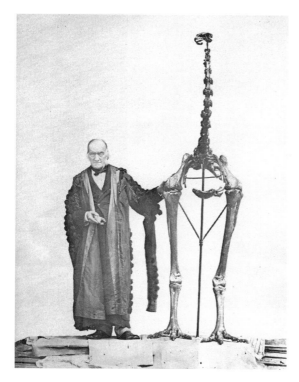

FIGURE 9. Richard Owen with a large moa skeleton, 1879. Here Owen is holding the bone fragment from which he originally described the moa forty years earlier. He stands next to a reconstruction of *Dinornis giganteus,* the largest of nearly two dozen moa species he introduced to science. From Richard Owen, *Memoirs on the Extinct Wingless Birds of New Zealand* (1879), vol. 2, plate 97.

to "very large birds" that had once roamed the countryside before being exterminated through overhunting. Polack decided that the bones belonged to some kind of large "emu" or similar bird of the genus *Struthio,* but he published his findings in a relatively obscure travel account and failed to produce the bones of the creature or to give it a name.[49]

The term "moa" first surfaced in 1838, the same year that Polack published his account of the species, when the missionary William Williams and the mission printer William Colenso traveled to the east coast after seeing a Maori translation of the New Testament through to publication. As Colenso later pointed out, "moa" was the Malay term for the cassowary (another flightless bird that inhabits Australia and New Guinea) and Polynesian for the domestic fowl.[50] The Maori that Williams and Colenso encountered told stories of a huge domestic rooster with the face of a man that inhabited a cave in the Whakapunake Mountain. While no one still living in the area claimed to have seen the bird firsthand, its bones were frequently found. A year later, Williams returned to the site with the Rev. Richard Taylor, and he began offering a reward for any remains of the moa that might be delivered to him. Each of these men—Williams, Colenso, and Taylor—subsequently published accounts of the species, leading to a bitter priority dispute that simmered for many years.[51]

In 1842, Williams and Colenso sent two large consignments of moa bones to the geologist and paleontologist William Buckland, one of Owen's colleagues in London. After Buckland granted Owen access to the specimens in the first package, he felt confident that he finally had enough material to formally name the new species. Owen proclaimed that the bones represented the remains of a new genus, which he dubbed *Dinornis,* from the Greek *deinos,* meaning "prodigious," "surprising," or "terrible."[52] He chose the name to make a splash, and he would recycle the prefix later that decade when he described the first dinosaurs.[53] The first species in this genus he dubbed *Dinornis nova ʒealandiae,* after the nation where it had been unearthed. Over four decades, he named and described more than twenty different kinds of moa in more than thirty papers, many of which were gathered into a two-volume anthology entitled *Memoirs on the Extinct Wingless Birds of New Zealand* (1879).[54]

Most of the bones that Owen studied had not yet fossilized. That fact, along with occasional reports of sightings, led to much speculation about whether the giant bird, which grew to as tall as twelve feet, might still roam the vast unexplored regions of that remote colony.[55] In his first notices on the moa, Owen hedged his bets by describing the species as one that had clearly once lived and might still exist somewhere in New Zealand. With increasing exploration and settlement, however, the likelihood that a bird that massive would continue to elude naturalists seemed increasingly slim. The failure to find any living examples of the mysterious bird coupled with the discovery of large caches of moa bones associated with human remains and artifacts—including apparent cooking and butchering sites—led to speculation that the Maori might be responsible for the extinction of many moa species.[56] Although the case for human culpability was not as clear-cut as it had been for the dodo, the suspicion that Maori had done the species in remained strong. As early as 1844, Owen himself began blaming humans for the presumed extinction of the moa: "It is not altogether improbable that the species of Dinornis were in existence when the Polynesian colony first set foot on the island; and, if so, such bulky and probably stupid birds, at first without the instinct and always without adequate means of escape and defence, would soon fall prey to the progenitors of the present Maoris."[57]

The German geologist Johann F. J. Haast, who came to New Zealand in 1858 as a British shipping company agent, was also fascinated with moas.[58] An indefatigable worker, he became instrumental in establishing the Canterbury Museum, at Christchurch, which began on a relatively modest scale in 1861 and opened the doors to its newly constructed building in 1870. In addition to figuring prominently in the exhibits at the museum, moa bones proved crucial to the growth of the collections; Haast engaged in a brisk trade in moa bones, which proved in great demand because, as one contemporary observer described it, "what Niagara is to ordinary waterfalls, the Moa was to all the bird-tribe."[59] Through sale and barter of this unique resource, Haast acquired biological organisms from around the world, and by 1873, the Canterbury Museum boasted an impressive fifty-six thousand specimens. Haast also published widely in the fields of natural history and archeology, including twenty-three papers on extinct birds.

Even if Maori were ultimately responsible for the extinction of the moa, over a century and a half after Owen published the first scientific account of the species, countless mysteries remain. What role did Maori fires, which transformed the habitat of much of the eastern South Island, play in the decline of the species? Did the introduction of poultry diseases, dogs, and rats that had accompanied the Maori contribute to the decline? Before the arrival of the Maori, New Zealand was a relatively small, faunistically impoverished island that contained only a handful of amphibian and reptile species, two mammals (both bats), and numerous birds. How did so many species of the moa come to inhabit an island that small and with so few other vertebrate species? Even more basic questions— like, Exactly how many species of moa were there? And when had each gone extinct?—continue to elicit comment and provoke controversy, despite the development of sophisticated new methods of analysis, like carbon dating. Writing in 1931, the longtime New Zealand resident T. Lindsay Buick noted that of all the ratites, past and present, "the Moa is indisputably the bird shrouded in the greatest mystery and steeped in the richest glamour."[60] In fact, more recent research has shown it to be only one of more than two dozen other endemic birds driven to extinction shortly after the arrival of the first humans to the islands.

IN SEARCH OF THE GREAT AUK

The same year that Owen first blamed the demise of the moa on the Maori, the last definite encounter with another flightless bird took place on Eldey, a small volcanic island surrounded by steep cliffs located ten miles off the coast of Iceland.[61] Prompted by a Reykjavik natural history dealer who had previously brokered the sale of nearly two-dozen auk skins and eggs, a local fisherman named Vilhjalmur Hakonarrson organized a party of fourteen men to row out to the island. Their aim was to capture specimens of this increasingly valuable bird, which had become a prized commodity as word of its rarity had spread through the natural history community. The three fishermen who landed on

FIGURE 10. Audubon engraving of the great auk, 1827–38. Audubon ventured to Labrador in 1833 with the hope of locating a specimen of this rare species to sketch. Although he found several other desirable species, he failed to discover a single great auk and was forced to use museum specimens as models. From John James Audubon, *Birds of America* (1827–38). Courtesy of the Darlington Memorial Library, University of Pittsburgh.

Eldey in June 1844 found only two great auks and managed to capture both; in the process they also smashed an auk egg they found. While numerous observers claimed to have spotted the species after 1844, these were not only the final two specimens to be collected but also the last incontrovertible evidence of the great auk's existence in the wild.

The great auk (or garefowl), known to scientists as *Pinguinus impennis*, originally ranged across the boreal and low-arctic waters of the north Atlantic.[62] It was a large bird, coming in at about thirty inches tall and weighing about eleven pounds, with a massive laterally flattened bill, black upperparts, white underparts, white patches in front of its eyes, and small wings that it used to help propel itself while diving for fish. The bird was the first to be called a penguin, which is probably derived from the Latin word *pinguis* (meaning fat); Europeans familiar with the north Atlantic species later applied the same name to an unrelated family of flightless birds in the Southern Hemisphere.

During the summer breeding season, great auks concentrated on a small number of offshore islands in the Gulf of St. Lawrence, east Newfoundland, Iceland, northern Scotland, and a handful of others sites. There they incubated a single egg on bare rocks. The most famous of these nesting sites was on Funk Island, off northeastern Newfoundland, which probably earned its name from the over-whelming aroma of bird excrement greeting those who dared to land there. At the time of first European contact, as many as one hundred thousand pairs of the great auk crowded onto the site, which was about a half a mile long and a quarter of a mile wide.[63] Isolated from the mainland and surrounded by steep cliffs, predators rarely ventured to this or other major auk nesting grounds. As had been the case of the dodo and the moa, the relative lack of predators meant that the great auk showed little fear of the humans.

Hungry sailors who began plying the north Atlantic waters in the sixteenth century feasted on the fat, fearless birds with abandon. In 1620, Captain Richard Whitbourne sang the virtues of the species, pointing out that they were "as bigge as Geese, and flye not, for they have but little short wings, and they multiply so infinitely, upon a certain flat Iland, that men drive them from thence upon a boord, into their boates by hundreds at a time, as if God had made the innocency of so poore a creature, to become such an admirable instrument for the sustena-tion of man."[64] In addition to being hunted down for its meat, the great auk was also aggressively pursued for its eggs, feathers, and oil.

By the end of the eighteenth century, the first calls of alarm about the plight of the species began to be sounded. In 1785, Captain George Cartwright predicted that if the practice of slaughtering the birds on Funk Island did not soon end, "the whole breed will be diminished to almost nothing."[65] A year later, the gover-nor of Newfoundland issued a proclamation protecting the island's great auk and its eggs, though it exempted birds captured for fish bait and seems to have been rarely enforced. Even as their numbers became thinned, persecution continued from individuals attracted to the islands by the other sea birds that congregated there in large numbers.[66]

In 1833, when John James Audubon traveled to Labrador to collect specimens, make observations, and complete drawings for his *Birds of America,* he landed on several immense seabird colonies in the Gulf of St. Lawrence.[67] There Audubon and his party witnessed several small, motley bands of men engaged in whole-sale bird and egg destruction for profit. He was so disgusted that he included a graphic account decrying the "eggers'" predations in the text that accompanied his plates. According to Audubon, on their first visit to a given site, the men en-gaged in this barbarous trade often destroyed every single egg they found. After repeating this process on as many islands as they could reach over the course of about a week, the party returned to each island to gather the freshly laid eggs. The men also massacred vast numbers of some species for their feathers. To add insult to injury, the eggers Audubon met up with failed to fulfill his request for specimens of rare great auk or eggs for his work. "So constant and persever-

FIGURE 11. John James Audubon in the field, ca. 1845–50. Featured here with his horse, dog, gun, and knapsack, the tools he used to obtain the specimens he drew for his monumental *Birds of America*. Audubon's widely reproduced images helped cultivate sympathy for the plight of vanishing birds. Courtesy of the Library, American Museum of Natural History, New York.

ing are their depredations," he lamented, that species that "were exceedingly abundant twenty years ago, have abandoned their ancient breeding places, and removed much farther north in search of peaceful security." "This war of extermination cannot last many more years," Audubon grimly predicted.[68] While he had no qualms about collecting specimens—sometimes dozens of examples of a single species—to produce his drawings, the large-scale commercial slaughter he witnessed in Labrador appalled him.

After three months of traveling in the region, Audubon managed to bring back seventy-three bird skins and to complete twenty-three large drawings for his magnum opus, despite the often cold, damp, and stormy conditions he and his party of five men encountered in the field.[69] One of the trip's greatest disappointments, however, was the failure to locate a single great auk. Local residents had spoken of "penguins" that they thought continued to reside on some of the islands in the region, but Audubon was unable to find any of them himself, and it now seems likely that by the time he arrived the birds had been lost from the area for many years. In the end, he was forced to base his great auk drawing on a skin purchased from a natural history dealer, a specimen that probably came from the island of Eldey, where the last pair of auks was collected in 1844.[70]

By the time of Audubon's visit, a run on great auk specimens had commenced, fueled by egg and skin collectors who vied keenly with one another to possess examples of the creature before it was altogether lost. Then, quite suddenly, in the mid-1840s, the supply of new specimens dried up, fueling an escalation in prices for the remaining birds and eggs. The price of stuffed specimen reached the considerable sum of £50 by the 1860s; twenty years later, auk eggs sold for as much as £110 and auk skins for up to £225.[71] Having been commodified in life, the great auk became even more valuable following its extermination. In all, about seventy-eight skins and about seventy-five eggs were gathered up before the species passed into oblivion.[72]

Among the many keen auk aficionados during the middle and later decades of the nineteenth century were two British naturalists, John Wolley and Alfred Newton.[73] With an passion that bordered on obsession, both men sought to amass specimens of the lost bird and its eggs, learn more about its life history, and reconstruct the story of its demise. Their curiosity ran so deep that in the early 1850s Wolley and Newton traveled to Iceland with the hope of learning what had transpired on Eldey in 1844. Somehow they managed to locate and interview several of the men involved in the slaughter of what seemed to be the last pair of great auks collected from the wild, though they never could track down the two specimens gathered that fateful day.[74] Wolley filled several notebooks with his auk research and ultimately hoped to publish a monograph on the species, but he died prematurely in 1859. Newton, who continued the research, remained among those who initially refused to accept the idea that the auks were gone. In a letter written in 1858, he still clung to a ray of hope:

> As to the extinction of the Great Auk, if it is extinct I think it has been mainly accomplished by human means. . . . Under the influence of the "Almighty Dollar" . . . those poor birds were persecuted, their eggs plundered and their necks broken to supply the demand which museums were then creating. And so the number dwindled, until in 1844, the only two then to be seen were taken, their egg broken (the shell left on the rock) and their skins shipped to Europe. I do not think there is any good evidence of the bird being seen since that time; but I confess I do not give it quite up, nor shall I for the next five or six years, though the places suitable for its breeding station must be very few in number.[75]

Newton shared Wolley's dream of producing a monograph on the garefowl, and after his collaborator died, he published the first in a series of articles on the species. He continued to amass information on the auk for more than thirty years, but never managed to complete the masterwork of which he dreamed.

As had been the case for the dodo, popular accounts of the great auk probably exerted more impact on the public's perception of the species than scientific ones. Charles Kingsley, a professor of history at Cambridge, naturalist, and friend of Newton, included the great auk in his well-known children's book, *The Water-Babies*, first published in 1863.[76] Kingsley depicted the last of the species standing all alone, crying tears of pure oil. Accounts of the sales of the auk and its egg also garnered media attention for the species, especially as the average selling price of specimens mounted. Eventually the great auk would be featured on prints, cigarette trading cards, and tobacco tins.[77]

It is often impossible to state with certainty exactly when a given species goes extinct. Lingering doubts about the status of the great auk probably blunted the potential impact of its loss, but the publicity surrounding both the decline of the great auk and its increasingly valuable skins and eggs contributed to raising awareness about human-induced population crashes in wildlife. The same was true of the dodo, where both the fact of extinction and human culpability seemed more straightforward. The case of the Maori and the moa proved more complex, but taken together, the loss of these three flightless island birds made a distinct impression not only within natural history circles but also in the broader culture. In the second half of the nineteenth century, all three species would be repeatedly deployed as reference points to document the threat of impending extinction of other wildlife.

DARWIN AND THE IMPORTANCE OF BEING INSULAR

While others were reconstructing the stories of the dodo, moa, and the auk, Charles Darwin and Alfred Russel Wallace struggled to understand the larger patterns of geographical distribution among island species. These cofounders of the theory of evolution by natural selection succeeded by combining broad reading in the scientific literature of their day with extensive firsthand experience with insular flora and fauna during their lengthy travels abroad. Both kinds of evidence proved crucial to the development of their ideas about how new species originated and distributed themselves across the globe. Prior to the publication of Darwin's *Origin of Species* in 1859, the most commonly held view was that an omnipotent, omniscient God had specially created plants and animals, which were therefore perfectly adapted for the environments in which they resided. Yet, this theistic explanation for the origin and distribution of life raised numerous questions when it came time to reconcile it with what had been learned

FIGURE 12. Watercolor of Charles Darwin, 1840. Drawn by George Richmond only a few years after Darwin returned from the voyage of the *Beagle*, a five-year expedition that proved central in his intellectual odyssey toward the theory of evolution by natural selection. Courtesy of the Darwin Heirlooms Trust.

about insular species. Why did islands contain so many unique species, like the dodo and the moa? Why did insular species seem so vulnerable to predation by humans and domesticated animals? And why did the number of species on islands seem to vary with their distance to neighboring mainlands? Darwin and Wallace's theory to explain how species came into being also offered compelling answers to these and other questions that increasingly puzzled naturalists. In the end, their evolutionary ideas ushered in a profound change in how not only the scientific community but also the public at large viewed the natural world. Far from being seen as an anomaly to be explained away, extinction became reconceptualized as central to the teeming diversity of life on earth.

Charles Darwin was the fifth child of a father who was a physician in Shrewsbury and a mother who came from the prominent, wealthy Wedgwood family.[78] As a child, he seemed more interested in collecting beetles than serious study of any sort. At the age of sixteen, he went off to Edinburgh University to train as a physician, but the sight of surgery performed without anesthesia so disturbed him that he left after two years; he spent the next three years at Cambridge University, where he hoped to train for the clergy. There he met the botanist John S. Henslow, who fired young Darwin's interest in natural history. Impressed with the potential of his otherwise unexceptional student, Henslow arranged for Darwin to become an unpaid naturalist on the *H.M.S. Beagle*, an Admiralty vessel charged with surveying the coast of South America and several Pacific islands. In the words of Gavin de Beer, the five years he spent on the voyage proved "the most important event in Darwin's intellectual life and in the history of biological science."[79] He left England late in 1831 with minimal formal training in science and the belief that creation was essentially static; he returned in 1837 as a seasoned naturalist convinced that the diversity of the natural world had resulted from gradual evolutionary change.

Fossils played a key role in the evolution of Darwin's ideas about species mutability. One of the books he took with him on the *Beagle* voyage was the first volume of Charles Lyell's *Principles of Geology*, and he received the second volume late in 1832, while in Montevideo. Darwin was immediately taken with Lyell's uniformitarianism, the belief that the earth has been shaped by the same terrestrial forces that were currently in operation and that those forces had always acted with similar levels of intensity.[80] This idea was in contradiction to the more widely held views of catastrophism, which stressed sudden, drastic, and more intense transformations in the earth's crust. In addition to applying Lyell's uniformitarian theory to numerous geological formations he visited during his five years on the *Beagle*, Darwin also absorbed many of his colleague's notions about the meaning of fossils, including his belief that extinction had been a regular part of the history of life on earth. Unlike many of his catastrophist colleagues, Lyell had blamed extinction on gradual environmental transformations.[81] As climatic conditions changed over time, species would be forced either to migrate or gradually perish as they became increasingly ill equipped to survive in the new environment.

The reality of extinction was brought home dramatically to Darwin when he found and excavated large quantities of fossil mammals in the vicinity of Bahía Blanca, on the eastern coast of South America. "I have been wonderfully lucky with fossil bones," he boasted after nearly a week of work at the first site in September 1832. "Some of the animals must have been of great dimensions: I am almost sure that many of them are quite new."[82] He was convinced that one of the massive jaws unearthed at the site belonged either to the megalonyx or a megatherium; a year later he found remains of the mastodon at another site. How and why had these large mammals become extinct? Darwin rejected the catastrophist explanations that Cuvier had proposed, variations of which remained common in his day. Rather, following Lyell's lead, he embraced the notion that extinction involved the gradual loss of individual species.

But there were problems with Lyell's idea that imperceptible changes in environment caused the destruction of these and other organisms. For example, the mastodon bones he located in 1833 were buried in a gully, in loam that had been deposited over shelly gravel. So the remains he exhumed had apparently come from mastodons that lived after the mollusks found at the site. Since many of the fossil shellfish he found appeared similar to species that were still alive, Darwin surmised that the climate could not have dramatically changed in the intervening years. If not killed off by climate change, could species be analogous to individuals, with a definite life span? he wondered.[83] In his published account of his *Beagle* voyage, issued in 1839, he once again resorted to this common analogy: "All that can be said with certainty, is that, as with the individual, so with the species, the hour of life has run its course, and is spent."[84]

By the time he published these words, Darwin was already toying with evolutionary explanations not only for fossils but also for the many perplexing facts of geographical distribution he observed during the voyage of the *Beagle*. Different regions of the globe often had different though similarly structured species, he noted, like the two forms of rhea he collected on the South American pampas, which resembled but were also clearly not the same as the African ostrich. At the same time, the flora and fauna of oceanic islands tended to resemble but also be distinct from species found on neighboring continents. Finally, adjacent islands that were nearly identical in climate and similar in physical features often held different assemblages of organisms. For Darwin, traditional creationist explanations of the origin of species failed to offer satisfying explanations for these observations.

The five weeks Darwin spent exploring the Galápagos Archipelago, six hundred miles west of mainland Ecuador, proved especially crucial in the development of his ideas about evolution.[85] Darwin prepared for his keenly anticipated visit by reading several previously published accounts of the islands. He became quickly enchanted with the biota he encountered there, including the islands' famous tortoises, iguanas, and boobies that exhibited no fear of humans. He quickly sensed that the flora and fauna bore a general resemblance to that of South America, but with important differences. In the process of collecting prodigious numbers of plants and animals on the four largest islands in the archipelago, he noted something unusual about the mockingbirds he encountered: "The specimens for Chatham [San Cristóbal] and Albemarle [Isabela] Isd. appear to be the same, but the other two are different. In each Isd. each kind is *exclusively* found; habits of all are indistinguishable." Darwin then remembered a comment from a British-born Ecuadorian official on the islands who claimed that local residents could tell from which of the individuals islands a tortoise had come by observing the form of its body, the shape of its scales, and its general size. During the long voyage home Darwin recognized that if these observations proved correct, they might "undermine the stability of species."[86]

With a new appreciation of the potential significance of his Galápagos material, Darwin lamented the fact that he had not been careful to note on which particular islands he had collected a long series of ground finches. Shortly after his return home, the British ornithologist John Gould named and described fourteen new species of finches based on the *Beagle* collections. Darwin then consulted the finch specimens that Captain Fitzroy and other *Beagle* crewmembers had collected and labeled in the hope of discovering the pattern of their geographical distribution, but he was never entirely successful at doing so. While the finches remained a muddle, Gould did confirm Darwin's hunch that the three different types of mockingbirds he collected were three separate species. Darwin knew that these three species were isolated on different islands; now he came to believe that they had descended from a common ancestor.[87]

Within a few months of his return to England, Darwin began a journal, his *Notebook on the Transmutation of Species,* in which he systematically gathered information relevant to the idea of species change. In a diary entry made at the time, he noted that identifying "species on the Galapagos Archipelago" provided the principal source of "all my views."[88] Although now convinced of the reality of evolution by common descent, he struggled for more than a year to come up with an explanation for how evolution occurred.

The eureka moment, when all the pieces finally fell into place, came in September 1838, after reading the grim speculations of the British political economist Thomas Robert Malthus.[89] Malthus had argued that the tendency for human populations to increase geometrically was inevitably checked by various forms of severe restraint: war, famine, disease, pestilence, and the like. While thinking through the implications of Malthus's insight for the natural world, Darwin came up with a mechanism that could drive the process of evolution: natural selection.

The basic idea seems relatively simple in hindsight. The reproductive capacity of organisms is incredibly high, yet most individuals fail to survive long enough to reproduce. Species also exhibit a great deal of variation, which results in some individuals being better adapted than others to the conditions of life in which they find themselves. These are the same individuals most likely to pass on their favorable traits to their progeny. Over the course of countless generations, these small heritable variations build on one another to produce major transformations in the physiology and structure of species. Under the constant pressure of natural selection, given enough time, not only new species, but also new genera, families, and orders will result. Using another metaphor, that of the wedge, Darwin attempted to capture a sense of the results of the fierce competition between species and the inevitability of extinction: "Take Europe, on an average every species must have the same number killed year with year by hawks, by cold &c—even one species of hawk decreasing must affect instantaneously all the rest—The final cause of all this wedging, must be to sort out proper structure, & adapt it to changes . . . One may say there is a force like a hundred thousand wedges trying [to] force every kind of adapted structure into the gaps in the oeconomy of nature, or rather forming gaps by thrusting out the weaker ones."[90]

After struggling to flesh out his ideas for several years, Darwin drew up a sketch of his theory in 1842 and a more developed version in 1844, but he refused to publish either essay for fear of the controversy his ideas would provoke. In the latter document, he argued that history demonstrated that "the disappearance of species from any one country has been slow—the species becoming rarer and rarer, locally extinct, and finally lost." While more recent species loss might ultimately be due to "man's direct agency" or his "indirect agency in altering the state of the country," the fossil record demonstrated a similar state of affairs: "decrease in numbers or rarity seems to be the high-road to extinction."[91] Moreover, naturalists thought nothing of the fact that one species of an existing genus was rare and another abundant, even if the exact cause of this phenomenon remained uncertain. Whether trying to explain the mysteries of species abundance, historical extinction, or prehistoric extinction, for Darwin natural selection provided the key: "We should always bear in mind that there is a recurrent struggle for life in every organism, and that in every country a destroying agency is always counteracting the geometrical tendency to increase for each species."[92]

Darwin was finally provoked to publish his evolutionary ideas after receiving a letter from Alfred Russel Wallace, who had independently arrived at the idea of evolution by natural selection. After receiving a draft of Wallace's paper in 1858, Darwin quickly wrote up his own results in *On the Origin of Species,* published a year later. Accounting for extinction loomed large in this, one of the most influential books of the nineteenth century. According to Darwin, "the extinction of old forms and the production of new forms are intimately linked together."[93] Because of the high reproductive potential of organisms, he argued, each area of the globe remained perpetually full. Yet, under the constant pressure of natural selection some forms would slowly but continually be modified, and as this occurred, they would out compete existing forms: "As new species in the course of time are formed by natural selection, others will become rarer and rarer, and finally extinct."[94] He had once been astonished at the fossil bones he unearthed in La Plata as he struggled to comprehend the force that had caused those species to perish. Now Darwin argued, "We need not marvel at extinction" for species loss was part of the regular order of nature, even if we could not pinpoint the precise events that allowed some species and even entire groups to survive across long stretches of time and forced others to fade away.[95]

WALLACE AND ISLAND BIOGEOGRAPHY

Darwin devoted an entire section of the *Origin of Species* to an explanation of species distribution on oceanic islands.[96] Here he argued that a combination of initial migration and subsequent descent with modification provided a convincing explanation for the patterns of island biogeography that had long puzzled naturalists, while the more prevalent idea of multiple independent creations failed to do so. Oceanic islands, he noted, possessed fewer total species, more endemic species, few or no representatives of some animals (e.g., reptiles, amphibians, and mammals), and trees or bushes belonging to orders that elsewhere include only herbaceous species. The "most important and striking" fact regarding island inhabitants, however, was "their affinity to those of the nearest mainland, without being actually the same species."[97] The example he used was the Galápagos Islands, where "almost every product of the land and the water bears the unmistakable stamp of the American continent." Twenty-five of the twenty-six species of land birds on the archipelago were endemic, yet in terms of their structure and behavior, they closely resembled those found on the South American coast. "Why should this be so?" Darwin asked, if these birds were created in the Galápagos and nowhere else, especially when the conditions there were so different from those in South America?[98] At the same time, the archipelago was populated by closely related but often distinct species, as, for example, the three mockingbirds of the Galápagos, each of which is confined to its own island. Again, Darwin argued that migration followed by gradual evolutionary change over time provided the best explanation for these and other distributional phenomena.

Darwin's cofounder of the theory of evolution by natural selection, Alfred Russel Wallace, also developed a deep interest in the problem of geographical distribution, especially on islands. Born the eighth of nine children to a family of modest means, Wallace became an apprentice in the surveying business of his oldest brother at the young age of fourteen, and for most of the next decade, he mapped and surveyed in Bedfordshire and Wales.[99] He relished the time he spent outdoors, which allowed him to indulge his growing passion for natural history, a passion he also nurtured by reading widely in the field. In 1848, he set out for

Brazil with his friend, the naturalist Henry Walter Bates, to collect specimens for the booming natural history market. The two young men honed their field skills, began speculating about the source of the bewildering array of life they found in the Brazilian rain forests, and gathered a variety of exotic specimens to send back to a London dealer who acted as their broker. Unfortunately, save for one shipment, all of Wallace's collections were lost when the ship on which he was sailing home caught fire. The four years he spent in South America was not a total financial loss, however, since he managed to publish two books based on his experiences, including a well-received narrative of his travels that established his reputation as naturalist on the make.[100]

In 1854, only two years after his return from the Amazon, Wallace set out again, this time for the Malay Archipelago. Over the next eight years, he hopped from island to island, collected more than one hundred thousand plant and animal specimens, and authored numerous scientific and popular articles on myriad subjects. One of the primary items on Wallace's desiderata list was the bird of paradise, an exquisitely beautiful, secretive, and rare bird that proved in strong demand among European collectors at the time. He was ecstatic, then, when he managed to capture a young king bird of paradise while on the island of Aru in 1857. His aesthetic interest in the species—which had long been an important dimension of natural history collecting—led him to openly speculate about the about the destructiveness of so-called civilized societies:

> I knew how few Europeans had ever beheld the perfect little organism I know gazed upon, and how imperfectly it was still known in Europe . . . I thought of the long ages of the past, during which the successive generations of this little creature had run their course—year by year—being born, living and dying amid these dark and gloomy woods, with no intelligent eye to gaze upon their loveliness—to all appearances such a wanton waste of beauty. Such ideas excite a feeling of melancholy. It seems sad that on the one hand such exquisite creatures should live out their lives and exhibit their charms only in these wild inhospitable regions, doomed for ages yet to come to hopeless barbarism; while on the other hand, should civilized man ever reach these distant lands, and bring moral, intellectual, and physical light into the recesses of these virgin forests, we may be sure that he will so disturb the nicely balanced relations of organic and inorganic nature as to cause the disappearance, and finally the extinction, of these very beings whose wonderful structure and beauty he alone is fitted to appreciate and enjoy. This consideration must surely tell us that all living things were *not* made for man.[101]

While important as an early example of lamenting the destructive changes that had accompanied European expansion, Wallace's revelation did not deter him from capturing a second example of the elusive species, nor did it diminish his ardor for collecting more generally.

In early 1855, while waiting out the rainy season in Borneo, Wallace composed his first significant statement of the evolutionary theory he had begun developing. In clear, straightforward language, he proposed a simple law: "Every species has come into existence coincident both in time and space with a pre-existing closely allied species."[102] Just as the fossil record revealed that the most closely related extinct species were in chronological proximity to one another, so too did studies of geographical distribution reveal that the most closely related living species were in nearby physical proximity to one another. Common descent explained both phenomena. He marshaled a variety of evidence to support his law, including Darwin's initial observations about the flora and fauna of the Galápagos published in *Voyage of the Beagle.*[103] What Wallace failed to come up with at this point, though, was a proper mechanism to drive the process of evolutionary change. Nonetheless, Wallace's 1855 paper provoked Lyell, whose geological writings had been a major influence on Wallace and Darwin, to begin keeping a new species notebook; Darwin, on the other hand, initially dismissed the paper. He did, however, come to appreciate its significance and entered into a correspondence with Wallace. In a letter dated 1857, Darwin mentioned casually that he had been at working on the question of how and why species and varieties differ from one another for more than two decades. He ended his letter, as he often did, with a plea for data: "One of the subjects on which I have been experimenting, & which cost me much trouble, is the means of distribution of all organic beings found on oceanic islands—& any facts on this subject would be received most gratefully."[104]

Later that year, while confined to bed with a debilitating malarial fever that had plagued him intermittently since his days on the Amazon, Wallace discovered his own answer to the riddle of evolution. Long after the momentous event, he would repeatedly claim that it was while contemplating Malthus's *Essay*—the very same book that had sparked Darwin's burst of insight—that the explanation "suddenly flashed" upon him: "on the whole the best fitted survive" in the constant struggle for existence, and over time this process would lead to gradual changes in organisms.[105] He fleshed out his argument in a manuscript that he sent to Darwin, who was shocked when he received it. So closely had Wallace's arguments and even his language paralleled his own, Darwin felt he was reading an abstract of the big book on natural selection he had been laboring over for years. After consulting with two of his closest colleagues—Lyell and the botanist Joseph D. Hooker—about how to proceed, in 1858, Darwin agreed to simultaneously publish an abstract from his 1844 essay, an excerpt from a letter sketching out his theory that he wrote in 1857, and Wallace's paper, without the latter's consent. The episode also pushed him to complete an abstract of his uncompleted longer book on natural selection, published in 1859 as *On the Origin of Species*.

Wallace continued to gather data on geographical distribution during his next four years in the Malay Archipelago. Analyzing that data, he discovered the boundary separating the faunas of Asia and Australia that ran through the region, what became known as "Wallace's Line." After returning home, he published a series of books that fleshed out his ideas about the relationship between the theory of natural selection and the distribution of organisms, including the *Malay Archipelago* (1869), *The Geographical Distribution of Animals* (1876), and *Island Life* (1880). The later two books, which synthesized and greatly extended the literature on biogeography (especially island biogeography) are considered his most important scientific contributions.

EVOLUTION'S IMPLICATIONS

Darwin and Wallace made sense of the strange creatures and the unusual distribution patterns on islands. Why was the dodo a flightless bird that appeared structurally similar to the ground dove and yet showed no fear of humans? Although neither author specifically addressed the issue, it is not hard to imagine how they might have answered the question based on their theories about evolution. The dodo had migrated to Mauritius at some distant point in the past, and in the absence of predators and other species that typically inhabited mainlands, over time it had grown in overall size while its wings diminished. A similar process of migration followed by modification had occurred with countless other species on countless other islands, thus explaining the high rates of endemism on those islands. And why had the dodo gone extinct? Because new arrivals to Mauritius—humans and their domesticated animals—ruthlessly preyed upon the bird while radically transforming the environment it depended on to survive. With the sudden influx of deadly new predators and the accompanying habitat destruction, the "ecologically naive" dodo simply could not evolve fast enough to survive.

Having traveled and read widely, Darwin and Wallace were among the first to appreciate the fact that, with minor variations, this same basic process that led to the extinction of the dodo had been occurring on hundreds of islands around the world since Europeans first set foot on their shores. They and their domesticated animals, pests, weeds, crops, trees, ornamentals, and diseases often exterminated individual insular species and overran their environments, which had previously been biotically impoverished. Not so long ago, the environmental historian Alfred Crosby coined an apt new term, "ecological imperialism," to describe the process whereby Europeans managed to stage a demographic takeover of indigenous peoples in many parts of the world with the aid of the plants, animals, and diseases they brought with them.[106] Writing a century prior to Crosby, Darwin and Wallace were among the first to understand how and why this process might take place.

Beyond their implications for island biogeography, Darwin's and Wallace's ideas about evolution had a profound impact on his biological colleagues and the broader culture. The evidence they gathered for the reality of evolution—from paleontology, geographical distribution studies, taxonomy, embryology, and morphology—was not only compelling but also synthesized the biological sciences in a way that had previously been impossible. Ironically, though, while the *Origin of Species* relatively quickly convinced scientists about the reality of biological evolution, until the Modern Synthesis of 1930s and 1940s, few of them accepted natural selection as the primary force driving evolution; for a variety of reasons, even Darwin himself began increasingly relying on other mechanisms in subsequent editions of the *Origin*.[107] Still evolutionary explanations of various sorts, which had already begun springing up in the social and the biological sciences, became pervasive after 1859.

The notion that struggle was ubiquitous in the human and the natural world also became increasingly commonplace. For some, like the American naturalists Henry F. Osborn, William T. Hornaday, and Aldo Leopold, the fact the native species had been the product of countless eons of evolutionary change provided a rationale for admiring and protecting them. As Darwin himself expressed it so eloquently in the concluding sentence of the *Origin:* "There is grandeur in this view of life, with its several powers, having been originally breathed into a few forms or into one; and that, whilst this planet has gone cycling according to the fixed law of gravity, from so simple a beginning endless forms most beautiful and most wonderful have been, and are being, evolved."[108]

Of course, evolutionary arguments were malleable enough that the could also be deployed into opposition to wildlife preservation initiatives. Rare species, like the California condor or the ivory-billed woodpecker, for example, would sometimes be portrayed as the dying relicts of another age, species that were on their way out before humans came on the scene, and thus beyond saving. In the most extreme cases, opponents of protecting wildlife would sometimes argue that extinction, even mass extinction of whatever cause, was a routine part of the history of life on earth, not something to be particularly concerned about.

The theory of evolution also had more subtle impacts on discussions about human-caused species extinction. Conspicuously absent from the *Origin* was any extended discussion of how humans might fit into the emerging evolutionary framework. Darwin remained silent on this issue until the final pages of his book, when he hinted that through his evolutionary ideas, "Light will be thrown on the origin of man and his history," a curiously passive but nonetheless suggestive passage.[109] Although in 1859 he was fearful of more directly stating that the same evolutionary process that had created the diversity of the natural world had produced humans, he had long accepted the notion himself and begun considering its many implications. For example, twenty years earlier, in his first *Transmutation Notebook*, he scribbled a powerful passage suggesting a bond between humans and other animals: "If we choose to let conjuncture run wild, then animals, our fellow brethren in pain, diseases, death and suffering and famine—our slaves in the most laborious works, our companions in our amusements—they may partake [of] our origin in one common ancestor—we may all be netted together."[110] A little over a decade after the *Origin*, Darwin did explicitly bring humans into his evolutionary framework beginning with *Descent of Man* (1871). There he demonstrated the structural, physiological, and behavioral continuity between humans and the great apes. If animals were indeed our kin, as Darwin had now made explicit, then we ought to take how we treat them more seriously. Darwin's ideas thus became one of many factors that helped produce the humane movement in Europe and the United States in the second half of the nineteenth century. Although humanitarians were mostly concerned about minimizing the suffering and pain of domesticated animals, as we shall see, they also occasionally engaged with wildlife issues.

BECAUSE OF THEIR relatively small sizes, impoverished biotas, and high rates of endemism, documenting the process of human-caused extinction on islands proved relatively easy. Not surprisingly, then, research on these sites promoted consciousness about the power of humans to destroy species in the middle decades of the nineteenth century. Could the same process of human-driven extinction be at work on continental species, which typically had much wider ranges of distribution, and thus often proved harder to locate, capture, and monitor? Not long after the appearance of Darwin's *Origin of Species*, the rapid decline of two North American species—the bison and the passenger pigeon—would definitively answer that question while giving rise to the first mass movement to protect endangered species.

CHAPTER THREE

SOUNDING THE ALARM ABOUT CONTINENT-WIDE WILDLIFE EXTINCTION

While the progress the century has wrought in respect to the development of the resources of our country is justly receiving so much attention, it may not be unfitting to notice briefly other attendant changes that are far less obvious, though no less real, than the transformation of hundreds of thousands of square miles of wilderness into "fruited fields," dotted with towns and cities, and intersected by a network of railways and telegraph lines. With the removal of a vast area of forest that has rendered possible the existence of millions of people where only a few thousand rude savages lived before, there has taken place a revolution in respect to the native animals and plants of this great region as great as has occurred in respect to the general aspect of the country. Not only has indigenous vegetation given place largely to introduced species, but the larger native animals have been in like manner supplanted by exotic ones. While these changes do not pass unnoticed by the naturalist, they are less apparent to the general observer.

J. A. ALLEN, 1876

CENTENNIAL CONCERNS

In 1876, the United States reached the centennial of its nationhood. It was a time to take stock of where the young nation had been and where it seemed to be heading. Many Americans viewed the occasion as an opportunity to celebrate the profound transformations of the previous century: a loose coalition of states had been forged into a Union, a once vast wilderness had been largely tamed, a reliable rail transportation system was knitting the expanding nation together,

FIGURE 13. Joel Asaph Allen, ca. 1876. A photograph of Allen taken at about the time he began both documenting and decrying the decline of North American birds and mammals. Courtesy of the Archives of the Ernst Mayr Library of the Museum of Comparative Zoology, Harvard University.

and industrialization had taken a firm hold. In a special centennial edition of the nation's history, the American author and editor Benson J. Lossing boasted of how much had been accomplished over the previous hundred years: "We may safely claim for our people and country a progress in all that constitutes a vigorous and prosperous nation during the century just passed, equal, if not superior to that of any other on the globe."[1] An advertisement for *The First Century of the Republic: A Review of American Progress* (1876) concurred, proclaiming that the nation's many achievements ought to "awaken a feeling of just pride in every American citizen."[2] Similar Panglossian sentiments rang out from the nation's pulpits, meeting halls, and legislative chambers.

Joel Asaph Allen, a young, ambitious, and painfully shy curator at Harvard's Museum of Comparative Zoology, offered a counterpoint to this centennial celebration of American progress. In a remarkable series of publications issued in 1876, he lamented the decline of North American wildlife, while suggesting that civilization's advancement also carried negative burdens that had yet to be fully comprehended.[3] Allen chronicled numerous native species that had either been lost or greatly diminished in number, he spoke briefly of the forces that threatened those that still managed to hang on, and he offered a series of concrete proposals that might stem their decline. While naturalists, sport hunters, and others had previously noted in passing how humans had driven individual species to extinction or near extinction, Allen's publications marked one of the earliest extended discussions of the larger process of human-induced wildlife extinction.

Allen is an important but largely unappreciated figure in late nineteenth- and early twentieth-century American biology.[4] He began collecting natural history specimens as a teenager on his family's farm in Springfield, Massachusetts. After several years at Wilbraham Academy, an education partly financed with the sale of his extensive bird collection, in 1862 he enrolled in Harvard's Lawrence Scientific School to study with the famed Swiss naturalist Louis Agassiz. Like most of Agassiz's early students, he soon became a curatorial assistant at the Museum of Comparative Zoology (MCZ). Only two years after his arrival, he secured a promotion to head of the bird and mammal department at that august institution. By the time he finally left the MCZ to accept a position at the American Museum of Natural History in 1885, Allen had secured a reputation as one of the nation's leading naturalists. Where most of his colleagues seemed content merely to catalog the young nation's flora and fauna, Allen searched for patterns in the geographical distribution and variation of organisms. As he began carefully analyzing multiple specimens of birds and mammals from across their range, he noticed that within a given species, extremities tended to be shorter in colder climates and longer in warmer climates. This ecological generalization is still known as Allen's rule.

In April 1871, Allen and two assistants headed west under the auspices of the MCZ.[5] For the next nine months, they traveled widely across the Great Plains and into the Rocky Mountains. In addition to helping Allen regain his failing health, a special object of the expedition was a series of American bison specimens for the MCZ. Railroads granted the party special rates, while the U.S. Army provided a base of operations. Among the thousands of specimens Allen shipped back to Cambridge were an extensive series of buffalo skins, skeletons, and skulls obtained in the vicinity of Fort Hays, Kansas.

Five years after he first ventured westward, Allen completed an exhaustive monograph, *The American Bisons, Living and Extinct.*[6] While the first part of his study consisted of technical descriptions of two lost and one living bison species, the second section told the troubling story of the sole surviving form of the American bison, *Bison bison.* This large, lumbering beast, which once inhabited almost half the North American continent, had been driven west of the Mississippi River by 1800. By the 1840s, Native Americans engaged in the fur trade and Euroamerican migrants heading to California placed increasing pressure on the species. The greatest period of decline, however, followed the completion of the first transcontinental railroad in 1869. The railroad's impact on the bison was multiple, Allen argued: it provided hunters with easier access to their quarry, offered a ready means of shipping bulky hides back to eastern markets, and promoted white settlement throughout the West. A year after the first transcontinental railroad began operation, eastern tanners discovered how to make buffalo hides into durable leather, thus increasing demand for the species. Though convinced that the bison remained abundant in the northern plains, Allen worried that at current rate of destruction, "the period of his *extinction* will soon be reached."[7]

Eager not to bury his alarming conclusions in an obscure academic treatise, that same year Allen issued a series of five more accessible articles detailing the decline of the buffalo and other North American vertebrates. The first of this series, "The North American Bison and Its Extermination," which appeared in the *Penn Monthly,* summarized the major findings of his monograph.[8] Like the longer study, it emphasized that the systematic destruction of the bison formed part of a disturbing pattern. "Wherever civilized man has held sway," Allen argued, larger mammals like the bison faced "annihilation." The only unusual thing about this particular case was the speed with which millions of bison had been destroyed, providing "one of the most remarkable instances of extermination recorded, or ever to be recorded, in the annals of zoology." "Another decade or two at its present rate of decrease," Allen predicted grimly, "will be sufficient for its total extermination."[9]

In the process of examining the natural history, travel, and exploration literature that had documented the bison's decline, Allen also noted changes in the status of other North American wildlife. In a second popular article entitled "The Extirpation of the Larger Indigenous Mammals of the United States," he decried the widespread decrease of the moose, gray wolf, panther, lynx, black bear, wolverine, caribou, elk, Virginia deer, and other large mammal populations since the beginning of white settlement.[10] That fact that so much wildlife destruction had been "reckless," "wanton," and wasteful particularly appalled him.[11] Allen questioned the commonly held assumption that humans owed no obligation to threatened species or to future human generations that might be deprived of their use and enjoyment. But, not surprisingly given the larger culture in which he found himself, he refused to condemn outright the continued westward march of civilization, which in his view had granted millions of American citizens the opportunity to flourish in an area that had once supported modest aboriginal populations.

As part of this same series of articles, Allen called attention to the plight of vanishing and lost birds. A brief notice for the *American Naturalist* pointed out that the great auk had once been abundant on Funk Island, until it had been mercilessly hunted down for feathers in the 1830s and 1840s.[12] A much longer article chronicled the decline of numerous avian species in his native Massachusetts, where at least four birds—the great auk, the wild turkey, the sandhill crane, and the whooping crane—seemed to have been exterminated since European contact.[13] Many others, including the raven, pileated woodpecker, passenger pigeon, heath hen, trumpeter swan, most of the wading and swimming birds, and nearly all the raptors also suffered marked population decreases. In a final article, Allen provided a more general account of the decline in bird populations throughout the United States.[14] Much of this paper was an expansion of his earlier article on Massachusetts birds, with the descriptions of former and current ranges suitably enlarged. Allen noted that settlement rapidly produces "a revolution in the haunts of many of the species as well as in their relative abundance."[15] While some smaller birds—like the robin, blue bird, and house wren—seemed to flourish in association with humans, others, especially larger species, frequently suffered population declines. Of the numerous birds Allen highlighted for more extensive discussion, the one most immediately threatened and the one to which he devoted most his attention was the beleaguered passenger pigeon.

Allen not only lamented the increasing rarity of many North American birds and mammals, he also suggested several measures to rescue them. First, he called for stringent state and federal legislation regulating hunting seasons and establishing bag limits. In the case of severely threatened species, Allen advocated outright prohibitions on hunting.[16] Since the emergence of recreational hunting as an elite activity in the 1830s and 1840s, sport hunting clubs, game protective associations, and related organizations had secured a long string of state laws to protect game species.[17] Now Allen hoped that America's sporting community might also be mobilized to lobby for legislation to rescue other forms of native wildlife. But he remained skeptical about sportsmen's enthusiasm for protecting nongame species, which in his words did "not appeal so strongly to their self-interest" as the creatures they sought to hunt.[18]

For some species, like the buffalo, Allen suggested setting aside federal land to serve as protective refuges. He noted that the czar of Russia pursued this approach in 1852, as part of an effort to save the last remaining European aurochs, the Old World cousin of the American bison.[19] No doubt Allen's suggestion was also related to the recent establishment of Yellowstone National Park, the first of its kind in the world. As a scientific member of the Northern Pacific Railroad

Expedition Party, he had visited the new two-million-acre park in 1873, one year after Congress had removed the site from the public domain. While the establishment of Yellowstone might have suggested the feasibility of national wildlife refuges, in actual practice this and other early national parks initially provided only minimal protection for native fauna. They were established primarily to protect the scenic wonders and natural curiosities contained within their boundaries, and during their early years they lacked sufficient staff even to achieve this limited goal.[20]

Beyond sportsmen, Allen hoped that recently established "associations for the 'Prevention of Cruelty to Animals'" might also aid the cause of wildlife protection.[21] While self-interested utilitarian motives informed the conservation activities of sport hunters, humanitarians were moved to action by a distaste for unnecessary animal pain and suffering.[22] The humane movement began in England in the 1820s and 1830s and was brought to the United States by the wealthy New York socialite Henry Bergh. In 1866, Bergh founded the American Society for the Prevention of Cruelty to Animals, the first in a series of a several dozen state and local humane societies organized over the next decade in the Northeast and Midwest. While humanitarians lobbied successfully for state anticruelty laws, as Allen pointed out, they focused primarily on minimizing the suffering of domestic animals. But they occasionally turned their sights on wild animals as well. Just two years earlier, for example, Bergh had received a series of letters decrying the slaughter of the bison that he widely circulated with the hope of securing legislation limiting hunting of the species.[23] At about this same time, Rep. Greenburg L. Fort, a Republican from Illinois, introduced a bill in Congress that would have made it illegal for any non-Indian to kill a female buffalo within the United States. Humanitarian sentiments loomed large in the debates about the proposal, which passed both houses of Congress in 1874, only to face a pocket veto from President Grant. Two years later, Fort's bill gained passage in the House, but, in the aftermath of the Custer's ill-fated campaign in the Black Hills, it failed to make it out of committee in the Senate.[24]

While sportsmen and humanitarians might serve as allies in the struggle to stem the decline of North American wildlife, they were unlikely to lead the fight. What was really needed, Allen believed, were new organizations devoted exclusively to the cause of wildlife protection. Nearly a decade later, he would be among those stepping forward to create such institutions. This chapter explores the circumstances that led to the creation of the Audubon bird protection movement at the end of the nineteenth century and the crucial role that naturalists played in forging that movement. Clearly, Allen was not working in a vacuum.

His centennial concerns about the growing problem of wildlife decline represented the culmination of a series of transformations in North America's physical and cultural landscape over the previous century.

NATURE'S NATION

The first Europeans to land on North American shores encountered a landscape teeming with wildlife. The superabundance of native flora and fauna, especially when compared to the more densely settled and biologically impoverished Old World, astounded early visitors and promoted the emergence of a persistent "myth of inexhaustibility."[25] The British author of *New England's Prospect* (1634), William Wood, for example, marveled at what seemed like a nearly boundless supply of plant and animal life in Massachusetts Bay Colony. During his four years there, he witnessed firsthand the annual spring migration of alewives "in such multitudes as is almost incredible, pressing up such shallow waters as will scarce permit them to swim"; he boasted that the supply of waterfowl was so bountiful that "some have killed a hundred geese in a week, fifty ducks at a shot, forty teals at another"; he noted how plentiful large and valuable trees were; and he spoke of great flocks of passenger pigeons with "neither a beginning nor ending, length or breadth of these millions and millions."[26] It would be easy to dismiss Wood's report if his visions of abundance were not confirmed by countless other explorers and settlers who ventured to the eastern shores of North America in the seventeenth century and further West well into the eighteenth and nineteenth centuries.

To this seemingly boundless landscape, Europeans brought nearly inexhaustible demands. From the beginning, immigrants like Wood felt driven by a divine mandate to transform the vast American wilderness—which they tended to view as chaotic, wasteful, threatening, even evil—into something that was orderly, productive, and good.[27] Felling trees, destroying predators, harvesting wildlife, draining swamps, introducing livestock, planting crops, building roads, and conquering Native Americans were all part of a holy mission to complete God's creation, to bring the shining beacon of civilization to an otherwise dark and foreboding countryside. The tree stump became the quintessential symbol of progress, and improving the land, settlers argued in a characteristically self-serving fashion, granted them the right to seize it from lazy aboriginals who contented themselves with living off the fat of the earth.[28] Where Native Americans tended to think in terms of various usufruct rights to the land vested in the entire community, Europeans asserted strong notions of private property, the right of an individual to own a bounded parcel of land and all that dwelt on it.

FIGURE 14. The superabundance of New World wildlife, 1618. Although some were undoubtedly exaggerated, early accounts like this seventeenth engraving invariably emphasized that North America was teeming with wildlife. From Theodor de Bry, ed., *America*, part 10 (1618). Courtesy of the Library of Congress, LC-USZ62-49747.

But most important of all, the new arrivals to the New World sought to turn a profit from America's natural resources—to extract valuable plants, animals, and minerals from the landscape—and to incorporate them into a web of commerce that stretched far and wide across the Atlantic and beyond. Native Americans could and did transform the natural world, but for the most part they engaged in only limited trade with neighboring tribes, while their exploitation of plants and animals was restricted by a set of deeply held spiritual beliefs that viewed all of creation as alive, imbued with will, purpose, and meaning. Europeans, in contrast, commodified the landscape, and in doing so, developed insatiable demands for the biota that resided on it.[29]

Given this set of attitudes and practices, newly settled areas of North America soon began experiencing dramatic transformations that proved difficult to ignore.[30] Once diverse habitats gave way to a more limited range of introduced crops, pesky weeds that seemed to sprout up everywhere, and vexing insect pests. Newly introduced domesticated animals roamed across the countryside in feral herds that consumed prodigious quantities of native vegetation, fostered shifts in plant species composition, and compacted the soil. Bounty programs decimated the populations of wolves and other predators. Timber became scarce around villages and towns, forcing their inhabitants to range farther and farther in search of the wood needed for dwellings, fences, and fuel. Deforestation, in turn, led to increased water runoff, wider temperature swings, and soil drying. And once bountiful fur and game species experienced sharp population declines.

Colonial officials responded to the local decrease of wildlife with a series of laws regulating the hunting of game species.[31] These were designed not only to reduce hunting pressures on declining wildlife populations but also to exert social control over those individuals who might be tempted to turn their back on civilization and adopt more primitive ways of life.[32] As early as 1698, for example, the Connecticut Assembly prohibited deer hunting between January 1 and July 15 each year.[33] Although several other states soon enacted similar legislation, these early laws gained spotty enforcement and faced frequent repeal. One thing the local depletion of wildlife failed to promote, however, was a serious challenge to the strong belief in progress, the continued commodification of nature, or the myth of inexhaustibility. Prevailing opinion considered the decline of valuable plants and animals to be a local phenomenon, little more than an inconvenience, rather than something to be lamented and fundamentally addressed.

That attitude first showed signs of change near the end of the eighteenth century and the first half of the nineteenth century. One factor fostering a nascent appreciation for the American landscape in something other than commodified forms was romanticism, a cultural movement that swept across Europe and then the United States as a reaction to the Enlightenment.[34] Where champions of the Enlightenment embraced reason, empiricism, and a mechanistic view of the universe, romantics revered imagination, intuition, mystery, and an organic worldview. Perhaps most importantly in this context, romantics relished regular human intercourse with the natural world, while raising difficult questions about the meaning and consequences of material progress. One inspiration for the movement, the French philosopher, writer, and political theorist Jean-Jacques Rousseau, for example, proclaimed the virtues of the "noble savage," the individual who lived a life close to nature, uncorrupted by the artificial trappings of civilization. Other romantics celebrated sublime scenery, dramatic landscapes like waterfalls, mountains, and chasms that induced feelings of awe and grandeur while revealing the profound power of a supreme being. Although the unabashed celebration of progress remained a dominant theme in American culture, romanticism offered an important counterpoint, one whose influence would steadily expand as the century wore on.

By the 1830s and 1840s, America produced an indigenous strand of romanticism known as transcendentalism.[35] Ralph Waldo Emerson became the leading exponent of this new philosophy, which portrayed all of nature as a font of the divine and saw proper interaction with the natural world as a means to achieve wisdom, strength, and renewal. Emerson's most enthusiastic disciple, Henry David Thoreau, combined a keen interest in the science of his day with the sensibilities of a philosopher-poet struggling to communicate the wonder, beauty, and mystery of the natural world.[36] Early on, he developed profound misgivings about modernity—technological development and the grasping commercial spirit that seemed to be closing tightly around him—not only because of its implications for humanity but also the threats it posed to nature itself. In a journal entry from 1856, Thoreau deployed a metaphor comparing wildlife with works of art, in this case poetry, to lament the decline of many native species around his home:

> When I consider that the nobler animals have been exterminated here,—the cougar, panther, lynx, wolverine, bear, moose, deer, the beaver, the turkey, etc., etc.,— I cannot but feel as if I live in a tamed, and as it were, emasculated country. . . . I take infinite pains to know all the phenomenon of spring, for instance, thinking that I have here the entire poem, and then, to my chagrin, I hear that it is but an imperfect copy that I possess and have read, that my ancestors have torn out many of the first leaves and grandest passages, and mutilated its pages. I should not like to think that some demigod had come before me and picked out some of the best stars. I wish to know an entire heaven and an entire earth.[37]

Although Thoreau was clearly referring to examples of local extermination, this passage represents one of the earliest impassioned expressions of concern about the consequences of anthropogenic extinction, and his comparison of wildlife with art would become increasingly commonplace by the end of the nineteenth century. A few years later, Thoreau suggested that each town should set aside a large area—five hundred to one thousand acres—as a "primitive forest" that would not only provide a home for endangered wildlife but also opportunities for "instruction and recreation."[38]

Nationalism provided a second source for valuing native flora and fauna beyond its economic worth as raw commodities. The nationalistic fervor that surrounded the Revolutionary War and infused Jefferson's *Notes on the State of Virginia* blossomed following the War of 1812. Increasingly, American naturalists struggled to gain control of the process of naming, describing, and classifying the young republic's native plants and animals.[39] Where they had once willingly submitted specimens to their European colleagues to amass in collections and write up for publication, they now founded new institutions—museums, scientific societies, and journals—to support a more indigenous, independent science. If the newly forged United States was blessed with an abundance of resources, "nature's nation" in the words of historian Perry Miller, then it was high time for American naturalists to assert their authority to catalog American biota.[40]

Alexander Wilson's nine-volume *American Ornithology* (1808–14) heralded the beginning of a series of elegant, large format, and heavily illustrated publications that proudly presented the young nation's biological richness to the world. The apogee of this ambitious, overtly nationalistic project came with John James Audubon's *Birds of America* (1827–38), which included exquisite life-sized color engravings of more than 435 North American species.[41] While some naturalists criticized the anthropomorphic, contorted poses of Audubon's birds, the project gained its creator wide acclaim. Cuvier, for example, commended his drawings as "the greatest monument yet erected by Art to Nature."[42] *Viviparous Quadrupeds of North America* (1845–54), which Audubon produced with the help of his two sons and the Charleston naturalist John Bachman, proved less successful artistically, but it still made a strong impact. The images from both publications were widely reproduced.[43]

Where Thoreau, Audubon, and other naturalists had made passing references to troubling changes in the environment, the first detailed indictment of the human power to transform the landscape came from the pen of the Vermont lawyer, businessman, philologist, farmer, congressman, and diplomat George Perkins Marsh.[44] A man of diverse interests, extraordinary ability, and prodigious girth, Marsh was, in the words of his biographer David Lowenthal, "the broadest scholar of his day."[45] Born in 1801, Marsh discovered the joys of nature at the age of eight, when an eye ailment prevented him from reading for four years. As an adult, he grew increasingly concerned about the environmental transformations he witnessed in his beloved Green Mountains. In 1857, after gaining an unpaid appointment as Vermont's Fish Commissioner, he issued a report bemoaning the human-induced changes that had decimated the inland fish stocks of his home state. Marsh worried not only about the loss of the native fish as a source of food and profit but also the recreation that angling provided. In an early version of the call for a strenuous engagement with nature that Theodore Roosevelt and his circle would later expand upon, Marsh charged that his fellow citizens were becoming "more effeminate, and less bold and spirited." "We have notoriously less physical hardihood and endurance" he continued, because "we are suffering . . . from a too close and absorbing attention to pecuniary interests."[46] Marsh advocated an aggressive breeding and restocking campaign that would not only restore Vermont's fish populations but also the prospects for healthful outdoor recreation.

Seven years later, Marsh published a remarkable book, *Man and Nature; or, Physical Geography as Modified by Human Action* (1864), that Lewis Mumford would later praise as "the fountainhead of the conservation movement."[47] Based on voracious reading in a wide variety of scientific and literary works, extensive travels through Europe and Asia, and changes he had witnessed firsthand in his home state of Vermont, Marsh became convinced that humans were radically transforming the natural world, with potentially devastating consequences. He marshaled a sweeping array of evidence to support his pessimistic thesis: "Man is everywhere a disturbing agent. Whenever he plants his foot, the harmonies of nature are turned to discords. The proportions and accommodations which insured the stability of existing arrangements are overthrown. Indigenous vegetable and animal species are extirpated, and supplanted by others of foreign origin, spontaneous production is forbidden or restricted, and the face of the earth is either laid bare or covered with a new and reluctant growth of vegetable forms, and with alien tribes of animal life."[48]

While the bulk of *Man and Nature* focused on the myriad problems associated with deforestation, Marsh also expressed concern about the widespread destruction of native flora and fauna. He warned, for example, about how modern transportation networks had brought "distant markets within the reach of professional hunters," thus causing wild birds to diminish "with a rapidity which justifies the fear that the last of them will soon follow the dodo and the wingless auk."[49] His basic framework was what we now term "ecological," though that term would not be coined for several more years. "All nature is linked together by invisible bonds," Marsh declared, "every organic creature . . . is necessary to the well being of some other."[50] Humans, he argued, tended to rent those connections asunder, thereby setting into motion a set of cascading effects that often proved irreversible. Marsh's pioneering, impassioned book, which went into three editions, gained widespread praise from scholars and the general public, but it remains unclear to what extent his contemporaries actually heeded his pessimistic message.

A SPORTING CHANCE

Romanticism, nationalism, and the recognition that humans could profoundly change the natural world provided frameworks for eliciting concern about human-caused extinction of wildlife in America. At the same time, the rise of recreational hunting and fishing offered a large, politically vocal constituency that sought to reverse the decline of numerous species, especially those pursued as game. While Southerners had long relished hunting for sport, by the early nineteenth century many Northerners had begun frowning on the activity, which they viewed as a form of wasteful idleness and a vestige of a more savage stage of human development.[51] By the middle of the century, however, recreational hunters managed to refashion their image. Drawing on the dual fonts of elite European sporting traditions and the mythology of frontier heroes like Daniel

Boone and Davy Crockett, American sport hunters began projecting themselves as hardy, self-reliant individuals, defenders of masculine virtue in an increasingly urban, industrial, commercial, and (in their view) effeminate world.[52] At the same time, an emerging sportsmen's code, which also drew heavily from European precedents, called on leisure hunters and fishermen to gain intimate knowledge about the quarry they pursued, to cultivate skill in using the equipment used in that pursuit, and to embrace a sense of fair play. Beginning in the 1830s, numerous books and periodicals codified the practices and values associated with sport hunting and fishing. By the time of the nation's centennial, at least three nationally circulated journals—*American Sportsman* (1871), *Forest and Stream* (1873), and *Field and Stream* (1874)—also helped build a sense of shared identity among wealthy and middle-class hunters and fishers. Two years later, *Forest and Stream* founding editor, Charles Hallock, identified more than 350 local, state, regional, and national organizations devoted to these activities.[53]

Sport hunters proved a potent force in wildlife conservation. While many factors—overhunting, the introduction of exotic species, and habitat destruction, to name a few—contributed to the increasingly obvious decline of many game species, sport hunters tended to pin most of the blame for declining wildlife populations on overzealous subsistence and commercial hunters, or "pot hunters" as they were then termed. Having offered a simplistic diagnosis of the problem, they offered an equally straightforward prescription: governmental regulation of access to game species. By the time of J. A. Allen's centennial warning about wildlife extinction, their persistent lobbying efforts had paid off in a long series of state laws that created game and fish commissions, established hunting seasons and bag limits, and mandated hunting and fishing licenses.[54] Those new state laws provided middle- and upper-class sport hunters with continued access to game species throughout much of the year, albeit for a modest price, while restricting legal hunting opportunities for those who were not wealthy enough to pay the required fees or who pursued game for profit. It is unclear to what extent these laws actually helped boost drooping wildlife populations, however, since they were only sporadically enforced and, as Allen pointed out, they applied only to a narrow range of wildlife: game species.

Anglers pursued a different approach, artificial propagation, as a panacea to restore declining fish stocks.[55] By the mid-nineteenth century, agriculture and industrialization both had exacted a heavy toll on America's fisheries. The damming of rivers for power, the dumping of industrial wastes, and the erosion of soil and disruption of water cycles that followed land clearing devastated the salmon and shad runs in rivers across the eastern and midwestern United States.[56] According to one early chronicler of the "fish culture" movement, "fisheries of

FIGURE 15. Seth Green demonstrating how to remove spawn from salmon trout, 1879. To restore America's depleted fish stocks, mid-nineteenth-century naturalists and fishers began promoting the idea of artificial propagation. Wildlife managers would soon embrace this approach as well. From R. Barnwell Roosevelt and Seth Green, *Fish Hatching and Fish Catching* (1879), 11.

all kinds had deteriorated, until they were on the point of extinction, and in fact, had been destroyed in some instances. . . . The time had arrived when, if our fish supply was to be saved at all, it had to be looked after."[57] A handful of naturalists, anglers, and entrepreneurs responded to this growing sense of crisis by initiating experiments with artificial fish propagation. Guided by reports from France, as early as 1853 two Cleveland physicians managed to fertilize eggs of the brook trout, a particularly popular species with anglers. Over the next decade, fish culture enthusiasts achieved success propagating numerous other aquatic species. They also presented bold claims about the superiority of artificial over natural methods of fish reproduction, claiming a success rate of as high as 80 to 90 percent, compared to only 1 percent in the wild. Widespread adoption of fish culture techniques would not only rescue numerous threatened species, they boasted, but also usher in a new golden age of piscine abundance.[58]

State legislatures responded to these wildly optimistic claims by commissioning a series of studies to investigate the condition of fisheries and the possibilities of fish culture within their boundaries. The most influential publication to emerge from this initial flurry of activity, written by none other than George Perkins Marsh, became an important manifesto for the emerging American fish culture movement.[59] Marsh began his 1857 report with a perceptive analysis of the ecological changes that Vermont's waterways had endured over the last century, when his home state's once abundant fisheries had been carelessly depleted. Overfishing, the "erection of sawmills, factories, and other industrial establishments," and the "general physical changes produced by the clearing and cultivation of the soil" had devastated the state's fish stocks.[60] While Marsh was remarkably prescient in recognizing the signs of ecological degradation, in many crucial ways he was also the product of his time and therefore reluctant to propose potential solutions that might challenge the economic and political status quo: "We cannot destroy our dams . . . ; we cannot wholly prevent the discharge of deleterious substances from our industrial establishments into our running waters; we cannot check the violence of our freshets or restore the flow of our brooks in the dry season; and we cannot repeal or modify the laws by which nature regulates the quantity of food she spontaneously supplies to her humbler creatures." Nor would protective laws "restore the ancient abundance of our public fisheries," Marsh argued, even if they were strictly enforced.[61] The only realistic solution was private initiative in the form of fish culture, which was "not only practicable, but may be made profitable, and . . . our fresh water may thus be made to produce a vast amount of excellent food."[62]

Institutionalization followed in the wake of these initial fish culture experiments. During the two decades following Marsh's report most New England and Mid-Atlantic states created fish commissions, and in 1870, a small but enthusiastic group of advocates established the American Fish Culturists' Association, soon to be known as the American Fisheries Society.[63] One of the new organization's first acts was to petition Congress to authorize "federal action in regard to the stocking of common waters of the United States."[64] In 1871, Congress responded by creating the U.S. Fish Commission and by appointing Spencer F. Baird, a widely respected naturalist who was also the assistant secretary of the Smithsonian Institution, to head the new agency.[65] Baird promoted Marsh's dream for the promise of fish culture through a series of federal hatcheries and a national fish distribution system that turned his fledgling agency into a "biological clearinghouse."[66] With the political support garnered through the broad distribution of hatchery-produced fish, he was also able to pursue less overtly utilitarian projects, like an inventory of North American fish. For well into the twentieth century, however, the main focus of the agency remained fish culture. That limited and strongly interventionist approach to the problem of declining wildlife would soon find support among advocates of captive breeding programs for endangered species.

IMMENSE BEYOND CONCEPTION

Sport hunters and anglers dealt mostly with game species that were facing local decline but were not necessarily in immediate danger of extinction across their entire ranges. It was the highly publicized and nearly simultaneously population crashes of two once profusely abundant North American species that first ignited widespread apprehension about the specter of extinction. The near loss of the bison and the demise of passenger pigeon presented a wake-up call to Americans, challenging the "myth of inexhaustibility" and placing the issue of wildlife extinction on the public agenda. Euroamerican greed has long been implicated in the annihilation of both species. Until recently, scholars have tended to point the finger of blame squarely on large-scale commercial hunting facilitated by newly constructed railroad networks that knitted together far-flung wildlife habitat and national and international markets. Recent research, however, suggests a more complex picture in which human and natural actors interacted with tragic consequences for both species.

The American bison (more popularly known as the buffalo) once occasionally ranged as far east of the Mississippi as the Atlantic Coast, but it was the Great Plains, especially the shortgrass-dominated western portion, that represented the species' population center and true home.[67] As many as 20 to 30 million bison once resided there, helping to shape the grassland habitat through nomadic browsing.[68] For much of the year, the species remained in relatively small bands, roaming in search of forage and shelter from extreme weather. During the summer rutting season, however, they aggregated into vast herds that routinely awed Euroamerican visitors to the region. In 1806, for example, Lewis and Clark encountered the species at the mouth of the White River in the Dakotas. Lewis wrote of the incident: "We discovered more that we had ever seen before at one time; and if it be not impossible to calculate the moving multitude, which darkened the whole plains, we are convinced that twenty thousand would be no exaggeration number."[69] Nearly thirty years later, the American ornithologist John Kirk Townsend reported on an expedition to the Columbia River in which he sighted the bison: "The whole plain, as far as the eye could discern, was covered by one enormous mass of buffalo."[70] Superlatives loom large in most early encounters with the plains bison during the summer season.

Perhaps the most famous brush with the species occurred in 1832 at Fort Pierre, in present-day South Dakota. Actually, "near encounter" is probably a better term for the incident, because it occurred a few days before the artist-ethnologist George Catlin arrived in the area, a site near the confluence of the Bad and Missouri Rivers that served as a major rendezvous point for the fur trade, so he only learned of the details secondhand.[71] When an "immense herd" of bison appeared on the river shore opposite the fort, several hundred Sioux Indians on horseback descended upon them. A general slaughter ensued, and the Sioux soon emerged with 1,400 fresh buffalo tongues, which they promptly traded for whiskey, leaving the rest of the animal to rot on the prairie. Under the

FIGURE 16. George Catlin, *Buffalo Hunt Surround—no. 9*, 1844. Early witnesses were awed by the teeming masses of bison they observed on the Great Plains. After learning of the massacre of a large herd outside Fort Pierre, South Dakota, in 1832, Catlin issued the first call to establish a national park that would preserve the bison as well as the Native Americans who relied on the species for survival. Courtesy of the Smithsonian American Art Museum, transfer from the National Museum of Natural History, Department of Ethnology, Smithsonian Institution.

combined sway of romanticism and nationalism, Catlin lamented the "profligate waste of the lives of these noble and useful animals" and predicated that their extinction was "near at hand." The species and its native pursuers were both "destined to fall before the deadly axe and desolating hands of cultivating man." Rather than accepting this apparent fate, Catlin called for both to be "preserved in their pristine beauty and wildness, in a *magnificent park*, where the world could see for ages to come, the native Indian in his classic attire, galloping his wild horses, with sinewy bow, and shield and lance, amid the fleeting herds of elks and buffaloes. What a beautiful and thrilling specimen for America to preserve and hold up to the view of her refined citizens and the world, in future ages! A *nation's Park*, containing man and beast, in all the wild and freshness of their nature's beauty."[72] Four more decades would pass before the federal government established the first national park at Yellowstone, and rather than welcoming Native Americans, officials did their best to bar them from their gates.[73]

As this episode suggests, Native American hunting provided one factor leading to an initial decline of the plains bison in the first half of the nineteenth century, although the overall scale of their activity remained modest at first. The bison was central to the life and culture of the Plains Indians, who had long relied on virtually every part of the species for food, clothing, shelter, tools, implements, weapons, drugs, glue, fuel, and ornaments. One scholar has suggested the large shaggy beast served as a kind of "tribal department store," a "builder's emporium, furniture mart, drugstore, and supermarket rolled into one."[74] When Native Americans began hunting for the fur trade in the 1820s and 1830s, however, they not only increased the quantity of bison they killed, but they also began seeking only its most marketable parts, especially its hide and tongue. In the middle decades of the century, California-bound white immigrants began regularly traveling through the region, a decade-long drought struck, and bovine diseases introduced by European livestock infected the herds, all of which produced additional downward pressures on the bison population.[75]

Beginning in the 1860s, hostilities between whites and natives increased dramatically, leading to ever more vocal calls to pacify Native Americans and confine them on reservations. Generals William T. Sherman and Philip Sheridan, both of whom had embraced strategies of destroying Confederate resources during the Civil War, recognized that as long as the Plains Indians enjoyed continued access to the bison, they could not be conquered. In 1875, for example, Sheridan is supposed to have discouraged the Texas state legislature from protecting the few bison that remained within its jurisdiction. Instead of trying to save the species, Sheridan reportedly argued, the lawmakers should not only thank the hunters but also award them a medal depicting a dead bison on one side and a dejected Indian on the other. The bison hunters had done more in the previous few years to "settle the vexed Indian question, than the entire regular army has done in the last thirty years. They are destroying the Indian's commissary. . . . Send them powder and lead, if you will; for the sake of lasting peace, let them kill, skin and sell until the buffaloes are exterminated."[76] Scholars have yet to uncover a smoking gun proving that the U.S. Army pursued an official policy of systematically destroying the bison as part of its campaign to pacify Native Americans. Historian David Smits has, however, found compelling evidence that numerous government officials knew that they would fail to triumph over the Plains Indians until the bison were gone. And he has shown that these same officials actively encouraged and supported commercial hunting of the species.[77]

The commercial bison trade received a boost in the late 1860s and the early 1870s with the completion of the transcontinental railroad, the manufacture of the first accurate, large-bore rifles, and the development of methods for tanning buffalo hides into leather that was in great demand for industrial belting.[78] After the Civil War, more than a thousand hide hunters descended on the plains hoping to cash in on the newly expanded market for bison leather. Depending on

FIGURE 17. Buffalo hides awaiting shipment, Dodge City, Kansas, 1874. With the spread of western railroads and the development of tanning methods that rendered their hides into pliable leather, demand for the bison soared following the American Civil War. Courtesy of the Kansas State Historical Society.

their skill level, a hunter could expect to bag anywhere from twenty-five to one hundred buffalo per day. The total slaughter ran to as many as 2 million or more bison per year.[79] In addition to overhunting, other bison casualties resulted from the introduction of Texas cattle fever in the southern part of its range, a series of abnormally severe winters, occasional grass fires, periodic drought, and most importantly, competition from introduced cattle. By the end of the 1870s, the southern plains bison had been reduced to a few hundred stragglers; a decade later, the same was true of bison in the northern plains. A species that had once blackened the landscape for as far as the eye could see was barely hanging on by a thread.

At nearly the same time as the bison, the passenger pigeon (*Ectopistes migratorius*) experienced a similar population crash from which it failed to recover.[80] This species, more than any other, has attained iconic status as a symbol of both the natural abundance of North America and the profligate waste of Euroamerican settlers. According to estimates by A. W. Schorger—the chemist, amateur naturalist, and author of a meticulously researched book on the species—as many as 3 to 5 billion passenger pigeons inhabited the eastern and central part of the continent around the time of European contact, a staggering number representing as much as a 25 to 40 percent of the entire avifauna of North America at the time.[81] Ironically, the apparent profusion of passenger pigeons when Europeans first arrived in the New World may have resulted from the crash of Native American populations in eastern North America during this same period.[82] Whatever the cause of their prodigious populations, we know that the species moved around in vast flocks both to exploit seasonally superabundant crops of mast—especially beechnuts, acorns, and chestnuts—and to overwhelm local predators at sites where they roosted and bred. That evolutionary strategy proved quite successful until Euroamericans, with their insatiable demands and access to far-flung markets, arrived on the scene.

The early descriptions of passenger pigeon flights, roosts, and nesting sites are mind-boggling. One afternoon in the early nineteenth century, the artist and naturalist Alexander Wilson witnessed a massive flock of pigeons flying overhead near the town of Frankfort, Kentucky. "I was suddenly struck with astonishment at a load rushing roar succeeded by instant darkness," he wrote of the encounter. "I took [it] for a tornado, about to overwhelm the house and everything round in destruction."[83] Wilson estimated that the densely packed flock, which was more than a mile wide and took longer than four hours to pass, contained more than 2.2 billion birds. "An almost inconceivable multitude," he admitted, "and yet probably far below the actual amount."[84] According to a later estimate, if each pigeon in Wilson's flight consumed about an eighth of a pint of mast each day, in a year, the entire flock would consume enough to fill "a warehouse 100 feet high, 100 feet wide, and 25 miles long."[85]

Numerous other writers corroborated Wilson's account. In 1813, for example, Audubon was traveling to Louisville, Kentucky, along the banks of the Ohio River, when he encountered a flight of passenger pigeons so large that "the noon-day was obscured as by an eclipse."[86] During this same period Audubon also visited a pigeon roosting site on the banks of the Green River that he estimated to be forty miles long and three miles wide. He was astonished at what he saw there: "The dung lay several inches deep, covering the whole extent of the roosting-place, like a bed of snow. Many trees two feet in diameter, I observed, were broken off at no great distance from the ground; and the branches of many of the largest and tallest had given way, as if the forest had been swept by a tornado. Every thing proved to me that the number of birds resorting to this part of the forest was immense beyond conception." The arrival of the birds in the early evening was also impressive, producing a noise that reminded Audubon of "a hard gale at sea, passing through the rigging of a close-reefed vessel."[87]

While awed by the massive flocks, naturalists learned only the rudiments about the passenger pigeon's life history and behavior before it fell victim to extinction. The species resembled an oversized mourning dove, a bird with which it would later be confused, and bred from April to June in deciduous forests of the northeastern United States and Upper Midwest. Vast nesting colonies stretched to as much as several hundred square miles, with new sites generally selected each year. Consistent with the evolutionary strategy of predator satiation, the stages of breeding at a particular site—from courtship and nest building to egg laying and hatching—occurred in an extraordinarily synchronous fashion. Generally, only a single egg would be laid in each nest, and only one brood was raised each year. Twelve or thirteen days after the egg was laid, the young would hatch. About two weeks later, the adults abandoned the site en masse, leaving the plump squabs to fend for themselves. Late in the summer, the species migrated to the southern portion of its range. Frequent movement from one site to another earned the species its scientific name, *Ectopistes migratorious,* which roughly translates to "wandering wanderer."

Humans living near temporary pigeon roosts and nesting sites had long utilized the bird for food and feathers. Native Americans and Euroamericans alike celebrated the periodic arrival of "pigeon years," when vast flocks descended upon an area with a sufficiently large mast crop to support the birds.[88] While deforestation pushed the species from the eastern seaboard by the mid-eighteenth century, it remained plentiful in the western part of its range for another century. In the early part of the nineteenth century, for example, Audubon witnessed residents of a community near Russellville, Kentucky, descend on a large pigeon roost in the area. Using pots of burning sulfur, pine-knot torches, long poles, and guns, they killed untold numbers of the birds. Once all the local wagons were groaning with the bodies of hapless pigeons, two farmers turned loose several hundred hogs to devour those that remained. While disturbed at the carnage he witnessed, Audubon believed that the species remained entirely safe from extinction: "Persons unacquainted with these birds might naturally conclude that such dreadful havoc would soon put an end to the species. But I have satisfied myself, by long observation, that nothing but the gradual diminution of our forests can accomplish this decrease."[89] Audubon was wrong, however; he not only greatly overestimated the reproductive potential of the passenger pigeon but also underestimated the impact of large-scale commercial exploitation of the species.

Market hunting of the pigeon, which began in earnest in the 1830s and 1840s, increased dramatically with the development of the railroad. In the second half of the nineteenth century, large urban dealers recruited vast networks of pigeon buyers and trappers, who were alerted to the movements of the species by railroad express agents using telegraphs.[90] When a roosting or nesting area was spotted, hundreds of professional netters descended on the site, where locals hoping to cash in on a visitation from the species would join them. The most efficient way of obtaining the birds was using large nets that could capture as many as several thousands pigeons each time they were sprung. Hunters used salt, grain, and blinded stool pigeons dropped from short platforms to lure the birds

FIGURE 18. Shooting passenger pigeons in northern Louisiana, 1875. Early accounts of the passenger pigeon routinely mention migrating flocks so vast they would blacken the skies for hours. Commercial hunting and habitat destruction quickly decimated the species in the second half of the nineteenth century. Smith Bennett, "Winter Sports in Northern Louisiana: Shooting Wild Pigeons," *Illustrated Sporting and Dramatic News* (July 3, 1875). Courtesy of the Marion duPont Sporting Collection, Special Collections, University of Virginia Library.

into their netting areas.[91] As late as the 1870s, market hunters shipped millions of pigeons from nesting sites in Michigan and Wisconsin. The vast majority were dead birds intended for food, but sportsmen also consumed large quantities of live pigeons to use as trap shooting targets.[92]

During the 1880s, sightings of the species became increasingly sporadic, while the number of pigeons shipped to market declined dramatically. Estimates of individual flock sizes that had once ranged in the hundreds of thousands and even the millions had now been reduced to thousands and hundreds. By the spring of 1888, the bird had become scarce enough that the Massachusetts ornithologist William Brewster ventured to Cadillac, Michigan, to investigate reports of pigeon sightings. While Brewster and his traveling companion located only scattered single pairs of nesting pigeons, they interviewed netters who claimed to have seen passing flocks varying from fifty birds to one that covered "at least eight acres." Based on these reports, Brewster concluded that the species was not yet teetering on the brink of extinction, as some had feared. But he predicted that without adequate protection "our Passenger Pigeons are preparing to follow the Great Auk and the American Bison."[93]

As early as 1862, the state of New York passed a law that prohibited disturbing pigeons at their nesting sites or discharging a gun within one mile of it perimeter.[94] Over the next several decades, numerous other states would enact similar laws, but because the public initially remained apathetic to the plight of the species, officials rarely enforced them. The Michigan legislature passed the only law giving complete protection to the pigeon in 1897, but by then, it was too late. While a decrease in the overall pigeon population greatly reduced the efficiency of hunters who sought the species, it may have actually increased the effectiveness of natural predators. Survival strategies consistent with predator satiation—for example, exposed nests and single egg clutches—worked quite well as long as the species bred in large flocks that overwhelmed the predators in any particular area. But as the size of flocks diminished, the remaining birds found themselves increasingly vulnerable to attack from other animals. At the same time, they no longer received the critical social cues that large flocks had once provided. Thus, while deforestation and overhunting brought the passenger pigeon to the brink of extinction, biological factors may have been responsible for the final blows. Claims of passenger pigeon sightings continued well into the twentieth century, but the last specimen taken from the wild was probably shot in Ohio in 1900.[95]

THE AUDUBON MOVEMENT TAKES FLIGHT

In early 1886, a decade after Allen first warned about the impending extinction of numerous American birds and mammals, the first inklings of a large-scale bird protection movement swept across the United States.[96] Over the next three decades, thousands of middle-class Americans took up the practice of birdwatching, devoured nature essays, joined Audubon Societies, and lobbied for legislation to protect nongame birds. They were moved to action by a variety of interrelated forces, including a romantic appreciation for nature, a humanitarian concern about the suffering of wild animals, and the aesthetic appeal of birds. Whatever their motivations, for the first time in this nation's history, a significant number of Americans began experiencing wild birds as something more than a resource to be consumed or a commodity to be sold. And as they became more aware of and emotionally connected with the nation's avifauna, they also grew increasingly haunted by threat of extinction. Through publications, lectures, and institutional connections, naturalists played a crucial role in organizing, leading, and sustaining the Audubon movement.

One early manifestation of that movement was a lengthy warning about the desperate plight of North American birds that first appeared in the journal *Science* in February 1886. This unusual, fifteen-page document was the product of the bird protection committee of the American Ornithologists' Union (AOU), an organization of largely technically oriented ornithologists that J. A. Allen, William Brewster, and Elliott Coues had founded three years earlier. Although first established one year after the creation of the AOU, the bird protection committee remained inactive until late 1885, when Allen, the union's president, convinced the amateur naturalist and oil machinery manufacturer George B. Sennett to take charge.[97] With most it members residing in the New York metropolitan area, over the next year, the committee held a series of weekly meetings at the American Museum of Natural History. With financial support from G. E. Gordon, the president of the American Humane Association, the committee distributed over one hundred thousand copies of its bird protection manifesto, which introduced many of the arguments that would become standard in the Audubon movement.

The introductory, longest, and most comprehensive essay in the group, "The Present Wholesale Destruction of Bird-Life in the United States," was penned by none other than Allen himself.[98] Echoing arguments he had first made a decade earlier, he castigated humans for severely disrupting "nature's balance." "The history of this country," Allen argued, "is the record of unparalleled destruction of the larger forms of life." Habitat destruction—through deforestation, the drainage of swamps and marshes, the transformation of "wildlands" into farms, and countless other changes accompanying European settlement—had devastated "the haunts and the means of subsistence of numerous forms of animal life" and resulted in "their extermination over vast areas." Market hunting, egg collecting, and indiscriminate shooting by "sportsmen," recent immigrants, and "colored people" each received brief coverage and strong condemnation. The principle target of Allen's wrath, however, was the millinery trade, which remained the nemesis of the bird protection movement for the next three decades. According to Allen, the recent fad for using bird feathers to decorate women's hats represented a threat to North American birds "many times exceeding all others together." Given this and other dangers facing native wildlife, Allen offered a grim prediction: "The fate of extermination, which to the shame of this country, has already practically overtaken the bison, and will sooner or later prove the fate of all of our larger game-mammals and not a few of our game birds will, if a halt be not speedily called by enlightened public opinion, overtake scores of our song-birds, and the majority of our graceful and harmless, if not somewhat less 'beneficial,' sea and shore birds."[99]

Why should anyone worry about the fact that "many species, and even genera, of birds are fast disappearing from our midst?" Deploying aesthetic and utilitarian arguments that would remain central to the bird protection movement for decades, Allen challenged the pervasive belief that nature was simply a boundless storehouse of economically valuable resources waiting to be exploited. In addition to whatever economic value they might possess, Allen proclaimed, birds also enjoyed an aesthetic value that rendered them "among the most graceful in movement and form, and the most beautiful and attractive in coloration, of nature's many gifts to man." At the same time, they were vivacious, charming creatures, with melodious voices. Allen seconded the opinion of a recent writer who claimed that the prospect of a countryside without songbirds was simply too horrible to contemplate—like "a garden without flowers, childhood without laughter, an orchard without blossoms, a sky without color, roses without perfume."[100]

For those who remained unmoved by an appeal to the living bird's aesthetic charms, Allen also offered a more practical rationale for bird protection: "The great mass of our smaller birds, numbering hundreds of species, are the natural checks upon undue multiplication of insect pests."[101] So-called economic ornithologists had been developing this basic argument for years. Having analyzed the food remains found in the stomachs of numerous avian species and swayed by the long-held assumption that nature remains in overall balance, naturalists hastily concluded that birds represented the primary agent in countering the potentially explosive growth of insect pests.[102] Allen remained confident the U.S. Department of Agriculture's recently created Division of Economic Ornithology and Mammalogy (the institutional precursor to the Bureau of the Biological Survey and the Fish and Wildlife Service) would confirm the central role that birds played in maintaining nature's balance.[103]

While the AOU's bird protection committee felt its primary mission was to educate the public about the problem of bird destruction, it also recognized the need for effective protective laws. To further that goal, it produced text for state legislation—the so-called AOU model law—aimed at protecting nongame birds.[104] Under the terms of this law, anyone who killed, purchased, or sold any nongame bird, its nest, or eggs would be subject to a fine of five dollars and up to ten days in jail for each offense. The only nongame species the committee specifically exempted from its protective umbrella was the English or European house sparrow, a recent introduction that faced increasingly strong condemnation from American ornithologists as it expanded its range across the United States. The model law also included a provision that authorized the collection of birds "for scientific purposes" through a special permit system.

Just as the AOU bird protection committee was moving into high gear, the patrician naturalist and editor George Bird Grinnell founded the first Audubon Society.[105] After earning his A.B. from Yale in 1870, Grinnell joined the army of students who amassed fossils for the preeminent American paleontologist O. C. Marsh, and three years later he became Marsh's assistant at the Peabody Museum. While J. A. Allen was writing his series of articles lamenting the decline of North American fauna, Grinnell was independently reaching a similar conclusion. As had been the case for Allen, westward travels—including stints as a naturalist on Custer's 1874 Black Hills Expedition and William Ludlow's 1875 reconnaissance of Yellowstone National Park—proved critical in the development of Grinnell's conservation consciousness. As he announced in a letter that accompanied his report, the latter experience convinced Grinnell that "the large game still so abundant in some localities will ere long be exterminated."[106] After earning his doctorate in paleontology in 1880, Grinnell became editor-in-chief and owner of *Forest and Stream,* a preeminent sporting and natural history periodical. He would occupy both positions for the next thirty-five years, thereby consolidating the magazine's position as a leading voice in condemning the commercial exploitation of wildlife.[107]

For a brief period, Grinnell also assumed a key leadership position in the fledgling Audubon movement before moving on to other conservation causes. In 1883, the same year he became a charter member of the AOU, he published the first in a series of editorials that sharply condemned the millinery trade's destruction of songbirds.[108] A year later, he became an active member on the AOU bird protection committee. In early February 1886, less than two weeks before that committee issued its first bulletin, Grinnell published an editorial proposing the formation of an "Audubon Society" dedicated to the "protection of wild birds and their eggs."[109] In a series of follow-up publications, he noted that recently the scale of bird destruction had expanded to the point it was "seriously threaten[ing] the existence of a number of our most useful species."[110] The choice of name for the proposed society seemed a natural one for Grinnell: not only had he once attended a school run by Audubon's widow but the famed artist's anthropomorphic and widely reproduced bird portraits often moved viewers to see wild creatures more sympathetically.

Membership in the new Audubon Society was open anyone who pledged not to kill nongame wild birds, destroy their nests, or wear their feathers. The national society was to be organized into a series of local chapters, to which Grinnell promised to provide circulars and other printed information promoting the cause. By the end of the first year, he boasted over three hundred local chapters and nearly eighteen thousand members.[111] In an effort to finance the increasingly burdensome workload associated with the enterprise, in 1887, Grinnell introduced *Audubon Magazine*, which featured news of the society, popular articles on birds, children's stories, diatribes against feather fashions, and a serialized version of one of the earliest field identification guides. One article chronicled the plight of the great auk, which, "like the Dodo of Mauritius and some other birds, has wholly ceased to exist because exterminated by the cruelty of man." If the predations of plume hunters remained unchecked, the anonymous author (probably Grinnell) warned, other birds would soon experience the same "melancholy fate" as these lost species.[112] In addition to ongoing concern about the cruelty of using feathers to decorate hats, Grinnell and the many supporters of the first Audubon Society clearly felt unease with the looming prospect of extinction.

After a burst of activity, however, the movement faltered. By December 1889, Grinnell claimed more than fifty thousand members for his Audubon Society, a remarkable accomplishment.[113] However, few of those who joined the movement proved committed enough to pay the modest subscription fee for its affiliated magazine, so he stopped publication with the January 1889 issue and abandoned the organization altogether soon thereafter.[114] The AOU bird protection committee experienced a similar fate. After securing the passage of versions of the model law in New York in 1886 and Pennsylvania in 1889, the committee languished. Many AOU members—resentful of the increasing restrictions on their ability to collect wild birds—seemed relieved at its apparent demise.

A NEW BEGINNING

The Audubon movement's more enduring revival came a few years later, just as the larger progressive reform movement was taking off and not long after the plight of the passenger pigeon and bison began reaching public consciousness. In February 1896, the Boston society dame Harriet Lawrence Hemenway and her cousin Minna B. Hall called the first meeting of the Massachusetts Audubon Society.[115] The stated purpose of the new organization was to discourage the use of bird feathers in hats and to otherwise "further the protection of wild birds."[116] After choosing the ornithologist William Brewster as president, the society began issuing a steady stream of pamphlets to bring its protectionist message before the public. Anxious reformers in dozens of other states quickly followed Hemenway and Hall's lead. As was the case in Massachusetts, male naturalists occupied most leadership positions within Audubon Societies, while females comprised the majority of the membership. The newly revived societies lobbied for protective legislation, hired wardens, promoted birdwatching, and championed nature study in the nation's schools.[117]

One force pushing state Audubon Societies to action was the naturalist and insurance salesman William Dutcher, who assumed chairmanship of the AOU bird protection committee late in 1895.[118] Dutcher's interest in natural history began as a sportsman in 1879, when he shot a bird on Long Island that he failed to recognize—a female Wilson's plover that turned out to be a new record for the area. Following that discovery, he began systematic study of the region's avifauna. His growing interest in ornithology led him to become associate member of the AOU at the time of its creation and treasurer of the organization between 1887 and 1903. After remobilizing the bird protection committee, Dutcher worked tirelessly to promote the creation of state Audubon Societies, to obtain passage of protective legislation, and to champion educational efforts aimed at publicizing the plight of birds. In November 1900, Dutcher called the first informal meeting of representatives of the various state Audubon Societies that had been organized up to that point. Five years later, he orchestrated the creation of the National Association of Audubon Societies, which was based out of New York City. He ran the new organization until 1910, when he suffered a debilitating stroke from which he never fully recovered.

Among the long string of conservation victories that Dutcher helped secure was the Lacey Act.[119] Signed into law on May 25, 1900, that pioneering wildlife legislation authorized the federal government to fund the restoration of wild bird populations, to regulate the importation of foreign animals, and most importantly, to prohibit the interstate shipment of "wild animals and birds" taken in violation of state laws. The final form of the bill, which emerged following four years of congressional consideration, was drafted in close consultation with the ornithologist T. S. Palmer, a bird protection committee member and chief assistant of the Bureau of the Biological Survey. The Audubon coalition of naturalists, humanitarians, and nature lovers fought for the bill's passage, but critical support also came from recreational hunters, especially G. O. Shields and his League of American Sportsman, who were working to eliminate the sale of all game.

Encouraged by the Lacey Act victory, in 1901 Dutcher and Palmer mounted an aggressive campaign to secure passage of the AOU model law.[120] Prior to that time, only five states had enacted satisfactory laws protecting nongame birds. Working closely with the AOU bird protection committee, state Audubon Societies, and sport hunting organizations, the two naturalists lobbied state legislatures up and down the eastern seaboard. As a result of their tireless efforts, by year's end, eleven states passed new or improved laws, and by 1903, twenty-nine states had adopted some version of the AOU model law.

Dutcher and Palmer also played a key role in the creation of the nation's first federal wildlife refuge.[121] On March 14, 1903, President Theodore Roosevelt signed an unusual executive order establishing Pelican Island—a four-acre site on Florida's Indian River that was home to a large nesting colony of brown pelicans—as a "preserve and breeding ground for native birds." The action came after the ornithologist and conservationist Frank M. Chapman began urging protection of the island's beleaguered inhabitants after several visits to the site. In 1902, Dutcher hired a warden to monitor the island during breeding season and filed paperwork to purchase the site from the federal government on behalf of the AOU bird protection committee. When the transaction became bogged down in bureaucratic red tape, a General Land Office employee suggested that Pelican Island might simply be declared a federal bird reservation. Roosevelt, who had long been interested in bird collecting and watching and a strong supporter of conservation initiatives, found the idea appealing. It also thrilled Dutcher and Palmer, who with Frank Bond, an Audubon official and General Land Office agent, began pressing Roosevelt to declare additional reservations. The president seemed quite willing to accommodate their requests, and before leaving office in 1909, he created fifty additional wildlife reservations, including several that were aimed at protecting big game animals.

In addition to his role in establishing the first national wildlife reservation, Chapman also proved central to numerous other bird conservation initiatives.[122] As a child, he had enjoyed exploring, hunting, and collecting birds in the countryside surrounding his home in Englewood Township, New Jersey, then a sparsely populated bedroom community for New York City. In 1888, the twenty-four-year-old Chapman abandoned a promising banking career to take up a position as an assistant to J. A. Allen, who served as an important father figure and mentor. He would remain at the American Museum of Natural History for the next half century, rising to the level of curator of the Department of Ornithology and establishing a reputation as one of the nation's preeminent ornithologists for his pioneering research on the biogeography and ecology of South American birds.

In addition to substantial accomplishment as a museum administrator and scientist, the indefatigable Chapman also pursued the equivalent of a second career as a bird popularizer and conservationist. His lectures, often illustrated with his own photographs and motion pictures, proved a hit with audiences across the nation, while his innovative field guides were best sellers. An active and early member of the AOU bird protection committee, Chapman helped found the

New York Audubon Society in 1897 and served as a longtime board member and chair of the National Association of Audubon Societies. Even more time consuming was his work for *Bird-Lore*, a popular bimonthly magazine of popular ornithology that he founded in 1899 and edited until 1934. Not only was the magazine brimming with useful information for bird enthusiasts of all ages, levels, and degrees of interest, it was also the official organ of the Audubon movement. As such, the magazine played a critical role in promoting a collective sense of identity among Audubon members across the nation.

Allen, Grinnell, Dutcher, Palmer, and Chapman were unusual only in the intensity of their devotion to the Audubon movement, which continued to expand well into the twentieth century. Along with many of their colleagues, these naturalists shared an abiding interest in science, a deep concern about the fate of American wildlife (especially birds), and a strong sense of obligation to act on that concern. Haunted by the specter of extinction, they sought to mobilize the public to save threatened species. Their campaign achieved remarkable success in securing a place for vanishing species on the nation's political and social agenda. The publicity drives they mounted, state and federal laws they promoted, and wildlife refuges they secured reversed the steep decline that many sea, shore, and wading birds had experienced at the hands of the millinery trade. Yet, for those species that had reached critically low levels, these measures remained insufficient to pull them back from the brink of extinction. To rescue the most endangered species, some conservationists began arguing for the more intensive, interventionist approach that fish culturalists had long embraced: artificial propagation.

CHAPTER FOUR

NATIONALISM, NOSTALGIA, AND THE CAMPAIGN TO SAVE THE BISON

*The wild buffalo is practically gone forever, and in a few more years . . .
nothing will remain of him save his old, well worn trails along the water-courses,
a few museum specimens, and regret for his fate. If his untimely end fails
even to point to a moral that shall benefit the surviving species of mammals* which
are now being slaughtered in a like manner, *it will be sad indeed.*

WILLIAM TEMPLE HORNADAY, 1886

*It takes millions of years to produce beautiful and wonderful varieties of animals
which we are so rapidly exterminating. Unless we can create a sentiment which will
check this slaughter, and devise laws for those who do not respect this sentiment,
the bones of our now common types will soon be as rare as those of the dodo and the
great auk; and man will be practically the sole survivor of a great world of life.*

WILLIAM TEMPLE HORNADAY, 1897

A CONSERVATION CONVERSION

In early 1886, William Temple Hornaday, the chief taxidermist at the U.S. National Museum, had bison on the brain. A decade before, the species that had come to symbolize the Great Plains had been all but obliterated from the southern portion of its once extensive range. Now increasingly frequent predictions about its demise in the north seemed to be coming to pass. After several months of correspondence about the status of the species, Hornaday reached a grim conclusion: fewer than three hundred bison remained throughout the entire United

States. Vast herds had once darkened the prairies with their thunderous, teeming masses, transforming the landscape during their seasonal migrations and filling those who witnessed them with a profound sense of awe; now the once-abundant species was teetering on the brink of extinction. "Could any war of extermination be more complete or far-reaching in its results!" Hornaday exclaimed.[1]

The impending demise of the bison greatly alarmed Hornaday, but most readers today are likely to find his initial response puzzling. When he discovered that the Smithsonian collection contained only a modest, motley assortment of old bison skins, skeletons, skulls, and mounted heads, Hornaday urged his boss, Secretary of the Smithsonian Spencer F. Baird, to sponsor a bison expedition while it remained possible. Taken with the idea, Baird urged Hornaday to collect enough specimens to supply not only the Smithsonian but also other smaller museums with examples of the increasingly rare beast. If they could still be found, Hornaday hoped to bring back as many as one hundred bison specimens, a prospect that apparently generated at least some pangs of guilt in the thirty-one-year-old taxidermist. "I am really ashamed to confess it, but we have been guilty of killing buffalo in the year of our Lord 1886," he later wrote in a widely circulated account of his expedition. "Under different circumstances nothing could have induced me to engage in such a mean, cruel, and utterly heartless enterprise as the hunting down of the last representatives of a vanishing race." But, Hornaday argued, he really had "no alternative." The species' days were clearly numbered, and it was far better for the remains of the final bison to be preserved for posterity than to "decay, body and soul, where they fell." It might strike us as odd today that Hornaday initially seemed more interested in preserving the remains of dead bison than of pursuing strategies to save the species as a living, breathing organism. But in doing so, he was following the dictates of natural history of his day, which was firmly rooted in collection and taxonomy. Never known for his modesty, Hornaday also believed that his taxidermy skills could render the bison "comparatively immortal," thereby helping to atone for humanity's reckless destruction of the species.[2]

That claim might sound like self-serving hubris, but at the time he approached Baird about funding the Smithsonian bison expedition, Hornaday had earned a reputation as one of America's leading taxidermists.[3] Born in Indiana in 1854 and raised in Iowa, he came of age amid a prairie that was still brimming with wildlife. Young William learned to hunt from an early age and soon became a crack shot with the rifle. During a visit with an older half-brother, he discovered a case of mounted ducks in a gun and fishing tackle store. The colorful specimens fired his imagination, and he entered Iowa State Agricultural College with the aim of becoming a professional naturalist, taught himself the rudiments of mounting animals, and gained an appointment as taxidermist at the college museum. Two years later, he left to join the staff of Ward's Natural Science Establishment, in Rochester, New York.

As America's leading natural history dealer, Ward's provided an ideal environment for Hornaday to cultivate his taxidermy and field skills.[4] The owner of the enterprise, Henry A. Ward, dispatched him on several collecting expeditions, including a round-the-world trip that Hornaday later chronicled in *Two Years in the Jungle* (1885). This first book established his reputation as a popular wildlife and adventure writer. While working for Ward, Hornaday also began developing what he termed "artistic groups" of mounted animals, an idea that one historian has called "the beginnings of the museological evolution of the habitat diorama in America."[5] As part of this important innovation in museum exhibits, he pioneered a sculptural approach to taxidermy that involved securing animal skins on carefully fashioned clay-covered manikins, thus rendering them more lifelike than traditional methods using bones and wooden armatures crudely stuffed with straw or rags. To promote his ideas and to further the profession of taxidermy, Hornaday and his colleagues founded the American Society of Taxidermists in 1880. Hornaday himself won the organization's first annual competition for a sensationalized orangutan group he entitled "Fight in the Tree-tops." The exhibit, which the U.S. National Museum purchased, earned him a position as chief taxidermist there. It is not clear why his attention turned to the bison in 1886, but he seems to have been motivated primarily by his desire to create a striking group display of this increasingly rare creature.

After securing pledges of a military escort and logistical support from the U.S. secretary of war, in May of 1886 Hornaday and two assistants set out to Miles City, Montana, near where a small herd of bison had reportedly been sighted. When the expedition party arrived, they encountered local residents who expressed skepticism about claims of live bison in the area and a landscape littered with bison remains. Row upon row of bleached skeletons provided an eerie reminder of what had been lost. Hornaday realized that any living bison he might find during this time of the year would probably be shedding its winter plumage, thus making it unsuitable as a museum specimen. But he hoped to find at least a few specimens that were still in reasonable shape, and he wanted to reconnoiter the area for a possible later expedition. Much to his delight, the expedition party managed to capture a month-old, light-colored calf, dubbed Sandy, which they transported back to Washington and put on display.

Hornaday and an assistant returned to Montana in late September 1886. With three cowboy guides and a four-man military escort, the party spent two months in the field before inclement weather forced their retreat. During this period,

FIGURE 19. William T. Hornaday and bison calf "Sandy," 1886. This young calf, which Hornaday captured alive during a western expedition to obtain bison specimens for the Smithsonian Institution, became a source of the inspiration for the U.S. National Zoo. Hornaday originally conceived of the institution as a way to preserve vanishing forms of North American wildlife. Courtesy of the Smithsonian Institution Archives, Record Unit 95, image no. 74-12338.

Hornaday's party secured twenty-five bison skins, one head skin, sixteen fresh and dry skeletons, fifty-one dry skulls, and two bison fetuses. On the final day in the field Hornaday even managed to bring down a massive old bull, "a truly magnificent specimen," that weighed about 1,600 pounds and measured five feet, eight inches at the shoulder, a full two inches taller than any other in the Smithsonian collection.[6] When Hornaday skinned the specimen, he discovered four old bullets lodged in its body.

Hornaday was ecstatic and soon after returning to Washington began designing a large exhibit incorporating the specimens as a proper "monument to the American bison."[7] Such an exhibit would provide an ideal opportunity to demonstrate the possibilities of the artistic groups he had been promoting while memorializing one of the West's most characteristic species. After receiving the requisite permissions, he selected six bison for the group, including not only the massive old bull, but also Sandy, the young calf that had lived on the mall in Washington for two months before its untimely death. He mounted the specimens using carefully constructed manikins for support, placed the group on a naturalistic setting that included material from the bison's habitat, and enclosed it in a massive mahogany and glass case. Among those who came to visit Hornaday while he was assembling the exhibit was General Phil Sheridan, who had once

FIGURE 20. William T. Hornaday's bison group, 1886. Hornaday created this impressive exhibit both to commemorate the bison, which was teetering on the brink of extinction, and to demonstrate his considerable taxidermy skills. The large male featured in the center would repeatedly be used as a model for stamps, coins, and insignia. Courtesy of the Smithsonian Institution Archives, Record Unit 95, image no. NHB-5470.

promoted the extermination of the bison as a means to pacify the Plains Indians; another enthusiastic visitor was Theodore Roosevelt, already a nationally recognized political figure. The two men spoke for an hour during their first meeting, found they had much in common, and became lifelong friends.[8]

Hornaday's bison group opened to much fanfare in early 1888. A local newspaper declared the exhibit a great success, a "picturesque . . . bit of the old wild west" transplanted to Washington.[9] Not surprisingly, Hornaday's boss, the director of the U.S. National Museum George Brown Goode, proclaimed his handiwork a "triumph of the taxidermist's art" that "surpassed in scientific accuracy and artistic design and treatment, anything of the kind yet produced."[10] Goode also confirmed plans for a series of similar exhibits of American mammals, each with its own case and natural accessories typical of the species' habitat. Hornaday's bison exhibit remained a favorite at the Smithsonian for the next seventy years, and the large male that towered over the other five specimens became the model for the buffalo-head nickel, a ten-dollar bill, several commemorative stamps, and the seal of the secretary of the interior.[11]

At some point after he returned from Montana, Hornaday realized that it was not enough simply to commemorate the bison in a museum exhibit, no matter how magnificent that exhibit might be. Experiencing a conversion from zealous collector to ardent protector, he grew deeply concerned about the plight of the few bison that remained and determined to do something to rescue them. One product of Hornaday's shift was a lengthy publication, *The Extermination of the American Bison*, that initially appeared in 1889 as part of the annual report of the U.S. National Museum and was later published separately. Reflecting Hornaday's background and shifting motivations, the report is a complex mixture of genres—part scientific treatise, part hunting account, and part call to action. Within its two-hundred-odd pages, he presented one of the longest monographs up to that point covering any single form of North American wildlife, including extensive details about the bison's life history, the factors that led to its extermination, the Smithsonian expedition, and his prized group exhibit. In characteristically hyperbolic fashion, Hornaday minced few words as he blamed the near extinction of the bison on "the descent of civilization, with all its elements of destructiveness, upon the whole of the country inhabited by that animal. From the Great Slave Lake to the Rio Grande the home of the buffalo was everywhere overrun by the man with a gun; and, as has ever been the case, the wild creatures were the first to go."[12] Unless something were done soon to stem the slaughter, numerous other species were likely to fall victim to this same destructive impulse.

A second product of Hornaday's conversion experience was a series of experiments with the captive breeding of vanishing species. Hornaday played a key role in the creation of the National Zoological Park in Washington, D.C., an institution he originally envisioned as a site to preserve and raise endangered North American species, especially mammals. At the time, naturalists and civic leaders had already opened some two dozen American zoos, mostly in northeastern cities, with the primary aim of providing entertainment and education for visitors. Zoo exhibits revealed the order and wonder of nature, offered tangible symbols of America's global reach, and provided a source of civic pride.[13] Hornaday considered these objectives important to garner the continued financial and political support zoos needed to survive, but he also thought they might play an important role in rescuing vanishing species. When his initial plans for the National Zoological Park were thwarted, he moved to New York State, where he was soon was recruited to help establish the New York Zoological Park (more popularly known as the Bronx Zoo). As the first director of that institution, Hornaday worked on techniques to raise and breed several declining mammals with mixed results. He enjoyed the most success with the bison, and in conjunction with the American Bison Society, he and his colleagues repopulated several western sites with the progeny from early experiments with this charismatic species.

He was not so fortunate with other species facing extinction. Some, like the pronghorn antelope, not only proved more difficult to ship and to maintain in captivity but also less appealing to the public. Other endangered species—like the passenger pigeon and the Carolina parakeet—seemed beyond the pale entirely. Both might have been saved through captive breeding programs if the same attention and care had been devoted to them as to the bison. But naturalists at the time seemed more interested in gathering up the last examples of these beleaguered birds for museum collections—just as Hornaday had done earlier with the bison—than in initiating systematic attempts at captive breeding. And nationalistic, nostalgic appeals to rescue the bison found more resonance with the American public than calls to save the passenger pigeon and the Carolina parakeet.

SAVING BISON, SAVING AMERICA

Shortly after his return from Montana in 1886, Hornaday issued a call for a Smithsonian-affiliated zoo that would shelter the breeding stock of endangered species and help educate the public about their plight.[14] The idea seems to have originated with the capture of Sandy, who remained on display in front of the U.S. National Museum for two months before the young calf finally perished. Several months after his return from his second trip to Montana, Hornaday approached Baird with the idea of creating a zoological park in Washington, but with Baird too ill to act, the idea initially languished. In October 1887, following Baird's death, Goode organized a Department of Living Animals as part of the U.S. National Museum, naming Hornaday as its curator. The ostensible rationale for the new department was to provide Smithsonian taxidermists with living models to observe, but Hornaday later wrote that he also viewed the experiment as a sort of trial balloon for his zoo idea. The first specimens for the new menagerie were gathered by Hornaday during a month-long western trip in the fall of 1887 and housed in a makeshift wooden structure just south of the original Smithsonian Building.

Upon his return, Hornaday began publicly promoting the idea of a full-fledged zoo in Washington, claiming that he was moved to action by the "fearful rapidity with which game is being killed in the West and the absolute certainty that in a few years many of the representative animals will be entirely extinct."[15] If these increasingly rare animals could be captured and successfully bred, Hornaday argued, they might be rescued from oblivion. Not surprisingly, Hornaday was especially interested in saving the beleaguered bison and urged Goode to procure specimens as soon as practicable: "It now seems necessary for *us* to assume the responsibility of forming and preserving a herd of live buffaloes which may, in a small measure, atone for the national disgrace that attaches to the heartless and senseless extermination of the species in the wild state."[16] Within a year, Hornaday had his wish, and six bison were among the growing collection of animals that graced the Washington Mall, where they attracted a throng of admiring visitors.

The popularity of the menagerie convinced Samuel Pierpont Langley, Baird's successor as secretary of the Smithsonian, to back Hornaday's proposal for a National Zoo in Washington. Once the appropriate legislation had been introduced into Congress, Langley authorized Hornaday to devote part of his work week to fleshing out and promoting the proposal. Hornaday played a central role in virtually every aspect of bringing his idea to fruition, from selecting the 166-acre site at Rock Creek to negotiating the purchase of land and from drafting legislation to testifying before Congress. He even prepared detailed plans for the grounds and animal accommodations. Throughout this arduous two-year process, he repeatedly stressed that one of the main purposes of such a zoo would be to provide a "suitable place in which to preserve representatives of our great game animals before they are all exterminated."[17] A little more than year after the final legislation authorizing the establishment of the National Zoological Park became law in April 22, 1889, however, Hornaday abandoned his cherished project, resigning from the Smithsonian after experiencing sharp differences with Langley over his role in the fledgling institution. In the absence of his visionary leadership, the new zoo quickly became a recreational site for the citizens of Washington, D.C., rather than a place where threatened North American species might be rescued from the fate of extinction.

Disillusioned at the experience, he spent the next six years as secretary of a real estate firm in Buffalo, New York, where he found the time to complete a novel, *A Man Who Became Savage,* and a bestselling nature book, *The American Natural History.* Hornaday was finally drawn back into the world of wildlife conservation in 1896, after receiving a letter from the paleontologist Henry Fairfield Osborn, a professor of zoology and dean at Columbia University, head of the Department of Vertebrate Paleontology at the American Museum of Natural History, and later longtime president of that institution. Osborn wondered if Hornaday might be interested in assuming directorship of a new zoo in New York. The organization behind the planned facility was the New York Zoological

Society, which itself was an outgrowth of the Boone and Crockett Club, a group of well-heeled hunters, explorers, and conservationists, most of whom resided in the metropolitan New York area.

As early as 1884, George Bird Grinnell had published an editorial in *Forest and Stream* calling for an active national association devoted to protecting game species that might complement the many local, regional, and state game associations established over the last several decades.[18] A year later, he met Theodore Roosevelt after publishing a mixed review of his first book. The two men immediately hit it off, found they shared a deep interest in the outdoors, and began meeting regularly to swap stories about their western adventures. One outgrowth of their conversations was a decision to launch the Boone and Crockett Club, named after two of America's most renowned frontiersmen, in December 1887. The main purposes of the new organization were to "promote manly sport with the rifle" and "to work for the preservation of the large game of this country."[19] Regular membership was limited to one hundred men who had successfully bagged at least one representative of each of the three kinds of North America's big-game species, though individuals who had made a significant contribution to wildlife conservation could also be invited to become associate or honorary members. One of the new organization's first major conservation campaigns met with success in 1894, when Congress voted to make it illegal to kill wildlife in Yellowstone Park, imposing a fine of up to $1,000 and a jail sentence of up to two years.[20]

Roosevelt and other members of the Boone and Crockett tended to be deeply ambivalent about the forces of modernity that were transforming the American landscape.[21] On the one hand, they benefited financially from the exploitation of the nation's natural resources—the Roosevelt family, for example, made its fortune in railroads, which consumed vast quantities of wood, coal, and iron while facilitating the commercial slaughter of wildlife throughout the United States. At the same time, these same individuals often lamented the decline of native animals, especially game mammals, that had resulted from overhunting, habitat destruction, and displacement by nonnative and domesticated species. They also supported Roosevelt's call for Americans to embrace the "strenuous life" and viewed the activity of hunting as a means to counter the ill effects of an increasingly modern, urban, industrial society. Vigorous pursuit of animals in the wild provided an especially appropriate way for Americans to reconnect with their pioneer origins, they argued, to reinvigorate the character and values of the American way of life that the historian Frederick Jackson Turner had highlighted in his famous 1893 address on "The Significance of the Frontier in

American History."[22] Not coincidentally, many supporters of game conservation during the period also tended to be vehemently anti-immigrant.[23]

A prominent example was the patrician lawyer, sportsman, naturalist, and eugenicist Madison Grant, who joined the Boone and Crockett Club in 1893.[24] Although he had recently graduated from Columbia University law school, Grant was always "more interested in the pursuit of game than of the law."[25] Not long after joining the society, Grant wrote Roosevelt, who remained president of the Boone and Crockett Club after moving to Washington to serve as Civil Service Commissioner, requesting that he appoint a committee to establish a zoological park in New York City.[26] After receiving Roosevelt's assent, Grant chaired a committee responsible for shepherding a bill through the state legislature that incorporated the New York Zoological Society and authorized the newly established institution to create a zoo. Boone and Crockett men dominated the executive committee of the new society, with Grant serving as secretary. Henry Fairfield Osborn also proved central in developing the policies and plans for the new zoological garden. Both men would continue to play leadership roles in the institution for the next several decades.[27]

In November 1895, Osborn and C. Grant La Farge (an architect and Boone and Crockett member) crafted a preliminary plan for the proposed zoo. They envisioned placing animals in enclosures that resembled their "natural surroundings" as much as possible rather than the cramped iron cages typical of American and European zoos at the time.[28] Two months later, the executive committee recommended hiring "an officer of practical experience and acknowledged scientific standing" to help locate the site for the new zoo, develop its policies, and oversee its construction.[29] After receiving favorable recommendations from George B. Goode and C. Hart Merriam, head of the Bureau of the Biological Survey, Osborn offered Hornaday the job.

Jumping at the opportunity to once again bring his vision for a modern zoo to fruition, Hornaday began work in April 1896. Of the several potential sites for the new facility under consideration, he immediately fell in love with South Bronx Park, which was blessed with ample space and a wide variety of relatively undisturbed habitats. In a telling metaphor, Hornaday wrote that if Noah were to arrive at the site "with his arkful of animals and turn them loose . . . each species would promptly find there its own suitable place." In addition to gathering a miscellaneous collection of animals displayed to entertain and educate "the general public, the zoologist, the sportsman, and every lover of nature," the new zoo would also be a modern-day Noah's ark that would rescue the many "native animals of North America" struggling for survival. Indeed, the first priority of the fledgling institution would be to collect "a liberal number" of the continent's notable animals threatened with extinction, for "nearly every wild quadruped, bird, reptile and fish is marked for destruction."[30] After an extensive tour of European zoos, Hornaday returned home to complete plans for, oversee construction of, and stock the new zoo, which finally opened to much fanfare in November 1899. As Hornaday and his colleagues originally envisioned it, then, the New York Zoological Park was to be a kind of hybrid institution, a mixture between a more traditional zoo stocked with a miscellaneous assortment of exotic creatures and a wild animal reserve for endangered North American species.

Obtaining a nucleus herd of bison to stock the twenty-acre range he had planned for the park's southeastern corner ranked high among Hornaday's priorities.[31] Following a series of inquiries, he negotiated to purchase four bison from Charles Goodnight in the Texas panhandle and three from Ed Hewens in Oklahoma. Hornaday also hoped to establish breeding herds of other large western mammals—antelope, caribou, mule deer, and other threatened species—but he experienced numerous difficulties achieving this goal. In its early years, for example, the zoo received sixteen pronghorn antelopes in two separate shipments, but all of them soon fell victim to disease or roaming dogs. Hornaday also initially struggled to maintain a self-perpetuating bison herd at the Bronx Zoo, one of the institution's most popular attractions in its early years. He blamed the frequent deaths of the species on gastroenteritis, which he thought was caused by eating rank grass, but the animals continued to die off regularly even after the grass was burned off, the terrain plowed, and the topsoil removed. Though soon forced to relinquish plans to raise large numbers of other North American game animals, Hornaday persevered with the bison. After repeated infusions of purchased or donated specimens, the zoo's bison herd eventually stabilized and then began to increase through the birth of new calves. The venture proved so successful it lead to an ambitious project to begin restocking the West with animals from the New York Zoological Society collection.

BACK HOME ON THE RANGE

The popular nature writer and lecturer Ernest Harold Baynes first conceived of the idea of establishing a national society dedicated to preserving the critically endangered bison.[32] The notion apparently came to him sometime after June 1904, when he took up residence on the edge of Austin Corbin's Blue Mountain Forest Game Reserve, in western New Hampshire. Among the many game species confined within the remote, 24,000-acre site were 160 bison, a herd that had grown from twenty-two animals introduced nearly three decades earlier to be-

FIGURE 21. Ernest H. Baynes with bison team, 1928. After training a pair of young bison to harness, Baynes began exhibiting them publicly as a way to gain support for the American Bison Society. From Raymond Gorges, *Ernest Harold Baynes: Naturalist and Crusader* (1928), 82.

come one of the largest remaining populations of the threatened species. Baynes was greatly taken with the bison he witnessed at the site and even managed to train two of Corbin's calves to pull a cart, a feat that greatly amused the locals who questioned how much control he was actually able to exert on the pair. One wag was reported to have commented that Baynes "hitches 'em up, and they take them where they damned please."[33] Soon Baynes began writing popular articles, presenting lectures, and warning governmental officials about the desperate plight of the species.[34]

In March 1905, after hearing Baynes speak in New York, Hornaday received authorization to offer the federal government twelve bison from the New York Zoological Society's modest herd.[35] The hope was to place these animals on the Wichita National Game Reserve in Oklahoma Territory that Roosevelt had established on a former Indian reservation recently opened to settlement.[36] Hornaday and his colleagues now admitted that zoo-confined animals, "even where the enclosures were as large as the New York Zoological Gardens," would inevitably fail to perpetuate the species over the long haul. They also feared that privately held herds could be sold at any time or crossed with cattle. The only way to maintain the bison as a purebred species "in full vigor for the next two hundred years, or more" would be by establishing a series of herds on public lands "in ranges so large and diversified that the animals will be wild and free." Under such conditions, Hornaday repeatedly asserted that the animals would not suffer any ill effects from inbreeding.[37] After securing permission from the society's executive committee, Hornaday offered a nucleus herd of twelve buffalo to the secretary of agriculture on the condition that Congress appropriate the funds to fence the area.

Baynes, who felt slighted by Hornaday's effort to introduce bison to the Wichita Game Reserve without consulting him, continued to push for the creation of a national society devoted to the species' preservation for "historical, sentimental, and practical reasons."[38] After several months of dragging their feet on the issue, Hornaday and Grant finally agreed to support the proposed organization. In December 1905, a small group of interested men (fourteen to be exact) gathered in the zoo's Lion House to found the American Bison Society. Theodore Roosevelt agreed to serve as honorary president of the new organization, Hornaday as president, and Baynes as secretary. Eventually, the society would grow to more than four hundred members, the vast majority of whom were wealthy males from the eastern United States.[39]

A publicity pamphlet the society issued shortly after its creation reveals much about its members' motivations. From the opening sentences, nationalism and nostalgia loom large in the four-page appeal: "The American Bison or Buffalo, our grandest native animal is in grave danger of becoming extinct; and it is the duty of the people of today to preserve, for future generations, this picturesque wild creature which has played so conspicuous a part in the history of America. We owe it to our descendants, that all possible effort shall now be made, looking to the perpetual increase and preservation of this noble animal, whose passing must otherwise be a matter of universal and lasting regret." While most remaining bison were now in private hands, the only way to ensure the long-term preservation of "our national animal" was through the creation of several government-owned herds in widely separated locations. The bison possessed definite commercial value, but it also has a "far better and stronger hold on the American people than can be estimated in dollars and cents." The species was "the most conspicuous that ever trod the soil of this continent," and its history was intimately interwoven with that of the American people. As such, it remained central to the nation's identity, and its extinction would not only represent "an irreparable loss to American fauna" but also a "disgrace to our country." In a relatively brief period, a creature that had once number in the "countless millions" had been reduced to a critically low point. Now was the time for action: "The least we can do now to partly atone for this ruthless slaughter, is to join in measures to prevent what must otherwise be the final result of perhaps the greatest wrong ever inflicted by man upon a valuable animal."[40]

FIGURE 22. Crating bison to ship west, 1907. Hornaday (*far left*) oversees shipment of the bison bred in the Bronx Zoo to the newly established Wichita National Game Reserve in Oklahoma Territory. This was the first in a series of federal bison reserves that the American Bison Society promoted as a way to repopulate the beleaguered species in the West. © Wildlife Conservation Society.

In other lectures and publications, representatives of the American Bison Society also stressed how the organization promoted and made possible continued beneficial contact with the natural world. For example, in 1914, the naturalist, museum director, and longtime vice president of the American Bison Society Franklin W. Hooper warned that modern Americans faced an increasing danger of losing "our respect and love for the animal life of the field and forest": "Blind to the great lessons of nature, we may so lose ourselves in the artificial maze and swirl of city life as to have no longer the stars to guide our course, the sturdy oak and the tall tapering pine to give us strength and inspiration, the flowers of the field to teach us beauty and humility, and birds of the air and the animals of the forest, companions of men for countless ages, to indicate to us how by diligence has come to pass the rising scale of life up to man." The American Bison Society was "bringing man back to nature," and one lesson to be found in observing wildlife like the bison was a "future wholesome and higher development of man; man not as master, but as servant, companion, and co-laborer of all created things."[41]

The first order of business for the newly founded society was completion of the Wichita bison reintroduction project.[42] Following a visit to the site, in February 1906, J. Alden Loring, a naturalist and former New York Zoological

Society curator, issued a report recommending an area of about twelve square miles on the western portion of the reserve as the most suitable range for the species.[43] Congress then unanimously appropriated the funds to erect a fence around this portion of the reserve, a project that was completed in 1907. In October of that year, Hornaday supervised the difficult process of loading fifteen bison into crates that were shipped cross-country from New York by rail and then hauled twelve miles by wagon to the newly established reserve. Ever mindful of the value of publicity, Hornaday made sure that photographers were on hand to document the move.[44]

The creation of the Wichita National Bison Herd, the first of several similar projects, was layered with irony. With the support of the American Bison Society, a New York zoo was exporting the buffalo back to its native western habitat in the Great Plains, to reservation lands that had been confiscated from the Apache, Comanche, and Kiowa Indians. The federal government, which only a few years earlier had actively encouraged the slaughter of the bison as a strategy to achieve final victory over the Plains Indians, was now appropriating money to rescue the beleaguered species. The same railroads that had facilitated the near extinction of the bison by linking hide-hunters to markets were now providing free transportation for live bison back to the West. And all this activity was being coordinated by a group of well-to-do easterners who viewed the species as a symbol of the untamed past that they hoped somehow to perpetuate. They were attempting to preserve the "wild" bison, however, by confining it to relatively small, fenced areas, where it was subject to nearly constant supervision and manipulation.

During this same period, members of the American Bison Society learned that Michel Pablo was trying to dispose of the bison herd that he had been grazing on Montana's Flathead Indian Reservation for more than two decades. Pablo and another half-Indian, Charles Allard, had apparently purchased fourteen bison in the early 1880s as an investment, and by 1907, the herd had grown to more than six hundred head, despite numerous sales in the interim.[45] When the federal government began opening up the Flathead Reservation to white settlement, Pablo realized that soon he would no longer enjoy access to free range. Congress refused to appropriate the money to buy the Pablo herd, so he approached the Canadian government hoping to negotiate a favorable lease for grazing land. Canadian officials responded by offering him $200 per head to purchase his bison. Once Pablo signed the contract, he began rounding up the creatures, which by all accounts proved a challenging ordeal. Between 1907 and 1912, he shipped 708 bison to an area known as Buffalo Park, near Wainwright, Alberta.

The sale of the Pablo herd to Canada stung the pride of Hornaday and other leaders of the American Bison Society. In response, the organization initiated a campaign to secure part of the Flathead Reservation as a federal bison preserve and to raise the funds needed to stock it. Overtly nationalistic appeals loomed large in letters urging Congress to authorize the purchase and fencing of suitable land on the site: "The loss of the Pablo-Allard herd, through its purchase last year by the Canadian government, was to all patriotic Americans a source of surprise and regret; but it is still possible for that loss to be made good, provided action is taken immediately."[46] Nationalism also featured prominently in letters soliciting donations to purchase animals for restocking (although at least one western recipient of the society's solicitation letter apparently mistook it for a fraudulent investment scheme!).[47] With the backing of President Roosevelt, in 1908, Congress authorized creation of the National Bison Range and provided the funds needed to fence the thirteen-thousand-acre area that was later chosen by representatives from the Biological Survey, the Forest Service, and the Bureau of Indian Affairs. With the more than $10,000 raised through popular subscription, mostly from wealthy easterners, Hornaday purchased thirty-four bison from Alicia Conrad in Montana, whose husband had earlier obtained his bison from Allard's heirs. The society also received several donations of bison from the Corbin and Goodnight herds.[48] However, Hornaday refused to negotiate with Pablo, who was still technically the owner of about seventy-five bison that had eluded capture. Apparently, he wanted nothing to do with the "half-breed Mexican-Flathead" who had sold bison to Canada.[49]

By 1910, Hornaday declared that the basic mission of the society had been accomplished. With the establishment of federal bison herds at Yellowstone, Wichita, and Montana, "the future of the American Bison, as a species" was "now secure."[50] He offered to resign his position as president but was convinced to stay on for a final year. Some of Hornaday's colleagues clearly felt more needed to done to ensure that the bison was truly safe, and in 1913, the society successfully lobbied for the establishment of a fourth federal herd at the Wind Cave National Park, in South Dakota. Once again, the New York Zoological Society provided a nucleus herd to stock the new reserve, shipping fourteen animals.[51] By that point, the society's annual bison census revealed a total of more than three thousand pure-blooded bison in North America, including five hundred "wild" bison, mostly in government hands. Madison Grant and Henry Fairfield Osborn, now president of the American Bison Society, began urging that the organization extend its activities to preservation of the pronghorn antelope, another western mammal whose range largely overlapped that of the bison, but they could never generate the same interest in this species.[52] Two years later, the society split in a vote over whether to disband entirely.[53] While enthusiasm for bison preservation waned in the face of apparent success, not until 1936 did the society quit collecting dues from its members. By that point, there were more than twenty-thousand bison in North America.

LOST OPPORTUNITIES

The success of the bison restoration efforts contrasted greatly with story of the passenger pigeon. At the annual meeting of the American Ornithologists' Union in 1909, the naturalist C. F. Hodge, of Clark University, wondered out loud: Has a "scientifically adequate search" been made to locate possible remnants of the once abundant species that might still be hanging on in the wild? When he received a resounding "no" in reply, he and William Beebe, a curator at the New York Zoological Society, raised a reward totaling more than $1,000 for anyone who could locate a nesting colony or even a single undisturbed nest of the species.[54] To aid in what they hoped would be an intensive nationwide search, the taxidermists and field-guide publishers Charles K. Reed and Chester Reed produced a widely distributed pamphlet with a color plate featuring the bird; Dutcher and the National Association of Audubon Societies also prepared colored plates of the passenger pigeon and mourning dove for circulation. In addition, Hodge spread word about the project through the popular press. Although the search was renewed for two subsequent seasons, none of the many reported sightings Hodge received panned out; most turned out to be the common mourning dove. Disappointed, by October 1912, he declared an end to the project. At that point the only passenger pigeons known to exist were an aging pair at the Cincinnati Zoo.

The passenger pigeon might have been saved through a systematic captive breeding program, but there were no sustained, organized attempts to do so, as there had been for the bison. There were, however, a handful individuals and institutions that dabbled with raising the species. In 1887, for example, the Milwaukee bird fancier David Whittaker obtained four passenger pigeons from a Native American who had captured them in northeastern Wisconsin. One of the birds died and another escaped, but the remaining two bred successfully in an outdoor cage he kept near his house. By the time the ornithologist Ruthven Deane visited Whittaker in 1895, the flock had grown to fifteen birds.[55] Over the next two years, he sold all of his passenger pigeons to the University of Chicago biologist Charles O. Whitman, who soon returned seven of them back to him. By 1909, all remnants of Whittaker's original group had perished.[56]

Whitman, who had raised domesticated pigeons as a boy, purchased Whittaker's birds primarily because he was interested in the light the species might shed on avian evolution and behavior.[57] The passenger pigeons were part of an

FIGURE 23. Charles O. Whitman in his columbarium, 1908. Whitman, who raised pigeons to study their evolution and behavior, seemed cavalier about the passenger pigeons that were part of his extensive collection. Although he once owned as many as fifteen of the endangered species, all of them had perished by the time this photograph was taken. Courtesy of the Special Collections Research Center, University of Chicago Library, ASAS-00316.

extensive bird colony—eventually numbering about 550 pigeons and doves, representing thirty species—that he kept at his home beginning around 1895. Nearly every summer for the next two decades, he shipped the entire colony by rail from Chicago to Cape Cod, Massachusetts, and back, so he could continue his studies without interruption while serving as director of the Marine Biological Laboratory at Woods Hole. Although he was undoubtedly aware of the passenger pigeon's endangered status, its preservation seemed less pressing to him than his scientific research. He failed to warn the public about the plight of the species, and, as the forced annual migration of his birds suggests, he seemed rather cavalier about the exceedingly rare specimens he possessed. Part of his response may have been related to his knowledge that his flock, all descendants from a single pair, seemed to be suffering from the ill effects of inbreeding. They would not survive over the long haul without the infusion of new blood, a prospect that seemed increasingly unlikely with each passing year. By 1907, all of Whitman's passenger pigeons were dead, save for two hybrids between that species and the common ring dove. Both were infertile males.

In 1902, Whitman did donate one female passenger pigeon to the Cincinnati

FIGURE 24. Martha, the last known passenger pigeon, 1912. Photograph by Enno Meyer. At one point in the late nineteenth century, the Cincinnati Zoo boasted as many as twenty passenger pigeons in its collections, some of which successfully reproduced. But the birds gradually died out, and after 1910, only a single example of this once prodigious species survived. From R. W. Shufeldt, "Published Figures and Plates of the Extinct Passenger Pigeon," *Scientific Monthly* 12, no. 5 (May 1921): 466.

Zoo, one of oldest continuously operating zoological parks in the United States.[58] Sometime around the time of its opening in 1875, the institution acquired several of the birds, although the exact numbers, dates, and sources are hopelessly confused in the meager, conflicting records that survive. We do know that the zoo enjoyed modest success in raising the birds, and that by 1881, it boasted twenty passenger pigeons on its grounds.[59] By 1907, however, the institution's flock had been reduced to three birds, two older males, and a single female, that had come from Whitman. One of these males died in 1909, leaving an elderly pair of passenger pigeons that zoo officials dubbed Martha and George Washington, after the nation's original First Couple.[60] When George died a year later, Martha was now utterly alone, the last passenger pigeon known to be living anywhere in the world. There in an eighteen-by-twenty-foot aviary with a sign revealing the species' precarious status, she lived out her final days under growing public scrutiny. As Christopher Cokinos has argued, it was the first time that a creature facing extinction became a celebrity, albeit a minor one.[61]

Sometime on the afternoon of September 1, 1914, Martha drew her final breath. Shortly after discovering her lifeless body, officials placed her in a large block of ice to stave off decomposition and shipped her by rail to the Smithsonian Institution. Three days later, the frozen bird arrived in Washington, D.C.,

FIGURE 25. John James Audubon portrait of Carolina parakeets, 1827–38. The boisterously polychromatic Carolina parakeet once ranged throughout Southeast and Midwest wetlands before falling victim to overhunting, habitat destruction, and the introduction of exotic species. Although parrots generally breed well in captivity, no sustained effort to capture and raise this particular species took place before it fell victim to extinction. From John James Audubon, *Birds of America* (1827–38). Courtesy of the Darlington Memorial Library, University of Pittsburgh.

where it was skinned, dissected, photographed, and preserved for posterity. The flamboyant ornithologist and anatomist R. W. Shufeldt, who was granted the honor of performing the autopsy on Martha, predicted grimly that it would not be the last time he would be called upon to undertake such a disquieting task: "In due course, the day will come when practically all the world's avifauna will have become utterly extinct. Such a fate is coming to pass now, with far greater rapidity than most people realize."[62] After a taxidermist mounted Martha's body in a lifelike pose, Smithsonian officials placed her on display in the U.S. National Museum as a warning to the many visitors who flocked to the institution: even a once profusely abundant bird like the passenger pigeon could fall victim to the juggernaut of modern civilization.

Four years later, the last known captive Carolina parakeet died at the Cincinnati Zoo. Known to scientists as *Cornuropsis carolinensis,* the species was a vivacious, stunningly beautiful bird with a bright green body and a head marked with brilliant yellow, orange, and red hues.[63] The Carolina parakeet once ranged widely across the southeastern and midwestern United States, where it thrived in wetland habitat, especially mature sycamores and cypress trees growing on riverbanks and in swamps. During the nineteenth century, extensive logging and the conversion of swamps into rice plantations played havoc with the home of the parakeet; the species also suffered persecution at the hands of farmers who blamed voracious flocks for destroying their valuable crops. The spread of the European honeybee may have further hastened the parakeet's decline by taking over the hollow trees the species preferred for nesting. By the end of the century, the brightly colored bird was also in demand for the pet bird trade, as decorations for hats, and as specimens for collectors. Avian diseases carried by domestic chickens and other animals may have also played a role in bringing down the species.

As early as 1831, Audubon noted that the Carolina parakeet "was rapidly diminishing in number."[64] Forty years later, in a report on an expedition he had recently made to Florida, J. A. Allen warned that the species was facing multiple threats: capture for the pet trade, wanton destruction by so-called sportsmen, and persecution from farmers who feared the bird would consume valuable crops. Already lost across much of its historic range, its ultimate fate seemed clear: "extermination."[65] By the end of the century, the parakeet's range had been reduced to the remote wetland regions of Florida and portions of South Carolina, leading the Smithsonian ornithologist Charles Bendire to echo Allen's grim prediction in 1895: "Civilization does not agree with these birds, and . . . nothing else than complete annihilation can be looked for. Like the Bison and the Passenger Pigeon, their days are numbered."[66] Yet, as had been the case for the passenger pigeon, naturalists gathered little data about the life history and behavior of the species before it was forced into oblivion. At the turn of the century, natural history remained focused on collection and classification, and as a result, ornithologists who encountered the bird in the field generally seemed more interested in possessing its skin and eggs than in trying to studying its behavior and biology. The more than eight hundred Carolina parakeet skins, skeletons, and eggs that gather dust in museum collections around the world today offer powerful testimony about the orientation of avian science at the time. While specimens are relatively abundant, we know precious little about how the bird survived in the wild.[67]

As it turns out, the Carolina parakeet might have been successfully bred if only a little more thought, care, and energy had gone into the process. While conducting years of painstaking historical research on the species, the biologist Daniel McKinley uncovered more than a dozen accounts of aviculturalists and zookeepers who boasted having the bird in their collections at some point from the mid-nineteenth to the early twentieth centuries, and several of them successfully reared young birds.[68] Among the most famous (and to the modern reader, troubling) examples were the birds that the Smithsonian ornithologist Robert Ridgway kept at his home in the Brookland neighborhood of Washington, D.C. There is no question that Ridgway was aware of the precarious status of the species when he collected approximately two dozen skins in southern Florida during the winter of 1896. When three of the birds he and his party shot received only mild wounds, from which they eventually recovered, Ridgway brought them home.[69] A male and female from this group even produced at least one brood of four eggs that hatched and grew to maturity. But then, one by one, Ridgway's parakeets perished. The female of the original pair died in 1902, after having laid many, mostly infertile eggs; the male died in 1903. One of their offspring, which Ridgway gave to the naturalist and Smithsonian colleague Paul Bartsch, apparently lived until 1914.[70]

By this point, even Hornaday had given up on the species. As early as 1902, he had received repeated offers from a Florida collector who had previously sold live birds to the New York Zoological Society. He refused to consider purchasing the Carolina parakeet because "the species is so nearly exterminated" and, in his view, there was no hope that it could be induced to breed in captivity. "I think it would not be right to encourage the capture of any of the few living specimens that now remain," Hornaday continued.[71] Nine years later, the society did accept the donation of two superannuated birds from the Cincinnati Zoo (which at the time had a flock numbering eight live Carolina parakeets), but both were dead by the spring of 1913.[72]

By 1916, only two of the birds were known to remain in captivity—a male and female, dubbed Incas and Lady Jane, both of which resided at the Cincinnati Zoo. Lady Jane died in 1917, and Incas in 1918.[73] Their passing received less notice than Martha's had only a few years earlier in part because apparently reliable sightings of the Carolina parakeet in Florida and South Carolina continued well into the 1920s and 1930s.[74] These reports offered a faint ray of hope that the species might still be rescued from the brink of extinction. In 1926, for example, the Chicago ophthalmologist and ornithologist Casey A. Wood offered an account of the work of the British aviculturalist Lord Tavistock, who had been advocating a systematic scheme of captive breeding of rare parakeets facing extinction using a system of large portable cages and a rigorous disinfection regime to minimize the chance of disease in his flocks. The hope was, in Wood's words, "to do for these vanishing species what has already been accomplished for the North American Bison." "Is not the Carolina parakeet a fit subject for such treatment," he asked, "if, perchance, a solitary breeding pair still survives the crass stupidity of our bird murderers?"[75] As it turns out, none of the surviving birds were captured, and by the late 1930s, the number of reports of the species in the wild greatly diminished.[76] No one knows exactly when and where the last Carolina parakeet perished, but given how conspicuous and mobile the species was, it is extremely unlikely that it continues to persist.

No one stepped forward to create a national organization, like the American Bison Society, to rally to the defense of the endangered passenger pigeon or the Carolina parakeet. While the Audubon movement provided regular updates on the status of both species, they were only two of the dozens of birds that movement sought to save. The declining status of passenger pigeon and the Carolina parakeet received wide publicity by the end of the nineteenth and the beginning of the twentieth centuries, but the two species failed to capture the imagination like the bison, which was not only one of the largest mammals in the United States but also viewed as central to America's identity as a nation. This was the charismatic creature that had once blackened the Great Plains during its seasonal migration, sustaining the lives of native peoples and greatly aiding the western migration of Euroamericans. The bison's continuing place in the nation's history and imagination would be commemorated in countless paintings, museum exhibits, coins, currency, and stamps.[77] The well-organized campaign to rescue this once conspicuous species came just as prominent Americans were beginning to worry about the future of a nation that seemed to be veering from its traditional values and ways of life. Industrialization and urbanization increasingly isolated Americans from regular contact with natural world, contact that was thought to be beneficial to body, mind, and spirit. Saving the bison represented a symbolic step in reaffirming a past way of life that seemed to be slipping away.

In addition to nationalism and nostalgia, biology may have also played a role in success of the campaign to save the bison. Like many ungulate species, the bison proved relatively easy to breed in captivity. Numerous ranchers had established successful, self-sustaining bison herds prior to the establishment of the American Bison Society, and indeed, these were the creatures that became the nucleus of the group that was acquired by the New York Zoological Society and eventually shipped back west to stock newly created bison preserves. While there were scattered successes with raising passenger pigeons and Carolina parakeets in captivity, their needs proved more difficult for their human captors to fulfill.

HORNADAY'S DIATRIBE

On the eve of the demise of the passenger pigeon and the Carolina parakeet and nearly a quarter century after he first sounded the alarm about the bison, Hornaday issued a lengthy diatribe about the extermination and preservation of

America's native birds and mammals. With its shrill tone, *Our Vanishing Wildlife* (1913) was vintage Hornaday, clearly designed to motivate its readers to action.[78] The four-hundred-page book was dedicated to William Dutcher, its foreword came from Henry Fairfield Osborn, and the preface offered lavish thanks to Madison Grant. The book's basic message was straightforward: a once profusely abundant population of North American wildlife was now lying in ruins. The great auk, Labrador duck, Pallas's cormorant, passenger pigeon, Eskimo curlew, Carolina parakeet, and five species of macaw and parakeet "had been totally exterminated in our own times."[79] Twenty-three additional North American birds were "threatened with early extinction."[80] Among the list of "Extinct and Nearly Extinct Species of Mammals" were several North American species and subspecies—the Arizona elk, the West Indian seal, the California elephant seal, and the California grizzly bear—along with several African, European, and Australian mammals. And at least two additional game mammals, the pronghorn antelope and the bighorn sheep, were clearly on their way out. Given this deplorable record of destruction, Hornaday argued, Americans needed to acknowledge that "the wild things of this earth are *not* ours to do with as we please. They have been given to us *in trust,* and we must account for them to the generations which will come after us and audit our accounts."[81]

Who or what was to blame for all this carnage? Hornaday presented a lengthy list of groups he considered responsible for the increasingly desperate plight of American wildlife. While praising "gentlemen sportsmen" as the "bone and sinew of wild life preservation," he strongly criticized all hunters who sought declining native species, especially when "extinction is impending."[82] He also had little patience for so-called pot hunters, who pursued game to put meat on the table when other food supplies were almost always available. All forms of commercial hunting, whether for meat or feathers, he condemned out of hand. As J. A. Allen had done twenty-five years earlier, Hornaday singled out two groups he considered especially destructive to wildlife. In terms that reflected the xenophobic circles with which he associated, Hornaday strongly denounced the slaughter of song birds by "Italians and other aliens from southern Europe," whom he referred to in biblical terms as a "pestilence that walketh at noonday."[83] Using equally sharp language, he castigated southern blacks for killing prodigious numbers of song birds, woodpeckers, and doves for food.[84] While Hornaday acknowledged the role that domestic cats, telegraph and telephone wires, predators, and diseases played in the decline of some forms of North American wildlife, he remained curiously silent on the issue of habitat destruction.

In the second half of his book, Hornaday chronicled efforts to rescue American wildlife from the brink of extinction. Here he praised the work of the federal government, which had established American bison in four national ranges, created fifty-eight bird refuges and five game preserves, protected the northern fur seal, and taken steps to conserve wildlife in Alaska. At the same time, game commissions and Audubon Societies had been responsible for the passage of much sound legislation on the state level. Beyond this, a series of New York–based national organizations—like the New York Zoological Society, the Boone and Crockett Club, the National Association of Audubon Societies, the Camp-Fire Club of America, the American Game Protective and Propagation Association, and the Wild Life Protective Association—had all done commendable work. But these organizations needed additional funding to continue their campaigns, while other cities needed to establish similar institutions to take up some of the conservation burden.

Hornaday offered several additional recommendations and warnings to his readers, including a lengthy list of new wildlife laws needed in each state as well as a call for new federal legislation protecting migratory birds and outlawing the sale of game. Beyond better laws, he strongly urged parents, educators, and zoologists to teach young children about the value of wildlife and its conservation.[85] Although America was now living through *"the middle of the period of Extermination,"* he remained convinced that most vanishing species, both game and nongame, could be saved, but only if they received "absolute protection from harassment and slaughter." "Recovery" would be "impossible," however, once a population had reached a critically low level where "the survivors" were "too few to cope with circumstances." The heath hen, passenger pigeon, whooping crane, sage grouse, trumpeter swan, wild turkey, upland plover, and perhaps even the antelope had already reached the point where they "will never come back to us, and nothing that we can do ever will bring them back."[86]

As he had done in a controversial article published fifteen years earlier, Hornaday railed against professional naturalists. As a group, he argued, naturalists tended to be apathetic about the plight of wildlife, "so intent upon the academic study of our continental fauna that they seem not to have cared a continental about the destruction of that fauna."[87] True, a handful of scientists at the Biological Survey, U.S. Forest Service, American Museum of Natural History, New York Zoological Society, Museum of Comparative Zoology, University of California, and a few other institutions had been active in wildlife conservation circles. Now it was time for scientists elsewhere to shoulder their fair share of the burden. In addition to offering a list of museums and scientific societies that he thought owed "service to wild life," Hornaday called for leading American universities—like Columbia, Cornell, Yale, and the University of Chicago—to hire at least one naturalist who would devote themselves entirely to the cause of wildlife conservation. "We don't want to hear about the 'behavior' of *protozoans*," he argued, "while our best song birds are being exterminated by negroes and poor whites."[88]

While many of the claims in *Our Vanishing Wildlife* were prone to exaggeration, there was also at least a kernel of truth in much of what Hornaday wrote. His indictment of scientists for their failure to engage with the issue of wildlife extinction offers an illuminating case in point. As the previous two chapters have tried to show, naturalists played a critical role in the turn-of-the-century wildlife conservation movement. The firsthand knowledge of vanishing fauna they gained while pursuing their scientific studies, along with the aesthetic and emotional attachment they developed with those species, pushed them to take active steps to stop the slaughter. Without the authoritative knowledge and leadership naturalists provided, the turn-of-the-century wildlife conservation movement would have soon faltered. Yet, for well into the twentieth century, the total number of naturalists who became active in conservation circles remained a small fraction of the biological community as a whole. Most naturalists seemed relatively indifferent to the plight of endangered species and more interested in safeguarding their prerogative to collect rare species than in rescuing them.

Those naturalists who did actively participated in the movement did not necessarily share Hornaday's strict protectionist views. For example, Joseph Grinnell, who established a leading program to study western fauna at Berkeley—a program that Hornaday had singled out for praise—also openly worried about the impact that restrictive wildlife laws were having on the practice of natural history. In 1915, only two years after the publication of *Our Vanishing Wildlife*, he issued an editorial in *Science* with the provocative title of "Conserve the Collector."[89] There he lamented the considerable decrease in the number of specimen collectors over the previous decade. An increased reliance on field identification, rather than the more traditional practice of securing specimens in hand, resulted in less precise and accurate data, thereby threatening "ornithology as a science." Moreover, as they forsook collecting, budding naturalists no longer gained the critical knowledge obtained from the pursuit and preparation of specimens in the field. Bird protection laws should include reasonable provisions for scientific collecting, Grinnell urged. While officials charged with enforcing those laws might rightly deny permits to collect "rare or disappearing species like the ivory-billed woodpecker or the Carolina parakeet," they should also more freely issue permits to collect other species.

Even when they did act on behalf of endangered species, there were clear limitations on what naturalists were willing to do or say. They devoted much more attention to mammals and birds facing extinction, especially game species, than to reptiles, amphibians, or insects. The focus of their concern tended to be larger, charismatic species that were not only easier to study but also attracted a larger number of interested researchers. They also tended to be economically important species. While naturalists were quick to point the finger of blame on the commercial exploitation of species, they refused to seriously question the larger political and economic system that promoted continued habitat destruction. American naturalists also tended to center their conservation gaze almost entirely on North American species. The nationalism that had informed American science since the early days of the republic still remained a potent force at the beginning of the twentieth century. Hornaday's *Our Vanishing Wildlife*, which included brief discussions of the lost and vanishing species on other continents, proved an exception to this otherwise general rule. Not until after World War I, with the United States' growing involvement with foreign affairs more generally, did a handful of American conservationists begin showing a more sustained interest in the plight of wild animals from other lands.

CHAPTER FIVE

GOING GLOBAL

THE AMERICAN COMMITTEE AND THE FIRST INVENTORY OF EXTINCTION

*Friends of wild life must act quickly, if we modern men are not to find
ourselves masters of a world shorn of much of its beauty at the very
stage when, through the development of machinery, we are obtaining
increasing leisure to cultivate an interest in the things of the spirit.*

HENRY CAREY, 1926

*Unless vigorous steps are taken in the next few years, we fear that
the rapid development and opening up of the great continent [of Africa]
will result in the permanent loss of some of the most interesting
productions of nature to be found anywhere in the world.*

JOHN C. PHILLIPS, 1935

COOLIDGE'S AFRICAN ADVENTURE

As he approached graduation at Harvard, Harold Jefferson Coolidge, Jr., experienced a burning desire to make a name for himself. His first opportunity came when he signed on as an assistant zoologist for the Harvard African Expedition of 1926 and 1927. One objective of this yearlong medical and biological survey of Liberia and the Belgium Congo was to secure a specimen of the mountain gorilla, a rare primate that had first been described twenty-five years previously.

Accompanied by fifteen Batwa pygmies and outfitted in stereotypical safari dress, Coolidge pursued the prized gorilla in the region surrounding Katana, in eastern Congo, where he had heard recent rumors of sightings. In a popular account of the adventure, Coolidge projected himself as an intrepid white hunter forced to surmount innumerable obstacles in the field: difficult terrain, high altitude, drenching rains, dense vegetation, a "large and utterly useless army" of native helpers, and a secretive, cunning, and potentially dangerous prey.[1] Two months of slogging up and down the area's steep slopes had resulted in only two fleeting encounters with his intended target, but no specimen, not even so much as a clear shot.

One day before he was scheduled to begin the journey to rejoin his American colleagues, the discovery of fresh gorilla tracks renewed Coolidge's hope for success. Soon he spotted several young gorillas frolicking high in the trees on the next slope. Coolidge and his men headed down the intervening ravine as quietly as they could, only to be greeted by a "fierce chest-drumming noise . . . undoubtedly a signal of danger from the male leader of the troop." For the next thirty minutes, both sides maintained a "most oppressive" silence that remained unbroken until one of his men issued a half-smothered cough.[2] Following a new round of chest drumming, the alarmed gorillas retreated, with Coolidge and his entourage in keen pursuit. Approaching the bottom of the ravine, Coolidge was startled by the sudden rush of a fleeing creature that seemed massive, about his height and nearly three times his weight. He quickly fired in the direction of the dark moving mass, which scampered across a stream and escaped into the bush.

A puddle of blood revealed that Coolidge's bullet had found its mark, but most of his men proved "too cold" and "too frightened" to continue the pursuit. Only the pygmy chief persevered in tracking the wounded animal through the dense jungle. When the two heard a "low, ominous growl" emitted from a thicket they were about to enter, the final native tracker also fled, while Coolidge braced himself for a possible charge from the creature, now only fifteen feet away, before firing again. Coolidge's second bullet also reached its target, but once again the beast managed to escape. Now alone in the jungle, Coolidge resumed searching until spotting the wounded gorilla leaning against a tree just twelve feet down the slope in front of him. After losing his footing in the mud, he began sliding down toward the fearsome beast. Coolidge fired once more, this time from a sitting position, fatally striking the gorilla in the heart.

Coolidge was delighted with the specimen, which he boasted was "far finer" than he had ever dreamed. His father paid to have the large male gorilla mounted for the Museum of Contemporary Zoology (MCZ), where it remains on exhibit to this day. While Coolidge expressed no concern about the precarious status of the vanishing mountain gorilla in his account of the adventure, he did reveal more than a little sympathy for his target, which he described in vividly anthropomorphic terms: "I had to take off my hat to this old king who handled

FIGURE 26. Harold J. Coolidge with gorilla specimen collected in Congo, 1927. Soon after his return to the United States, Coolidge became a founder of the American Committee for International Wild Life Protection. Two decades later he would also play a central role in creating the Survival Service Commission of the International Union for the Conservation of Nature. Courtesy of Miles Coolidge.

his troops so well, covered the rear of his retreat, and handled his medicine like a man. . . . That night in some unknown valley of a remote Congo mountain probably the gorillas held a meeting to mourn the loss of their great sovereign and elect his successor."[3]

The product of a prominent New England family, Coolidge had entered Harvard in 1923 with the hope of becoming a diplomat.[4] Gradually, however, he gravitated toward natural history, and in the summer following his sophomore year, he and Charles Day ventured to Alaska to collect brown bear specimens for the U.S. Bureau of the Biological Survey. The next fall he discovered the MCZ's famous "eateria," where museum staff and visitors regularly gathered to socialize and consume exotic game. Following his return from Africa, Coolidge graduated with a bachelor's of science degree and spent the next year studying zoology at Corpus Christi College, Cambridge. At the suggestion of the Yale primatologist Robert Yerkes, he decided to specialize in the great apes. Although he never earned a Ph.D., which had already become a standard entry requirement for an academic career in science, in 1928 Coolidge became an assistant mammalogist at the MCZ, a position he was to occupy for the next two decades. There he com-

FIGURE 27. John C. Phillips, 1928. Trained as a scientist and physician, Phillips only briefly practiced medicine before devoting himself to conservation, natural history, and sport hunting. As founders and early leaders of the American Committee, he and Harold J. Coolidge set the organization's agenda. Courtesy of the Archives of the Ernst Mayr Library of the Museum of Comparative Zoology, Harvard University.

pleted a major taxonomic revision of the genus *Gorilla,* discovered the pygmy chimpanzee, led a major expedition to Asia, and began regularly coteaching a course called "The Evolution of Animal Sociology."

While studying in England, Coolidge attempted to contact the American conservationist T. Gilbert Pearson, who had ventured abroad to preside over the first general meeting of the International Committee for Bird Protection. The two failed to meet up, but Coolidge sent Pearson a letter in which he revealed an ambitious idea. There ought to be, he wrote, "an International Committee for the Protection of the Flora and Fauna of the World and for the Advancement of Science. This committee to be made up of leading Scientists (such as directors of large museums) and prominent men interested in Preservation (by Preservation I mean also Protection)."[5] The organization could act as an information clearinghouse on international wildlife conservation issues, fund fieldwork for naturalists studying endangered species, lobby for the creation of national parks around the world, and compile a blacklist of wildlife poachers. Pearson's response to Coolidge's idea has not survived, but soon after beginning work at the MCZ, he found a sympathetic ear in the person of the physician, sportsman, and naturalist, John C. Phillips.

Nearly three decades older than Coolidge, Phillips was also from a family with deep New England roots, but one whose financial prospects seemed a good deal more solid.[6] A modest, self-effacing man with a well-developed sense of humor, he became keenly interested in outdoor pursuits, especially hunting and fishing, at a young age. Phillips earned a bachelor's degree in science from Harvard's Lawrence Scientific School in 1899 and an M.D. from Harvard Medical School in 1904. But save for a couple of years at Boston City Hospital immediately following graduation and a brief stint in a military field hospital during World War I, Phillips never practiced medicine. Instead he traveled widely—to Greenland, Mexico, the Middle East, the Far East, and Africa—to hunt and study wildlife, especially game birds and mammals. In addition to amassing trophies for his home, he donated numerous specimens to the MCZ, where he served as a research curator of birds. Between 1922 and 1926, he published his magnum opus, a copiously illustrated, four-volume *Natural History of Ducks*. Phillips was also active in numerous state and national conservation organizations and a vice president of the Boone and Crockett Club.

Early in 1930, Coolidge and Phillips successfully launched an organization that bore a remarkable resemblance to the one Coolidge had outlined in his 1928 letter to Pearson: the American Committee for International Wild Life Protection.[7] This was a small, informal group whose membership was limited to representatives from a handful of zoological societies, museums, and conservation organizations, most of whom served as official advisors. Much of the work of the American Committee was transacted through a small executive committee, which met once or twice a year, with Phillips serving at its first president and Coolidge as its secretary. Although the group was a spin-off from the Boone and Crockett Club, its leaders were quick to point out that they were not primarily an organization of sportsmen. Rather, they hoped to use the authority of science to further their main objective: "to promote the preservation of rare species in all parts of the world."[8]

In actual practice, the American Committee proved more regionally focused than this ambitious mission statement suggests. For its first several years, the organization devoted most attention to African wildlife conservation. Every member of its first executive committee had spent time in Africa hunting game or collecting specimens—a distinction that was not always easy to make in actual practice—and five of the organization's first seven publications dealt with that continent. Toward the end of the 1930s, however, for reasons outlined in the next chapter, the American Committee increasingly set its sights on Latin America.

The specter of extinction clearly haunted members of the American Committee. Yet, when it came to threatened species, they found themselves in a quandary. On the one hand, like many other naturalists from this period, they sought to "preserve" rare animals as specimens to be amassed in museums for exhibition and scientific study. On the other hand, they also sought to "preserve" living examples of those species, ideally in their native habitats. These differing notions of preservation came into increasing conflict as vanishing wildlife populations moved closer and closer to oblivion. In the mid-1930s, the American Committee confronted this basic tension by passing a controversial resolution calling for scientific institutions—primarily natural history museums and zoological parks—to refrain from collecting rare species. They also began an ambitious project to create an authoritative inventory of the world's vanishing species—first of mammals and then of birds. Given the limited state of knowledge about threatened species at the time, the task turned out to be much more difficult than originally anticipated, and over the next two decades, the effort to complete this catalog consumed much of the committee's time and energy. Nonetheless, the resulting publications—which finally began appearing in 1943—not only rank as one of the American Committee's most enduring legacies but also became an important model for later official lists of endangered species.

CROSSING BOUNDARIES

Efforts to protect migratory species provided the first inkling of a more international conservation consciousness in the United States. Many forms of wildlife refuse to stay put, regularly crossing national boundaries during their seasonal movements. Beginning in the early twentieth century, the process of trying to resolve competing ownership claims involving migratory species and to regulate their taking as their numbers crashed led the U.S. government to enter into several international treaty negotiations.

The first (and least successful) of those treaties involved fishing rights in the waters between the United States and Canada, a subject that had long provoked tension between the two nations.[9] By the end of the nineteenth century, increasingly efficient harvest methods and continued resistance to any form of regulation brought the disagreement to a head. As early as the 1890s, the U.S. secretary of state initiated talks about how to address the problem of diminishing fish stocks, talks that led to the appointment of a special commission of scientists (one American and one Canadian) to gather relevant data and formulate recommendations. Four years in the making, that initial report nonetheless failed to break diplomatic deadlock on the issue. In 1908, officials succeeded in negotiating a treaty that called for another scientific commission to draw up regulations designed to make the boundary water fisheries more sustainable. In the end, however, the regulations floundered in Congress. Though clearly in their long-term interest, fishermen strongly opposed the regulations, while the fish themselves lacked the charisma needed to elicit sympathy from the public or even most prominent conservationists.

Efforts to reach an agreement to save the North Pacific fur seal proved more successful.[10] This aesthetically appealing and economically valuable mammal was born on American islands in the Bering Sea but spent most of its life in international waters. In the 1880s, Canadians began pursuing fur seals on the high seas, an act that was within their legal rights but which also led to appalling waste (as many as 80 percent of the seals shot were never recovered) and a cruel death by starvation for young seals whose mothers were killed. Scientists tended to blame pelagic hunting for the precipitous turn-of-the-century decline of the North Pacific fur seal. If something were not done soon, they warned, the species might soon go the way of the passenger pigeon. Alarmed at this prospect, conservation-minded naturalists—like David Starr Jordan, William T. Hornaday, and Henry W. Elliot—rallied behind the cause of seal protection and gained widespread public sympathy for the appealing but increasingly beleaguered species. By 1911, Canadian and U.S. officials reached an agreement with their counterparts in Japan and Russia to ban the pursuit of fur seals on the high seas and to divide the profits from the more sustainable practice of harvesting on land. In the words of environmental historian Kurkpatick Dorsey, that landmark treaty "snatched the northern fur seal from the jaws of extinction," and the herd quickly rebounded.[11]

The Migratory Bird Treaty sought to protect another form of wildlife that regularly crossed international borders.[12] As we saw in chapter 3, by the end of the nineteenth century, Americans had grown increasingly concerned about the decline of wild birds, many of which seemed to be teetering on the brink of extinction. The Audubon movement succeeded in gaining limited protection for some vanishing species, particularly nongame birds, through a series of state laws, while sport hunters secured passage of state laws regulating the taking of game birds. The problem was these laws lacked uniformity from one state to the next, and they were not always adequately enforced. Federal legislation seemed a logical solution, but long-standing tradition and a recent Supreme Court decision (*Geer v. Connecticut*, 1896) held that wild game was the property of the state. The Lacey Act, passed in 1900, represented the federal government's tentative entry into the arena of wildlife regulation, but it merely outlawed the interstate shipment of wildlife taken in violation of any state law, without challenging the idea of state ownership.

Beginning in 1904, supporters of federal protection of wild birds introduced the first in a long series of bills that were initially aimed at safeguarding migratory waterfowl and wading birds. In 1912, T. Gilbert Pearson, the acting head of the National Association of Audubon Societies, urged Congress to offer federal protection for *all* migratory birds. The next year, Congress responded with the Migratory Bird Act, though opponents of the legislation continued to argue that it represented a blatant violation of states' rights. To make sure the new law survived constitutional muster, supporters moved to have its major provisions introduced into a treaty with Great Britain (on behalf of Canada) and then, in 1918, successfully shepherded the enabling legislation through the House and the Senate. The treaty, which declared a permanent closed season on most nongame migratory birds and authorized the federal government to set seasons on game species, survived judicial review within two years after its passage.

Beyond treaties, the next stage in the development of a more international conservation consciousness was the creation of institutions devoted to the issue. In 1922, T. Gilbert Pearson set out for Europe to learn more about avian conservation initiatives on the other side of the Atlantic and to explore the possibility of international cooperation.[13] He met with the wealthy aviculturalist and ornithologist Jean Delacour, the president of the French League for the Protection of Birds whose elegant five-hundred-acre estate was stocked with hundreds of birds from around the world, and with the Dutch naturalist and conservationist Peter G. Van Tienhoven, who would later establish the International Office for the Protection of Nature.[14] After preliminary discussion with these and other individuals, he invited Delacour, Van Tienhoven, and eight other colleagues to London to discuss the idea of forming an international society devoted to bird preservation. The group enthusiastically supported the idea and voted to found the International Committee for Bird Protection (ICBP), with Pearson as its chairman. Membership was to be composed of individuals nominated to represent various societies with an interest in bird protection. The members from a given country were organized into national sections, with a chair serving as a node of communication between the section and Pearson.

The stated goal of the ICBP was "co-ordinating and encouraging the preservation of birds."[15] That was to be accomplished through a variety of means. First, through correspondence, informal meetings (held in 1923 and 1925), a series of more formal congresses (the first of which was held in 1928), the publication of an irregular bulletin (first issued in 1927), and the exchange of other publications, the society served as an information clearinghouse on the status of bird protection within its member countries. In addition, Pearson used the ICBP as a vehicle to promote the cause of bird protection in nations where it seemed lacking. For example, prior to his second trip to Europe in 1923, Pearson wrote Percy

Lowe, an ornithologist at the British Museum of Natural History and head of the ICBP's British section, to ask if he knew of any "worthwhile man or men in Italy whom we might approach with the possible hope of getting a bird organization started there." Pearson hoped that promoting conservation sentiment in Italy would not only benefit the birds of that particularly nation, but would also serve as a positive influence for the "tens of thousands of Italian workers" who immigrated to America each year. According to Pearson, "Every blooming one of them gets hold of a gun and they today as a class constitute the most tremendous human force of destruction operating against our songbirds."[16] During his trip, Pearson found four Italian organizations willing to nominate members to the ICBP.

Although a loose-knit organization, the ICBP remained active on several fronts. The organization's Declaration of Principles, passed in 1923, emphasized that wild birds were critical to maintaining the balance of nature, that they were important as objects of scientific study, that they exerted a "great aesthetic influence on all right-minded people," and that they served utilitarian functions as sources of food, destroyers of unwanted rodents and injurious insects, and as objects of pursuit in field sports. Despite these many benefits to the environment and humanity, in many countries bird populations were diminishing "at an alarming rate," and many "interesting and valuable species" had already been "exterminated from the earth."[17] Consistent with these general principles, the ICBP urged the passage of legislation protecting insect-eating birds, called for elimination of the feather trade, promoted the creation of bird sanctuaries, and called for the end of hunting during breeding season and while birds were raising their young. The organization also promoted international treaties to protect migratory birds and minimize oil pollution of navigable waters.[18]

During its first few decades, the ICBP faced numerous challenges that greatly limited its effectiveness as an agent for international avian conservation. One major problem was the geographical, cultural, and linguistic diversity of its membership. By 1935, the organization boasted 203 members representing 130 societies and 26 countries.[19] While it was true that the diverse membership of the ICBP agreed on the need for greater protection of birds and the most active members tended to be in Europe, members often differed on the specifics about how to achieve that goal. The problem of diversity within the organization was exacerbated by Pearson, who could communicate only in English, a severe disadvantage in such a polyglot organization. To make matters worse, Pearson arranged for publication of the first ICBP Bulletin only in English. The result was that, until the Second World War, the organization remained more important as an information clearinghouse than as a force for significant change.

A growing interest in African big game provided a second context for a modest upsurge in concern about international wildlife in the United States. Since the colonial period, Americans had enjoyed viewing exotic African creatures in circuses, menageries, and traveling shows. The first lion, for example, was displayed in North America as early as 1716; the first elephant in 1796.[20] Although difficult to obtain, African species also proved popular in the American zoos that began opening their gates in the second half of the nineteenth century, while stuffed examples of these creatures figured prominently in public natural history museums. The first inklings of anxiety about the plight of African wildlife, however, did not emerge until the early twentieth century.

The specific episode that first raised American consciousness about the fate of African big game was an expedition that Theodore Roosevelt mounted immediately after leaving the White House in March 1909.[21] Determined both to make a scientific contribution and to fend off potential criticism of his African big game hunting, Roosevelt negotiated with Smithsonian officials to sponsor the expedition. Private donations (largely from Andrew Carnegie) financed most of the trip, while the ex-president covered the rest using a hefty advance he negotiated for published accounts of the adventure. With his eighteen-year-old-son Kermit in tow, Roosevelt arrived in Mombassa, Kenya, in April 1909. Also accompanying him were three Smithsonian-recommended field naturalists, two white guides, and more than 260 African grooms, gun bearers, tent men, askari, and porters.[22] It was reputed to be the largest hunting party ever to set out in Kenya.[23] By the time the expedition ended, the two Roosevelts had bagged no less than 17 lions, 11 elephants, 14 rhinos, 8 buffaloes, and numerous other species, more than 512 mammals in all.[24]

When the *New York Times* expressed concern about a report that Roosevelt shot eighteen antelope and gazelles along with two wildebeests in a single day, the ex-president felt compelled to respond. All the African animals he killed, save for a handful needed for food, were being preserved for the Smithsonian Institution and thus represented an important contribution to science.[25] The outpouring of publicity surrounding the expedition not only brought attention to the plight of African wildlife but also spawned a host of imitators who regularly sought museum sponsorship for African safaris.[26]

One hunter who had advised Roosevelt in planning his trip—the taxidermist, naturalist, sculptor, conservationist, and inventor Carl Akeley—also met up with him briefly in Africa. Through his publications, movies, and especially his magnificent museum dioramas, Akeley played a key role in bringing representations

FIGURE 28. Kermit and Teddy Roosevelt with African hunting trophy, 1909. Roosevelt's safari garnered specimens for the Smithsonian Institution, publicity for the plight of African wildlife, and criticism of what some considered to be overzealous hunting. Courtesy of the Smithsonian Institution Archives, Record Unit 7179, image no. SIA2008-2356.

of the "dark continent" to America's shores for decades after his death in 1926.[27] He first gained notoriety in 1886, when as an employee of Ward's Natural Science Establishment, he helped mount P. T. Barnum's famed elephant, Jumbo, after a locomotive accidentally struck and killed the beast. After leaving Ward's, Akeley found employment as a taxidermist at the Milwaukee Museum, the Field Museum of Chicago, and beginning in 1909, the American Museum of Natural History in New York. While on his third collecting trip to Africa in 1910, a charging elephant nearly killed him. During the lengthy period of recovery that followed, he developed a dream that would soon become an obsession: the great African Hall at the American Museum of Natural History. Anxious to capture a sense of the sublime grandeur of the African continent before the march of civilization destroyed it, Akeley envisioned a massive exhibition that would serve as an enduring monument to the continent's "fast vanishing wildlife." "Two hundred years from now," he predicted grimly, "naturalists and scientists will find in such museum exhibits as African Hall the only existent records of some of the animals which today we are able to photograph and study in their forest environment."[28] The hall would consist of a large open space surrounded by forty meticulously appointed dioramas featuring a variety of scenes of "primeval

Africa."[29] The centerpiece of the exhibit was to be "Akeley's taxidermic tour de force," a raised platform featuring an elephant group that included a cow collected by Roosevelt and a calf collected by his son.[30] Henry Fairfield Osborn, the president of the American Museum, loved Akeley's idea and sold it to the museum's board of directors. Until his death in 1926, Akeley struggled to secure the specimens and funds needed to transform his ambitious dream into a reality. Though he managed to place several wildlife groupings on exhibit, not until ten years after his death did a scaled-down version of the African Hall (with twenty-eight rather than forty dioramas) finally open to the public.

During his fourth expedition to Africa in 1921–22, Akeley traveled to the Kivu region of Belgium Congo to collect mountain gorilla specimens for one of his habitat groups.[31] He succeeded in obtaining two large males, two females, a youngster, and the first motion pictures ever made of the species using a camera of his own invention. In the process of observing the species in the wild, however, he not only developed an increasing sense of kinship with the gorilla but also fear about its future. After returning to New York, Akeley wrote several articles emphasizing the sorry plight of the creature and went on a lecture tour featuring movies from his expedition. He also drew up plans for a wildlife sanctuary in the area where he had hunted and presented them to John C. Merriam, the paleontologist and conservation-minded head of the Carnegie Institution of Washington, who in turn passed them on to the Belgium ambassador to the United States.[32] Following prodding from Akeley, in 1925, King Albert I created the eponymous Parc National Albert. The decree setting aside the area, the first modern national park on the African continent, declared that it was time to "take steps to preserve the remaining gorillas from extermination."[33]

To drum up funding for the African Hall while cultivating public support for conservation on the continent, Akeley convinced the pioneering wildlife filmmaking duo of Martin and Osa Johnson to become formally affiliated with the American Museum of Natural History.[34] In 1923, Osborn and Akeley also persuaded museum officials both to sponsor the Johnson's extended trip to Africa and to invest in a series of three films about the continent. The hope was that the profits from these motion pictures would provide a cash infusion for the African Hall fund. At the time, public museums generally avoided money-making ventures of this sort, and for a variety of reasons, the Johnsons managed to finish only one commercially viable film during their five-year stint with the museum. Nonetheless, *Simba, King of the Beasts: Saga of the African Veldt*, released in 1928, was a financial (if not a critical) success, providing compelling visual scenes of Africa to compliment the increasing number of written accounts of the continent.

By the mid-1920s, even periodicals aimed at scientifically oriented naturalists devoted increasing attention to the plight of African wildlife. In May 1926, for example, the *Journal of Mammalogy* included a stirring article entitled "Saving the Animal Life of Africa—a New Method and a Last Chance," authored by the Philadelphia lawyer, conservationist, and nature writer Henry Carey.[35] Once abundant species, like the white rhinoceros of South Africa, the blauwbok antelope, and the quagga, Carey lamented, were now gone forever. Others, like the okapi, the white-tailed gnu, and the white rhinos of the Upper Nile were poised "on the verge of the abyss." In support of his claim about the general decline of African wildlife, Carey quoted a pessimistic passage from Osborn, who had written, "We paleontologists alone realize that in Africa the remnants of all the royal families of the Age of Mammals are making their last stand, that their backs are up against the pitiless wall of what we call civilization. Human rights are triumphing over animals rights, and it would be hard to determine which rights are really superior or most worthy to survive."[36]

Ultimately, Carey seemed most concerned about the aesthetic, philosophical, and spiritual consequences of this mass extinction. The destruction of African wildlife was not only threatening many species, he argued, but also "shrinking our own intellectual and spiritual resources." American conservationists had long expressed cultural concerns about the loss of game species in the United States. Carey extended those concerns to Africa, while boldly declaring that continent's wildlife to belonged to the entire world: "Man is stripping a vast playground of the beautiful creatures that gave it life. . . . We need a place in the world where we can recapture 'The lost arts of wildcraft' in such surroundings, because 'the self dependent life of the wilderness nomad brings bodily habits and mental processes back to normal, by exercise of muscles and lobes that otherwise might atrophy from want of use.'"[37] To achieve this end, Carey called for the creation of a permanent international game commission to be set up under the auspices of the League of Nations.

By the second half of the 1920s, then, a long series of lecturers, publications, films, and exhibits had begun exposing American audiences to the beauty, wonder, and vulnerability of African wildlife. New Yorkers were thus well primed when, in November 1929, C. W. Hobley, secretary of the Society for the Preservation of the Wild Fauna of the Empire (SPWFE), came to the United States to appeal for funds.[38] Established in 1903 and known by the sobriquet "the penitent butchers," the SPWFE was an organization of British sport hunters, naturalists, and governmental officials concerned about the desperate plight of wildlife in Britain's colonial possessions in Africa and Asia.[39] The society published a periodical, lobbied for the establishment and enforcement of protective legislation, sponsored scientific studies of endangered species, promoted a protective ethos among settlers, and pushed for the creation of secure wildlife refuges. Two months after Hobley's visit, New Yorkers also heard from the American sportsman, adventurer, and soldier of fortune Frederick R. Burnham, who reported a "shocking decrease" in game in areas of Africa he knew well from his days as a military scout.[40] Hobley's plea and Burnham's bleak report generated nearly $10,000 in pledges for the SPWFE. They also led Osborn to suggest the formation of "an American auxiliary" to the SPWFE that would "assist both in propaganda and in raising funds."[41]

LAUNCHING THE AMERICAN COMMITTEE

In the midst of this flurry of activity, Coolidge and Phillips seized the opportunity to establish the American Committee for International Wild Life Protection. After gaining support from Kermit Roosevelt, by then a prominent conservationist and a member of the Boone and Crockett Club's executive committee, and from Madison Grant, the organization's president, Coolidge introduced a motion that he and Phillips's had authored at the club's business meeting early in 1930.[42] That motion highlighted the conservation work being done in Europe and proposed that the Boone and Crockett Club establish a committee of its own to deal with international wildlife affairs. Although initially created by and closely affiliated with the Boone and Crockett Club, from the beginning the new group was supposed to broadly represent "American sympathy and interest in international wild life protection."[43]

During the first executive committee meeting in May 1930, the committee voted to make its headquarters in Harvard's Museum of Comparative Zoology and established a small advisory committee, which initially consisted of nine individuals chosen to represent various museums and conservation organizations across the United States.[44] Included on the first advisory committee were many of America's best-known naturalists, conservationists, and sportsmen, including Thomas Barbour, director of the Museum of Comparative Zoology; Charles M. B. Cadwalader, director of Academy of Natural Sciences of Philadelphia; Joseph Grinnell, representing the American Society of Mammalogists; Alexander Wetmore, representing the Smithsonian Institution; and Stanley Field, director of the Field Museum.

At this initial meeting, the executive committee also hammered out the goals of the new organization, declaring that it intended (1) to promote the creation of national parks or reserves and new game laws; (2) to gather and disseminate "correct information on matters of international wildlife conservation"; and

(3) to promote sportsmanship among Americans hunting abroad while eliminating abuse of any special permits they may have been granted to collect protected animals.[45] The committee made it clear that it intended to complement and support rather than compete with the SPWFE and the International Office for the Protection of Nature in Brussels. Writing to Hobley at about the time of this first formal meeting, Coolidge stressed that the new American Committee hoped to work closely with its European counterparts, with whom they shared a deep sense of responsibility for the future of threatened flora and fauna: "I realize as you do that Anglo Saxon people will have to bear a large portion of the burden, financial and otherwise in this great conservation work."[46] The committee was also anxious to differentiate itself from the ICBP. While recruiting Wetmore to serve on the advisory committee of the American Committee, Phillips argued that "we will not conflict in any way with Pearson's work on bird protection in Europe, as we are more nearly concerned with extinction in the Colonial possessions of England, France, Belgium, etc."[47]

In its mission statement, the committee indicated it was "especially interested in preserving much depleted forms of wild life, particularly . . . the larger game mammals such as the Giant Sable in Angola, the White Rhinoceros, the Gorilla, etc."[48] The list of threatened species highlighted here is telling. During its first several years, the American Committee engaged with efforts to protect many vanishing forms of wildlife—ranging from the koala bear in Australia and the wisent of Europe to great whales that traversed international waters and the chinchilla in South America. But it primarily grappled with the threat facing African big game mammals. And while the committee focused on the conservation of game animals, it remained anxious to show that the dictates of sport hunting did not set its agenda. As Phillips argued in a letter to Hobley, "primarily we are not an organization of sportsman, not the least bit interested in the welfare of the sportsman, and almost all our Advisory Board consists of representatives of museums and zoological gardens. We are interested primarily in saving rare and vanishing species for future generations, not for recreation especially."[49]

One of the first issues the American Committee wrestled with was the destruction of African wildlife undertaken as part of tsetse fly control campaigns. The insect serves as carrier for a trypanosome that causes nagana in cattle and sleeping sickness in humans, both of which seemed to be on the rise following the imposition of European rule in Africa.[50] Although scientists at the time argued about what role, if any, game played in the apparent increase of these dreaded diseases, during the early twentieth century, officials in Central and South Africa began periodic wildlife culling programs aimed at stemming their spread. Responding to a plea from Hobley, in 1930 the newly created American Committee protested the "excessive slaughter" of game animals that had been undertaken in Zululand with the apparent approval of the South African government.[51] The organization also commissioned Robert Strong, the head of the Harvard School of Tropical Medicine, to undertake a systematic review of research on the relationship between game and the spread of tsetse fly–borne diseases. Strong and his two coauthors found no consistent correlation between the abundance of game and the number of tsetse flies present in a given area, but they did find at least one species of the fly that seemed to decline when large game animals were "considerably reduced."[52]

Another issue that periodically emerged during the committee's early years was how to respond to cinematic nature faking. Numerous feature films released in the United States during the 1930s purported to depict the behavior of African (and occasionally Asian) animals in their native habitat. As it turns out, many of these movies were partly or entirely staged using captive creatures that were starved or otherwise manipulated in order to obtain the most gruesome and spectacular footage possible. A notorious example released in 1930, *Ingagi*, even depicted gorillas carrying off native women. By 1933, the American Committee had investigated no less than six films containing "unnatural natural history" of this sort. While these movies claimed to be "genuine celluloid documents of African expeditions," many of them were actually made in Hollywood using animals from private zoos.[53] However, attempting to point out their inaccuracies to a gullible American public could backfire. When the American Committee exposed the fallacies of *Ingagi*, for example, the creators of the film sued the organization for $3 million. To add insult to injury, before dropping the suit, they used the ensuing publicity to drum up interest in the film. In the end, the committee decided to refrain from further exposés of fraudulent wildlife films, deciding instead to report them to the Better Business Bureau and the National Humane Society.

One of the most significant events in the American Committee's early history was its contribution to the London Convention for the Protection of African Flora and Fauna.[54] The frenzy of African colonization in the second half of the nineteenth century brought with it unprecedented pressures upon the continent's native wildlife, with habitat destruction, displacement by domesticated species, enclosure, disease, and hunting for sport and profit each taking their toll. What had happened to the bison in North America seemed to be being repeated with numerous species throughout Africa. With as much as two-thirds of the continent's game in territory under its control, in 1899, the British circulated a proposal to the other European colonial powers that expressed a desire to stem the slaughter. The resulting convention, signed on May 19, 1900, applied to a

2,500-mile swath that cut through the heart of the continent, between latitude 20° north and a line extending across the north boundary of Southwest Africa (Nambia) through the Zambezi River. It provided a list of recommendations offering decreasing levels of protection to a list of animals (schedules 1–4) and promoted the destruction of animals deemed noxious (schedule 5). Other provisions called for the establishment of reserves (where hunting could take place only by permit), closed seasons, a licensing system, and a permit system for the export of certain threatened species.[55] Signatories agreed to put these and other measures into effect within a year after the convention came into force, though numerous clauses permitted delays in implementation. Even with this wide latitude, the convention never took effect, although several governments adopted portions of its recommendations.

By the early 1930s, British conservationists were once again clamoring for greater protection of African wildlife. In December 1932, Hobley informed Coolidge that his organization was anxious to negotiate a revised version of the convention.[56] Coolidge responded by lobbying various British and American officials to get John Phillips appointed as an advisor to the conference drafting the convention, an action that he later called "the most important single thing outside of our publications that this committee has done."[57] In preparation for the conference, the American Committee also published a lengthy pamphlet on *Africa Game Protection* that provided a brief listing and map of the 117 protected areas on the continent and notes on its game species "nearing extinction" or "needing additional protection."[58] A press release distributed on the eve of Phillips's arrival in England declared that a new treaty, if ratified and enforced, would "play a vital role in preserving from extinction species such as the gorilla and the white rhinoceros."[59]

The 1933 conference reaffirmed many of the provisions of the 1900 agreement, but also extended it in significant ways.[60] The new convention was geographically more expansive, for example, covering the entire continent of Africa rather than just its central two-thirds. To replace the cumbersome system of five categories of animal schedules, the 1933 convention established only two: class A, species requiring special protection; and class B, those needing some protection. (Forty years later a similar two-part classification scheme would be at the heart of the U.S. Endangered Species Act.) Gone entirely was the attempt to categorize some animals as noxious, deserving of systematic destruction. One of the most important new developments in the 1933 agreement was language calling for the establishment of permanent national parks, with secure boundaries that could be modified only by legislative action and strict limitations on the activities allowed within them. As with the 1900 convention, it turned out that few nations actually ratified the treaty, but it did provide a framework for much of the African wildlife regulation that followed in its wake.

John Phillips returned from the conference brimming with ideas that would become central to the future of the American Committee. As one indication of the high hopes he and the committee had for the London Convention, the committee published its full text, including an illustrated version of the two annexes of special protected species and a revised listing of African national parks and reserves. In his forward to the volume, Phillips urged all participating governments to ratify the new convention as quickly as possible: "Unless vigorous steps are taken in the next few years, we fear that the rapid development and opening up of the great continent will result in the permanent loss of some of the most interesting productions of nature to be found anywhere in the world."[61] The struggle to produce the two schedules of African species needing special protection highlighted the need for better data on the status of threatened wildlife around the world and pushed forward a major long-term project to compile such an inventory. At the same time, the success in negotiating the London Convention suggested the possibility of a similar agreement in the Western hemisphere, with the United States taking the lead, much as Britain had done in Africa. While the protection of African wildlife continued to be a priority of the American Committee, during the mid-1930s, the organization increasingly set its sights on the conservation of Central and South American wildlife. Before making this shift, however, it engaged with an issue that revealed a basic tension between wildlife preservation and scientific study.

COLLECTING CONFLICTS

In April 1934, Hastings William Sackville Russell, the Marquess of Tavistock, wrote a letter to the editor of *The Auk* to complain about the "ruthless and excessive destruction of rare birds by the American Whitney Expedition." One of his colleagues had recently returned from the Pacific Islands with reports that the collectors associated with this expedition had "apparently exterminated" the small lory found only in the interior of Viti Levu, killing forty-seven specimens, while their permit allowed them to take only five. Moreover, due to expedition's excessive collecting, the "rare" masked parrot (*Prosopeia personata*) could no longer be found in the wild, while the Norfolk Island parakeet (*Cyanoramphus cooki*) had been "decimated or exterminated." Tavistock feared that "great mischief has been done on Antipodes Island and other islands as well." An avid aviculturalist, Tavistock hoped the reports were exaggerated, but they seemed reliable. He seemed particularly distressed by the "slaughter of rare birds of the parrot family" because, if a few birds of each species had been trapped, they could have been bred in captivity, thereby producing an abundance of museum specimens. For some time, he had been encouraging his colleagues in California to begin breeding "species threatened with extinction in a wild state." He had even sent a few examples of the Norfolk Island parakeet to the West Coast, where they now appeared to be reproducing successfully.[62]

Frank Chapman, curator of the bird department at the American Museum of Natural History (the institution that sponsored the Whitney South Seas Expedition), issued a series of responses to Tavistock's troubling charges. In his initial reply, published with Tavistock's letter, Chapman countered that the "small lory" that Tavistock mentioned was not only found on Viti Levu, but also on several other islands in the Fijis as well. The Whitney party collected twelve specimens of the bird (not forty-seven) during its weeklong visit, proof that the species was not particularly rare. Of the masked parrot, twenty-six (not eighteen) had been collected, but the fact that they all had come from a small portion of a relatively large and unexplored island suggested that "this species is neither rare nor threatened with extinction." The expedition party had taken only two specimens of the Norfolk Island parakeet. The Marquess's information was inaccurate, Chapman asserted, and in the future, anyone who had concerns about collecting done under the auspices of the American Museum would do well to contact him directly: "We have nothing to conceal, and if excess of zeal should lead one of our collectors to violate the ethics of the profession we should be among the first to admit and regret it."[63]

Chapman seemed inclined to drop the matter at that point, but rumors about the Whitney Expedition's overzealous collecting of rare species continued to circulate. With prodding from the American Committee, then in the midst of formulating guidelines for the collection of threatened species, Chapman published a lengthy rebuttal in *Science* magazine.[64] There he stressed the strictly scientific character of the Whitney expedition, pointing out that thus far it had not only resulted in important additions to the museum's ornithological collections but also in forty-four published papers. Chapman emphatically denied that the collections, made over the course of fourteen years from six hundred separate islands and more than one thousand localities, had posed a serious threat to any birds. He admitted that on Chatham Island the expedition party had collected twice as many individuals of three species as their permit allowed. But this was inadvertent, he claimed, because several members collected independently of one another and thus did not realize their error until they returned to headquarters. Moreover, the "excess" specimens had been donated to the New Zealand Dominion Museum.[65]

Like most naturalists of his day, Chapman firmly believed that scientific collecting posed little or no actual threat to wildlife, even in cases when rare species were taken. In terms of their impact on wildlife populations, other factors—habitat destruction, commercial hunting, systematic persecution, and the introduction of exotic species and disease—proved much more destructive than collectors. But what Chapman failed to acknowledge was that in addition to its potential implications on animal populations themselves, collecting rare wild animals also had a symbolic dimension that grew more pronounced as a species declined in number. Seeking out animals known to be rare gave the impression that naturalists were more interested in the pursuit of science than in protecting vanishing wildlife.

The issue of defining the boundaries of legitimate scientific collecting was one that the American Committee had struggled with from the beginning. Early in 1931, Coolidge contacted William K. Gregory, a curator in the Department of Comparative Anatomy at the American Museum of Natural History and a close friend of Osborn, about three live juvenile gorillas that Martin Johnson had recently captured or purchased while filming in the highlands of the Belgium Congo.[66] When officials in Brussels caught wind of what had transpired, they complained that Johnson had failed to obtain the necessary permit to take the specimens, which were protected by law throughout the colony. Although this particular expedition relied solely on private funds, Coolidge feared that foreign officials would associate his illegal activities with the American Museum, which had sponsored some of Johnson's earlier travels. As a scientist who specialized in the great apes, Coolidge relished the prospect of having three live mountain gorillas available for study in American zoos; but as secretary of the American Committee, he was also quite concerned that these specimens had apparently been procured without proper permits. He was particularly troubled by the chilling effect that Johnson's activities might have on future American efforts to collect gorillas in the area. In the end, after much discussion, Johnson gained permission to export the three young gorillas and the issue died down.

Three years later, Tavistock's charges concerning overcollecting by the Whitney expedition forced the issue of taking rare species back in the limelight. In the spring of 1934, the American Committee asked Alexander Wetmore—an ornithologist, assistant secretary at the Smithsonian Institution, and member of the committee's advisory board—to author a resolution on the issue. Wetmore, in turn asked his staff in the bird department to draft an initial version that strongly condemned the "increasing collecting of rare species" and urged that "larger museums" pressure the government to restrict unnecessary collecting.[67] Fearful that this draft was too restrictive to gain wide acceptance within the zoological community, Wetmore jettisoned it and presented an entirely new version at the American Committee's annual meeting later that year. Now the first paragraph blamed "progressive exploitation of natural resources" for the "alarming reduction in numbers of even extermination in numerous forms both of animals and plants." The second paragraph affirmed the right of museums, zoological gardens, and arboretums to gather flora and fauna for "scientific study," but called for "discretion" in the taking of "rare species" so that "collecting in the name of science may not lead to actual extermination." The final paragraph urged organizations and private collectors to refrain from activities that would "injure the continued existence of any form."[68] Another paragraph penciled in at the meeting asked signers of this resolution to do everything in their power "to prevent the extermination of threatened species," but this proposed modification failed to gain support. The initial plan was to circulate the toned-down version of this resolution to various biological institutions for signatures and then issue the resolution and supporter list as a printed circular.[69]

The American Committee advisory board, which now included representatives from fourteen institutions, remained divided over the resolution. Joseph Grinnell, the founding director of the Museum of Vertebrate Zoology at Berkeley, who two decades earlier had publicly lamented the decline of collecting in ornithology, strongly supported the idea.[70] He thought the resolution would make the officials of America's museums and zoological societies "and especially the sportsmen who frequently operate under the auspices of these institutions, more conscious of their responsibility not to pursue rare animals anywhere near to the point of extinction."[71] However, Charles M. B. Cadwalader, the director of the Academy of Natural Sciences in Philadelphia, feared that it would open up institutions to "unnecessary criticism and unfavorable publicity" if a sponsored collector should "accidentally shoot one of the few remaining forms" of a rare species. He also thought it better for each museum or zoological society to subscribe to a resolution along the lines being proposed and that these would then be deposited with the committee. This procedure would allow institutions to maintain their autonomy and prevent them from being publicly associated with others of "definitely lower standards."[72]

Coolidge then hashed out the issue with the American Committee's executive committee, which now included Wetmore, Cadwalader, and five other members. In the end, the executive committee decided to circulate a condensed version of the resolution to forty-five American museums and zoos, asking for support in the form of a letter that would be kept on file. The final paragraph, urging collectors not to carry on their activities in such a manner as not to threaten "the continued existence of any form" was dropped. Although thirty-three institutions had signed on to the resolution by the end of the year, it was unclear how much effect it had on their collecting activities.[73]

The resolution did lead to calls to condemn collecting abuses. One telling case is the American Ornithologists' Union (AOU), which had long been dominated by technically oriented ornithologists, many of whom resented outside interference in their activities. In 1930, as the newly elected president of the AOU, Joseph Grinnell appointed a new bird protection committee whose chair proposed creating a "white list" of "rare and vanishing" birds for which collecting permits would not be issued, an idea the full committee rejected. Five years later, at the AOU meeting following the circulation of the American Committee collecting resolution, the retired businessman and ornithologist S. Prentiss Baldwin presented a resolution calling on AOU members to refrain from "collecting of skins and eggs of rare and apparently vanishing birds." Through various behind-the-scenes machinations, opponents managed to squelch the resolution. When a similar measure failed to gain passage the next year, the National Association of Audubon Societies took the AOU to task. Finally, in 1937, the AOU passed a resolution that opposed "any scientific or other collecting or investigational activities which may in any way endanger or adversely affect the status of seriously depleted species by molestation, invasion of territory or otherwise."[74]

Ironically, Coolidge himself later found himself under attack for overzealous collecting of a rare species. In 1937, he led the Asiatic Primate Expedition to Siam (Thailand), Borneo, and Sumatra. The goal of the expedition—which also included the primate morphologist Adolph Schultz, the psychologist and anthropologist C. Ray Carpenter, and three other participants—was to observe, photograph, and collect gibbons in the hope of discovering more about their evolutionary relationship with humans.[75] When Van Tienhoven, of the International Office for the Protection of Nature, and Henry Maurice, of the SPWFE, heard troubling rumors that Schultz had killed two hundred gibbons during the expedition, they wrote to W. Reid Blair, the new secretary of the American Committee. Blair passed on their queries to Coolidge, who authored a lengthy letter to Maurice revealing just how touchy the subject of taking rare species remained in conservation circles. "No one in this country that I can think of," Coolidge asserted, "has a stronger feeling [than I do] in regard to the importance of protecting vanishing species and the promotion of national parks and game preserves in all parts of the world." Yet, in his opinion the Asiatic Primate Expedition's toll, a total of 175 gibbon specimens in Northern Siam and twelve specimens in North Borneo, proved entirely justified. The "considered judgment" of the expedition party, which included three of the world's authorities on primates, was that the "vast gibbon population" in the region would not be harmed by the collecting ac-

FIGURE 29. Sherwood Washburn (*left*) and Harold J. Coolidge (*right*) collecting gibbons in Siam (now Thailand), 1937. Not long after the American Committee adopted a policy calling for restrictions on the collection of rare and endangered species, European conservationists accused the Asiatic Primate Expedition (which Coolidge had organized and financed) with overzealous collecting of gibbons. Courtesy of Miles Coolidge.

tivities of its members. Moreover, the collections promised to reveal valuable information about the taxonomy and morphology of the gibbon that would nicely complement expedition field studies of their psychology and sociology.[76]

In his reply to Van Tienhoven, Coolidge reiterated one of the basic premises of the American Committee—the claim that detailed, firsthand knowledge of native species in their natural environment provided crucial insight into the challenges those species faced in the wild: "Both John Phillips and I had a definite feeling that a person who has hunted or collected in the field can, in many ways, evaluate certain wild life problems better than one who has not the advantage of field experience. I mention this because I think the question of gibbon skeletons is a case in point."[77] What Coolidge failed to reveal in either response is the fact one of his collectors killed several gibbons in close vicinity to a Buddhist temple, a site where local residents considered all life to be sacred, thus creating great resentment in the region.[78] Here was a case when the symbolic dimensions of specimen collecting loomed particularly large.

Of course, naturalists, sportsmen, and collectors needed access to reliable information about the status of potentially rare species if they hoped to avoid contributing to their demise. At this point, a layer of irony further clouded the debate about overcollecting rare and threatened species. It turns out that naturalists desiring to fill gaps in their collections not only routinely sought out specimens of vanishing organisms in the wild and extinct specimens from other collectors, they also created the first systematic, worldwide inventories of those species.

The British naturalist and banker Walter Rothschild offers a vivid case in point. An avid collector since his youth, Rothschild possessed both an obsessive desire to amass natural history specimens and the prodigious wealth to fully indulge his passion.[79] His family estate at Tring Park housed the world's largest private museum ever assembled; at its height, the famous Rothschild collection contained 280,000 bird skins, 3,400 mammals, millions of insects, and an assortment of other invertebrates, reptiles, and fish.[80] Other statistics associated with the enterprise are equally staggering. Shortly after his parents built a large museum to house his burgeoning collection in 1889, Rothschild hired two German curators, the ornithologist Ernst Hartert and the entomologist Karl Jordan, who between them would describe five thousand new species and author more than a thousand books and papers based on the specimens at Tring. According to Miriam Rothschild (herself an accomplished naturalist and the author of a fascinating biography of her uncle), between 1890 and 1908, Lord Rothschild dispatched more than four hundred bird collectors to the far corners of the globe in an attempt to fill in his ambitious desiderata list. He was particularly enamored with rare species of birds and spared no expense tracking them down in the wild and attempting to purchase extinct species in other private collections. His interest in the subject culminated in the publication of a lavishly illustrated book, *Extinct Birds* (1907), which represents the first attempt to compile a thorough inventory of birds thought to be lost.[81]

By the early decades of the twentieth century, Harvard's Museum of Comparative Zoology was also racing to obtain specimens of as many of the world's birds as possible. With encouragement from Thomas Barbour (who in 1927 would become director of the MCZ), and a contingent of supporters that included John C. Phillips, John E. Thayer, and several other ornithologist-patrons, around 1920, curator Outram Bangs announced an ambitious program to acquire examples of every known avian genera for the MCZ bird collection.[82] To monitor progress in his quest, in 1923, Bangs recruited the ornithologist James Peters to compile a card catalog of the museum's collection and want list.

This initiative soon blossomed into a major, multivolume project, *Check-list of Birds of the World* (1931–87), which Peters spent the rest of his career compiling and other ornithologists finally completed more than three decades after his death.[83] Another spin-off project from the original card file was a catalog of rare and vanishing species that Thomas Barbour and John Phillips began maintaining.[84] Based on the references in this catalog, in 1926 Phillips published a list of nearly 150 "extinct and vanishing birds of the Western Hemisphere," a third of which he considered "certainly" or "probably" extinct.[85]

From the time of the American Committee's foundation, Phillips, Coolidge, and Barbour had noted the need for a comprehensive study of the world's vanishing forms of mammals and birds.[86] As early as 1932, the committee began maintaining a simple card index of threatened mammals around the globe.[87] A year later, Phillips returned from the London Convention negotiations with a renewed conviction that an authoritative inventory of the world's extinct and vanishing mammals was crucial. Such a report would not only prove important in its own right but would also help the committee set conservation priorities. With this information in hand, the hope was that the committee could "draw up proposals for the protection of vanishing species in their natural habitat through the establishing of adequate national parks and reservations."[88]

Despite renewed interest in the idea, two challenges initially kept the project from getting off the ground. First was the issue of financing. The economic depression that continued to grip the nation limited the American Committee's ability to raise the funds needed to commission such a study, while repeated attempts to secure foundation funding initially came to naught.[89] The other problem was locating someone qualified, willing, and available to compile such an inventory, which would require not only considerable scientific background but also the language skills needed to slog through hundreds of widely scattered publications.

During the summer of 1935, Phillips finally found someone who seemed perfect for the job. Francis Harper was a talented but temperamental zoologist with a keen memory, an accessible writing style, and a Ph.D. from Cornell.[90] Despite his qualifications, the forty-eight-year-old was also an outspoken perfectionist who seemed unable to hold down a permanent position of any sort. Immediately before beginning work for the American Committee, he spent several years summarizing scientific papers for *Biological Abstracts*, a task he considered "severe drudgery," even if it did provide him with constant exposure to the latest research and regular practice in translating foreign languages. When Phillips approached him about the extinct and vanishing mammals project, Harper expressed interest but refused to commit.[91]

FIGURE 30. Francis Harper in the field, 1934. Given his training and background, Harper seemed a perfect fit for the job of compiling a volume on lost and vanishing mammals of the world. However, his slow work pace proved a source of growing tension between him and the American Committee. Courtesy of Special Collections, Spencer Research Library, University of Kansas Libraries.

Confident that Harper would eventually sign on, Phillips and Coolidge began drawing up plans for the project. The members gathered at the American Committee's annual meeting approved a modest annual budget that included a special research allocation of $1,000, enough to pay Harper a salary for four months.[92] Early estimates suggested that the project could be completed in "a year and perhaps longer." In his correspondence with Harper, Phillips stressed that the resulting publication should provide a widely accessible but authoritative compilation, and he urged emphasis on the utilitarian arguments for wildlife preservation. The work should highlight "humanistic appeal, economic possibilities and possible uses for scientific experiments or use in preparing serums," he argued. "We want to assemble all the arguments we can and avoid as much as possible appeal to pure sentimentality; though sentiment is, I suppose, the reason behind most of our efforts."[93] Consistent with this general orientation, the report would stress "those forms of important interest and direct benefit to mankind."[94] Cadwalader, a member of the American Committee board, granted Harper permission to work at the Academy of Natural Sciences in Philadelphia, which possessed a library he knew quite well from his work for *Biological Abstracts*. Coolidge and Phillips felt their belief in the project's value to conservation gained vindication when they attended a meeting of the North American Wildlife Conference in February 1936. There Aldo Leopold circulated a draft paper calling for a "conservation inventory of threatened species" that seemed quite similar to the project they were undertaking.[95]

Harper finally began work for the American Committee in May 1936. For the next three years, he not only scanned a seemingly endless stream of relevant publications but also widely circulated a form letter asking for up-to-date information on the status of lost or vanishing mammals.[96] After laboring on the project for a year, he reported at the American Society of Mammalogists annual meeting that he had identified approximately 390 species and subspecies that warranted inclusion in his inventory and that as many of fifty-five of these had already gone extinct. Although he had written up only a handful of forms, already patterns were beginning to emerge. "In practically every case," Harper asserted, "civilized man has been responsible, either directly or indirectly, for the extinction of these species. It is only too evident that the process of extermination is being accelerated as time goes on. The general advance of settlement, clearing of land, firearms in the hands of primitive Africans and Asiatics, introduction of exotic pests—all these contributory causes of extinction are directly chargeable to civilized white man, chiefly in the last century. It is a terrible indictment." Perhaps surprisingly, the continent "ahead of all others in its record of destruction" was North America, where twenty-four of fifty-five forms had been lost and government agencies were making a "determined effort to wipe out, with poison, a large portion of our native mammalian fauna." Even if the many subspecies of grizzly bear on this list were reduced to a single form, fourteen North American species of mammal were gone forever. "This is America's shame," Harper admonished his colleagues. How was native fauna to be preserved from the onslaught of civilization? While there were no easy answers, Harper believed the most expeditious solution was the creation of "a sufficient number of inviolate sanctuaries."[97]

In response to Harper's claim about the abysmal record of fauna destruction in North America, Phillips warned that geographical comparisons of extinction rates were misleading, particularly since the mammals of other continents were less well known. He also urged Harper to avoid "critical attitudes and propaganda" in his forthcoming book.[98] Still, the apparently high rate of human-caused extinction in North America was an issue that had dogged the American Committee since its founding. For example, in 1936, when writing to the director of the South African Museum about a campaign to protect the mountain zebra, Coolidge noted that his own nation's spotty conservation record potentially comprised the American Committee's ability to successfully lobby officials of other countries: "We are fully conscious of the plight of many of our own disappearing species. There are only six hundred Grizzlies left in the United States. There are only six of the Sierra Nevada race of our Mountain Sheep that still survive. The Trumpeter Swan and the Ivory-Billed Woodpecker are in a very precarious condition. As you know the Carolina Paroquet and the Heath Hen are already extinct. If we had only made a better showing ourselves we would be in a stronger position in approaching your Government. Nevertheless I do hope that some good may come from our action."[99]

The slow pace of the project proved another source of tension between Harper and Phillips. In May 1937, a year after Harper began working for the American Committee, Phillips reported that he told Harper he must begin writing up his final report by the end of July and have it completed by the end of the year, but he also added, "I don't really expect this to happen."[100] Several months later, when Phillips expressed his concerns directly to Harper, he responded that he was "as anxious as anybody to wind the work up in the shortest reasonable time."[101] By December 1937, the committee had expended nearly $5,000 on the Harper project, but still no end was in sight. Following a trip to Philadelphia, Barbour reported back to Coolidge that Harper was doing "a good job though not a very speedy one. I don't suppose, however, that anyone, not having a job to turn to when this task is finished, is very likely to hasten its conclusion. Human nature ain't made that way."[102]

By this point, Coolidge and Phillips were both experiencing health problems and anxious to relieve themselves of the constant pressure of fund-raising on behalf of the American Committee. They entered negotiations with the New York Zoological Society, which had expressed interested in expanding its conservation work following receipt of the Permanent Wildlife Fund. The hope was that if the home office of the committee relocated to New York, the organization would be in line for a portion of the interest from the $130,000 fund that William Hornaday had bequeathed to the society following his death in May of that year.[103] While zoo officials found the prospect of housing the American Committee appealing, they wanted nothing to do with the Harper report.[104] A disappointed Phillips agreed to oversee the report's completion but died of a heart attack November 1938, nine months after the American Committee's official move to New York. A committee consisting of Coolidge, Wetmore, and Cadwalader then assumed responsibility for monitoring Harper's progress.

They began to fear that Harper might never finish the project. In December 1938, he reported that he had written up about 318 Old World forms but still had about eighty more to go.[105] Coolidge warned him that "the financial situation looks desperate" and the committee could not continue his salary after Janu-

FIGURE 31. Glover M. Allen at the Museum of Comparative Zoology. When it looked like Francis Harper might never complete his report on lost and vanishing mammals of the world, the American Committee recruited Allen, a renowned mammalogist, ornithologist, and Museum of Comparative Zoology curator, to finish the portion on New World mammals. Courtesy of the Archives of the Ernst Mayr Library of the Museum of Comparative Zoology, Harvard University.

ary 1939.[106] In response, Harper urged the publication subcommittee to release the Old World material before it became outdated.[107] After initially balking at the idea, they reconsidered when Glover M. Allen, a curator of mammalogy at the MCZ, agreed to assume responsibility for a second volume on New World and marine mammals. Since Allen was already on the payroll of the MCZ, recruiting him to the project would not only speed the appearance of a significant portion of the inventory but would also save the American Committee a great deal of money. The publication subcommittee also granted Harper a final three-month extension to finish writing up the Old World material. He still had twenty-five to thirty forms to complete after this final deadline, in mid-April 1939, three years after he began working on the project.[108] Allen, on the other hand, made rapid progress, finishing not only his volume by January 1940, but also seventeen Old World forms that Harper failed to complete.[109] When negotiations with several academic publishers fell through, the American Committee raised the funds to publish the volumes themselves. [110]

Allen's *Extinct and Vanishing Mammals of the Western Hemisphere, with All the Marine Species of All the Oceans* finally appeared in 1942, the same year that its author died suddenly from a heart attack, and was dedicated to John C. Phillips.[111]

In his forward to the book, Coolidge likened the project to a "clearing of a new trail into a virgin forest" and alluded to the war as he urged an expansion of efforts to protect endangered species: "In these dark days, when even man is fighting to save himself from extinction, let us hope that this book, inspired by a great sportsman and conservationist and written by a great naturalist 'whose gentleness and purity of spirit were beyond all praise,' will in some measure be helpful in promoting a wider understanding to the need for preserving the wildlife heritage of the American republics for the enjoyment and wise use of generations to come."[112] Allen's introduction was extremely brief and curiously devoid of detailed analysis of overall trends. He did claim that as of the date of publication "few modern species of New World or marine mammals" had been "directly exterminated by man." But he also predicted that economically valuable species were "likely to be endangered increasingly with the more intensive [harvesting] methods of modern times and the growth of populations."[113] The body of his more than six-hundred-page inventory treated nearly three hundred species and subspecies of North American and West Indian mammals (including eighty-six forms of grizzly and black bear), twenty-five forms of South American mammals, and fifty-four of oceanic mammals.

Following a series of additional delays, Harper's *Extinct and Vanishing Mammals of the Old World* finally appeared three years after Allen's volume and nearly a decade after the project began.[114] Once again Coolidge offered a foreword to the book in which he expressed hope that the two volumes would "serve as a foundation . . . on which will be built future plans for the preservation of vanishing species of mammals in their native habitats." He thought this goal might be accomplished within a "framework of international cooperation," such as the London Convention or the newly negotiated Inter-American Convention. He also proposed regarding "threatened species" as "a sort of international trust by the country under whose jurisdiction it may fall."[115]

Harper's introduction was not only much longer than Allen's but also grappled with the larger implications of his findings. During the previous two thousand years, Harper noted, the "world has lost, through extinction, about 106 forms of mammals," about seventy-seven of which were full species. The rate of extinction seemed to be steadily increasing, with the previous century alone witnessing the extinction of 67 percent of the 106 extinct forms and the previous fifty years witnessing the extinction of 38 percent of the forms. The majority of mammalian extinctions occurred in North America (twenty-seven) or the West Indies (forty-one), while the least number occurred in Asia (one) and South America (three). Europe, Africa, and Australia fell in between these extremes with six, nine, and eleven extinctions, respectively. Harper noted that insular faunas seemed particularly susceptible to loss and that at present there were about six hundred vanishing or threatened forms of mammals around the world. After briefly reviewing the causes of wildlife decline in several major regions, Harper concluded that a single agent was responsible for the vast majority of recent extinctions: "The primary factor in the depletion of the world's mammalian faunas is civilized man, operating either directly through excessive hunting and poisoning, or indirectly through invading or destroying natural habitats, placing fire arms in the hands of primitive peoples, or subjecting the primitive faunas of Australia and of various islands to the introduction of aggressive foreign mammals, including the fox, mongoose, cat, rat, mouse, and rabbit. Except in the West Indies, comparatively few species seem to have died out within the past 2,000 years from natural causes, such as evolutionary senility, disease, or climatic change."[116]

Reviewers at the time typically praised both volumes while expressing a hope that they would serve as a wake-up call for saving endangered mammals around the world. For example, W. L. McAtee, a naturalist with the U.S. Fish and Wildlife Service, commended the reports for providing widely accessible account of the human record of wildlife destruction: "May it prove more than a pious hope that these books will stimulate action before it is too late for the preservation of some of the earth's most interesting creatures."[117] The British ecologist Charles Elton was even more effusive, calling the works "godsends to the ecologist interested in the fate of the world's mammals." Elton recognized that, if the political will for conservation could be ever mustered, much more research would be needed to successfully "protect and manage" the world's disappearing species. But in providing a ready compilation of virtually all that was known about extinct and vanishing mammals, these volumes represented an important first step.[118]

There were, of course, limitations to the Harper-Allen reports. Perhaps the most obvious of these was that they treated only a single class of charismatic species and only the most economically important examples of that class. In 1945, George Sprague Myers, an ichthyologist and professor of biology at Stanford, asked whether the committee might sponsor a companion volume on extinct and vanishing fish.[119] Although doubtful if Myers was the right person to take on the project, Wetmore expressed enthusiasm about the general idea and even suggested that it be expanded to include reptiles and amphibians as well as fish.[120] Still smarting from the delays surrounding the Harper report, though, the executive committee voted not to pursue the project, citing the difficulty in finding someone with the drive to complete it and the uncertainty about its cost as reasons for its decision. An inventory on extinct and rare birds was "more needed" the committee argued, and MCZ curator James Greenway had been maintaining a file on the subject for many years.[121] Jean Delacour, a member of the American

Committee since his move to the United States at the beginning of World War II, agreed to chair a committee on the proposed bird volume and a year later reported that Greenway had finished a draft of the volume. Not until 1954, however, was the final version completed, and four more years would pass before the *Extinct and Vanishing Birds of the World* finally made its appearance.[122]

These three reports represent the most enduring legacy of the American Committee for International Wild Life Protection. At a time when wildlife conservation in the United States remained firmly focused on the plight of native species, Coolidge, Phillips, and a handful of their colleagues called attention to the growing dangers facing birds and mammals around the globe. Haunted by the specter of extinction, they recognized that effective wildlife protection needed to begin with the construction of authoritative knowledge about the status of vanishing species. Their pioneering efforts to systematically document what would later be known as the "extinction crisis" were not only widely acclaimed at the time, but also became the inspiration for the "Red Books" that the International Union for the Conservation of Nature began compiling in the 1960s and the official list of Rare and Endangered Wildlife of the United States that the Fish and Wildlife Service initiated shortly thereafter.

SHIFTING FOCUS

Even more so than many other conservation organizations in the United States at the time, the American Committee sought to use science to rescue endangered species. Most of its members were naturalists with extensive experience in the field, often in far-flung locations around the world. Not surprisingly, then, they sought to use the methods of natural history to craft a global inventory of species that seemed to be facing extinction, and they regularly appealed to the authority of science when deploying arguments in the defense of those species. As the contentious debate about limiting the collection of rare forms of wildlife attests, at times their commitment to the values and institutions of science even threatened to overshadow their sense of responsibility to protect vanishing species.

Yet, despite repeated appeals to the seemingly timeless authority of science and a prescient interest in international wildlife issues, in fundamental ways the American Committee was also a product of its times. Like many social, intellectual, and political leaders at the end of the nineteenth and the beginning of the twentieth centuries, Coolidge, Phillips, and other members of the committee not only shared a sense of racial superiority but also often displayed a condescending paternalism when dealing with non-Western peoples. Just as imperialists had long justified appropriating other peoples and other lands on the grounds that bringing the benefits of civilization to unwashed savages was part of the "White Man's Burden," in public and in private, American Committee members declared wildlife preservation to be an "Anglo-Saxon" duty. They regularly condemned native uses of wildlife, even as they vigorously defended the rights of white settlers, sportsmen, and (especially) scientists to hunt those species. Rarely did they acknowledge the role that imperialism and Western-driven economic development played in threatening native wildlife in Africa and Asia. One important exception was Harper's report, which blamed "civilized man" for the extinction of numerous species and pointed out that the rate of extinction in North America was much higher than on other continents.

The American Committee was also a product of its times in terms of its basic agenda. The organization began during a period of increasing American interest in Africa and its wildlife, and all of its initial members had experience hunting or collecting specimens (the distinction was not always easy to make) on that continent. By the end of the 1930s, however, a series of events fostered a shift in the committee's geographical focus. The centennial of Charles Darwin's famous visit to the Galápagos Islands raised consciousness about the plight of endemic species on that archipelago. A dramatic decline in migratory waterfowl fueled concern about the many species that spent part of the year south of the U.S. border. At the same time, the establishment of Roosevelt's Good Neighbor Policy highlighted the United States' long-term economic, military, and political entanglements to the south. The advent of World War II, which greatly limited communication and travel outside of the Western hemisphere, proved to be the final straw that pushed the American Committee to begin paying greater attention to Latin America.

CHAPTER SIX

THE LATIN AMERICAN TURN

NATURE PROTECTION IN THE WESTERN HEMISPHERE

The perpetuation of the wild life of the world is a duty which nations in the forefront of civilization cannot ignore with due regard to the needs of mankind! . . . We must all remember that we are trustees for the generations of the future, and no more species should be allowed to go the way of the Passenger Pigeon, the Heath Hen, the South African Quagga, the Antarctic Wolf of the Falklands, and the Dodo.

HAROLD J. COOLIDGE, 1939

LOOKING SOUTHWARD

In 1920, the ornithologist Alexander Wetmore left Washington, D.C., for an extended ornithological reconnaissance of Argentina, Chile, Paraguay, and Uruguay. His trip came after the U.S. Senate had passed an unusual resolution calling for a treaty to protect North American birds that spent part of the year below the Rio Grande.[1] A thirty-two-year-old assistant biologist with the Bureau of the Biological Survey, Wetmore was a logical choice not only to assess the current status of migratory birds in that region of the world but also to gage the prospects for an international agreement aimed at safeguarding them.

Wetmore had experienced a passion for birds since the age of five, when his mother gave him with a copy of Frank Chapman's *Handbook of Birds in Eastern North America*.[2] By the age of thirteen, he authored his first publication, a paper entitled "My Experience with the Red-Headed Woodpecker" that appeared, fittingly enough, in Chapman's *Bird-Lore*. After earning his B.A. from the Univer-

FIGURE 32. Alexander Wetmore in Riacho Pilagá, Formosa, Argentina, 1920. Wetmore later became the foremost expert on Central and South American birds and one of the forces pushing the American Committee to become more involved with Latin American conservation. Courtesy of the Smithsonian Institution Archives, Record Unit 7006, image no. SIA2008-2355.

sity of Kansas, in 1912, he joined the Biological Survey as a field agent. By dint of hard work, determination, and dedication, he received regular promotions, while earning his M.A. (1916) and Ph.D. (1920) from George Washington University. Nearly three decades later, an admiring colleague nicely captured a sense of who he was: "The quiet soft-spoken Wetmore is a striking figure, whether speaking before a scientific meeting, or collecting a birds in the heart of a tropical forest. . . . About him there is an air of quiet modesty, but of hidden strength, emphasized by his clean-shaven jaw, wide firm mouth, and slightly uptilted nose. His deep, drawling voice and earnest manner command respect, and he seldom speaks at length unless he has something worth telling. Although he likes the company of men, particularly scientists, he is happiest when he is with the birds."[3]

His early years provided a solid foundation for an extraordinary scientific career. Before his death in 1978, he was to author 6 major books and more than 700 scientific papers, describe 189 new species and subspecies, collect more than 25,000 skins, and prepare more than 4,000 anatomical specimens. Colleagues recognized his exceptional achievement in avian paleontology, biogeography, and systematics by electing him to membership in countless societies, by granting him numerous awards, and by naming no less than fifty-six new biological organisms after him, a group he fondly referred to as his "private zoo."[4] He managed to accomplish so much while assuming a heavy administrative burden. In 1924, he transferred from the Biological Survey to serve a brief stint as superintendent of the National Zoological Park before accepting a position as assistant secretary (1924–44) then secretary (1944–52) of the Smithsonian Institution.

Wetmore had already begun to show signs of promise when he was recruited to survey the state of South American birds in 1920. He finished his yearlong tour with the impression that the "protection of birds in these countries is about the same stage as it was in the United States thirty years ago." While Argentina led the four nations he visited in the number of game laws, the public largely ignored them. Moreover, agricultural settlement was increasing apace, and many recent colonists came from the "south of Europe, where sentiment for the protection of birds, particularly the smaller species, is wanting." Plume birds faced exploitation from commercial hunters, and in one restaurant, Wetmore had even been served the increasingly rare upland plover. The situation was only a little rosier in Uruguay, where support for avian conservation seemed stronger, but in Paraguay there was "not much bird protection" and in Chile "everything is shot."

Given what he had observed, Wetmore reached a pessimistic conclusion: in the near future, treaties between the United States and Argentina, Paraguay, or Chile would accomplish little in the way of protecting migratory birds. Even if the governments of these nations agreed to negotiate a treaty, they lacked the commitment needed to put it into operation. While a protective treaty might be secured with Uruguay, few North American birds actually migrated there. "After a year in the field," Wetmore concluded, "it is my candid opinion that no legislation will serve to preserve some of these birds . . . or in most cases prevent their decrease. . . . Personally I consider the outlook for some of our birds as decidedly gloomy."[5]

Wetmore's dim prognosis was one of several factors that led U.S. officials not to immediately press for a treaty protecting birds that migrated to Central and South America. In Mexico, the winter home to many species of North American waterfowl, continued political upheaval and efforts to nationalize the oil industry chilled relations with the United States, leading to the severance of diplomatic ties. Here and throughout most of Central America, the repeated intervention of U.S. troops fostered an atmosphere of suspicion and hostility rather than the mutual trust necessary to successfully conclude an agreement protecting birds that crisscrossed national boundaries.[6] The general lack of conservation sentiment and protective legislation in most Latin American nations represented the final straw. U.S. naturalists and governmental officials seemed to agree with Wetmore, who felt that even if a migratory bird treaty could be arranged with one or more Latin American countries, the region's social and political climate provided little hope that it would be enforced. E. W. Nelson, the former chief of the Bureau of the Biological Survey, spoke for many of his colleagues when he admitted privately in 1924 that "I am distinctly opposed to entering into negotiations for a treaty on migratory birds merely for the purpose of having a treaty when at the same time we know that such a treaty would be completely ineffective."[7]

Prospects for agreement on this and other conservation fronts improved with the implementation of the Good Neighbor Policy in the 1930s, as heavy-handed U.S. military intervention in Latin America gave way to more subtle forms of influence and persuasion. Accompanying this foreign policy shift was a crash in North American waterfowl populations, the result of drought, habitat destruction, and overhunting. Finally, thanks to a large-scale cooperative banding program, scientists at the Biological Survey began to draw more precise maps delineating the migration patterns of North American birds. The net effect of these political, ecological, and scientific transformations was a growing interest in the plight of Latin American wildlife among U.S. conservationists. Although this interest would never seriously challenge the attention devoted to species considered native to their own continent, it was nonetheless substantial.

While national and international events played a role in pushing the American Committee toward work in Latin America, Wetmore himself also proved central to this reorientation. Through extensive work in the field, research on museum specimens, and a long series of publications, he established a reputation as a renowned authority on Central and South American birds. He also earned respect as a pragmatic conservationist deeply concerned about the plight of birds and other wildlife at home and abroad. An active member of the American Committee from its founding, Wetmore became a board member in 1934, where he helped orchestrate the organization's shift in focus from African to Latin American conservation. The culmination of that transformation was the Convention on Nature Protection and Wildlife Preservation in the Western Hemisphere (1940), a landmark treaty that nineteen American republics would ultimately ratify. The American Committee proposed, drafted, and generated support for this agreement, while Wetmore chaired the U.S. Committee of Experts responsible for gathering information in support of the proposed treaty, reviewing the American Committee draft, and passing that draft on for final approval by the Inter-American Committee of Experts, which he also chaired.

The first inkling of the American Committee's turn toward Latin American conservation revolved around the Galápagos Islands, named after the area's massive land tortoises. As we have already seen, island species of all sorts have long attracted the interest of naturalist and conservationists. Because insular faunas tended to exhibit a great deal of endemism—species unique to a given area—they shed important light on the process of evolution; at the same time, those distinctive species often proved quite vulnerable to attack by humans and the many pests they introduced, thus eliciting concern from those concerned about wildlife extinction. Nowhere were these two tendencies—high rates of endemism and vulnerability—more evident than in the Galápagos Islands, six hundred miles west of mainland Ecuador.

Those islands also occupy a special place in the history of science.[8] Charles Darwin was not the first naturalist to venture to the area, but his visit proved the most famous, and the specimens he collected there in the mid-1830s played a critical role in the formulation of his ideas about evolution by natural selection. By the end of the nineteenth and the beginning of the twentieth centuries, collectors from around the world were flocking to the remote archipelago to secure examples of its unique tortoises, iguanas, birds, and other flora and fauna. In the decade between 1897 and 1908, for example, Walter Rothschild, Stanford University, and the California Academy of Sciences each sponsored major Galápagos expeditions that returned home with thousands of specimens.[9] During the early 1920s, the number of visitors to the region further increased after the New York Zoological Society's flamboyant curator of birds, William Beebe, made a highly publicized visit. In a series of popular articles and a book, *The Galápagos: World's End,* Beebe portrayed the islands as an enchanting cradle of evolution, a "place of infinite variety and charm."[10] With the appearance of Beebe's book, the completion of the Panama Canal, and the continuation of a post–World War I yachting craze, a rash of wealthy easterners began venturing to the area. Many of those came simply "to sightsee or fish" but some also hosted a "veneer of naturalists, having a zoo or museum sponsor and publishing accounts of the adventure."[11] Whatever their motivation, many Galápagos visitors ended up destroying native species.

By the 1920s, American scientists realized that the flora and fauna of the region faced growing danger. Once again, a naturalist associated with the New York Zoological Society played a leading role in sounding the alarm. In 1924, Charles H. Townsend, the longtime director of the society's New York Aquarium, authored a short article warning of the "Impending Extinction of the Ga-

FIGURE 33. Galápagos tortoise from Pinzón Island, 1891. Photograph by C. H. Townsend. The Galápagos tortoise, the largest living land tortoise, once inhabited nine islands in the archipelago. The species served as an important food source for ships plying the region, leading to a dramatic decline in its population and complete loss on some of the islands. From Charles H. Townsend, "Impending Extinction of the Galápagos Tortoises," *Zoological Society Bulletin* 27, no. 2 (1924): 56.

lápagos Tortoises," one of the first publications to express concern about the threatened status of a species other than a bird or mammal.[12] An enthusiastic naturalist from his youth, Townsend was a onetime employee of Ward's Natural Science Establishment, where he met William T. Hornaday, Frederic A. Lucas, and others who would occupy important positions in American natural history institutions.[13] Following his departure from Ward's, he spent several years at the Academy of Natural Sciences in Philadelphia, where by his own admission he "dabbled in ornithology and other zoologies, with great personal satisfaction but to little scientific effect."[14] Over the next two decades, he held a variety of governmental posts, including a stint on the *Albatross* (1886–89), which cruised the Galápagos while undertaking deep-sea explorations in the Pacific. The Pribilof fur seals proved one of his special interests; for years, Townsend made annual inspections of the seal rookeries for the U.S. government, and he served on the Fur Seal Advisory Board, which played an important role in the passage of the North Pacific Fur Seal Convention of 1911.

In his 1924 article, Townsend warned that the Galápagos tortoise (*Geochelone nigra*) had already disappeared from the "smaller and more accessible islands" in the region, although a modest number of them probably still existed on Isabela, the largest island. What was now urgently needed was "not the collecting of more specimens for Museum purposes, but the preservation of such animals as may be living." For many years, settlers in the Seychelles, north of Madagascar, had successfully raised and bred the Aldabra tortoise in semidomestication. If some of the remaining Galápagos tortoises were transported to suitable climates and provided with adequate care, this "valuable animal" might also be saved.[15]

To learn more about the plight of the Galápagos tortoise, Townsend consulted the logbooks of whaling ships that had visited the region. His research confirmed that the animal provided an important source of fresh meat for the whalers, "serving a role not unlike that of the bison to the settlers of the great plains."[16] Although it was backbreaking work to round up the large, unwieldy reptiles, they survived well in the holds of ships and were considered fine eating, so sailors "took on board all they could get without too great of delay—sometimes a hundred or more."[17] The seventy-nine ships whose records he consulted, undoubtedly a small fraction of the total that had plied the region's waters in the middle decades of the nineteenth century, carried away a shocking thirteen thousand tortoises: "It was the abundance of the tortoises that attracted food-seeking ships for more than three centuries, an attraction that persisted until the exhaustion of the supply" in the mid-nineteenth century.[18] More recently, over-collecting and introduced animals that had gone feral—rats, goats, pigs, cats, and dogs—were wreaking havoc with the tortoises that remained.

Four years after he first sounded the alarm about the decline of the Galápagos tortoise, Townsend led a New York Zoological Society expedition aimed at transplanting the species. Over the course of about a month in 1928, Townsend's party secured 180 live tortoises on Isabela Island, individuals ranging in size from young hatchlings to large specimens nearing breeding age.[19] He distributed them to just over half a dozen zoos and aquariums in the Panama Canal Zone, Bermuda, and the southern United States in an attempt to establish captive breeding populations. Townsend hoped his efforts would rescue the Galápagos tortoise by establishing the reptile as a domesticated food source. These hopes were dashed, however, when they failed to breed well in captivity, though some of these specimens captured in 1928 remain alive to this day.[20]

Two years later, Townsend returned to the Galápagos as director of the Astor Expedition, also sponsored by the New York Zoological Society. Included among the party that Victor Astor invited to join him on his "large and speedy yacht" were not only Townsend but also Kermit Roosevelt, American Museum curator James Chapin, and eight other naturalists and adventurers. The focus of the trip was exploration of Santa Cruz Island, the second largest island in the

FIGURE 34. Charles H. Townsend in the Galápagos, 1927. Townsend was one of the first naturalists to call attention to the decline of the famed tortoises that inhabit the archipelago. Here he his sampling one of the species of spineless cactus favored by the reptiles. From Charles H. Townsend, "The Galápagos Islands Revisited," *Bulletin: New York Zoological Society* 31, no. 5 (1928): 160.

Galápagos archipelago but, because of its "rough volcanic terrain" and "tangled vegetation," the least well known. In two weeks, the expedition party captured so many live birds and iguanas that the upper deck of the ship resembled a "small menagerie during the return voyage." However, it managed to secure only eight tortoises, all of which were deposited in the Bronx Zoo.[21] At the end of his published report on the expedition, Townsend turned to the issue of preserving the wildlife that was struggling to survive on the Galápagos. Declaring that "the extinction of an animal useful to man is always deplorable," Townsend called for legal protection of the remaining tortoises and flightless birds, the elimination of sealing throughout the islands, the transplantation of land iguanas to islands where pigs and dogs had yet to gain a foothold, and the "control of pests."[22]

Other American naturalists also grew increasingly anxious about the plight of Galápagos wildlife. Among those was the ornithologist Harry S. Swarth, a long-time curator at the Museum of Vertebrate Zoology at Berkeley, where he collaborated with Joseph Grinnell on numerous publications dealing with the birds of California and Baja California.[23] In 1927, Swarth transferred to the California Academy of Sciences to study the institution's extensive collection of Galápagos land birds, nearly six thousand specimens obtained during an expedition more than two decades earlier. Four years later, he published a lengthy monograph on the region's avifauna that focused on the island's puzzling finches, the same species that had once vexed Darwin. His previous research on North American birds failed to prepare him for the extreme variation he found in the Galápagos finches he so meticulously measured and compared. To sort out the complex taxonomic challenges such variation posed, Swarth called for additional field-work, including selective collecting and "carefully directed observations of birds amidst natural surroundings."[24]

In 1932, he received the opportunity to venture to the Galápagos as part of the Templeton Crocker expedition.[25] That visit convinced him that something had to be done soon to save the islands' unique flora and fauna, and within a year after his return he began circulating a proposal that he hoped would achieve that end. Swarth advocated designating portions of the archipelago as a "reserve" that would serve as a "sanctuary for the indigenous fauna and flora," as a "monument to Charles Darwin," and as an "out-door biological laboratory." The Galápagos Islands were "one of the most amazing *natural laboratories* of evolutionary processes on earth," Swarth declared. To protect the remnants of this unique landscape, he called for an end to settlement and a hunting ban on several of the uninhabited islands as well as a prohibition on the destruction or disturbance of many reptiles, birds, and mammals throughout the archipelago. Finally, he recommended a campaign to exterminate the feral dogs and cats that were de-stroying native fauna.[26] Variations on Swarth's basic proposal would continue to circulate for the next several years.

While searching for kindred spirits who might share his concern about the increasingly desperate wildlife situation in the Galápagos, Swarth stumbled onto Robert T. Moore, a research associate at the California Institute of Technology.[27] A 1904 M.A. graduate from Harvard, Moore had previously pursued a variety of business ventures, including breeding foxes, mining, and lumbering. In 1927, he accompanied the first of two ornithological expeditions to Ecuador on behalf of Cal Tech. From that point, he began agitating with various Ecuadorian officials for preservation of the Galápagos's unique wildlife. In the summer of 1932, he gained support from V. M. Egas, the former consul of the Ecuadorian Republic to Los Angeles, with whom he worked closely to draft legislation and push for its passage. In 1933, when Moore's efforts came to the attention of the American Committee, members invited him to serve on their advisory board.[28] A year later Moore agreed to chair a special Galápagos committee that the American Committee established, a position he was to hold for the next four years.

FIGURE 35. Robert T. Moore examining his bird collection. Moore, a business-man and naturalist with a keen interest in Latin American birds, served as the American Committee's point person on conservation in the Galápagos during the mid-1930s. Eventually, he would amass a collection of more than sixty-five thousand birds, most of which were from Mexico. Courtesy of John C. Hafner of the Moore Laboratory of Zoology.

CAUTIOUS CONSERVATION

While the American Committee welcomed Moore into the fold, initially it re-sponded cautiously to calls for creating a reservation and a biological station on the Galápagos. Committee members feared that the area was too remote to serve as a proper biological laboratory and attempting to establish one there might dilute a recent campaign to support a research station on Barro Colorado Island, in the Panama Canal Zone.[29] At the same time, they initially worried that some sort of commercial enterprise might be behind the calls for setting aside part of the Galápagos as a reserve.[30]

Convinced of the sincerity of the Moore and Egas proposal, however, in May 1934 the American Committee passed a formal resolution approving a draft "Sci-entific Station at the Galápagos Islands Act" that the two had authored. "We are joining with the other scientific institutions throughout the world," the resolution proclaimed, "in recommending to the government of the Republic of Ecuador, the creation of certain reservations . . . in the Galápagos Islands and the preservation of the rare zoological species, which exist only in this Archipelago, and which have been made famous by the visit of Charles Darwin." Like Swarth's earlier recommendation, the proposed legislation called for setting aside several of the islands in perpetuity as "inviolate refuges for all forms of zoological life," while calling for the permanent protection throughout the archipelago of many of the same species that Swarth had originally highlighted, plus eighty additional birds from the checklist included in his monograph. In addition, the legislation empowered the president of Ecuador to contract with a reputable scientific institution to oversee the preservation and breeding of all zoological species in the reservation, to establish a "Darwin Memorial Zoological Laboratory" on San Cristóbal or Isabela Island, and to hire the wardens needed to enforce the provisions of the act. Writing to the U.S. Minister in Ecuador, Coolidge indicated that his committee considered it "extremely urgent to have some definitive action taken by the government of the republic of Ecuador to preserve the endemic zoological species found on the Galápagos islands, and especially those species which have been very much reduced to a point where there is a danger of their extinction."[31] While generally supportive of the resolution, Wetmore remained characteristically gloomy about its prospects: "The Ecuadorians may do this as a gesture but I am somewhat pessimistic regarding anything coming of it."[32]

Following modifications that Ecuadorian officials requested, in August 1934 the chief executive of the Republic of Ecuador, Abelardo Montalvo, issued a decree that echoed many of the provisions in the original Moore/Egas proposal.[33] Montalvo declared several Galápagos islands to be protective refuges for resident and migratory animal life, prohibited the destruction of a pruned down list of species considered to be "in real danger of extinction," and required newly arrived visitors to land on San Cristóbal Island, where they would be required to sign a document agreeing to abide by the laws protecting the archipelago's wildlife. Violations of the decree carried heavy fines, but a specific enforcement mechanism was lacking. In an article announcing the decree, Moore pointed out that "legislation without enforcement is usually of little value" and until wardens were hired and properly equipped, "poaching may continue with more or less impunity." The decree did make it possible, however, to enforce the Vandegrift Tariff of 1930, a U.S. law that made it illegal to import any wild mammal or bird without a certificate from appropriate authorities indicating that the species had not be obtained in violation of the laws of the country from which it was exported.

The most enigmatic figure to promote Galápagos conservation during this period was the American travel writer, explorer, and adventurer Victor Wolfgang von Hagen.[34] Born in St. Louis and educated in Europe, von Hagen was only nineteen years old when he ventured on his first expedition to Africa in

1927. Six years later, he was in San Francisco promoting what he called the Darwin Memorial Expedition to the Galápagos Islands.[35] The original plan called for von Hagen and fifteen associates to spend two and a half years traveling up and down the coast of Central and South America, where they would undertake archeological, biological, and ethnological studies. To commemorate the one hundredth anniversary of Darwin's famous visit to the archipelago, the expedition party planned to erect a permanent monument on San Cristóbal Island, a bronze bust of Darwin with a short inscription written by his only surviving son, Leonard Darwin, who had given his blessing to the endeavor. Before raising the monument, von Hagen planned to travel to Quito, where he hoped to gain authorization to establish a permanent biological station on Santa Cruz Island that would promote conservation of and research on the unique fauna of the region. To finance his expedition, von Hagen charged admission to the docked ship he hoped to use, sold subscriptions to a planned journal of the expedition, and recruited individuals to join him on the venture. To fund the biological station and warden, von Hagen proposed convincing the Ecuadorian government to authorize the release of a special series of postage stamps to commemorate the Darwin-Galápagos centennial. A dealer in New York had already agreed to buy the entire issue, and the net profits from their sales would be placed in trust. If that plan failed, von Hagen would "appeal to the wealthy yachtsmen who have visited the Galápagos in the past years" for the needed funds.[36]

Von Hagen's plans fell through when he failed to raise enough money to mount his expedition. His financial prospects brightened considerably in 1933, however, when he married a well-to-do San Francisco heiress. Von Hagen and his new bride traveled to Honduras, where they managed to locate the elusive quetzal, the royal bird of the Aztecs that some feared was extinct. Although this rare bird had never been photographed before, much less captured alive, he managed to ship back nine live specimens to the Bronx Zoo.[37]

Von Hagen and his wife then traveled to Ecuador, where they spent two and one half years.[38] In Guayaquil, he joined forces with several faculty at the University of Guayas to form a National Scientific Corporation for the Study and Protection of the Riquezas Naturales (Natural Wealth) of the Galápagos Islands, an organization whose goal was to promote the establishment of biological research station on Santa Cruz Island. Von Hagen hoped that a group staffed almost entirely by Ecuadorian nationals would allay fears about outside interests interfering in the internal affairs of this proud South American republic.[39] The issue remained a sensitive one in Ecuador because of concerns that its territorial claims to the islands might be challenged and the United States might seize them for their strategic importance in defending the Panama Canal.[40]

FIGURE 36. Victor Wolfgang von Hagen observing the quetzal in Honduras, ca. 1937. The enigmatic von Hagen was the first naturalist to photograph and capture wild quetzals. He was also an important promoter of conservation on the Galápagos Islands. From V. Wolfgang von Hagen, "Capturing the Royal Birds of the Aztecs," *Travel* 72 (December 1938): 38.

Meanwhile, American Committee members grew increasingly concerned about political developments in Ecuador and increasingly suspicious of von Hagen. The small South American nation had long suffered from political instability, but a newly enacted constitution (the thirteenth in just over a century), the formation of a series of new political parties, and the havoc of the Great Depression ushered in a turbulent period that was unusual even by Latin American standards.[41] As one measure of that ongoing political turmoil, fifteen different individuals served as chief executive of the troubled nation during the 1930s, and not one of them managed to complete a single term of office.

Worried about the lack of enforcement mechanism in the original decree and fearful that future chief executives might reject it, Moore and Egas began working on revised legislation.[42] Following a personal appeal from von Hagen, who had authored his own proposal, the newly installed Chief Executive Frederico Páez, issued a new decree in May 1936.[43] Its preamble declared that the fauna of the "Archipielago de Colón" was "in danger of being totally extinguished by the depredations of unscrupulous travelers and tourists," and if this state of affairs continued, it would represent "an irreparable loss to Science."[44] Páez declared a number of Galápagos Islands as "National Reserve Parks," but eliminated the specific list of protected species found in the earlier decree. The Páez decree also authorized the creation of a provisional five-person committee charged with selecting a permanent board of directors that would oversee the protection of the flora and fauna of the islands, establish research stations, and find suitable national and foreign scientific institutions to sponsor those stations. Von Hagen issued a bilingual pamphlet reproducing the contents of the decree under the sponsorship of the Comité Provisional para la Protección de la Fauna del Archipiélago de Colón. Although he did not explicitly claim as much in this pamphlet, the implication was that he had secured appointment to the official provisional committee created by Páez's decree.

Unsure what to make of the new decree and increasingly leery of von Hagen, American Committee members were initially at a loss as to how to respond. As we have already seen, when the committee first learned of the Darwin Memorial Expedition several years earlier, they feared that it was ultimately a moneymaking venture. Concerns about von Hagen's motives heightened in 1935, when the U.S. Department of State began investigating his efforts to obtain British funding for a research station in the Galápagos and when he protested an American naval vessel's planned visit to the archipelago.[45] Writing to Julian Huxley, who had inquired about von Hagen that same year, Coolidge answered diplomatically that he thought von Hagen was a "fair naturalist" who "seemed honest" but that he was also a "soldier of fortune" who appeared motivated by "personal publicity."[46] Especially troubling were statements von Hagen had made during a visit to Washington in 1937, including claims to Smithsonian curators that they would need a permit from him before they could land on any of the Galápagos Islands.[47] At this point, Ecuadorian officials began denying that von Hagen had authorization to act in any official capacity, and they launched an investigation into his background.[48] A Better Business Bureau report revealed that, in the early 1930s, he had failed to return rental equipment and had issued several bad checks.[49] Even more damning, in 1932, he faced arrest in Mexico City for using the apparently forged signature of a prominent governmental official to obtain financial backing for a planned expedition to Monte Alban, in southern Mexico. With their suspicions about von Hagen confirmed, American Committee members began circulating a photostat of the newspaper article describing his arrest with the hope of eroding any support he might still enjoy.[50]

The situation grew more tense in the spring of 1937, when von Hagen traveled to England to seek backing from the British Association for the Advancement of Science's (BAAS) Galápagos Committee, formed in 1935. Claiming to be a member of the official provisional committee named in the 1936 decree, von Hagen said he would post the various islands that had been designated as reserves and turn over the money raised from the sale of Ecuadorian stamps to the BAAS Committee so it could establish a biological station and warden on the islands. When O. J. R. Howarth, secretary of the BAAS committee, wrote to the American Committee asking for support for the plan, Moore drafted a reply in which he expressed profound doubt that von Hagen had any formal connection to the Ecuadorian government and described his previous failed efforts in San Francisco to obtain sponsorship from the University of California, Stanford University, and the California Academy of Sciences.[51]

The BAAS Committee then began circulating a new proposal that would commemorate Darwin's visit while preserving the unique biological organism on the islands. To achieve those goals, the committee proposed a fundraising campaign to establish the Darwin Memorial Biological Station on Santa Cruz Island, where a suitable house had already been located. The hope was that a "warden possessing biological qualifications appointed and supervised by the committee" would review applications to undertake scientific research on the islands and report any violations of the decrees protecting Galápagos wildlife. The committee estimated that a capital fund of between £40,000 and £50,000 would be needed to establish, equip, and staff the proposed station. Before Moore could formally respond to this initial proposal, Howarth sent two mildly revised versions of the BAAS Committee plan seeking support from the American Committee.[52]

Moore then sent a letter to the Galápagos Committee of the American Committee requesting input on the most recent British proposal. There he warned that "report after report" confirmed the continued destruction of the fauna on the islands; unless something were done soon, "immediate extinction of species is threatened." "The British *have a program* and are ready to act," Moore continued, and he strongly urged his fellow committee members to approve the British initiative with only minor changes.[53] The American Committee became skittish, however, when State Department officials objected to "any foreign nations having a hand in the collecting of funds" for such a project. Representatives from the State Department promised to take up the matter of Galápagos conservation at the next Pan American Conference.[54] Although the British Galápagos Committee continued fund-raising for the project, World War II soon intervened. Not until nearly two decades later were the plans for a Darwin Research Station that were first developed in the 1930s finally brought to fruition.[55]

Meanwhile, the situation in the Galápagos continued to deteriorate. In 1937, the Swedish zoologist Rolf Blomberg offered a gloomy assessment following a recent visit to the islands. Especially distressing was an incident during his return to the mainland, when the sailors on an Ecuadorian gunboat killed an "enormous number of sea-lions" for their valuable hides, even though the area was supposedly protected. Two years later, David Lack, a British ornithologist who would soon author a major study on Darwin's Galápagos finches that was destined to become a classic, issued a similarly bleak report on wildlife conditions in the islands. The Ecuadorian government had passed regulations that required visiting expeditions to obtain permission on the mainland before landing in the Galápagos, mandated fees for collecting larger animals, and instituted a strict limit for the number of specimens of each species that could be legally taken. Yet, the number of private yachters, settlers, and soldiers on the islands had continued to mount, wreaking havoc on the wildlife. As a result, the "undoubtedly well intentioned" regulations proved entirely "ineffective."[56]

In 1941, only four days after the bombing of Pearl Harbor, U.S. troops from the Canal Zone occupied the islands with the "grudging consent of the Ecuadorian government," thus revealing the limitations of Roosevelt's Good Neighbor Policy. Over the next year, U.S. forces built an airbase on Baltra and erected more than two hundred buildings for the one thousand or so troops stationed there. The result was ecological devastation, including the loss of the islands' iguana population, which has never recovered.[57] During the war, the American Committee repeatedly pressured the U.S. military to protect the unique flora and fauna of the islands and even considered issuing a booklet "to arouse interest in the Galápagos and the conservation of its wildlife."[58] They abandoned that plan, however, after military officials advised them to drop the matter.[59]

BIRDS WITHOUT BORDERS

Concern about the plight of endangered Galápagos fauna in the years surrounding the centennial of Darwin's visit to the region provided one entry point for U.S. naturalists to begin engaging more actively with Latin American conservation. Fears about avian species that migrated southward represented another. As early as 1913, when conservationists first discussed the prospect of using a treaty to shield recently enacted federal migratory bird legislation from potential judicial override, they considered opening negotiations with Mexico. That nation was experiencing deep political turmoil, however, and the Huerta administration, which came to power following a coup that same year, proved openly hostile to conservation.[60] Negotiations with British officials on behalf of Canada, with whom the United States enjoyed relatively cordial relations, seemed a much surer route to success.

Once Canada had been brought on board, interest in Latin American wildlife protection grew again. But those who argued for extending treaty protection to migratory birds that wintered in Central and South America faced a critical challenge: a lack of reliable data not only about the species they sought to protect but also of existing game laws in the region. Writing to Pearson in 1920, Biological Survey Chief Edward W. Nelson argued that despite the recent Sen-

FIGURE 37. E. A. Goldman and Indian guide, Mount Turnbull, Arizona, 1916. During many trips to Mexico, Goldman became acquainted not only with the fauna of the region but also many prominent Mexican naturalists and conservationists. Both proved important during the negotiations that led to the 1936 Convention for the Protection of Migratory Birds and Game Mammals. Courtesy of the National Archives and Records Administration, Record Group 22-WB, Box 47, no. 16974.

ate resolution, there was no chance for the negotiation of appropriate treaties until his agency could complete the necessary background investigations.[61] As we have seen, in 1920, he sent Alexander Wetmore on a year-long reconnaissance to begin addressing this lacuna. While Wetmore expressed pessimism about the prospects for a meaningful treaty with all but one of the four Latin American nations he visited, his report did provide a better sense of where various species resided and their current status.

The Bureau of the Biological Survey also continued to sponsor exploration in Central America, particularly in Mexico. A central figure here was E. A. Goldman, who quit his job as a California vineyard foreman in 1891 to go exploring with Nelson.[62] Nelson became an important mentor for Goldman, who had long shown an interest in natural history, despite a lack of formal training. For the next fourteen years, the two naturalists roamed widely across the Mexican countryside on behalf of the Biological Survey, collecting specimens, making observations, and writing up reports. Both became widely recognized experts on the fauna of Mexico, particularly mammals, and both went on to long careers at the Bureau of the Biological Survey, with Nelson becoming chief of the agency in 1916. Goldman continued to make regular visits to the region over the next three decades.

In addition to dispatching naturalists southward to learn more about the status of North American birds that migrated to Central and South America, the Bureau of the Biological Survey also initiated a large-scale bird banding project. That project involved placing small aluminum bands on the legs of birds—either fledgling young or trapped adults—with a brief message asking those who later trapped or shot the bird to contact the agency. Wetmore initiated the agency's first banding studies during the early years of World War I, while researching an outbreak of duck sickness in the Bear River Marshes of Great Salt Lake.[63] In 1920, faced with the task of enforcing the recently enacted Migratory Bird Treaty Act, Nelson arranged for his agency to take over the work of the American Bird Banding Association, which had been active since 1909. He then hired Frederick C. Lincoln, a curator of ornithology from the Colorado Museum of Natural History, to oversee a corps of several hundred volunteer banders, a number that soon increased to about two thousand.[64] The hope was that analysis of return records would place management of waterfowl populations on a sounder scientific basis. While sifting through the growing stream of migration data flowing into his office, Lincoln developed the concept of flyways, ancestral migration routes, rarely more than fifty to one hundred miles wide, that large numbers of bird species seemed to follow each year. According to Lincoln, North America contained four major flyways—the Atlantic, the Mississippi, the Central, and the Pacific—and the birds that utilized one of them rarely, if ever, used the others.[65] While return records from Latin America remained relatively rare, they did help fill in the considerable gaps in knowledge about waterfowl movements in the Western Hemisphere.

Once the United States restored formal relations with the Mexican government in 1923, pressure increased to implement the Senate resolution calling for migratory bird treaties with Latin American nations.[66] A year later, Biological Survey naturalists drew up a rough draft of a treaty for State Department review, while United States and Mexican officials initiated informal discussions on the subject. Negotiations quickly reached an impasse, however, when the Mexican government insisted on linking bird and fisheries conservation in a single treaty. They expressed particular concern about the depredations of commercial fishermen from California, who were developing increasingly efficient means to capture fish off of Mexican waters while avoiding custom duties on their catches. Negotiations continued in fits and starts over the next decade, but the issue of linkage between bird and fish regulation proved an insurmountable obstacle to progress, despite strong pressure from American conservationists, who continued to argue that a migratory bird treaty would strengthen the 1916 agreement with Canada while cultivating greater conservation sentiment among the Mexican people.

Finally, in 1934, a series of events renewed hopes for a wildlife treaty with Mexico. On the Mexican side, the most important change was the election of President Lázaro Cárdenas, a populist reformer who made conservation "one of the top priorities of his administration."[67] Among his notable accomplishments in this area, Cárdenas established the first autonomous conservation agency in Mexico—the Department of Forestry, Fish, and Game (founded in 1935)—and increased the number of Mexican national parks from two to forty-two. On the U.S. side, the adoption of the Good Neighbor Policy and the appointment of the sportsman and political cartoonist Jay "Ding" Darling to head the Bureau of the Biological Survey also brightened prospects for an agreement.[68] Though not a biologist, Darling proved a capable leader who was not only deeply committed to waterfowl conservation but also quite adept at wresting funds from President Franklin Roosevelt amid an increasingly severe economic depression. Even before being sworn into office, Darling began prodding American officials to act on the proposed treaty with Mexico. His initial optimism quickly faded, however, when he discovered that Cárdenas appeared uninterested in rapid resolution.[69]

In an effort to break the continuing deadlock, in January 1935 Darling gained State Department approval for a plan to dispatch Goldman to Mexico City. But even with his intimate knowledge of the region's wildlife and personal relationships with many of the principals involved in the discussions, Goldman quickly grew discouraged.[70] He and other U.S. officials in Mexico City faced numerous challenges in their effort to nail down the terms for a treaty.[71] The small but well-connected community of Mexican sportsmen feared the proposed agreement was a veiled attempt to curtail recreational hunting, the high turnover rate among Mexican officials disrupted the continuity of the discussions, and Mexican negotiators remained committed to the concept of "patrimony," the idea that wildlife properly belonged to one nation or another. This idea stood in opposition to the U.S. claim that ownership of migratory wildlife could not be rightfully claimed by any particular nation.[72] Finally, and perhaps most importantly, the Mexican government remained more interested in a treaty covering fisheries—which had immediately obvious economic implications—than in one protecting migratory birds, while powerful California fishermen blocked forward motion on the fisheries front.[73] A frustrated Darling lashed out: "The commercial fisheries of California are denuding the waters off Point San Lucas and the Lower Gulf of California as ruthlessly as the lumber barons raped the forests of North America. If Mexico were big enough, she'd be justified in knocking our block off."[74]

The deadlock finally broke in September 1935, when Juan Zinzer, head of the game division of the newly created Department of Forestry, Fish, and Game, introduced a proposal for a bird treaty that looked much like the one the Americans had originally suggested.[75] Although several sticking points remained, negotiations moved quickly from then on, as both parties seemed anxious to trumpet their achievement at the first North American Wild Life Conference in Washington, D.C., scheduled for several months hence. Goldman returned to Mexico City in early 1936 and soon finished ironing out the remaining sticking points. Officials from both governments signed the treaty on February 7, 1936, and later that day, Darling received a thunderous round of applause when he announced the fruitful conclusion of treaty negotiations to the delegates attending the conference.[76] Finally, nearly two decades after the idea of a bird treaty with Mexico had first been seriously raised, success seemed at hand.

The initial euphoria quickly dissipated, however, when the specifics of the secretly negotiated treaty began to circulate. The Convention for the Protection of Migratory Birds and Game Mammals did declare it "right and proper" to prevent from being "exterminated" the birds that moved seasonally between the United States and Mexico.[77] To achieve this lofty goal, the pact called for the establishment of closed seasons for migratory birds, the end of spring hunting for waterfowl, the creation of safe "refuge zones" where hunting would be banned, the introduction of strong restrictions on the killing of insectivorous birds, and the prohibition of hunting from aircraft. But the treaty also declared that the mechanisms used to protect migratory species should permit "the utilization of said birds rationally for the purposes of sport, food, commerce and industry."[78]

Not surprisingly, some U.S. conservationists squawked at a convention that seemed to sanction the commercial use of avian species, especially when the wildlife protectionists had been struggling for decades to end the American trade in wild birds. When he learned of the clause the day before the treaty was signed, Assistant Secretary of Agriculture Rexford Tugwell fired off a letter urging the Departments of State and Agriculture not to approve it in its current form. The U.S. government, Tugwell asserted, had always claimed that "it is impossible to preserve from extermination or undue depletion any species of migratory birds when that species is permitted to be utilized *commercially* or *industrially*."[79] The strident conservationist Rosalie Edge proved even more blunt, declaring in one of her characteristically blistering pamphlets that the treaty was an example of "double-crossing conservationists and migratory birds."[80] Rather than building on the success of the earlier convention with Canada, Edge and others viewed the treaty with Mexico as a giant step backward. Despite its shortcomings, however, the agreement did highlight the possibility of successful negotiations regarding wildlife between the United States and a Latin American nation, and in doing so, it helped push the American Committee into high gear in the region.

THE LIMA RESOLUTION

In February 1935, Coolidge wrote to Julian Huxley praising Lord Onslow, president of the Society for the Preservation of the Wild Flora of the Empire, for his broad conservation vision. He was particularly impressed with Onslow's recent call for a conference to negotiate an Asiatic wildlife protection treaty modeled on the convention that had been concluded for Africa two years previously. In the same letter, Coolidge expressed pessimism regarding the possibility of a similar "Pan American plan," at least in the near future: "Perhaps in a few years the picture will have changed, and if the other conferences have proved successful, our Latin American neighbors may wake up. I think they are rather more likely to do this if they think it is fashionable rather than for the good that the conservation may do them. In many ways they are like the Chinese with little regard for human life and less for animals and birds."[81] Coolidge's comment was revealing not only of the condescending attitude that he and many of his colleagues shared toward Latin Americans, but also about the slim chance they saw for conservation initiatives in that region of the world.

A year later, however, with the successful negotiation of the treaty with Mexico to protect migratory birds, attitudes began to shift. Among the many research projects listed in a funding proposal that the American Committee prepared early in 1936 was a plan for "preliminary studies in preparation for a Pan American Conference or Conventions similar to the London Convention in the interest of promoting National Parks and the preservation of rare species in the New World."[82] This proposal suggested that this research should be undertaken in cooperation with the Pan American Institute of Geography and History, which had been established in 1928 to promote inter-American cooperation in the areas suggested by its title. In October 1935, the institute had passed a series of resolutions calling for "preservation of Nature Monuments and historic places" in Latin America. A primary mover behind these resolutions was undoubtedly Wallace W. Atwood, a geographer and geomorphologist who had served as president of the National Parks Association (1929–33) and president of the Pan American Institute of Geography and History (1932–35).[83] The American Committee funding proposal also called for a study of the "racial attitude toward nature protection in the various parts of the world." The proposal remained vague on exactly what such a study might entail, but it did suggest that consciousness about wildlife protection might be raised in South American countries by naming protected natural areas after national heroes, who were widely revered.[84]

In May and June 1936, Phillips visited Europe to confer with his conservation-minded colleagues abroad. He came home with a renewed sense that the American Committee needed to expand its limited protective campaign in the Galápagos and begin taking more seriously the possibility of conservation in the rest of the Western Hemisphere.[85] Following his return, Phillips wrote to his brother, William Phillips, a well-connected State Department official, and to L. S. Rowe, the director general of the Pan American Union, asking if the issue of "preserving primitive areas and historic sites" in Central and South America might gain some form of "official recognition" at the next Pan American Conference, set for Buenos Aires.[86]

While it proved too late to get the issue of conservation onto the agenda for the upcoming Buenos Aires meeting, the program for the 1938 meeting (scheduled for Lima, Peru) had not yet been set. In August 1936, Phillips encouraged Wetmore to contact Assistant Secretary of State Sumner Welles "while the matter is still hot" and find out if the State Department would accept suggestions regarding conservation from the American Committee.[87] Initial signs proved encouraging, but for the next year, Phillips and Coolidge remained too preoccupied with supervising the Harper Report, fund-raising for the American Committee, and leading the Asiatic Primate Expedition to give the matter much attention. Nearly a year later, Phillips called on Atwood, encouraged him to drum up support for the idea of conservation at the next meeting of the Pan American Institute of Geography and History, and suggested Wetmore's name as a possible delegate.[88] In a follow-up letter after the meeting, Phillips also pointed out that from his recent visit with prominent Europeans interested in "international nature work" he had "gathered the impression that they thought it was up to us over here to start some real movement in South America . . . it's up to you in the United States to stir up the Latin American countries, to the end that you can eventually call an all-American Convention on nature protection and conservation of natural resources."[89]

Within a year, the Pan American Union and the Department of State granted approval for the American Committee to draft a resolution for introduction at the Lima meeting, scheduled for December 1938.[90] Initially, the committee tried to get Moore or Wetmore to attend the Lima meeting as an official delegate, but neither was able to make the long journey to Peru. Instead, the archeologist Alfred Kidder, Jr., who was Thomas Barbour's son-in-law and already undertaking excavations in the region, agreed to present the resolution on behalf of the

American Committee.[91] To provide the Department of State with background information on the resolution, the committee also assembled a large packet of documents that highlighted the success of the London Convention and other international treaties, while outlining the many conservation challenges in Latin America.[92]

The resolution approved in Lima was largely modeled on the American Committee draft.[93] The preamble declared the "American Republics" to be "richly endowed with natural scenery, with indigenous wild animal and plant life, and with unusual geologic formations, which are of national and international importance." Reflecting the ongoing concern with the notion of plenitude, the preamble also proclaimed that these nations are "desirous of protecting and preserving in their natural habitat representatives of *all species and races of their native flora and fauna* including migratory birds, in sufficient numbers, and over areas extensive enough to assure them from becoming extinct through any agency within man's control." To achieve this goal, the resolution urged nations in the Western Hemisphere to adopt appropriate measures to promote "Nature Protection and Wild Life Preservation," established an Inter-American Committee of Experts charged with drafting a convention on international cooperation among the American republics regarding "preservation of fauna and flora in their natural habitat," and called for the Pan American Union to "take the necessary steps to carry out the above provisions," whether through a special conference, at the next scheduled meeting, or simply opening them for signature.

MAINTAINING MOMENTUM

Having played a key role in the success of the Lima resolution, Coolidge moved into high gear to secure its implementation. In February 1939, he traveled to Washington to meet with Wetmore, L. S. Rowe, William Manger (counselor to the Pan American Union), and various State Department officials.[94] He emerged from those meetings convinced that "the Pan American Union is counting on our American Committee to play an extremely important part in helping them to work out this convention."[95] To draft an initial version of the agreement, Coolidge suggested appointing an informal technical advisory committee, which would include representatives from the American Committee, the National Association of Audubon Societies, the Bureau of the Biological Survey, the National Park Service, and the U.S. Forest Service. He expressed confidence that this committee's draft would be "very close" to the version that the Pan American Union planned to circulate to the Inter-American Committee of Experts, which would be charged with writing up the final version.[96]

Coolidge did his best to spin reports about the resolution and the upcoming convention in the media, lest they suggest that the United States was trying to force the initiative upon unwilling Latin American nations. Writing to W. Reid Blair, the current executive secretary of the American Committee, he complained about newspaper coverage that credited the committee for the resolution and proclaimed that the United States was a "prime mover in this hemisphere" when it came to "conservation as a whole." While Coolidge considered both statements to be true, he also thought it important not to give Central and South American nations the impression that "Americans want to put over on them an international treaty, no matter how altruistic the motives of such a treaty may be."[97] Similarly, he suggested to the American Museum ornithologist Robert C. Murphy that when it came to building support for conservation in Latin America, a "holier than thou attitude" must be avoided.[98] Coolidge thought Murphy's recently published map, "S.O.S. for a Continent," might prove a useful tool in this regard because it openly acknowledged the conservation shortcomings of the United States and Canada. That map highlighted the fate of the great auk, the heath hen, and other North American species and landscapes that had suffered at the hands of humans. The short introduction to the map was particularly unflinching in its critique: "Owing to rapid settlement and reckless exploitation, North America has suffered more from improper land use and the upsetting of Nature's balance than has any other continent within so short a span of time."[99]

Coolidge also penned an article of his own for the Pan American *Bulletin*, which represented an impassioned, if somewhat scattered, plea on behalf of nature protection in the Americas.[100] As a detailed attempt to summarize contemporary arguments for conservation, it is worth exploring at some length. He began by declaring that "conservation of natural resources, especially wild plant and animal life[,] is a matter of vital concern to every one of the American republics!" Now more than ever, strong leadership was needed to "save many important vanishing species of animals, birds and plants from total extinction at the hand of man."[101] The current push for nature protection was a "natural reaction" to the excessively "materialistic developments of the last decades" and a protest against "all needless destruction and unreasonable prodigality in land cultivation, and works for the preservation of natural beauty, virgin landscapes, and natural objects of unusual significance as being of fundamental value for the present and coming generations."[102]

Coolidge then presented a brief enumeration of the many arguments supporting nature protection, a list that had become increasingly familiar by the middle of the twentieth century.[103] As was often the case, economic considerations came first. Wild flora and fauna represented "an immeasurable capital, a capital that ought not to be ruthlessly spent, but rather stewarded with care." At the same time, nature parks provided valuable recreational and educational opportunities, especially "for the inhabitants of large cities." In addition, nature protection had an aesthetic dimension, while continued access to natural objects and systems remained crucial to biological research. Wild nature even had important medical implications, since it provided many of the drugs used to treat illnesses and the animals used in laboratory experiments. And, finally, biological organisms served as an important source of inspiration for technological innovation: airplane design owed much to the study of birds, while ship builders drew insights from the shape of fish. One item curiously absent from Coolidge's list was the role that individual wildlife species played in maintaining the ecological integrity of biological communities, an argument that was becoming increasingly important by the time he was writing.

In the second half of his article, Coolidge first turned to a brief discussion of a handful of plants and animals suffering precipitous declines. "The most seriously threatened victims of civilization," he declared, "are the large animals with a limited range and slow rate of multiplication, and the species that have commercial value. . . . Nothing seems to make us fully realize the startling fact that we are witnessing the end of the age of mammals which is being brought about by our own efforts." [104] A handful of international organizations had sounded the alarm about these and other vanishing species, while several governments had created national parks—like Yellowstone, the Parc National Albert, and the Kruger National Park—that contributed to preserving them. Coolidge praised the London African Convention of 1933 as "perhaps the most significant event in the whole development of wild life protection through international agreement," summarized a few of its major provisions, and briefly highlighted the success of wildlife treaties in which the United States had become involved. [105] He concluded by declaring emphatically that "the perpetuation of the wild life of the world is a duty which nations in the forefront of civilization cannot ignore with due regards to the needs of mankind!" [106]

In April 1939, a special committee appointed by the governing board of the Pan American Union agreed on a plan to implement the Lima resolution on nature protection and wildlife preservation. [107] The special committee ruled that the Inter-American Committee of Experts should consist of a single delegate from each of the member countries, that it should convene at the Pan American Union building in Washington in April 1940 (to run concurrently with the Eighth American Scientific Congress), and that the Pan American Union should formulate a "draft convention" as the basis for discussion at that meeting. [108] The special committee also released a questionnaire (that Coolidge helped author) requesting information on game laws, protected natural areas, and the status of vanishing fauna or flora in each of the American republics.

With a clear mandate to act, Coolidge arranged a meeting between the Pan American Committee of the American Committee—which included himself (as chair), Alexander Wetmore, William G. Sheldon, and Frederic Walcott—and representatives from the Pan American Union, the National Association of Audubon Societies, and two relevant governmental agencies. At the first meeting in May 1939, Coolidge, Wetmore, and Sheldon were joined by Manger, T. Gilbert Pearson, F. C. Lincoln (representing the Bureau of the Biological Survey), and Victor Cahalane, C. C. Presnall, and C. P. Russell (representing the National Park Service).[109] The purpose of this initial meeting was to begin blocking out the form for the proposed convention. Following a series of suggestions from Coolidge, the assembled naturalists and conservationists agreed to sections dealing with migratory birds (which incorporated features from the Canadian and Mexican treaties), national parks (which was modeled on the London Convention), the identification and protection of vanishing species, and various enforcement mechanisms. They decided to exclude fisheries and "other controversial topics of economic importance" from the proposed treaty. They also agreed to divide up the burden of crafting specific clauses for these and other sections, which were then circulated within the committee. The Pan American Committee completed a final draft of the convention during their second meeting in early October 1939.[110]

Even as plans for a Pan American convention on nature protection were moving forward, the Galápagos situation remained politically charged and divisive. Huxley and Moore continued to press the American Committee for stronger action in the archipelago. The committee's unwillingness to support the British Galápagos Committee plan to establish a research station on the islands and the decision to appoint a separate Pan American Committee irritated Moore, who tendered his resignation from the American Committee's Galápagos Committee in the summer of 1939.[111] In a series of exchanges with Julian Huxley, Coolidge argued that criticism in Ecuador should not be stirred up at this juncture because doing so might prejudice the delegate that nation would be sending to the upcoming Inter-American Committee of Experts meeting, thereby jeopardizing the proposed convention.[112] He reassured Huxley that the Galápagos Islands would come up for discussion during those meetings. When Huxley asked if representatives from Canada and governments with colonial possessions in the Western Hemisphere would be represented as advisors at those meetings, Coolidge dodged the question. He did, however, send a copy of the draft convention to Henry Maurice, president of the Society for the Preservation of Wild Fauna of the Empire, to gain his input and support. "It is most important that there should be no feeling among our neighbors to the South, that this is something which the United States is trying to put over on them," Coolidge reiterated. "It may be that your diplomatic representatives could do a little quiet missionary work in the interest of promoting New World cooperation for National Parks and the protection of migratory birds. Your government has taken the lead in dealing with these problems in Africa, and you have plans for the same sort of convention for Asia. I think that we are the natural ones to take such a leadership as far as the New World is concerned."[113]

Another delicate situation involved dealing with T. Gilbert Pearson and the International Committee for Bird Preservation. Phillips had repeatedly complained to Audubon Society President John Baker about Pearson's ineffective administration.[114] He should step down from chairmanship of the organization, Phillips argued, because he lacked the social graces, the diplomatic tact, and the language skills necessary to serve in this important capacity.[115] Baker tended to agree with this assessment and by the middle of 1936 was actively working to terminate Pearson's connection with Audubon, a move that ultimately failed.[116] A year and a half later, when Phillips and Coolidge were searching for a way to shed the continued financial burden the American Committee represented, Baker suggested merging the American Committee and the International Committee as an Audubon department under new personnel. Although Phillips seemed interested in this proposal, the American Committee instead decided to move its offices to the New York Zoological Society early in 1938.[117]

AN UNCONVENTIONAL CONVENTION

After securing appointment as the official U.S. delegate to the Inter-American Committee of Experts, Wetmore conferred regularly with Coolidge. In one exchange, Coolidge suggested that it might be advantageous for Wetmore to criticize various aspects of the convention at the meeting as a strategy to convince other American republics that the agreement was not "one more case of a put up job by interests in the United States."[118] Coolidge and Wetmore also corresponded with Latin American naturalists and conservationists to address concerns and to bring as many of them as possible on board with the agreement. In the end, the extensive preparations paid off: the three-day gathering proceeded as planned in May 1940. In addition to the United States, seventeen Central and South American republics sent delegates, with only Honduras, Guatemala, and El Salvador missing. The final convention, approved unanimously by the Inter-American Committee of Experts, emerged little changed from the draft the American Committee had prepared earlier.[119]

The preamble for the Convention on Nature Protection and Wild Life Preservation in the Western Hemisphere (sometimes referred to as the Pan American Convention) began with essentially the same wording as the Lima Resolution, but the position of the first two paragraphs was transposed.[120] Twelve articles supporting the goal of preventing extinction of all species in the region followed. The first laid out a set of common definitions for the various forms of protected areas mentioned in the treaty—national parks, national reserves, nature monuments, and strict wilderness reserves—and provided a list of birds considered migratory. A particularly interesting item in this section was the idea of declaring not only regions or objects as inviolate "nature monuments" (a traditional use of this term) but also extending the designation to "a species of flora or fauna."

Article 2 called for the contracting governments to set aside protected natural areas as quickly as they could, while Articles 3 and 4 established a set of common rules for the management and perpetuation of those areas. A call for the "protection and preservation" of threatened organisms outside officially designated natural areas, along with similar measures designated "natural scenery, striking geological formations, and regions and natural objects of aesthetic interest or historic or scientific value," appeared in Article 5. The next article urged the contracting governments to support scientists engaged in research and field study in the American republics, while calling for exceptions in protective measures to allow for scientific research. Article 7 covered migratory birds and directed the contracting governments to adopt "appropriate measures for the protection" of those species of "economic or aesthetic value or to prevent the threatened extinction of any given species." Curiously, the language that many American conservationists found objectionable in the Migratory Bird Treaty with Mexico, which sanctioned utilization of migratory birds for "the purpose of sports as well as food, commerce, and industry," also found its way into this hemispheric treaty. An annex to the convention, mentioned in Article 8, provided a list of threatened species "declared to be of special urgency and importance" which were to be "protected as completely as possible." Article 9 set up a system to regulate trade of protected fauna and flora, including the creation of export certificates and prohibition on the importation of species protected by the country of origin. The final three articles established procedures for sharing information about nature protection, ratifying the convention, and withdrawing from it. The convention would go into effect three months after the deposit of not less than five ratifications with the Pan American Union.

The Pan American Convention was born in a wave of optimism. In an article for *Science* announcing that eight nations had signed the treaty on the first day it was opened for endorsement on October 12, 1940, Coolidge called the agreement the "third great cooperative step taken by the United States to further wild-life protection by international treaty on the American Continent": "Science, art and literature have forged strong ties between the American Republics. This convention should add still another tie through its parks, reserves and monuments, which will help to bring about a common interest in the great masterpieces of creation in the Americas."[121] Newton Drury, the director of the National Park Service, lauded the convention as "the spark that may arouse to crusading vigor the preservation of superlative examples of nature throughout the Americas."[122] William G. Sheldon, a member of the committee that had helped author the treaty, also offered praise in *American Wildlife*, highlighting the fact that the clouds of world war made it especially important "to foster and develop greater cultural understanding between the North and South Americas." The "common heritage of great mountain ranges, rich forests and all the various species of flora and fauna found therein" provided a strong bond and a "stabilizing force" that, if nurtured could contribute to "cultural hemispheric unification." Sheldon looked forward to the day when similar treaties would be negotiated "in all parts of the globe."[123] The same sentiment was expressed by Lord Onslow, who wrote to Coolidge expressing his hope that "some day, though I fear it will be a long time ahead, there will be some sort of general agreement all over the World for the protection of Flora and Fauna."[124]

Continuing the active role he had assumed from the beginning, Coolidge lobbied vigorously to gain ratification for the new treaty. On the day it was opened for signature at the Pan American Union building in Washington, the United States and seven other nations formally endorsed the measure. Not long after the signing ceremony, Coolidge wrote to William Phillips urging him to use his Washington connections to push for "immediate ratification of this most important Treaty."[125] When Coolidge learned of the State Department's reluctance to forward the convention to the Senate without the annex of threatened species mentioned in Article 8, Wetmore assured him that the Pan American Committee could quickly generate an appropriate list.[126] The version that went to the Senate included the woodland caribou, sea otter, manatee, trumpeter swan, whooping crane, Eskimo curlew, Hudsonian godwit, Puerto Rican parrot, and the ivory-billed woodpecker. Early in January 1941, Coolidge wrote to the editor of the *New York Times* urging him to publish a story on the convention and indicating that he had been "working for almost three years with several others quietly and indirectly to bring this Convention into existence"[127] He wrote a similar note to the editor of the *Journal of Mammalogy* and one to Fairfield Osborn, urging him to publish an appropriate notice of the convention in the New York Zoological Society's *Bulletin*.[128] Coolidge's own announcement in *Science*, which appeared in November 1940, was also part of the successful ratification campaign.[129] On April 7, 1941, the Senate passed the resolution authorizing ratification without debate. The treaty came into effect just over a year later, after Guatemala, Venezuela, El Salvador, and Haiti joined the United States in ratifying it.

Coolidge also scrambled for the funding and personnel needed to implement the treaty. Shortly after the Inter-American Committee of Experts met in May 1940, he approached the ornithologist and former editor of the National Audubon Society's *Bird-Lore*, William Vogt.[130] For the previous year, Vogt had served as consultant for the Peruvian Guano Administration, where his job was to study the population dynamics of the sea birds that nested on islands off the coast of Peru. As he described it, the goal of the project was to "increase the increment of excrement" from these birds, which was valuable for use as fertilizer. Coolidge thought Vogt would make an excellent point person for promoting the Convention on Nature Protection and Wild Life Preservation throughout Latin America.[131] He authored a multiyear proposal to finance Vogt and circulated it to Laurance Rockefeller, who had joined the American Committee's Executive Committee when the organization moved its headquarters to New York two years earlier. Coolidge hoped that Rockefeller would secure funding for his proposal through the Rockefeller Foundation or some other private agency.[132] While it failed to fund Coolidge's proposal, the foundation did provide a grant to the Committee for Inter-American Artistic and Cultural Relations, which hired T. H. Goodspeed, a professor of botany at the University of California, to lecture on scientific and conservation topics in Colombia, Peru, Chile, Argentina, Uruguay, and Brazil.[133] Vogt, on the other hand, secured a position as associate director of the Division of Science and Education at the Office of the Coordinator of Inter-American Affairs.

Coolidge continued to argue that a "trained biologist" ought to travel throughout Latin America to "gather information and offer advice on all problems dealing with conservation matters." He also maintained that this expert should serve in a newly constituted Conservation Section of the Pan-American Union, the mission of which was to "aid and encourage the conservation movement" within the region.[134] At their annual meeting in December 1942, the American Committee passed a resolution supporting the latter notion, and the next year, the Pan American Union obliged by creating a Conservation Section with Vogt as its head.[135] He spent the next seven years traveling in Central and South America and issuing reports on the relationship between resources and human population for the governments of Mexico, Costa Rica, El Salvador, and Venezuela. Vogt's Latin American studies became the basis for a Spanish-language textbook, *El Hombre y la Tierra* (1944), and a bestselling book, *Road*

to Survival (1948).[136] The latter was a neo-Malthusian diatribe that warned how increasing populations were outstripping the food supply; it was translated into nine languages and read by millions throughout the world.

CONVENTIONAL WISDOM

Numerous factors converged to produce the Convention on Nature Protection and Wildlife Preservation in the Western Hemisphere. The successful negotiations that produced the London Convention of 1933, which provided a useful framework for African conservation, offered a beacon of hope that a similar agreement covering North and South America might also be concluded. At the same time, the Convention for the Protection of Migratory Birds and Game Mammals, which U.S. and Mexican officials negotiated just three years later, suggested that at least some Latin American nations might now be ready to sign on to a broad, multilateral conservation treaty. Roosevelt's decision to implement the Good Neighbor Policy eased tensions between the United States and nations to the south, providing a shot in the arm to the Pan American movement that had begun ever so halting several decades earlier. These critical developments coincided with the outbreak of World War II, which not only limited European influence in the region but also bolstered a sense of hemispheric unity in that face of fighting in Europe and the Pacific. Clearly, the time seemed ripe for the Pan American Convention, but auspicious timing alone can only go so far in explaining the completion of this landmark agreement.

More than anything, the treaty owed its existence to a small group of foresighted naturalists and conservationists affiliated with the American Committee for International Wild Life Protection. Haunted by the specter of extinction, this small nongovernmental organization had formed in 1930 with a primary focus on African wildlife conservation. Within several years, however, it had shifted its sights to Latin America. Although he did not live to see the Pan American Convention concluded, John C. Phillips was the first to appreciate the possibility of a comprehensive hemispheric treaty, having been inspired to suggest the idea after attending the London Convention negotiations. Undoubtedly, his family connections within the Department of State played an important role in gaining federal approval to begin the process of crafting the treaty. No one knew more about the deteriorating situation for wildlife across Latin America than the world's leading authority on Central and South American birds, Alexander Wetmore, who chaired both the U.S. and the Inter-American Committees of Experts charged with finalizing the treaty. And no single individual did more to encourage, frame, and promote the Convention on Nature Protection and Wildlife Protection in the Western Hemisphere than Harold J. Coolidge.

All three were acutely aware of the difficult challenges involved in negotiating and implementing a conservation treaty in the face of ongoing political, economic, and social instability in the region. Through their extensive efforts to protect endangered wildlife in the Galápagos Islands, American Committee members had learned firsthand how changing administrations, foreign entanglements, and differential priorities often hampered efforts to rescue Latin America's rare and vanishing species. Despite this experience, Coolidge pushed for a broad, binding approach to the proposed hemispheric agreement, while Wetmore thought a more limited, flexible approach would prove more prudent. The final version of the treaty, which eleven nations had ratified by 1947, offered a mixture of both viewpoints.[137]

Has the Convention achieved its aim of advancing the cause of nature protection and wildlife preservation in the Western Hemisphere? The handful of scholars who have studied its implementation disagree on the legacy of this landmark agreement. On the one hand, the environmental lawyers Kathleen Rogers and James A. Moore praise the treaty as "ahead of its time in the strength and breadth of its commitments to protecting plants, animals, and their habitats."

But they also conclude that rather than living up to its initial promise, the agreement has languished, remaining "dormant and largely unimplemented since its inception."[138] The historian Kurkpatrick Dorsey seconds this generally negative assessment, while questioning the depth of the signatories' commitment to the values the treaty represented.[139] On the other hand, in a recent, detailed study using both U.S. and Latin American sources, historian Keri Lewis proves more optimistic about the impact of the treaty. She argues that this "comprehensive, flexible, and malleable agreement . . . laid the foundation for effective nature protection across the Americas." Latin American officials, for example, have repeatedly turned to the treaty for support and guidance in establishing protected areas, while the provisions calling for cooperation between scientists, conservationists, nongovernmental organizations, and governmental officials have bolstered nature protection throughout the Americas. The case of Costa Rica, Lewis notes, provides a particularly striking example of the treaty's power to promote conservation; shortly after ratifying the treaty in 1967, the government of that Central American nation initiated a national park system and began shifting its economic base from agriculture to ecotourism.[140]

Whether or not the treaty has achieved its full potential, it remains an important and underappreciated document that would have not been possible without the prodding and expertise of the American Committee. At a time when most conservationists in the United States remained firmly focused on trials and tribulations of North American wildlife, the committee recognized the importance of a more international approach to the problem of extinction. Three decades later, that approach would manifest itself even more forcefully in the Endangered Species Act of 1973, which cited the Convention on Nature Protection and Wildlife Preservation in the Western Hemisphere as a source of authority for the federal government to begin more aggressively regulating the taking of vanishing wildlife.

One thing the convention failed to do, however, was reference the science of ecology. Given the timing of the agreement, this is a rather curious omission. The naturalists and conservationists who drew up the treaty may have been innovative in their embrace of an international approach to wildlife conservation, but they remained largely disconnected from the emerging field of biology that took the study of organisms and their relations to their natural environment as its central subject matter. By the time the Convention on Nature Protection and Wildlife Preservation in the Western Hemisphere gained ratification, that science had already begun to make its mark on domestic conservation policy in the United States.

CHAPTER SEVEN

ENTER ECOLOGY

PRESERVING NATURE'S LIVING LABORATORY

*Biologists, above all others, should be in a position to appreciate the loss to
science which results from the destruction either of single natural species or of natural
associations of species. They are in a unique position to give advice as to what
particular species and associations are of greatest importance to science, and as to
which ones are in most urgent need of protection.*

FRANCIS B. SUMNER, 1921

*In general, from a philosophical and practical viewpoint, the unmodified assemblage
of organisms is more valuable than the isolated rare species.*

VICTOR SHELFORD, 1933

A PLEA FOR PRESERVATION

In March 1920, the American naturalist Francis B. Sumner published a remarkable essay in the *Scientific Monthly*. The goal of his lengthy, impassioned plea
was to convince his biological colleagues to undertake a "serious effort to rescue
a few fragments of vanishing nature." According to Sumner, the natural world
faced a relentless assault at human hands. While several authors had recently
highlighted the need for "prompt and drastic action to save our native fauna,
especially the birds and mammals," Sumner feared that those warnings about impending extinction remained overly narrow. "Forests are vanishing," he declared
with alarm, "brush land [is] being cleared, swamps drained and desert irrigated."
No place seemed entirely safe from the juggernaut of civilization, as even "the
remotest mountain fastnesses and the wildest solitudes of the desert" were being

opened up to exploitation by the mass adoption of the automobile. While Sumner admitted that it was impossible to "arrest the march of progress" entirely, he proposed a sweeping expansion of the "conservation of our fauna and flora and natural scenery to an extent hitherto not contemplated by our people or our government." More specifically, he called for conservationists (and conservation policy) to move beyond a preoccupation with individual species and begin paying more attention to biological communities and habitat: "Large tracts of land, representing every type of physiography and of plant association, ought to be set aside as permanent preserves, and properly protected against fire, and against every type of depredation. Here would be included desert and chaparral, swamp land and seashore, mountain and prairie."

As practitioners of the "science of life," Sumner continued, biologists should staff the frontlines of an aggressive campaign to rescue at least some patches of nature from oblivion. But instead of promoting preservation, he found too many of his colleagues unmoved by the ecological devastation occurring around them: "One might conclude . . . that there are zoologists, and some of them occupying high positions, who would not be greatly disturbed if the entire natural fauna and flora of the earth, with a few specified exceptions, should be destroyed overnight." As long as they managed to obtain the limited repertoire of animals needed for their research—the fruit fly, sea urchin, mouse, and a few other organisms that had become standard experimental models—laboratory-oriented biologists seemed largely indifferent to "the worldwide assault upon living nature."[1] In a follow-up article the next year entitled "The Responsibility of Biologists in the Matter of Preserving Natural Conditions," Sumner again urged his colleagues to step forward to protect both endangered species and endangered biological communities.[2]

Sumner offered two general arguments in support of his plea for setting aside generous tracts of land. On the one hand was the aesthetic argument, a line of reasoning that romantic-oriented preservationists had long leaned on to make their case. By aesthetics, he meant not just "the appreciation of natural scenery" as traditionally construed, but also the "deep-rooted feeling of revolt . . . against the noise and distraction, the artificiality and sordidness, the contracted horizon and stifled individuality, which are dominating features of life in a great city." For Sumner, the aesthetic appeal of nature was similar to but also "vastly more compelling, than that of either music or poetry." Beyond a sentimental appeal to aesthetics were more hard-nosed scientific considerations. If biologists, especially field-oriented biologists, hoped to continue their research into the future, they needed to secure "from destruction the greatest possible number of living species of animals and plants" and to do so as "far as possible

FIGURE 39. Francis Sumner in the field. Trained as an ecologist, Sumner sought to combine the rigor of laboratory methods with extensive work in the field. He also sought to engage his colleagues in a campaign to broaden the focus of conservation from declining individual species to endangered landscapes. Courtesy of the Scripps Institution of Oceanography Archives, University of California, San Diego Libraries.

in their natural habitats and in their natural relations to one another." Here Sumner was not just talking about preserving research opportunities for old-school naturalists with their "traditional passion for naming species," but especially those biologists whose scientific interests centered on discerning larger relationships in nature. Chief among these were modern ecologists, concerned with what Sumner characterized as the "totality of animal and plant life in particular regions."[3]

Though perhaps more outspoken than his contemporaries, Sumner represented a new breed of Ph.D.-trained biologist who came of age in late nineteenth- and early twentieth-century America.[4] Following his birth in Connecticut in 1874, he spent the first ten years of his life on a small, isolated farm on the outskirts of Oakland, California. There he received formal instruction from his father, a schoolteacher and author of geography textbooks who encouraged his young son's interest in raising reptiles and collecting shells, birds' eggs, and insects. At the age of sixteen, he entered the University of Minnesota, where he earned his B.S. degree in 1894. Following graduation, Sumner spent the next few years on a trip to South America (undertaken in an effort to regain his frail health), an expedition to the Egyptian Sudan, and graduate work at Columbia University. His mentors included the embryologist Edmund B. Wilson, the paleontologist and museum administrator Henry Fairfield Osborn, and the ichthyologist Bashford Dean.

Sumner's 1901 doctoral thesis on the embryological development of fish typi-
fied the research completed in the graduate biology programs launched in America at the close of the nineteenth century. As historian of science Robert Kohler
has ably documented, by the early twentieth century, a reaction against this kind
of laboratory-based morphology, embryology, and cytology had begun to set
in among some American naturalists. Biological reformers sought to carve out
intellectual and institutional space for a "new natural history," an approach that
blended the rigor of laboratory-based studies that had come to predominate in
academic institutions with the older, field-based natural history.[5]

Sumner's subsequent research career exemplified this trend. Following
lengthy stints teaching at the College of the City of New York and as laboratory
director at the U.S. Fish Commission in Woods Hole, Massachusetts, he accepted
a position at the Scripps Institution for Biological Research (later renamed the
Scripps Institution of Oceanographic Research) in 1913, where he was to remain
for the rest of his scientific career. There Sumner completed a long-term study
of the deer mouse (*Peromyscus*) that secured his scientific fame. His murine research combined extensive fieldwork at multiple sites with the breeding of various subspecies (geographic races) under controlled laboratory conditions. Witnessing the impact of development on the once remote western locations where
he captured his specimens undoubtedly contributed to Sumner's growing sense
of alarm about human-induced transformations of the natural world. Extensive
contact with Joseph Grinnell, the outspoken Berkeley naturalist who regularly
decried the turn-of-the-century ecological devastation occurring in California,
further fueled his emerging environmental consciousness. No ivory-tower biologist, Sumner believed that scientists had a solemn duty to apply their knowledge
in the social and political sphere.[6]

He discovered several kindred spirits who shared his concerns, especially
among those naturalists who identified with the emerging discipline of ecology.
Although first coined in the 1860s, the term "ecology" failed to catch on until
later in the nineteenth century, when American botanists adopted it to describe
the broad study of the relationship between living organisms and their environment. By 1915, there was sufficient interest to found a separate institution, the
Ecological Society of America (ESA), devoted to promoting the nascent field.
One of the central players in this new institution was Victor Shelford, a Chicago-
trained Ph.D. whose research and teaching helped put animal ecology on the
map. Shortly after helping found the ESA, Shelford established the Committee
for the Preservation of Natural Conditions within the fledgling organization.
With Shelford's leadership and constant prodding, over the next three decades
this committee compiled an extensive inventory of natural sites needing protec-

tion, it lobbied for state and federal protection of those sites, and it developed a new rationale for the preservation of nature. Beyond the aesthetic, economic, and recreational concerns that had long been central to conservation discourse, Shelford and his colleagues argued that individual species were crucial components of biological communities. Only through sustained ecological research on these communities could scientists hope to develop models of how healthy landscapes functioned in nature. Indeed, it was in his capacity as a cochair of Shelford's committee that Sumner published his 1921 article challenging his biological colleagues to become more active in the preservation of nature.

Today the term "ecology" has come to mean both the scientific study of the interrelationship between organisms and their environment and (more popularly) the broader social and political movement aimed at shielding the natural world from the full brunt of civilization. This chapter shows how Sumner, Grinnell, Shelford, and a handful of their ecologically oriented colleagues pushed to expand ecology beyond its original narrow moorings in academic biology. During the period between the First and Second World Wars, ecologists provided both a new set of lenses for viewing the natural world and a new set of arguments for nature preservation. At the same time, they highlighted the role of habitat protection in saving endangered species. One naturalist, Aldo Leopold, went further by trying to incorporate ecological ideas into a radical new ethic guiding human relations with the natural world. While ecologists achieved only limited success within conservation circles during the interwar years, ecological approaches and arguments would subsequently become central to the postwar environmental movement.

THE NEW NATURAL HISTORY

The science of ecology emerged during a period of reform fervor in American biology. Although natural history had always encompassed a diverse variety of practices, methods, and approaches, during the late eighteenth and early nineteenth century, taxonomy—the naming, description, and classification of organisms—began to predominate within the scientific study of flora and fauna.[7] By the 1880s and 1890s, however, a new laboratory-based embryology, cytology, and physiology eclipsed the descriptive, field- and museum-oriented, and taxonomically focused natural history.[8] The laboratory revolution in American biology began at about the same time that universities established the first doctoral programs in the United States, beginning with Johns Hopkins University in 1876. Where naturalists had once learned their craft through self-study, amassing collections, fieldwork, informal apprenticeships, and (less often) formal study at the undergraduate level, over the ensuing decades graduate study in an academic laboratory increasingly became the entry point for serious research (much less careers) in the biological sciences. By 1910, the Ph.D. had become a standard credential for individuals hoping to fill a professional position not only in biology, but most other scientific disciplines as well.[9]

Even as experimental and laboratory-based biology gained in prestige and visibility in American institutions of higher learning, however, a backlash began to develop. By the early twentieth century, reformers began calling for a "new" or "scientific" natural history, a revitalization of that older tradition that sought to combine the rigor, control, and replicability of laboratory work with the excitement, immediacy, and close contact with nature that field research offered.[10] One of Shelford's classmates, the ecologist Charles C. Adams, published one manifesto of this reform movement in 1917. There he argued that a new generation of more diversely trained scientists filled the breach that had opened up between laboratory and field studies. "The new natural history," he continued, "is working on a higher level, with a broader outlook, and has a saner and closer contact with nature than was possible by either the laboratory or the older field method alone. It takes the laboratory into the field and brings the field problems into the laboratory as never before." Not surprisingly, for Adams the prime exemplar of the new natural history was his own chosen field of ecology.[11]

The term "ecology" had first been coined in 1866 by Ernst Haeckel, the inveterate neologist and leading German disciple of Darwin.[12] Finding inspiration in the *Origin of Species*, Haeckel defined ecology as "the science of the relations of living organisms to the external world."[13] Darwin's big book was brimming with examples of how living organisms interacted with one another in what he described as the "complex web of life." In one passage that bore a resemblance to the old nursery rhyme "The House That Jack Built," Darwin even discussed what would later be dubbed a food chain when he showed how the amount of clover growing in the area around his home was indirectly related to the number of cats (since cats devoured mice, which in turn consumed humble-bees, which fertilized clover).[14] Though a number of naturalists pursued research that would later be characterized as "ecological," the term itself failed to catch on until the until the early 1890s, when a group of American botanists interested in studying adaptation, geographical distribution, and physiology in plants picked it up.[15]

Once planted, the seed of ecology grew and blossomed in the fertile soil of modern American universities. One prominent ecological research school flourished at the University of Nebraska, under the supervision of the botanist Charles E. Bessey.[16] His most renowned student was Frederic E. Clements, a prickly figure whose methodological and theoretical innovations dominated

American ecology from the 1910s until World War II.[17] In 1897, Clements and one of his classmates, Roscoe Pound, developed the idea of the quadrat—a one- to five-square-meter patch of ground divided by evenly spaced grid lines. They hoped that meticulously conducted inventories of the plants growing in these carefully chosen plots would bring more precision and rigor into ecological re- search than the more casual sense impressions typically used to survey vegeta- tion at the time. Strongly influenced by German phytogeographers, Clements also developed a dynamic approach to plant communities. He argued that those communities—which he viewed as a kind of superorganism—were subject to regular, predictable patterns of invasion from various species in a process known as succession. The final stage of this process he called the climax community, supposedly the most stable, diverse, and hence the most natural plant formation for a given set of environmental conditions.

By the early twentieth century, the leading training ground for American ecologists was the University of Chicago, which boasted one of the most in- novative and productive biology faculties of any U.S. university at the time.[18] There the quiet but inspiring Henry Cowles trained a generation of students about plant succession through regular field trips to the sand dunes that bordered the shores of Lake Michigan.[19] The site provided a vivid demonstration of the process of succession, where "a pattern of ecological development *in space* paral- leled the development of vegetation in time."[20] While repeatedly examining the varied flora that inhabited dunes of differing age, Cowles and his students also gained firsthand knowledge of how humans could radically transform a once re- mote, wild, and scientifically valuable location. In the early part of the twentieth century, the construction of a massive U.S. Steel plant in the area around Gary, Indiana, provided a striking example of how development could despoil a once beautiful, productive research site, in this case by transforming it into a sprawl- ing industrial complex surrounded by a bustling new town.[21]

Initially, most ecological research centered on plants and plant communities, which—being relatively limited in number, visible, and static—were easier to study than more numerous, less conspicuous, and more mobile animals. In the 1910s and 1920s, however, several ecologists began laying the groundwork for the study of animal ecology. Prominent among those were two of Cowles's stu- dents at Chicago: Charles C. Adams and Victor Shelford, who each published pioneering books on the subject in 1913. Adams's *Guide to the Study of Animal Ecology* was a compact bibliographic and methodological textbook that empha- sized survey work.[22] Shelford's *Animal Communities in Temperate North America* was a more ambitious and ultimately more influential book that was based on ten solid years of fieldwork.[23] Among those inspired by Shelford's study was the British naturalist Charles Elton, who managed to synthesize animal ecology into a coherent subdiscipline.[24] Following research on fur-bearing animals in the Arctic, where the food relations between organisms proved much easier to unravel than in the more biologically rich, diverse temperate zones and tropics, Elton published his widely read *Animal Ecology* (1927). Declaring ecology to be a form of "scientific natural history," Elton first raised to prominence several key ecological concepts, including food chains, food webs, and the pyramid of numbers (i.e., the idea that the number of individual animals and their total weight decreased at each level of the food chain).

Ecology was (and has remained) a notoriously diverse field, with practitioners who pursue a bewildering variety of approaches and methods. Beyond the division between plant and animal ecology, there were also fundamental differences between ecologists who tended to focus either on terrestrial or aquatic species, saltwater or freshwater environments, individual species or entire communities, and basic or applied research. An apparently exasperated Cowles admitted in 1903 that "the field of ecology is chaos. Ecologists are not agreed even as to fundamental principles or motives; indeed, no one at this time, least of all the present speaker, is prepared to define or delimit ecology."[25] While Clements offered a theoretical framework that many plant ecologists found useful, it failed to provide much role for animals in the process of succession and its applicability to aquatic environments was equally questionable.

What ecologists lacked in uniformity of approach to their subject, however, they made up for in enthusiasm and numbers. By 1914, enough American scientists identified with some aspect of ecological research to contemplate establishing a national society.[26] Undoubtedly, the founding of a British Ecological Society a year previously helped plant the idea in the mind of Robert H. Wolcott, a professor of zoology at the University of Nebraska. In March 1914, Wolcott wrote to Shelford suggesting the formation of an ecological society that would primarily be devoted to guided field excursions, like the kind that Cowles regularly led with his students around Chicago. That exchange led Cowles to call for a meeting during the annual gathering of the American Association for the Advancement of Science (AAAS) later that year, where attendees agreed to create a small committee to further explore the idea. At the next meeting of the AAAS, in 1915, about fifty scientists voted to found the Ecological Society of America and elected Shelford as its first president.

During the next three decades, both the society and the larger discipline experienced impressive growth. By 1917, membership in the fledgling organization had reached 307 scientists, with nearly a third of those identifying plant ecology as their field of major interest, a nearly equal portion animal ecology, and about

15 percent forestry or entomology.[27] Not surprisingly given the position of the Chicago school of ecology, thirty-two of the members at the time were from Illinois; Washington, D.C., with its many federal bureaus, was second, with thirty representatives. By 1930, the ESA had grown to 653 members.[28] As another indicator of the scale of ecological activity during this period, the word "ecology" appeared in the titles of at least 27 dissertations and 133 American books during the 1920s; by the 1930s, those numbers had grown to 33 dissertations and 201 books.[29] Clearly, ecology had gained a firm foothold within American biology.

INSTITUTIONALIZING ECOLOGICAL ACTIVISM

In 1917, two years after the founding of the ESA, Shelford wrote to the new president of the organization, Ellsworth Huntington, suggesting the formation of a committee for the "Preservation of Natural Conditions for Ecological Study."[30] What prompted him to make this recommendation remains unclear, but the establishment of the National Park Service in 1916, which placed national parks and monuments under a single, unified management structure, may have influenced the timing of his proposal. Increasing mobilization for World War I, which led to calls to open up those parks for resource exploitation, may have also played a role.[31] Whatever the source of the idea, Huntington responded enthusiastically and suggested that the new committee not only grapple with relevant legislation aimed at preserving natural conditions but also create an inventory of "typical areas which ought to be preserved in various parts of the country." Shelford jumped at the opportunity, and for the next several decades, he worked tirelessly on behalf of this committee and its successors.

At the time he wrote Huntington, Shelford was one of America's leading ecologists.[32] Raised in rural western New York, he enrolled in West Virginia University as a premedical student at the age of twenty-two. There, his uncle, William Rumsey, a Cornell-trained entomologist affiliated with the Agricultural Experiment Station at Morgantown, took him on long hikes and exposed him to serious botanizing and insect collecting for the first time. After several semesters, Shelford transferred to the University of Chicago, where he received his B.S. in 1903 and his Ph.D. four years later. At Chicago, he fell under the spell of several gifted teachers, including the geneticist Charles Davenport, who nudged him in the direction of animal ecology; the developmental biologist Charles M. Child, who inspired him with long rambles through the countryside and taught him experimental techniques that would become central to his research; and especially the ecologist Henry C. Cowles, who introduced him to physiological ecology and enchanted him with regular field trips to the dunes of Lake Michigan.

FIGURE 40. Victor Shelford in the field, Reelfoot Lake, Tennessee, 1937. Photograph by Eugene Odum. Shelford not only made important contributions to animal ecology but was also the primary force behind the Ecological Society of America's engagement with conservation issues for more than three decades. Courtesy of the University of Illinois Archives.

After graduation, Shelford remained an instructor at Chicago for seven years before accepting a faculty position at the University of Illinois, from which he was forced to step down in 1946 due to a compulsory retirement policy. Many of Shelford's colleagues considered him to be the "father of animal ecology in America," an accolade gained through his wide-ranging research and the more than two dozen doctoral students who worked under his supervision.[33] Despite a heavy schedule of teaching, writing, and fieldwork at Illinois, Shelford managed to devote countless hours to conservation.

Articulating a scientific rationale for nature preservation provided one initial focus for Shelford's committee, which numbered seventeen ecologists by the end of its first year. In 1913, Charles C. Adams had stressed that ecologists needed continued access to relatively undisturbed natural areas so they could gain an understanding of the "normal processes of nature": "We need what might be called a bionomic baseline, an idea of conditions which existed before man came upon the scene, the conditions which would again supervene if the human inhabitants were withdrawn." He warned that as civilization increasingly encroached upon "habitats and associations," "the chances of preserving adequate records before their complete extinction" was becoming ever more remote.[34] Still, at this point, Adams failed to issue a formal call for political action to save the sites that remained. Similarly, in *Animal Communities in North America,* published that same year, Shelford himself stressed the need for a more complete knowledge of nature to inform policies about how to manage it properly: "We must know nature, not a part but the whole, if we wish to treat the simplest everyday problem of our relations to animals intelligently and justly.[35]

Willard G. Van Name, a curator of invertebrate zoology at the American Museum of Natural History and well-known gadfly in conservation circles, took the next step in a blistering editorial published in *Science* in 1919.[36] There he chided his fellow naturalists for failing to seriously engage with the issue of nature preservation, especially when rapid postwar economic expansion and improved transportation networks posed a grave threat to nearly "every part of the globe." Zoologists, botanists, and ecologists should play a more visible role in the struggle for the "preservation from extinction or destruction of hundreds of interesting species of animals and plants and the many places of unusual scientific interest that are being sacrificed for the selfish interests of a few." Biologists needed to become more politically active not only to ensure continued access to the specimens and sites they needed for their research, but also for moral and aesthetic reasons. Using language intended to resonate with an American audience still fired with postwar patriotism, Van Name compared the careless destruction of nature to the wartime destruction of culture: "We may be shocked and indignant at the vandalism of the Huns of ancient and modern times in respect to art and the results of human industry, but we ourselves act no better toward natural objects of unique interest, value and beauty, and more intelligent generations in the future who will find themselves deprived of much that it was our duty to preserve for them will not doubt regard us with the same kind of feeling as we look upon the despoilers of Belgium, France, and Serbia." Francis Sumner's declarations that naturalist had a duty to engage in nature preservation appeared soon after Van Name's critical editorial.

These publications, and the arguments they articulated, were all available to Shelford's Committee for the Preservation of Natural Conditions, which had grown to seventy members by the time they issued their first separately published report in 1921. Funded with a small grant from the National Research Council, *Preservation of Natural Conditions* was a heavily illustrated but only marginally effective pamphlet. Rather than offering a coherent narrative, the text presented a hodgepodge of quotations and previously published material only loosely strung together. The section offering "Some Reasons for Preserving Natural Areas," for example, included a variety of statements articulating the need to act quickly to save natural communities and habitat for aesthetic, recreational, cultural, economic, and scientific reasons. Here, among many similar declarations, was one of Francis Sumner's earlier calls for conservation: "The main thing is to recognize (as many biologists do at present) the need of concerted action. Without this, we shall certainly lose the greater part of material upon which our sciences of ecology, geographical distribution, taxonomy, etc. are based."[37] The pamphlet also warned that when it came to protecting undisturbed land, "Early action is always important, for the destruction of natural areas is continuous and progressive."[38] The site of Gary, Indiana, offered a compelling example. Fifteen years earlier, the biologically rich area had reverted to the state because of delinquent taxes. But rather than preserving it for recreation and scientific study, state officials had allowed much of the land to fall into private hands. Now "tall stacks and a cloud of smoke" marred the very landscape where so many Chicago students had first been introduced to the process of ecological succession.[39] To drive home the point, the committee's pamphlet included numerous before and after pictures depicting the young, bustling town of Gary.

Before the preservation committee could effectively act to save potentially endangered research areas, however, it needed a better sense of where the most important and imperiled sites were located. Creating an inventory of these areas proved one of the most time-consuming tasks the committee accomplished during its initial decade. As one of his first acts as chair of the preservation committee, Shelford had drawn up a natural area description card, which included space for identifying climax vegetation, animal populations, topography, and the potential objections to preserving a particular location. Over the next several years, he widely circulated the cards among his colleagues, urging them to complete them and to "DO IT NOW."[40] By the end of 1921, Shelford reported that he and his assistants had compiled most of their report, which was likely to contain descriptions of about a thousand areas that were either already preserved or worth saving.[41] The final report, which did not appear until 1926, represented the labor not only of its general editor, Shelford, but also seven associate and special editors, a "publication editor" (the ecologist Forrest Shreve), and more than seventy-five individual contributors.

Emblazoned with an American bison on the cover and touted as the "crowning achievement of the Committee on the Preservation of Natural Conditions," *Naturalist's Guide to the Americas* represented an impressive accomplishment.[42] In the book's preface, Shelford offered a brief history of the project while admitting that its coverage of invertebrates and lower vertebrates was unduly lacking. The first section contained nineteen short essays exploring the uses, values, and policy issues associated with natural areas. In one of these, Shelford argued for the importance of ecology and the study of natural habitats more generally to various branches of biology. He predicted that soon not just ecologists but all biologists—both pure and applied—would "require preserves of natural conditions in connection with their various scientific interests."[43] The largest section of the book consisted of chapters on each state, Canadian province, Latin American country north of Brazil, and large West Indian island. The individual chapters included a brief account of the physical conditions, climate, and biota of each area under discussion. The authors placed particular emphasis on undisturbed places they considered "natural" or "virgin." To aid naturalists desiring to visit these sites, some chapters included practical advice on how to reach them and the available accommodations. A handy list of organizations interested in the preservation of natural conditions and two indices rounded out the ambitious publication.

SAVING NATURE IN NATIONAL PARKS

With this monumental project behind them, Shelford and his preservation committee mobilized to protect the areas with natural conditions ecologists needed to conduct their research. National parks and forests, which were already under federal management, seemed particularly strategic sites to fulfill this aspect of the committee's mission. A key challenge, though, was the fact that neither the National Park Service nor the Forest Service considered the preservation of relatively undisturbed nature as a primary goal. Despite a legislative mandate to manage national parks in a way that left them "unimpaired for the enjoyment of future generations," the National Park Service seemed more interested in protecting scenic vistas and maximizing recreation opportunities for visitors than in perpetuating intact biological communities.[44] Similarly, the Forest Service promoted a multiple-use management policy for the land under its jurisdiction, a policy that promoted timber extraction, grazing, and recreation.[45] Consistent with their overall goals, both agencies routinely effected drastic manipulations of the landscapes under their charge—including building roads, suppressing fires, destroying predators, managing game, and introducing nonnative species— actions that threatened to disrupt or decimate the natural communities that once inhabited those sites.

The ESA preservation committee pursued multiple strategies to safeguard examples of federally owned land from development. First, it pushed to have appropriate new sites incorporated into the national park system. Second, it lobbied to have minimally disturbed research areas carved out of existing national parks and forests, areas where human intervention and public access would both be extremely limited. Finally, the committee sought to minimize intrusions in existing national parks and forests. During its nearly three decades of existence, the committee enjoyed limited success in all these areas.

The campaign to preserve Alaska's Glacier Bay as a national park—an area that was both scenically compelling and ecologically significant—represented the committee's first major entry into the political arena.[46] Dominated by a ramified fjord some sixty miles long, the site first became widely known following John Muir's visit in 1879. By the 1880s and 1890s, steamships regularly brought awe-struck tourists to gaze at the giant ice sheets that regularly calved into the bay from the three-hundred-foot-tall Muir's Glacier, the largest and most imposing of nine tidewater glaciers that bordered the region. As those glaciers slowly receded, the land behind them was laid bare, allowing plants to colonize. Thus, in a more compressed time scale than the Indiana Dunes, the site provided a vivid demonstration of the process of ecological succession.

In 1916, the ecologist William S. Cooper made his first trip to Glacier Bay.[47] An avid mountain climber, Cooper vacillated between an interest in the formal science of ecology and a more popular style of nature writing reminiscent of Muir. After dabbling in graduate study at Johns Hopkins, he transferred to the University of Chicago, where he imbibed in the gospel of dynamic ecology under Henry Cowles and accompanied his mentor on regular pilgrimages to the Indiana Dunes. Following graduation from Chicago in 1911, Cooper began searching for his own ecological promised land, a research site "where vegetational change and development were proceeding so rapidly that they could be studied with fair completeness in the span of the lifetime."[48] He found exactly what he was looking for after reading Muir's description of Glacier Bay in *Travels in Alaska* (1912). During the first of several trips to the region, Cooper installed a series of permanent one-meter quadrats to track how the area's vegetation changed over time.[49]

In 1922, when Cooper approached the ESA to explore the possibility of gaining formal protection for the area that had become central to his research, he found himself appointed chair of a subcommittee charged with exploring the idea.[50] Another member of that subcommittee, Robert Griggs, who had recently participated in a drive to have Katmai National Monument established in Alaska, suggested that a national monument designation might prove much easier to obtain than one for a national park. The former could be accomplished simply through presidential proclamation, while the latter required an act of Congress. The next year, the ESA formally approved the report Cooper's subcommittee presented to the membership. Included in that document was a resolution calling

FIGURE 41. William S. Cooper at Taku Inlet, Alaska, 1916. After beginning research in Glacier Bay, Alaska, a region that provided a vivid demonstration of the process of ecological succession, Cooper led a successful campaign to have his study site declared a National Monument. Courtesy of the University of Minnesota Archives, University of Minnesota–Twin Cities.

for the "preservation" of Glacier Bay as a "National Monument for permanent scientific research and education, and for the use and enjoyment of people."[51] The committee then mounted a massive letter-writing campaign in support of the proposal. Beyond encouraging the other twenty-seven organizations represented on the Council on National Parks, Forests and Wildlife to champion the idea, the committee drummed up support using the list of local and national conservation organizations that had been compiled for the *Naturalist's Guide*. Impressed with this outpouring of support, in 1925, President Calvin Coolidge declared a 1,820-square-mile area as Glacier Bay National Monument. The proclamation accompanying Coolidge's executive order specifically noted the ESA's enthusiastic encouragement of the project and indicated that the new monument presented a "unique opportunity for the scientific study of glacial behavior and the resulting movements and development of flora and fauna."[52] In a letter to the editor of *Science* published nearly two decades later, Shelford singled out the Glacier Bay campaign as one of the most enduring achievements of the preservation committee.[53]

Shelford and the ESA preservation committee also participated, though less centrally, in the drive to establish Everglades National Park.[54] The Everglades was a massive, biologically rich south Florida wetland famous for its subtropical flora and profusion of birdlife. The lifeblood of this sawgrass-dominated landscape—once aptly described as a "river of grass"—was the periodic sheet flow of water from Lake Okeechobee to Florida Bay in a band that averaged about forty miles wide. During the first several decades of the twentieth century, a series of roads, drainage works, and flood control projects disrupted the supply of water, gravely threatening the unique site. In 1928, a group of concerned Florida residents organized the Tropic Everglades National Park Association to promote preservation of the area as a national park. They met with stiff opposition not only from individuals who wanted to develop the land in question but also from unexpected quarters: conservationists who felt the site represented an unworthy candidate for that designation because it lacked the monumental grandeur of existing national parks. Typical of this latter viewpoint was William Temple Hornaday, who dismissed the Everglades as a mere "swamp" that was "a long ways from being fit to elevate into a national park, to put alongside the magnificent array of scenic wonderlands that the American people have elevated into that glorious class."[55]

As early as 1930, the ESA weighed in on the controversy when it passed a resolution calling for the "preservation" of the Everglades in its "essentially natural condition." The society endorsed the idea of a national park as what it termed a "nature preserve" or "museum of nature."[56] After a congressionally authorized committee and the National Parks Association each investigated the area and endorsed the project, Congress finally approved the proposal in 1934. The enabling legislation authorized the creation of Everglades National Park once title to sufficient acreage had been obtained by the state of Florida. It also declared that the area "shall be permanently preserved as a wilderness" and that no park development intended to facilitate visitor access should "interfere with the preservation intact of the unique flora and fauna and the essential primitive conditions." In the words of the landscape architect Ernest Coe, the tireless leader of the campaign to save the Everglades, Congress's action represented a turning point in conservation history. For the first time a concern about preserving "natural ecological relations" loomed large in the creation of a national park.[57]

CREATING NATURE SANCTUARIES

While pushing for the establishment of new national parks, the ESA and its preservation committee also sought to have areas within existing national parks and national forests set aside as more-or-less inviolate nature sanctuaries. As early as 1921, the ESA passed a resolution condemning the introduction of nonnative plants and fish into national parks, both of which had become standard practices in a federal agency dominated by landscape architects rather than biologists. The rationale for this resolution centered on the claim that the National Park Service had a duty to "pass on to future generations for scientific study and education natural areas on which the native flora and fauna may be found undisturbed by outside agencies."[58] Four years later, the ecologist and longtime preservation committee member Barrington Moore amplified this general argument. In a brief chapter for a Boone and Crockett Club–sponsored publication on hunting and conservation, Moore lamented the fact that political support for national parks relied too heavily on the recreational benefits these areas provided. Rather than catering simply to tourists, however, national parks should be maintained as a kind of "nature's laboratory." The knowledge gained through the study of areas where the "balance of nature" remained intact was not only important for furthering the science of biology, but also for providing knowledge needed to maximize agricultural production.[59] That same year, the ESA preservation committee called for the "preservation of definite 'wilderness areas' in our National Parks and Forests" and created a separate subcommittee to study the idea.[60]

The committee's interest in maintaining natural conditions on federal lands received a considerable boost in 1929, when the biologist George M. Wright offered both to fund and help conduct an in-depth study of wildlife in national parks.[61] A former student of Joseph Grinnell, Wright had not only conducted extensive fieldwork in Yellowstone, but also long promoted a more ecologically sensitive management policy there and in other national parks. Among Grinnell's many other students and associates who pursued careers in the National Park Service were two biologists who joined Wright on the survey project: Ben Thompson and Joseph Dixon.[62] The team's first publication, *Fauna of the National Parks of the United States* (1933), was a landmark study that called for the park service not only to maintain existing natural conditions within national parks, but also to restore degraded park fauna to a "pristine state." While recognizing that nature was in a constant state of flux, Wright, Thompson, and Dixon nonetheless identified the "period between the arrival of the first whites and the entrenchment of civilization" as an appropriate baseline for "representing the original or primitive condition that it is desired to maintain."[63] Impressed with the research that went into the creation of the *Fauna of the National Parks,* in 1933 the National Park Service created a Wildlife Division, with Wright as the chief and Dixon and Thompson as staff biologists. One year later, the Park Service Director Arno Cammerer declared the report's findings as official policy of his agency, though they would never be fully implemented.[64]

Cammerer's declaration notwithstanding, the ESA preservation committee

FIGURE 42. George M. Wright (*left*), Ben H. Thompson (*middle*), and Joseph S. Dixon (*right*) in the field at Mono Lake, California, 1929. The authors of *Fauna of the National Parks of the United States* (1933), a groundbreaking report that called for managing national parks in a more natural state. All three had connections with the University of California at Berkeley, and all three were hired for the newly created Wildlife Division at the National Park Service. Courtesy of Pamela Wright Lloyd and the George Wright Society.

recognized that absolute preservation of all national parks was incompatible with the park service's primary goal of providing recreation opportunities for park visitors. As a result, the committee advocated establishing "research reserves" within the national parks, areas that would be maintained in as natural a condition as possible and open only to authorized scientists through a permit system. In the spring of 1931, Cammerer issued a research reserve policy to "preserve permanently" selected natural areas in parks "in as nearly as possible unmodified condition free from external influences."[65] By 1942, the park service had established twenty-eight research reserves in ten national parks. These ranged in size from about thirty-two thousand acres in Glacier Bay National Park to about seventy-five acres in Great Smoky Mountains National Park.[66]

The preservation committee also lobbied for the establishment of nature sanctuaries in national forests. As early as 1921, Aldo Leopold had called for the designation of a wilderness area within Gila National Forest.[67] He seemed particularly concerned about the impact of road building and mass adoption of the automobile, which made previously remote and relatively undisturbed areas in the American Southwest accessible to unprecedented numbers of visitors. Initially, Leopold focused more on preserving opportunities for backcountry recreation than on maintaining intact biological communities, though he would soon become an eloquent spokesman for a more ecological approach to conservation. One year later, one of Leopold's colleagues, the forester G. A. Pearson, issued a call for the "Preservation of Natural Areas in the National Forests" in *Ecology*. There Pearson pointed out that the policy of "highest use" that governed the management of national forests "recognizes that although forests as a whole should be devoted primarily to timber production, specific areas may serve the public interest in better ways." One appropriate use for some remote sites would be to preserve them as "so-called natural areas where plant and animal life and natural features in general may remain undisturbed by human activities."[68] In 1928, the ESA passed a resolution declaring that "preservation of natural conditions for scientific, educational[,] aesthetic[,] and economic reasons becomes more urgent and insistent with the encroachment of civilization." More specifically, the society urged the Forest Service to create "representative sample areas within their boundaries, of such areas as will best pass on to future generations unimpaired the native plants and animals, as well as other features of unusual scientific and educational value."[69]

The Forest Service responded to this call by amending its official manual in 1929. Newly adopted regulation L-20 required the chief of the federal agency to establish a "series of areas of National Forest land known as Research Reserves, sufficient in number and extent adequately to illustrate or typify virgin conditions of forest growth in each forest region, to be retained, so far as practicable, in a virgin or unmodified condition for the purposes of science, research and education." Under this new regulation, natural conditions were also to be largely maintained in "Primitive Areas," which were set up primarily for public "education, inspiration, and recreation."[70] The sites within the National Forests formally designated as Research Reserves tended to be relatively small in size. By 1940, the Forest Service had established forty-one such areas, with an aggregate total of only about 47,549 acres out of nearly 175 million acres in the National Forest system as a whole. The wilderness areas carved out of national forest land, on the other hand, tended to be much larger. By 1940, thirty areas of over one hundred thousand acres had been formerly designated as wilderness, a total of nearly 12 million acres.[71]

The plethora of new natural area designations provided at least nominal protection to numerous sites, but it also led to much confusion about what was being protected and how those sites should be properly managed. To bring more clarity to the issue, Shelford organized a conference on nature sanctuaries in conjunction with the 1931 meeting of the ESA. The twelve naturalists, ecologists, and conservationists who attended the conference reached consensus on the importance of establishing "as nature sanctuaries or nature reserves, areas of natural vegetation containing as nearly as possible all the animal species known to have occurred in the areas within historical times." Preserved areas should be surrounded by buffer zones, only slightly modified areas that might be devoted to "experiments, recreation or game culture, etc." These buffer zones should be "left alone without management" except in the case of an emergency, when "control measures" might be undertaken "after most careful consideration and determination as to their practical necessity."[72]

A year later, Shelford published a lengthy statement on the "Preservation of Natural Biotic Communities" to flesh out the consensus reached at the nature sanctuary conference. In the introduction, he argued that while a desire to "preserve at least some of the original vegetation and wild animals" was nearly universal around the world, little had been done to actually achieve that goal. Vegetation other than forests had generally not escaped destruction, while the larger wildlife in most nations faced widespread persecution. National parks tended to be created from relatively undisturbed areas, but even they did not provide absolute protection; rather, a "large amount of modification" had gone on in those areas, both before and after they were supposedly protected.[73]

Shelford and his committee also decried the trend toward "specialization on particular objects or particular organisms" in teaching and research. Outside of "modern ecology and geography," biologists rarely attempted to deal with "the entire life of natural areas." Nature study in schools tended to suffer from the same shortcoming. As a result, the growing sentiment for protecting individual species had not yet translated into a proper concern about the ongoing degradation of entire communities. Yet, Shelford continued, "from a philosophical and practical standpoint, the unmodified assemblage of organisms is commonly more valuable than the isolated rare species." Both threatened individual organisms and extensive biological communities could be safeguarded in sufficiently large nature sanctuaries.[74]

The body of the report discussed the definition, size, and classification of "natural areas containing original plant and animal life." Here Shelford admitted that "just what original nature in any area was like from a biological viewpoint, is not known and never can be known with any great certainty." "'Nature' and 'natural areas' are purely relative terms," Shelford continued, terms that "can have significance only as averages, because the outstanding phenomenon of bi-

otic communities is fluctuations in numbers of constituent organisms or reproductive stages of organisms over a period of one to thirty or more years." Indeed,
a nature sanctuary represented a place where "these fluctuations are allowed free
play."[75] After running through a list of the various classes of nature sanctuary
established in North America, Shelford reported that the ESA preservation committee recommended subdividing larger reserves into a sanctuary proper, a buffer zone of "partial protection," and an area managed for human use in cases
when this was one of the aims of the reserve. These areas should be designed to
provide the "best conditions for roaming animals within the buffer zone and the
sanctuary." Among the animals most needing protection in nature sanctuaries
were "the carnivores," which tended to be unpopular with farmers, ranchers,
and sportsmen.[76] In this and other efforts to protect predators, the preservation
committee found itself embroiled in one of the most divisive issues in wildlife
conservation circles prior to the Second World War.

PROTECTING PREDATORS

Euroamericans had long feared large predators—especially the gray wolf (Canis
lupus)—with an intensity that now seems disproportionate to the actual physical
or economic threat these species posed.[77] As early as the seventeenth century,
colonial governments in America began offering bounties on wolves in the hope
of minimizing livestock losses to creatures they considered both evil and raven-
ous. Even without the economic incentives that local and state bounty programs
provided, however, settlers rarely passed up the opportunity to kill these and
other large predatory species using pits, circle drives, guns, and poisoned meat.
As a result of ongoing persecution, by the 1850s the wolf had become rare in the
East.[78] The settlers who poured into the West following the Civil War carried
with them a keen desire to exterminate not only the wolf, but also its smaller,
more numerous, and secretive western cousin, the coyote (Canis latrans). In addition to farmers and ranchers, sportsmen reviled these and other predators,
which they blamed for the apparent decline in game populations over much of
the United States. Even early wildlife conservationists generally shared a dim
view of the gray wolf. In 1904, for example, William Temple Hornaday condemned the species as the most "despicable" creature in all of North America:
"There is no depth of meanness, treachery, or cruelty to which they do not cheer-
fully descend."[79]

Buckling to mounting pressure from western ranchers, in 1915 Congress appropriated funds for the Bureau of the Biological Survey to begin an all-out war
on the predators, especially the coyote. The Biological Survey was an agency

FIGURE 43. Coyote pelts gathered in Wyoming during November 1921. Beginning in 1915, the U.S. Bureau of the Biological Survey initiated a large-scale campaign to reduce the coyote population in the West. Critics soon charged that the agency was trying to eradicate the species, an accusation that federal officials vehemently denied. Courtesy of the National Archives and Records Administration, Record Group 22-WB, Box 55, Neg. no. B3575M.

within the Department of Agriculture that had first been established in 1885 and subsequently endured a series of reorganizations and name changes.[80] For decades after its founding, the agency was primarily staffed by old-school naturalists, most of whom had learned their science through self-study and apprenticeships rather than formal training. Although most of its early personnel during were interested in studying the taxonomy, geographical distribution, and life history of birds and mammals, particularly game species, from the beginning the agency also regularly conducted applied research in wildlife. Following the passage of the Lacey Act (1900) and the creation of the first federal wildlife refuges, the Biological Survey assumed a growing burden of enforcement and administrative duties as well. The agency's budget and staff remained modest, however, until Congress provided funds to expand its limited experiments in predator control in the early years of World War I. By 1931, three-fourths of the Biological Survey's budget was devoted to its burgeoning predator control program.[81]

Naturalists were the first to decry the federal government's widespread campaign to eradicate predators. Joseph Grinnell became an early critic of the popular program, though he refrained from publicly engaging in the issue to shield his institution from the potential wrath of angry state legislators.[82] Open opposition to the Biological Survey's activities first surfaced at the 1924 annual meeting of American Society of Mammalogists, a professional organization that survey employees had created only a few years earlier. There, the ecologist Charles C. Adams and Grinnell's Berkeley colleague, Lee R. Dice, raised pointed questions about the goals and methods of the government's large-scale predator control program, while Survey mammalogists E. A. Goldman and W. B. Bell strongly defended the work of their agency.[83]

The mammalogists' continuing split on the issue of predator control is revealed in the deliberations of a small committee appointed to explore the issue after the 1924 meeting. No one was terribly surprised four years later when the five-member committee failed to reach consensus and issued two separate reports. On the one hand, Vernon Bailey and Goldman, both longtime staff members at the Biological Survey, argued that predators deserved preservation but only in select national parks and isolated parts of the public domain. On the other hand, Dixon, Adams, and Edmund Heller, director of the Milwaukee Museum, charged that under mounting pressure from western livestock interests, the Biological Survey was conducting "an eradication campaign against western wildlife that could not be defended on scientific or economic grounds." Stung by the ongoing criticism, Biological Survey Chief Paul Redington responded by claiming that his agency was merely "hastening the inevitable," since large predators were doomed wherever civilization held sway. Only specially designated sanctuaries provided any hope of preserving these species. While defending the need for control work, Redington vehemently denied the charge that his agency was intent on "exterminating" wildlife, though the term remained prominent in Biological Survey reports and discussions for several more years. A. Brazier Howell, another of Grinnell's colleagues, responded with a petition signed by 148 scientists that strongly condemned the predator-control program for threatening "the very existence of all carnivorous animals, including those valuable species which constitute the chief check upon injurious rodents and are a vital element of our fauna."[84]

In 1930, when the Biological Survey asked Congress for $1 million per year over the next decade to expand its campaign against predators, the ESA preservation committee finally entered the fray. Its initial foray came in the form of a tepid resolution the committee authored and the ESA membership approved at its annual meeting in late December 1930. While the preamble declared the organization to be "gravely concerned" about the proposed increase in appropriations for predatory mammal control, the resolution itself expressed sympathy for predator and rodent control measures, but only when they had been shown to be scientifically necessary and in the interest of the *"general public."* The ecologists argued that the ten-year plan under consideration in Congress failed to meet either criteria. The second part of the resolution called for dramatically increased funding for scientific research in wildlife management. There should be a "marked increase in the research functions of the Biological Survey, so that all its wild life activities may be conducted on the basis of science and intelligence, and with full regard for the interest of the general public."[85]

Concerned that in the politically charged atmosphere of the predator debate even this mildly worded resolution might create undue friction, at the same meeting, the ESA moved to split its old preservation committee. Henceforth, a newly created Committee for the Study of Plant and Animal Communities would have responsibility for selecting areas for preservation, studying management policies, and promoting scientific investigation. Government-employed naturalists, who were barred by law from lobbying and who wanted to avoid embarrassing the agencies for which they worked, could remain active in this large committee. A much smaller, five-person Committee on the Preservation of Natural Conditions would be responsible for lobbying Congress, state legislatures, and governmental agencies. The division proved largely symbolic since Victor Shelford initially chaired both committees.

As we have already seen, the issue of predators also loomed large in the discussions about nature sanctuaries that the preservation and study committees held over the next several years. The report on the "Preservation of Natural Biotic Communities" that the study committee published in 1933 defined "First Class Nature Sanctuaries" as "any area of original vegetation, containing all the animal species historically known to have occurred in the area (except primitive man), and thought to be present in sufficient numbers to maintain themselves." In establishing protection for such areas, "carnivores" were the animals "requiring first and most careful consideration." In particular, the range of these predators—up to fifty square miles for the wolf, twenty square miles for the coyote, and twenty square miles for the mountain lion—should be taken into account. "These animals are slated for general extermination" and can only be adequately protected in "larger well-buffered parks or remote wilderness areas of the national forests."[86]

In 1935, Shelford authored and the ESA passed another pair of resolutions dealing with the contentious issue of predator control. The first called on the federal government and individual states to fund "thorough studies of the life histories and ecology of the flesh-eaters, to serve as a basis for their scientific management." The second proclaimed "the larger wild animals, especially carnivores" as valuable in the "economy of nature, and therefore of importance in the proper development in the science of ecology, which supplies information to be used in interpreting the past and predicting the future of biological events." The resolution reaffirmed the society's commitment to the establishment of "buffered sanctuaries" that would include all natural flora and fauna. A series of particular recommendations where such sanctuaries might be established in national parks and forests followed.[87]

While the ESA made inroads into its goal of having nature sanctuaries established on federal lands, it failed to effect fundamental change in the Biological Survey's predator control program. As historian Thomas Dunlap has documented, serious reevaluation of that program would not occur until the ecological ideas that Shelford and his committee had promoted began diffusing into the broader public in the post–World War II era. Even then, the issue remained contentious.

THE EDUCATION OF ALDO LEOPOLD

At the same time that Shelford and his colleagues fought for the creation of nature sanctuaries to preserve not only beleaguered species like the gray wolf but also intact biological communities and the opportunity for ecological study, at least one naturalist pushed to forge the science of ecology into a new conservation ethic. Although trained as a forester in the mold of Gifford Pinchot, over the course of his career Aldo Leopold moved beyond his narrow utilitarian perspective to articulate a radical new way of framing the relationship between humans and the natural world. Ecology proved central to Leopold's notion of the "land ethic"—the idea that human interactions with the natural world should aim to maintain the "integrity, stability, and beauty of the biotic community."[88] While today this idea is most closely associated with *A Sand County Almanac* (published posthumously in 1949), Leopold first began fleshing it out in a series of articles and essays that appeared in the 1930s and early 1940s.

Leopold was born in 1887, to an outdoor-loving family of German extraction.[89] From a young age, he developed a passion for hunting, exploration, and wildlife watching in the area surrounding his boyhood home in Burlington, Iowa, located along a major migratory flyway. When it came time to select a vocation, Leopold settled on forestry, one of a few suitable options then available for avid outdoor enthusiasts. After securing a master's degree in forestry from Yale in 1909, he joined the Forest Service and began work in New Mexico and Arizona territories. He cruised timber until 1913, when a misdiagnosed bout with nephritis nearly cost him his life. During his long period of recuperation, Leopold avidly read William Temple Hornaday's recently published diatribe, *Our Vanishing*

FIGURE 44. Aldo Leopold weighing woodcock, 1944. Photo by Robert McCabe. A pioneer in the field of wildlife management and conservation, Leopold also sought to develop an ecologically informed ethic governing humanity's relationship with the natural world. Courtesy of the University of Wisconsin–Madison Archives.

Wildlife. A personal visit from Hornaday two years later further galvanized his interest in endangered wildlife. From that point on, Leopold pursued numerous opportunities for work in recreation management and game protection within the Forest Service. It was during this period, for example, that he suggested an area at the headwaters of the Gila River be managed as a wilderness area or "national hunting ground." In 1924, the same year the Forest Service accepted this recommendation, he left the southwest for Madison, Wisconsin, where he would remain for the rest of his life.

Following a brief stint as an administrator at the U.S. Forest Products Laboratory, Leopold turned to wildlife research. Soon he was he was helping to professionalize the emerging field of game management. After conducting game surveys of several midwestern states, he published *Game Management* (1933), a widely adopted textbook that his biographer Curt Meine has praised as "the most extensive collection on wildlife conservation yet assembled, a masterful synthesis of management theory and techniques."[90] His pathbreaking research led to an appointment as professor of game management at the University of Wisconsin. It was the first such position in the United States, and over the remainder of his career, he influenced a generation of students. He also participated in the formation of the Wildlife Society, an organization of professional wildlife managers, in 1936.[91]

Although *Game Management* was clearly oriented toward utilitarian conservation and it focused on game species rather than wildlife more broadly, Leopold's outlook had clearly begun to be shaped by ecological thinking. As early as 1930, he had begun referring to ecology as the "rock bottom of game management."[92] A year later, he met the British ecologist Charles Elton at the Matamek Conference on cyclic biological phenomena in Quebec, a meeting that Leopold called "the best thing of its kind that I have ever attended."[93] Indeed, it was Elton's research that had inspired the gathering in the first place. The two naturalists immediately hit it off and became lifelong friends.

Part of Leopold's growing interest in the science of ecology stemmed from a notorious failure of utilitarian wildlife conservation: the Kaibab deer crisis.[94] This classic episode, recounted in countless textbooks, involved an aggressive, misguided predator-control campaign undertaken in an effort to increase the supply of mule deer in the Kaibab National Forest, an island-like area on the North Rim of the Grand Canyon surrounded by steep canyons and desert. Beginning in 1906, the Biological Survey supervised the systematic destruction of wolves, coyotes, mountain lions, and bobcats on the one-thousand-square-mile site. Soon the predator-control program suffered from too much success: an explosively growing deer herd that was severely overgrazing the area. By the mid-1920s, disease and starvation struck the burgeoning herd, forcing a precipitous (and, to the Biological Survey, embarrassing) crash in its population. Leopold followed the situation closely, and ultimately, he would associate the Kaibab deer crisis with other human-induced wildlife irruptions that resulted from a fundamental failure to appreciate the interconnectedness of biological communities.

In "The Conservation Ethic," a landmark essay published in 1933, Leopold took the first step toward incorporating the science of ecology into an ethic dealing with "man's relationship to the land and to the non-human animals and plants that grow upon it."[95] Over eons of time, the "human community" had developed strongly held notions of right and wrong, first in the context of relationships between individuals and then in relationships between individuals and society. It was now time to take the next step: to develop a "land-relation" ethic that acknowledged the central role ecology had played (and continued to play) in human history. Among other things, the science of ecology offered the opportunity for a more informed and robust response to the problem of wildlife extinction: "Why do species become extinct?," Leopold asked rhetorically. "Because they first become rare. Why do they become rare? Because of shrinkage in the particular environments which their particular adaptations enable them to inhabit. Can such shrinkage be controlled? Yes, once the specifications are known. How known? Through ecological research. How controlled? By modifying the environment with those same tools and skills already used in agriculture and forestry."[96]

Two years later, Leopold became a founding member of the Wilderness Society, a small but politically active organization devoted to the preservation of large tracts of wild nature.[97] Preservationists who treasured the aesthetic and recreational opportunities that wilderness areas provided initially dominated the group. As environmental historian Paul Sutter has shown, the Wilderness Society's founding members shared a particular concern about the impact of the automobile, which was opening up even the most remote wild areas to tourists, road building, and development. Leopold shared his colleagues' concern about the effects of the automobile, but he also pushed them to consider the ecological dimensions of wilderness preservation. Wilderness provided a baseline for measuring the experiment in civilization, he argued: "The long and short of the matter is that all land-use technologies—agriculture, forestry, watersheds, erosion, game, and range management—are encountering unexpected and baffling obstacles which show clearly that despite the superficial advances in technique, we do not yet understand and cannot yet control the long-time interrelations of animals, plants, and mother earth." Any such tinkering should be ecologically informed and undertaken with "an intelligent humility toward man's place in nature."[98]

In 1939, Leopold accepted an invitation to deliver the plenary address to a joint meeting of the Ecological Society of America and the Society of American Foresters. The topic he chose, "A Biotic View of the Land," reflected his ongoing struggle to develop a new way of perceiving nature that incorporated the latest findings from ecology.[99] According to Leopold, ecological research had demonstrated that previous attempts to label some species as "useful" and to dismiss others as "harmful" were doomed to failure because they failed to recognize the interdependence of organisms within biological communities: "The only sure conclusion is that the biota as a whole is useful, and biota includes not only plants and animals, but soils and waters as well."[100]

Drawing heavily from Elton's *Animal Ecology,* Leopold argued that the "biotic pyramid" offered a particularly useful metaphor for proper thinking about the land. He favored this idea over the older notion of "the balance of nature," which continued to dominate conservation discourse even though ecologists had begun to challenge it as overly static. Soil formed the pyramid's base, a plant layer rested on the soil, an insect layer formed on top of the plants, and so on up through the various groups of fish, reptiles, birds, and mammals. Predators formed its apex. Each successive layer decreased in overall abundance of its population. Energy, which plants captured from the sun, flowed up and down through the pyramid as organisms provided food and other services for those above and below it. Each individual species, humans included, formed a link in numerous individual food chains, so the overall pyramid was "a tangle of chains so complex as to seem disorderly." Closer examination, however, revealed "a highly organized structure" whose "functioning depends on the cooperation of all its diverse links."[101]

Recent human manipulations of nature represented a profound threat to this finely tuned structure. The pyramid of life had once been "low and squat," while the individual food chains that characterized it remained "short and simple." Over eons of time, evolution had slowly added layer after layer, thus elaborating its structure. The end result was a highly complex, highly interdependent edifice that functioned smoothly as an energy circuit. Humans equipped with technology, however, were now causing transformations of "unprecedented violence, rapidity, and scope." Predators were being "lopped of the cap of the pyramid," while "food chains, for the first time in history" were being "made shorter rather than longer." Poor agricultural practices, the proliferation of pollution, and the worldwide redistribution of organisms produced other transformations in the pyramid. While some biotas adjusted to this human-induced violence, others proved more fragile.[102] "Vanishing species, the preservation of which we now regard as an aesthetic luxury," might prove essential to the maintenance of the pyramid as a whole, Leopold warned. To understand the full effects of man-made changes in the land, therefore, it was critical "to preserve samples of original biota as standards against which to measure the effects of violence."[103]

Two years, later Leopold further elaborated on this idea in a short essay entitled "Wilderness as a Land Laboratory."[104] Here he argued that if humans hoped to understand how to more effectively maintain the health of the land and to respond appropriately to obvious signs of ecological ailment—like declining soil fertility and rodent irruptions—it was critical to preserve intact, fully functioning landscapes: "A science of land health needs, first of all, a base-datum of normality, a picture of how healthy land maintains itself as an organism." Wilderness areas—which paleontologists had shown existed for "immensely long periods" with only rare losses to its component species—provided the ideal "land-laboratory." Each "biological province" needed its own wilderness sites "for comparative studies of used and unused land." Although Leopold put a finer point on it, this was essentially the argument that Shelford and other ecologists had been making for two decades.

Leopold's most famous essay, first drafted in 1944 at the urging of one of his students, shows how far his ideas had developed since his early years with the Forest Service.[105] "Thinking Like a Mountain" presents the poignant story of his encounter with a mother wolf and several of her grown pups while cruising timber in the American Southwest. "In those days," Leopold recalled, "we never passed up a chance to kill a wolf," so he and his party quickly emptied their rifles into the pack, fatally wounding the mother and striking one of her pups. His response to witnessing the mother wolf's death highlighted his sense that land-use policy needed to adopt a more ecological perspective: "We reached the old wolf in time to watch a fierce green fire dying in her eyes. I realized then, and have known ever since, that there was something new to me in those eyes—something known only to her and to the mountain. I was young then, and full of trigger-itch; I thought that because few wolves meant more deer, that no wolves would mean hunters' paradise. But after seeing the fierce green fire die, I sensed that neither the wolf nor the mountain agreed with such a view."[106] Though a powerful mea culpa, Leopold's story should not be taken literally; he experienced no one single flash of ecological insight. Rather, the story represents the distillation of his ideas about the ecological importance of intact biological communities, an idea that he had been groping toward for more than a decade.

ECOLOGICAL ACTIVISM AND ITS DISCONTENTS

Although the science of ecology made significant inroads in American conservation discourse and practice during the 1920s and the 1930s, it also experienced a number of disappointing setbacks as World War II loomed on the horizon. The preservation of landscapes for ecological research, for example, had always been fraught with difficulty. Decisions to safeguard a particular area could easily be reversed, and once a site had been radically transformed through logging, road building, or development, the biotic community it contained would inevitably be compromised.

As an example of the tenuousness of protection, in 1936 a windstorm blew down many trees in Andrews Bald, one of several research reserves that had recently been established in Great Smoky Mountains National Park. The park's superintendent and most of his staff wanted the area immediately cleaned up.[107] They argued the downed trees were not only unsightly but also presented an unacceptable fire hazard. Wildlife biologists, on the other hand, thought the fallen trees should remain untouched so the ecological effect of windstorms on grassy balds—open, mountaintop areas of grasses that had long been of scientific interest—could be adequately studied. The Park Service's acting director responded to the controversy by not simply granting permission for the trees to be cleared but also by abolishing the research reserve on the site. By the time World War II began, the entire research reserve program seems to have become moribund, and those areas began to be routinely altered in the same way as other parts of national parks: through such practices as fire suppression, insect control, grazing, and fish stocking.[108]

Wildlife research and conservation programs could also be scaled back or entirely eliminated. By 1935, the National Park Service employed twenty-seven biologists in its newly created Wildlife Division.[109] Central to the division's success was its founder and chief, George Wright, a charismatic leader who managed to gain a hearing for a more ecological approach to park management in a federal agency firmly committed to the proposition that parks existed for the recreational benefit of people. When Wright died in an automobile accident in 1936, the Wildlife Division languished. By 1938, the number of biologists had dwindled to only ten. A year later, as part of an overall reorganization of conservation-related agencies within the Roosevelt administration completed at the urging of Secretary of the Interior Harold Ickes, the remaining biologists in the Wildlife Division were transferred to the Biological Survey, which had been reconstituted as the Fish and Wildlife Service. Although they retained a connection with the National Park Service, their influence diminished. And even though the decision was reversed following World War II, it would be two more decades before scientifically based resource management in the national parks would experience a renaissance.

Obviously, the biologists affiliated with the National Park Service did not remain completely inactive. For example, in the late 1930s and early 1940s, the naturalist Daniel B. Beard, who would later become the first supervisor of Everglades National Park, conceived of and saw through publication a popular volume on North America's endangered wildlife. Completed with the aid of a five-member committee that also included Ben Thompson, *Fading Trails* (1942) began with an introductory chapter documenting what seemed like a limitless supply of American animals before European contact, an "abundance of wildlife never recorded in the history of any other continent."[110] The second chapter, "Civilization's Heavy Heel," revealed how the depredations of "commercial market hunters" and the "change of environment by the hand of man" had resulted in a dramatic reduction of North American wildlife like the bison and the demise of the passenger pigeon.[111] The bulk of the book consisted of vignettes treating about three dozen animals that had "approached the brink of extinction" and now stood "ready to follow a fading trail down into the twilight."[112] Like most other wildlife conservation literature at the time, the volume stressed birds and mammals rather than the full panoply of endangered flora and fauna, and it focused on individual species rather intact biological communities.[113] It did, however, devote attention to what one reviewer called "the ecological problems involved in the perpetuation of certain birds and mammals."[114]

Private preservationist-oriented organizations were no less immune to changes in priority or internal policy shifts than governmental agencies like the National Park Service. Perhaps the most dramatic example was the ESA preservation committee itself, which had long served as the leading voice for ecologically informed nature preservation. At least twice before, in 1928 and 1935, the committee's lobbying activities, controversial conservation stands, and ongoing funding demands had prompted reviews of its activities.[115] On both occasions, ESA officials decided to continue its work, but never with the financial backing Shelford thought necessary to do the job properly. In an effort both to secure that support and to counter growing dissatisfaction with the preservation committee among the ESA's leadership, in 1943 Victor Shelford published an account of the committee's achievements that included yet another plea for adequate funding. "With wartime and post-war pressure to destroy nature mounting," Shelford warned, "it is well for those interested in its preservation of nature for scientific purposes to look over the machinery by which some of it may possibly be saved."[116] He then circulated a reprint of his account with a ballot asking ESA members what they thought of the committee and its activities. Among the approximately 690 members of the ESA, about four hundred replied, with 85 percent of those indicating that the work of the committee was as important as the society's annual meetings and journals. Encouraged by the strong show of support, Shelford then called for an amendment to the society's constitution that would guarantee the funds needed to continue its activities.

Given the favorable response from his recent survey, Shelford and his supporters were stunned when the ESA's executive committee proposed discontinuing the preservation committee at the next annual meeting later that year. Opposition seems to have stemmed from the feeling that it was improper for a scientific society to act as a political pressure group.[117] Attendees at the business meeting decided to put the issue before the full membership in the form of a referendum. Despite Shelford's pleas, in the summer of 1945 the ESA membership overwhelming supported an amendment to the society's bylaws that continued the preservation committee but barred it from taking "direct action designed to influence legislation," including presenting resolutions or taking part in "political, pressure group, or propaganda activity."[118] The next year, the society decided to abolish the preservation committee altogether. Just over two decades after Francis Sumner had issued an impassioned plea for naturalists to become more involved in the preservation of nature, the ESA abandoned its long-standing institutional commitment to political activism. Ironically, that repudiation came just as the physicists who had created the atomic bomb began calling on their colleagues to assume a greater sense of social responsibility for their scientific research.[119]

Disappointed but undaunted, Shelford and a number of his supporters decided to organize a new group, the Ecologists' Union, whose objective was the "preservation of natural biotic communities, and encouragement of scientific research in preserved areas."[120] By December of 1946, 158 ecologists had joined the fledgling organization, which continued to lobby Congress and federal agencies on behalf of preservation causes. One Ecologists' Union officer, the botanist and former Shelford student George Fell, soon argued for an expansion of the society's membership beyond scientists. He and his wife also put together a list of one hundred natural areas that they felt must be preserved. In 1950, members of the Ecologists' Union voted to reconstitute the group as the Nature Conservancy, which has since grown into the largest and most successful nature preservation organization in the world.

Although the ESA abandoned its institutional commitment to political activism on behalf of nature, by the 1930s and 1940s ecological ideas had begun slowly making their way into conservation discourse. Increasingly, wildlife supporters described threatened species not only as aesthetically interesting and useful objects but also as vital components of ecological communities. Indeed, this transformation is evident in a revealing exchange between the ecologist and former Shelford student Charles Kendeigh and John Baker, the executive director of the National Audubon Society. In 1944, Kendeigh wrote to Baker arguing that the Audubon Society was primarily interested in the "preservation of species," while the ESA was concerned with the "preservation of biotic communities." Baker immediately fired back: "While you are right that our Society is vitally concerned with the preservation of species, we have come to recognize that that objective is probably only to be obtained thru [sic] our being vitally concerned with the preservation of habitat and therefore, soil, water and plants as well as wildlife."[121] While the depth of Baker's commitment to ecological ideas might be questioned, National Audubon's ongoing campaign to repair the tarnished reputation of the predatory birds provides evidence that Shelford and his colleagues had enjoyed at least limited success.

CHAPTER EIGHT

RECONSIDERING RAPTORS DURING THE INTERWAR YEARS

It seems true wisdom to preserve even apparently injurious species from wanton destruction. What moral right has man to decree the extermination of any bird which at worst merely reduces the number of some of its fellows? As biologists can we believe that the earth and all its inhabitants exist solely for the benefit of man?

JAMES CHAPIN, 1932

The time to protect a species is while it is still common. The way to prevent the extinction of a species is never to let it become rare.

ROSALIE EDGE, 1934

SAVING THE BALD EAGLE

In 1940—as American participation in World War II loomed on the horizon—the Bald Eagle Protection Act finally became law. This step came more than a century and a half after the Continental Congress had chosen this majestic bird (known to science as *Heliaeetus leucocephalus*) as the centerpiece for the first official seal of the United States. As the only eagle with a range restricted to North America, the species soon became commonly known as the American eagle and widely adopted as an emblem of the young nation's freedom, power, and sovereignty. Yet, despite the nationalistic symbolism associated with this quintessential example of charismatic megafauna, the move to grant federal protection to the bald eagle was long in coming and remained controversial even when it finally passed.[1]

FIGURE 45. Lithograph of the bald eagle, 1893. Drawn by Robert Ridgway. Despite the majestic symbolism associate with the species, like other raptors, it was often subject to systematic persecution at the hands of farmers, ranchers, sport hunters, and game managers. Not until the Bald Eagle Protection Act of 1940 did the bald eagle finally receive federal protection. From Albert K. Fisher, *The Hawks and Owls of the United States in Their Relation to Agriculture* (1893), 96.

Although the ominous clouds of military conflict clearly played an important role in rallying support for bald eagle protection, its success was not simply a manifestation of war-fueled patriotism. The act, which countered a tradition of treating the eagle and other birds of prey as pariahs, represented the culmination of a long campaign. For two decades preceding its passage, raptor enthusiasts had struggled to repair the tarnished reputation of predatory birds, to fight bounty laws aimed at reducing their numbers, and to secure legislation to protect them from continued harassment at the hands of farmers, ranchers, sportsmen, and others who considered them "vermin," fit only for systematic obliteration from the landscape. Although the Bald Eagle Protection Act fell short of achieving its supporters' ambitious aims, it nonetheless stands as a landmark piece of federal wildlife legislation that has yet to gain the historical attention it deserves.[2]

Persecution of the bald eagle and other raptors was part of a much broader antipredator campaign waged by private citizens, sportsmen's clubs, arms and ammunition manufacturers, and local, state, and federal officials. As we saw in the previous chapter, by the 1920s, the Bureau of the Biological Survey's growing involvement in that campaign provoked a schism in the wildlife conservation community. On one side of the debate were those who viewed predators as a significant threat to game and livestock. For them, systematic predator control seemed a rational, efficient means to protect vulnerable animals, a way for humans to improve upon a disorderly and sometimes dangerous nature. On the other side of this contentious issue were the scientists, humanitarians, and nature lovers who initially opposed predator control largely because of their romantic commitments or concerns about minimizing suffering in wild animals and who later turned to the science of ecology to find support for their position.

Much has been written about the long-standing controversies surrounding mammalian predators, especially wolves and coyotes.[3] This chapter explores an analogous debate over avian predators that raged in the interwar years and culminated in the Bald Eagle Protection Act of 1940. What this story reveals is an active, increasingly vocal community of dedicated bird enthusiasts—primarily scientific ornithologists and serious amateur birdwatchers —who challenged private and governmental conservation authorities in an attempt to gain protection for all manner of hawks and owls, including the bald eagle.

Several concerns motivated these bird enthusiasts to action. First was their aesthetic interest in bird life of all kinds, including predatory species. In private correspondence and public debate, raptor enthusiasts repeatedly declared that the substantial pleasure they gained from viewing predatory birds in the wild deserved as much consideration as the sportsmen's and farmers' interest in destroying them. For example, in the mid-1920s, when a prominent wildlife defender declared the "wanton annihilation of a species" to be a "real crime," he used language that metaphorically linked wild animals to artistic masterpieces: "The destruction of a great work of art calls forth genuine condemnation, in spite of the fact that it may conceivably be reproduced, or even excelled. But how about the creature that has been millions of years in the making, which, once gone, is gone forever?"[4] When asked by a friend, "What do you *do* when you look at a bird?" a birdwatcher from this period responded, "What do you do when you see a great painting?"[5] The rise of birdwatching in the late nineteenth and early twentieth centuries produced a growing constituency of middle- and upper-class Americans who regularly ventured out into the nation's fields and forests to experience nature in nonconsumptive ways. Having developed strong emotional and aesthetic bonds with avian species, birdwatchers increasingly challenged the many forces that seemed allied against those species.

A profound unease with the implications of human-induced extinction also moved bird enthusiasts to action. At least since the publication of George Perkins Marsh's *Man and Nature* (1863), conservation-minded individuals had been warning that the New World's natural resources were not limitless, as had once been widely assumed. Not until the end of the nineteenth century, however, with the near demise of the once ubiquitous bison, did the message that humans might drive species to oblivion finally begin to receive a broader hearing. The subsequent loss of other species—the last passenger pigeon in 1914, the last Carolina parakeet only four years later, and all but a handful of heath hens by the 1920s—provided other sobering reminders that humans could drive once abundant wildlife populations over the brink of oblivion. The language of eradication that the most strident control advocates deployed proved especially troubling for predator control critics. In the eyes of the individuals who defended predators, to willfully destroy a species, even one that many considered a pest, represented a deplorable act against present and future generations.[6]

Finally, bird enthusiasts repeatedly claimed that predator control advocates had failed to make a convincing scientific case proving that raptors were necessarily harmful to wildlife or livestock populations. Initially, bird-of-prey advocates looked to the science of economic ornithology to find support for their position. Through analysis of the stomach contents of various avian species combined with field observations of their feeding habits, economic ornithologists sought to determine their diets and make authoritative judgments about their worthiness for protection. They tended to do so within a framework that explicitly designated species as either "good" (and therefore worthy of protection) or "bad" (and deserving of persecution) based on whether they were directly beneficial or harmful to human interests.[7] By the mid-1930s, critics of avian predator control also began to turn to arguments from the newly emerging science of ecology to bolster their position. As we have already seen, the new ecological framework portrayed all species as crucial components of biological communities.

The campaign to protect avian predators represents an important episode in the history of American wildlife conservation. In addition to shedding light on the origins of the Emergency Conservation Committee (ECC), a radical wildlife organization that, among other accomplishments, forced a series of reforms within the National Association of Audubon Societies (NAAS), the campaign reveals the extent to which concern about human-induced extinction had become central to wildlife discourse by the 1920s and 1930s. Long before the passage of the Endangered Species Act, a significant number of Americans began to be haunted by the specter of human-induced extinction and to take steps to address the problem.

The continued vulnerability of predatory birds received graphic confirmation in 1917, only a year after Congress passed the enabling legislation for the Migratory Bird Treaty. As we have already seen, that landmark agreement granted U.S. and Canadian officials the authority to regulate the hunting of migratory game birds and declared a permanent closed season on most nongame birds. But it excluded birds of prey from its purview, for treaty supporters feared that protection for raptors, even if they were migratory, would diminish support for the already controversial measure.[8] Taking advantage of this loophole, the Alaskan Territorial Legislature began offering a fifty-cent bounty on the bald eagle. The move came in response to concerns about rising food prices during World War I and to pleas from salmon fishers and fox farmers, who claimed that eagles posed a threat to their livelihood.

Once stateside bird enthusiasts caught wind of the legislation, they responded with alarm. For example, in 1920, a group of naturalists at the American Museum of Natural History in New York petitioned the American Ornithologists' Union (AOU)—the nation's most prestigious ornithological society—to take a public stand against the Alaska eagle bounty. According to the petitioners, the legislation placed the bald eagle in "serious danger of extinction." Yet the AOU, the National Association of Audubon Societies, and the Bureau of the Biological Survey had failed to raise a voice in protest, even though more than five thousand birds had already fallen victim to the bounty and the Bureau was on record as opposing bounties on predatory birds. A. K. Fisher, head of the AOU bird protection committee and longtime employee of the Bureau of the Biological Survey, responded that eagles posed a genuine threat to young blue foxes just after they had left the den and to salmon on their spawning grounds. Though he regretted the necessity of controlling these and other predators, Fisher argued that it was simply a case of "wild life and economics run amuck."[9]

Fisher's response to the petition contained more than a little irony since his 1893 book, *The Hawks and Owls of the United States in Their Relation to Agriculture,* had long served as a source of information and inspiration for wildlife conservationists hoping to rehabilitate the negative image of predatory birds. In 1885, C. Hart Merriam convinced Fisher, his friend and medical school classmate, to join him at a newly created federal wildlife agency, the Division of Economic Ornithology, which eventually became known as the Bureau of the Biological Survey. Under Merriam's leadership, this organization quickly established itself as the primary center for the practice of economic ornithology and mammalogy in the United States. Fisher's book, which was based on stomach-content analysis of some 2,700 birds of prey, was the agency's third major publication. While quite willing to condemn a handful of predatory birds as largely harmful, most species Fisher declared to be either beneficial or neutral in their overall effects. The bald eagle he placed in this later class. In a letter of transmittal accompanying the book, Merriam proclaimed that Fisher's research conclusively demonstrated that "a class of birds commonly looked upon as enemies to the farmer, and indiscriminately destroyed whenever occasion offers, really rank among his best friends, and with few exceptions should be preserved, and encouraged to take abode in the neighborhood of his home." Fisher's work also showed "the folly of offering bounties for the destruction of hawks and owls."[10]

While bird enthusiasts continued to decry the Alaskan eagle bounty in editorials and magazine articles, they initially failed to interest any major private or public conservation organization to take up their cause, even after the bounty was raised from fifty cents to one dollar in 1923.[11] Undaunted, a handful of bird-of-prey advocates kept pressure on the National Association of Audubon Societies and the Bureau of the Biological Survey, while working behind the scenes to construct a network of like-minded supporters. Their goal was not simply to repeal the troublesome bald eagle bounty in Alaska, but also to arouse wider public sympathy for all avian predators, which continued to be subjected to local and state bounties, routinely exempted from bird protection laws, and persecuted by farmers, sportsmen, and even bird sanctuary organizers.

A series of events in the mid-1920s galvanized concern about predatory birds. During this period, critics of the Bureau of the Biological Survey's expanding predator-control program became increasingly vocal in their opposition to federal campaigns designed to eradicate mammalian predators in the West. At the 1924 meeting of the American Society of Mammalogists, for example, Biological Survey naturalists found their agency's predator policies under attack, despite the fact that they had dominated the organization since its founding five years previously.[12] This initial skirmish signaled the beginning of a protracted battle that continued to divide professional mammalogists for the next several decades. It also emboldened sympathizers of predatory birds to speak out.

The predator-control campaigns that sport hunters mounted in an effort to bolster game populations further fueled the concerns of bird-of-prey enthusiasts. Despite decades of state and federal game protection, the prospects for many game species failed to improve by the mid-1930s. Many ducks and upland game birds even seemed to be declining in population rather than increasing.[13] While the causes of these declines were complex, sport hunters often focused their ire on the wildlife that preyed on game. Sportsmen's clubs, state game officials, and arms and ammunition manufacturers routinely offered bounties on mammalian

The Hawk   Owl—Crow
Pest Must Go!

The New Jersey Fish and Game Commission is befriending farmers and poultry raisers as well as sportsmen in starting out to eradicate the ravaging pestilence of vermin.

The only good hawk, owl, crow, roving house cat, weasel, fox, bobcat wolf or snake is a dead one. Let's up and at 'em and make them all good.

The last passenger pigeon on earth died September, 1914, six years before the passage of the New Jersey law against killing passenger pigeons. Post mortem activities became undertakers, not the friends of useful wild life.

There were no horses in America when this continent was discovered—only fossil remains. There will soon be nothing left of our song birds, game birds and game animals but fossil remains if FOSSILS remain in successful control of vermin.

If you want to be shown how to lend a hand—"do your bit"—in this holy was to rescue God's innocent and useful wild creatures from the ravages of viper trash and the ignorant preachments of those who raise neither pigs nor potatoes, chickens nor children, and to make New Jersey a happier hunting ground than Indian ever dreamed of, ask

REV. NOEL J. ALLEN,
Lecturer and Vermin Control Expert
New Jersey Fish and Game Commission
State House, Trenton, N. J.

THE HAWK-OWL-CROW-CAT PEST MUST GO

FIGURE 46. Antiraptor propaganda, 1933. Issued by the New Jersey Game Commission, this advertising was part of a campaign to "eradicate the ravaging pestilence of vermin," including hawks, owls, crows, and cats. From *Bulletin of the Hawk and Owl Society* 3 (1933): 7.

and avian predators, sponsored predator hunting contests, and established game propagation farms on which predators were systematically removed through trapping, hunting, and poisoning. Shooting hawks and owls during migration, when their populations tended to concentrate along seasonal flyways, also became an increasingly popular pastime in rural areas. The rationale behind these actions is revealed by the anonymous author of a 1936 pamphlet issued by the More Game Birds in American Foundation, a national organization of sport hunters that had been founded six years previously with the goal of increasing the supply of game birds through systematic propagation. While birds of prey had once played an important role in maintaining the balance of nature, that balance had been upset when human hunters arrived on the scene. Now the only effective way to restore game populations, the pamphlet argued, was to remove the population pressure that natural predators represented: "The old saying, 'Let the gamekeeper take care of vermin and the game will take care of itself,' does not tell the whole story, but it contains a great deal of truth."[14]

Raptors even failed to find safe haven in bird sanctuaries. In the late nineteenth and early twentieth centuries, hundreds of local Audubon Societies and bird clubs sprang up across North America.[15] In an effort to promote birdwatching, cultivate sympathy for conservation, and provide habitat for increasingly beleaguered birds, these organizations frequently established protected areas were hunting was prohibited. Yet, birds of prey often found themselves barred from their gates. For example, the reformed hunter, popular lecturer, and self-styled naturalist Jack Miner established a particularly famous bird sanctuary on the outskirts of his brick-and-drain-tile factory in Kingsville, Ontario. By the mid-1920s, when North American duck and geese populations already showed signs of serious decline, Miner's sanctuary attracted throngs of migratory waterfowl and an equal number of amazed human visitors. Though he had relinquished a youthful interest in hunting and professed to an abiding friendship with wildlife, Miner routinely referred to predators, particularly avian predators, as "cannibals": "With natural menaces—vermin like crows, hawks, and rattlesnakes—I've no patience."[16]

The National Association of Audubon Societies' failure to more vigorously defend hawks and owls also disturbed bird-of-prey advocates. They hoped the oldest, largest, and best-financed bird protection organization in North America would take a leadership role in a broad campaign to rehabilitate the reputation of predatory birds. Instead, bird-of-prey defenders became increasingly frustrated with the leadership of T. Gilbert Pearson, who had headed the organization since 1910. They criticized Pearson as timid in his general approach to wildlife conservation, far too cozy with sportsmen, excessively stingy with Audubon funds, and largely indifferent to the increasingly desperate plight of birds of prey. Their fears intensified in late 1924 when Pearson vigorously defended his philosophy of "conservative conservation" in an address at the National Association's annual meeting. There he attacked "extremists" whose legislative proposals reflected a limited knowledge of wildlife, including those who called for new restrictions on hunting or even its elimination. He also reaffirmed his support for sportsmen, whom he called the "largest effective force for the preservation of game in this country." [17] A year later, when Pearson refused to condemn a hawk destruction campaign organized by the American Game Protective Association, raptor enthusiasts felt they could remain silent no more.

SAVE THESE BIRDS!

The ornithologist Waldron DeWitt Miller played a leading role in the campaign to gain a more sympathetic hearing for birds of prey. Keenly interested in birds and wildflowers since childhood, in 1903, the twenty-four-year-old insurance agent found employment as an assistant in the bird department at the American Museum of Natural History in New York.[18] His new job provided him with access to a world-class bird collection, an extensive ornithological library, and colleagues who guided him through the intricacies of avian taxonomy. With long hours at the museum and regular forays into the countryside surrounding his home, he eventually established a reputation as one of America's premier ornithologists. Unlike most of his colleagues at the time, however, he never showed much interest in specimen collecting. His generally sympathetic attitude toward birds also found early expression in the New Jersey Audubon Society, an organization for which he served as an incorporator and vice president from the time of its creation in 1897 until his death in 1929.

Miller held a particular passion for birds of prey, and as early as 1914, he began systematically collecting data on their diets. Over the next decade, these studies confirmed what he had long suspected: hawks and owls were unjustly persecuted, declining in number, and clearly in need of greater protection. A fellow curator in the bird department and close personal friend later described Miller's attitude toward birds of prey in these words: "It seems true wisdom to preserve even apparently injurious species from wanton destruction. What moral right has man to decree the extermination of any bird which at worst merely reduces the number of some of its fellows? As biologists can we believe that the earth and all its inhabitants exist solely for the benefit of man?"[19]

Miller found numerous kindred spirits who supported his agenda. One of the most important of these was a maverick coworker, Willard G. Van Name, an as-

FIGURE 47. Waldron D. Miller in the field, 1928. A renowned ornithologist and curator at the American Museum of Natural History, Miller was also a strong advocate for birds of prey. The blistering pamphlets he coauthored helped raise consciousness about the plight of raptors and provoke reform of the National Association of Audubon Societies. From *Waldron De Witt Miller: In Memoriam* (1929): 7. Courtesy of the Department of Ornithology, American Museum of Natural History.

sociate curator in the invertebrate department at the American Museum who was seven years his senior.[20] With his pessimistic attitude toward human nature and a propensity for pointed critique, Van Name had already established a reputation for being a "quixotic, truculent curiosity" within conservation circles. He shared Miller's interest in birdwatching, his love of wild creatures, his impatience with existing conservation organizations, and his zeal for reform. The two undoubtedly served as the moving force behind the 1920 petition calling on the AOU to condemn the Alaskan eagle bounty. Both were also active in the Linnaean Society of New York, a regional bird study society that had been established in the late 1870s. There they found a largely sympathetic membership of serious amateur ornithologists and birdwatchers who experienced great aesthetic and emotional pleasure in viewing birds of all kinds, including birds of prey. Throughout the 1920s and 1930s, the organization remained a hotbed of agitation on behalf of predatory birds.[21]

Miller not only helped mobilize the Linnaean Society but also pushed other bird organizations, like the Delaware Valley Ornithological Club (DVOC), into action on the issue of predatory birds. In late 1925 and early 1926, both groups passed resolutions urging the National Association of Audubon Societies to devote more money and effort to the cause of bird-of-prey protection. The exact wording of the resolutions varied, but the message proved virtually identical. Although hawk populations were plummeting in the eastern United States, their persecution seemed to be increasing. These disturbing developments had occurred even though A. K. Fisher had long ago demonstrated that most hawks were either neutral or beneficial to agricultural interests. Both organizations asserted that "nature lovers" had as much right to view birds of prey for the aesthetic pleasure it provided them as sport hunters had to kill these species in an misguided effort to protect game. The Linnaean Society resolution argued that the "interests of game conservation do not require the extermination of such raptorial birds as prey upon it." Echoing this sentiment, the DVOC resolution urged Pearson to take "immediate action" to prevent or discourage the antipredator campaigns by "so-called sportsmen."[22]

Miller also orchestrated a letter-writing campaign to accompany the resolutions. At his prodding or on their own initiative, dozens of bird enthusiasts urged Pearson to give the protection of birds of prey a higher priority within his organization. For example, in a letter dated February 21, 1926, and apparently written at Miller's urging, the ornithologist Aretas A. Saunders commended Pearson for the success that the National Association had enjoyed with its campaigns to protect gulls, terns, egrets, and other birds that once had been "threatened with extinction." Now, however, it was time to turn attention to another class of endangered birds. Saunders's letter reveals how the specter of extinction increasingly haunted naturalists and nature lovers alike during this period: "The extinction of any native species, even a harmful one, is in my opinion a crime. While the economic argument might be necessary for some minds, I do not think bird-lovers need to hesitate to try to protect any and every species of native bird, just because it is a living creature with as much right to its place in the world as man. The hawk kills because he must do so or starve. Man kills cruelly and wantonly, or to gratify his greed, or to have stuffed trophy as a token of his skill and needless cruelty. So here is hoping the National Association will not sit idly by and allow selfish greed or ignorant prejudice destroy to the verge of extinction some of our largest and most magnificent birds."[23]

The onslaught caught Pearson off guard. Initially, he suspected that Van Name was behind the avalanche of letters flowing into his office, many of which he found "insulting and abusive." He also clearly resented the implication that his organization had abandoned its duty to come to the aid of birds of prey. In his replies, Pearson strongly defended his leadership on the issue, pointing out that he had not only written numerous editorials but also delivered many addresses on the subject of predatory birds, that the first Audubon educational leaflets covered these species, and that the organization had recently begun systematically gathering data on their current status. He also reminded his correspondents that the National Association of Audubon Societies (and its less-formal predecessor, the National Committee of the Audubon Societies) had been responsible for securing passage of a uniform bird protection law in forty states. While the model form of that law made it a crime to harm most nongame species, Pearson argued that "in many states we found it absolutely impossible to pass a law protecting song and insectivorous birds unless the hawk clause was amended to include all hawks in the non-protected list." He felt he could only push so hard on the issue of protecting birds of prey without appearing a "crank" and potentially undermining the organization's effectiveness on other issues.[24]

Underwhelmed by Pearson's response, Miller began laying plans for a circular on hawk protection designed to keep pressure on the Audubon president. He worked closely with Henry Carey, a Philadelphia lawyer, naturalist, and author of the Delaware Valley Ornithological Club's resolution on predatory birds, to write the text, gather illustrations, develop a mailing list, recruit cosigners, and raise the necessary funds.[25] By the time of the publication of *Save These Birds!* in April 1926, Miller and Carey had persuaded several prominent naturalists, artists, and nature writers—including Thorton Burgess, Louis Agassiz Fuertes, Ernest Thompson Seton, Gerald Thayer, and several DVOC officers—to formally endorse it.

Mailed to some two thousand individuals and organizations, Miller's circular began by pointing out that early writers had once considered birds of prey to be the most majestic and beautiful of avian species, universal symbols of "power and might." Lately, however, these "splendid birds" had fallen on hard times as sportsmen and farmers had begun portraying them as "vermin," fit only for extermination. Miller decried campaigns aimed at reducing the populations of birds of prey, presented scientific evidence for the benefit of hawks and owls (drawn largely from A. K. Fisher's earlier study), and called on readers to raise the issue with local bird protection organizations and the National Association of Audubon Societies: "Is it not high time that the nature-lovers of this country asserted their right to a share in its wildlife? Are we not justified in demanding our rights, and those of our successors, to the enjoyment of these marvelous creatures?"[26]

Save These Birds! prompted an increase in the volume of mail urging Pearson to devote more attention to birds of prey and resulted in a new round of pro-bird-of-prey resolutions from the Nuttall Ornithological Club and the American Ornithologists' Union. Much to Pearson's consternation, several of those who wrote him mistakenly thought that his organization had issued the circular. He assured one such correspondent that nothing could be further from the truth; rather, it was the product of "a little group of people who have apparently been displaying as much zeal in their efforts to discredit this Association and the writer, as they have to protect Hawks."[27]

Whatever his attitude toward those behind the circular, Pearson felt pressure to respond publicly. He did so during an address delivered at the 1926 meeting of the National Association. There he once again defended his record as president, claiming that he had always stood for "a policy of sane and conservative conservation" while endeavoring to steer clear of "false sentimentalism." Following a request from the Audubon board of directors, Pearson announced that he had called on Alaskan officials to repeal the eagle bounty law "until such time as a careful investigation of its food habits can be made."[28]

The Audubon board also persuaded Pearson to venture to Alaska so he could gain firsthand information about the impact of the eagle bounty. Following a brief visit in the summer of 1927, he reversed his earlier position calling for its repeal. Pearson now argued that bald eagles remained more abundant in Alaska than statesiders could possibly imagine and that at times they did prey on healthy fish and game. Moreover, while the payment of more than forty thousand bounties may have reduced the eagle population in the more accessible coastal regions of Alaska, the species certainly did not face "any immediate danger of extermination" within the territory as a whole. Rather than a fight to repeal the eagle bounty, Pearson argued for a broad educational campaign in Alaska to cultivate sympathy for all forms of bird life.[29]

A CRISIS IN CONSERVATION

Clearly frustrated by Pearson's backsliding on the Alaskan eagle bounty, Miller, Van Name, and Davis Quinn felt obligated to respond. In June 1929, they issued a scathing sixteen-page pamphlet entitled *A Crisis in Conservation: Serious Danger of Extinction of Many North American Birds.* Without referring to the National Association of Audubon Societies by name, the three naturalists minced no words as they charged the nation's largest and most widely recognized wildlife conservation organization with "neglect, indifference, and incompetence." Although this well-financed institution had help secure protection for most "song and insectivorous" birds, it was now sitting by idly while dozens of North American birds—including the bald eagle—were rushing headlong toward extinction. Among numerous other charges, Miller, Van Name, and Quinn castigated Audubon leaders for being slow to publicize the Alaska bounty. They also took them to task for issuing "anti-eagle propaganda" that gave the impression that the species was "pretty destructive of game and fish" while failing to mention its threatened status.[30]

A Crisis in Conservation set into motion a series of events that rocked the National Association of Audubon Societies to its core. One recipient of the harsh indictment was Rosalie Edge, whose copy arrived while she was vacationing in Paris. A well-heeled New Yorker and former women's suffrage activist with "sharp, restless blue eyes," Edge had recently met Van Name while on one of her regular birding jaunts in Central Park. As an ardent birdwatcher, an active participant in the Linnaean Society, and an Audubon Society life member, she was greatly disturbed by what she read. Later, Edge recalled how strongly the pamphlet impressed upon her "the tragedy of those beautiful birds, disappearing through the neglect and indifference of those who had at their disposal wealth beyond avarice with which those creatures might be saved." After returning to New York, she attended the 1929 annual Audubon Society meeting to demand a response.[31]

Unsatisfied with the reply she received, Edge met with Van Name, who had been barred from issuing further publications without prior approval from museum authorities (several of whom served on the Audubon board of directors). In part as a way to circumvent this ban, the two launched the Emergency Conservation Committee, a radical organization that remained a gadfly in wildlife conservation circles for the next three decades. Davis Quinn, a young bird-of-prey enthusiast of no particular renown, and Irving Brant, a midwestern journalist who had earlier charged that gun companies had gained control of the Audubon association, also agreed to serve as officers in the new organization. Undoubtedly, Miller would have also become involved if he had not been killed in a tragic accident—a bus struck his motorcycle while he was birding in the New Jersey countryside—shortly following the publication of *A Crisis in Conservation*.[32]

Although the Emergency Conservation Committee is probably best remembered today for the reforms it provoked within the National Association of Audubon Societies—including the forced resignation of Pearson in 1934—from the beginning the organization proved firmly committed to securing greater protection for predatory birds. A series of provocative pamphlets castigating the National Association of Audubon Societies, the Bureau of the Biological Survey (condemned in one ECC publication as the "Bureau of Destruction and Extermination"), and other American conservation institutions soon became the hallmark of the new organization.[33] The first three ECC pamphlets all treated birds of prey. After a reprint of *A Crisis in Conservation*, two publications with an equally strident tone quickly followed. *Framing the Birds of Prey*, written by Davis Quinn, decried the "fanatical and economically harmful campaign of extermination being waged against the hawks and owls." According to Quinn, ammunition manufacturers hoping to bolster sales and sportsmen attempting to stave off a reduction in bag limits or a shortening of open seasons fostered the increasingly common perception that birds of prey were simply "vermin." If something were not done quickly to save these "beautiful" birds, they would soon be "creatures of the past."[34] *The Bald Eagle: Danger of Its Extinction by the Alaskan Bounty* lambasted the Bureau of the Biological Survey for failing to exercise its authority to overturn the eagle bounty in Alaska territory and the National Association for neglecting to "carry on any earnest or persevering work to save the Bald Eagle."[35]

The furor over *A Crisis in Conservation* also provoked Pearson into action. At the time of its publication in June 1929, he was still basking in the glory of one his greatest conservation victories: the passage of the Norbeck-Andresen Migratory Bird Conservation Bill. By the end of the 1920s, the "duck crisis" was deepening as North American waterfowl populations continued to plummet due to drought, habitat destruction, and overhunting. Although he had repeatedly defended sportsmen and refused to sanction further restrictions on their activities, now Pearson had reluctantly concluded that a reduction in waterfowl bag limits was necessary. With backing from a newly created National Committee on Wild-Life Legislation, which he chaired, Pearson also threw his support behind a federal refuge bill for migratory birds that Senator Peter Norbeck of South Dakota had introduced into Congress in 1928. Using NAAS staff and working closely with several other prominent conservation organizations, Pearson coordinated a massive publicity campaign to garner support for the bill, which made it through Congress and was signed into law early in 1929. Previous federal refuge bills had been roundly criticized for allowing hunting in the very areas supposedly being created to restore waterfowl populations; the Norbeck-Andresen Act, however, authorized the expenditure of up to $8 million in federal funds to purchase "inviolate sanctuaries" for waterfowl along migratory flyways. John Burnham, the president of the American Game Protective Association, congratulated Pearson on the victory, exclaiming "you deserve all the glory that is coming to you."[36]

Coming on the heels of the Norbeck-Andresen Act, Pearson felt blindsided by the turmoil that *A Crisis in Conservation* ignited. Fearful of eroding support within the National Association of Audubon Societies in the wake of the controversy, he began devoting more attention to birds of prey. In a carefully worded address delivered to the American Game Conference late in 1929, Pearson defended the right of gamekeepers and farmers to protect themselves from avian predators. But he also argued that they had an obligation to consider the views of those who opposed the wholesale killing of "useful hawks," birds that rendered valuable services as "destroyers of rats, mice, and various insects." "For the sake of these great, handsome birds themselves, and in the spirit of fair-play," Pearson concluded, game officials should discourage bounties on birds of prey and help his organization educate the public on how to identify those species protected by law.[37] Pearson's address was hardly the ringing endorsement of hawks and owls that his critics had been urging him to make, but it sent a signal that Audubon planned to give greater attention to the issue.

A further encouraging sign came three weeks later, when Pearson authored a Bald Eagle Protection Act and convinced Senator Norbeck and Rep. August P. Andresen of Minnesota to introduce it in Congress. At the time, only five states specifically mentioned the bald eagle in their bird protection laws, while thirty-nine states protected it by inference, three states included the species on their unprotected list, and one state (Wyoming) had no laws covering nongame birds. Even in those states where the bald eagle received explicit or implicit protection by law, prosecutions for killing them proved extremely rare. And Alaska continued to pay bounties on "this magnificent emblem of our country," which, according to an editorial Pearson published in *Bird-Lore*, "was becoming a very rare species over large areas of its range." Although Norbeck initially rejected Pearson's request that Alaska be specifically included in the bill, Pearson stood firm on the issue, and the territory appeared in the draft presented to Congress.[38]

As he had done for the Norbeck-Andresen Act a year earlier, Pearson unleashed an impressive publicity drive to drum up support for his eagle bill. He and his staff dispatched editorials to newspapers across the country stressing the importance of the legislation; they sought backing from hundreds of major conservation organizations, bird clubs, sportsmen's groups, and even the Benevolent Order of Eagles; and they urged countless American bird lovers to write their congressional representatives urging its passage. By January 1930, Pearson reported that the campaign was "keeping his office-force after hours and working at top-speed," although privately he admitted that the bill had little chance to gain passage.[39]

Later that month, in hearings before the House Committee on Agriculture and Forestry, Pearson joined Theodore S. Palmer, an ornithologist, longtime Biological Survey staff member, and first Audubon vice president, to testify on behalf of the bill. During nearly two hours before the committee, Pearson reviewed the status of the bald eagle, which he claimed was "nearly bordering on extinction in many places," while Palmer argued that the eagle bill would "encourage patriotism" and provided historical background on how the species had become the "national bird." In response to questioning, both naturalists denied that the bald eagle was a migratory species, a point with which their critics took exception. Although only one individual, the delegate to Congress from Alaska Territory, offered opposition to the bill and more than a hundred supporters sent letters strongly urging its passage, the House Committee on Agriculture reported unfavorably on the measure. Committee members claimed that because the bald eagle was not considered a migratory bird, the act would unconstitutionally interfere with the right of states to regulate wildlife within their borders. They also argued that the Depression was an inauspicious time to enact legislation protecting such a clearly "destructive bird." Norbeck had better luck in the Senate, where he steered Pearson's bill through with only a relatively minor amendment authorizing the destruction of any bald eagle "found in the act of killing domestic fowl, wild or tame lambs or fawns or foxes on fox-farms." The measure died, however, when it failed to gain passage in the House of Representatives.[40]

REHABILITATING RAPTORS

Disappointed but undaunted, in January 1932 a small, enthusiastic group of predatory bird supporters (most of whom were also Linnaean Society members) decided to found the Hawk and Owl Society. The primary force behind the new organization was its secretary, Warren F. Eaton, a thirty-two-year-old, Harvard-educated textile and dry goods wholesaler, dedicated birdwatcher, and president of the Linnaean Society of New York. An energetic, dapper young man with a wiry physique, Eaton "gloried in birds of prey, his adoration amounting almost to an obsession." The purpose of the Hawk and Owl Society was to foster "greater popular appreciation of the aesthetic, scientific, and economic value of hawks and owls"; to combat the steady stream of anti-bird-of-prey propaganda; and to work for the passage and enforcement of protective laws for avian predators. The group also came out in firm opposition to the ethos of eradication that seemed to inform many predator-control campaigns, arguing that "No species should be exterminated or extirpated from *any* part of its habitat." [41] Within a few months of its founding, Eaton had received expressions of support from more than twenty bird clubs, state and local Audubon Societies, and scientific organizations sympathetic to its goals. Not surprisingly, the Delaware Valley Ornithological Club, the Linnaean Society of New York, the Nuttall Ornithological Club, and the bird protection committee of the AOU were among those offering resolutions welcoming the new organization.

Under Eaton's leadership, the Hawk and Owl Society quickly became an important clearinghouse for the latest information on birds of prey and their protection. Through a voluminous correspondence, the publication of a modest irregularly issued bulletin, the wide distribution of article reprints on predatory birds, and the creation of local committees, the new organization not only linked together hawk and owl supporters throughout the nation but also promulgated authoritative information on their status. In addition, Eaton and other members of the Hawk and Owl Society regularly delivered pro-bird-of-prey speeches, pushed game authorities to enforce existing laws, lobbied for additional protective legislation, and used various other means to "exert influence in favor of raptors."[42]

Beyond engaging in successful legislative campaigns in several states, the Hawk and Owl Society also played a role in the creation of Hawk Mountain Sanctuary, the world's first refuge for birds of prey. After World War I, slaughtering hawks during their fall migration became an increasingly popular pastime in rural communities along Blue Mountain (also known as Kittatinny Ridge), a corduroy ridge that begins in southeastern New York and runs to southern Pennsylvania. The killing proved particularly intense along the rock outcrops above Drehersville, Pennsylvania, where at the height of the migration season, as many as four hundred shooters vied to bag the thousands of raptors that soared passed each day. So many hawks were killed and left to rot where they fell that "a bad odor hung over the place" during the fall migration season.[43]

After reading about the annual massacre in an ornithological journal, Henry H. Collins, Jr., and Richard H. Pough—both avid Philadelphia birdwatchers, Delaware Valley Ornithological Club affiliates, and members of the Hawk and Owl Society—visited the area during the fall migration of 1932. Disgusted by what they saw, the two published graphic accounts of the grisly ritual in the *Hawk and Owl Society Bulletin* and *Bird-Lore*.[44] In October 1933, Pough presented an illustrated lecture on the subject to a joint meeting of the Hawk and Owl Society, the Linnaean Society of New York, and the National Association of Audubon Societies. He had spoken with realtors, he announced, and the entire mountain could be purchased for a reasonable price. Following Pough's presentation, the meeting broke into enthusiastic applause when Pearson announced that Audubon would buy the mountain and stop the slaughter.[45] By June 1934, when it looked like Pearson was not going to follow through on his promise, Rosalie Edge secured a lease on the property, some 1,398 acres, with an option to buy it for $3,500. Though her preemptive action ruffled some feathers within the Hawk and Owl Society and Audubon, Edge managed to obtain financial support from both organizations to help enforce a ban on shooting within the borders of the newly established Hawk Mountain Sanctuary.

Although he failed to purchase Hawk Mountain, Pearson did devote increasing attention to bird-of-prey protection in the early years of the Great Depression.

FIGURE 48. (*above*) Rows of raptors killed at Hawk Mountain, 1931. Photograph by Richard Pough. During the early 1930s, slaughtering migrating hawks remained a popular pastime along parts of the Blue Mountains, which stretch from southeastern New York to Pennsylvania. This photograph represents the kill at a single stand near Drehersville, Pennsylvania. Courtesy of Hawk Mountain Sanctuary.

FIGURE 49. (*left*) Rosalie Edge with hawk. A birdwatcher and cofounder of the Emergency Conservation Committee, Edge served as an important gadfly in wildlife conservation circles during the 1930s and 1940s. She also secured the purchase of Hawk Mountain Sanctuary. Courtesy of Hawk Mountain Sanctuary.

For example, in 1932, he lodged a protest with Montana state game officials after they announced a "Common Enemy Contest" that targeted hawks and owls. In response to anti-eagle stories that began circulating in newspapers across the nation, he widely distributed a press release defending the species. While admitting that bald eagles regularly feasted on carrion and occasionally robbed fish from ospreys, Pearson argued that those biological realities should not detract from the grandeur of America's national emblem, this "symbol of valor and of power."[46] *Bird-Lore* also began to include regular reports of various state campaigns to extend legal protection to birds of prey and to repeal bounty laws targeting them. The title and portions of the text of Pearson's 1933 address to the International Association of Game, Fish and Conservation Commissioners—"Evils That Lurk in the Bounty System"—sounded more like an Emergency Conservation Committee pamphlet than the bland statements he had typically issued in the past. There he decried the results of state and private bounty campaigns and pointed out that twenty-six states protected "useful Hawks" by law, while sixteen failed to protect them. According to Pearson, bounty programs invariably led to the needless destruction of hawks because these species were notoriously difficult to identify correctly.[47]

NAAS involvement with hawk and owl protection became even more pronounced early in 1933, when John H. Baker gained election as chairman of the Audubon board of directors. A strong-willed Harvard graduate, successful Wall Street investment banker, and ardent birdwatcher since childhood, Baker came to the attention of the Audubon board while serving as president of the Linnaean Society of New York. They recruited him in the hope of bringing an end to the controversy that continued to plague the organization under Pearson's reign. The ongoing conflict not only threatened Audubon's authority, but also caused a precipitous decline in membership, which fell from 8,400 in 1929 to 3,400 by 1933. Baker wasted no time recruiting allies to oust the increasingly ineffective Pearson, who was finally forced to tender his resignation in October 1934. Baker then assumed the position of executive director of Audubon, with responsibility for the day-to-day administration of the organization.[48]

Signs of change abounded within the association. For example, in 1934, the Audubon board adopted a strongly worded resolution lifted almost verbatim from the Hawk and Owl Society mission statement. The organization put itself on record as opposing the extermination of any species, advocating protection for all "beneficial" hawks as well as any that were "rare," and condemning all bounty laws and anti-hawk campaigns. The only exception in the board's generally pro-predatory bird resolution was in the case of individual birds caught in the act of damaging property. Audubon also committed itself to "create greater popular appreciation of the aesthetic, scientific, and economic value of Hawks and Owls," "to combat the constant propaganda which encourages their destruction," and "to work for the enactment of laws for their protection."[49]

The resolution marked the beginning of an active "Hawk and Owl Preservation Campaign" at Audubon. Baker invited Eaton to join the Audubon staff, a move that, in effect, merged the Hawk and Owl Society into the larger and more established bird conservation organization. In addition, the National Association published a lavishly illustrated book, *The Hawks of North America* (1935), a project that had begun under Pearson's administration. An anonymous donor subsidized publication of the volume with the agreement that any profits from its sales would be used to help fund Audubon's bird-of-prey campaign. The organization also issued a large color poster urging the young men in Civilian Conservation Corps camps not to engage in the "promiscuous killing of eagles, hawks, and owls." By the end of 1935, Baker claimed that Audubon had distributed more than eighty thousand pieces of printed material relating to hawks and owls.[50]

There were also subtle changes in the way Audubon pitched bird-of-prey protection to the public. Baker and many of his younger staff felt it was important to place wildlife conservation on more secure scientific foundations. In addition to funding long-term research on the status of several endangered species— including the bald eagle, California condor, and ivory-billed woodpecker— Audubon officials began resorting to ecological arguments in their literature intended for public consumption. For the first time, Audubon publications routinely cited the work of a new generation of ecologists, wildlife biologists, and game managers who had taken up the systematic study of predators and their roles in biological communities. For example, in a 1935 article for *Bird-Lore*, Eaton summarized the research of Paul L. Errington, an associate of Aldo Leopold who had just completed a five-year study of the bob-white quail. Errington's research demonstrated that predators and game birds had "existed together for centuries before the balance of nature was disturbed by man." Now, however, the most important factor in the survival of game species was the degree to which humans had modified their environment. Adequate food supplies and suitable cover proved much more important than natural predation in determining the number of quail that survived in a given year.[51]

The same article quoted extensively from Leopold, who was laying down the foundations of modern wildlife management through the introduction of ecological concepts and research methods. Leopold had recently warned that hawks and owls were becoming "almost as scarce as ducks": "Apparently we are about to repeat what has already happened in England, exterminate our breeding raptors in the alleged interest of game." While predator-control efforts had made serious inroads into bird of prey populations, "research has piled more and more evidence that such control is usually futile." There were no easy answers to this problem. Legal protection of harmless raptor species would fail without public support and adequate enforcement. The only solution was the "slow and painful process of teaching sportsmen and farmers the *ecology of predation*." [52] This new, more ecological approach to conservation did not immediately supplant earlier arguments based on aesthetic and utilitarian grounds. But it would become increasingly important in the post–World War II era, as ideas from academic ecology began to diffuse into the wider culture.

While sympathetic to Leopold's call for more education about the role of predators in biological communities, Audubon officials also continued to value protective legislation. In 1935, when Rep. Jennings Randolph of West Virginia introduced a new eagle protection bill into the U.S. House of Representatives, Audubon once again unleashed its publicity machine to promote the measure. The Emergency Conservation Committee also lobbied aggressively for the legislation, issuing a second edition of their earlier eagle pamphlet under the title *Save the Bald Eagle: Shall We Allow Our National Emblem to Become Extinct?* Randolph's bill, apparently introduced without Audubon input, sought to "preserve from extinction the American eagle, emblem of the sovereignty of the United States of America" using the enforcement provisions of the Migratory Bird Treaty Act. As had been the case with a similar bill introduced five years previously, Senator Norbeck successfully shepherded companion legislation through the Senate, but the House refused to pass the measure. Clearly, there was much more educational work to be done. [53]

A LIMITED VICTORY

Audubon resolve on the issue of birds of prey was put to a test a year later (1936), when Eaton died suddenly from complications following surgery for appendicitis. He had been an indefatigable champion for birds of prey, first as "the works" behind the Hawk and Owl Society and later as head of the Hawk and Owl Division at Audubon, and there were fears that the organization would no longer prioritize raptors without his leadership. Those concerns quickly diminished, however, when Baker hired the Philadelphia engineer and businessman Richard H. Pough—one of the primary instigators behind the creation of Hawk Mountain Sanctuary—to replace Eaton. Like most of those involved in the ongoing campaign to rehabilitate the image of avian predators, the thirty-two-year-old Pough had discovered the joys of birdwatching as a youngster; over the years, his determined pursuit of the activity had led him to join numerous bird clubs and to visit forty-five of the forty-eight states. After joining Audubon, he continued the aggressive education campaign that Eaton had begun. In addition to regular reports on the status of birds of prey in *Bird-Lore,* Pough widely distributed a new poster that included views of how hawks appeared from the air on one side and graphic representations of the their diets on the other.[54]

The growth of falconry—the practice of taming wild birds of prey to use for hunting—also promoted greater public appreciation of these species.[55] American interest in reviving this ancient sport—which had long been popular among the British upper classes—began in the 1920s, when the Cornell bird artist and ornithologist Louis Agassiz Fuertes published an article on the subject in *National Geographic* magazine.[56] By 1941, more than one hundred devotees joined together to form the Falconer's Association of North America, which soon began publishing its own short-lived periodical, the *American Falconer.* Hundreds of additional falconers joined local organizations devoted to the sport or pursued it on their own. Like birdwatching, falconry had a strongly aesthetic dimension; one prominent enthusiast called it a "glorified form of bird worship." Supporters of the activity also claimed that the "overwhelming majority" of its practitioners were ardent conservationists who were "keenly interested in the future of hawks and falcons."[57] However, critics of the practice—like Rosalie Edge of the Emergency Conservation Committee—condemned most falconers as both cruel and a "new and very serious factor in the progressing extermination of our birds of prey." Whatever their direct impact on raptor populations, falconers published widely circulated accounts of their sport, accounts that invariably highlighted the grandeur and majesty of the species they trained. Even Edge recognized the "good a few falconers have accomplished in introducing their birds to prejudiced people who have never perhaps seen a hawk."[58]

The most renowned American falconers from this period were John J. and Frank C. Craighead, twins from Washington, D.C. In 1930, at the age fourteen, the Craighead brothers stumbled onto an old copy of *National Geographic* containing Fuertes's article on falconry and then witnessed a trained hawk in person. Captivated by what they had read and seen, the two began searching for suitable young birds to begin working with themselves. Soon they had secured four nestling Cooper's hawks from high in a pin oak tree along the Potomac River. They kept two of the birds for themselves, gave a pair to their buddies, and spent the next several months learning to care for and train their young charges. This initial experiment in falconry proved such a success that within a few years the Craighead brothers began traveling widely to capture, photograph, and film a remarkable variety of birds of prey. They published enthusiastic reports of their adventures in a series of popular articles that they expanded into a book, *Hawks*

FIGURE 50. Frank (*far left*) and John (*far right*) Craighead and friends with trained hawks, ca. 1937. Beginning in the 1920s, the age-old practice of falconry experienced a modest revival in the United States, thereby helping to cultivate a broader sympathy for raptors. The twin brothers Frank and John Craighead were well-known falconers who went on to distinguished careers in wildlife biology. Courtesy of the Craighead Collection.

in the Hand (1939), before heading off to graduate school in wildlife biology at the University of Michigan. Their publications invariably presented hawks and owls in extremely sympathetic terms. Rather than the nameless destructive "vermin" so often vilified by hunters, farmers, and ranchers, the Craigheads stressed the charm and distinct personality of their birds. "Hawks," they were fond of saying, "vary individually as much as people."[59]

While there continued to be setbacks, Audubon's ongoing educational campaign, the continued growth of interest in birdwatching, and the modest expansion of falconry eventually produced legislative results. By 1939, for example, thirty-eight states had enacted laws protecting the bald eagle, and Audubon officials were actively lobbying for similar protection in eight of the remaining states.[60] One year later, responding to mounting public pressure and the upsurge of patriotism on the eve of America's involvement in World War II, Congress finally passed the Bald Eagle Protection Act. This particular version had been introduced at the urging of Maud G. Phillips, the founder of the Blue Cross Society, a regional humanitarian organization based out of Springfield, Massachusetts. In 1939, she convinced two of her congressional representatives, Senator David Walsh and Rep. Charles R. Clason, to sponsor her bill and then testified on its behalf. Phillips's legislation declared that the bald eagle, the "symbol of American ideals of freedom," was clearly "threatened with extinction" and therefore in dire need of federal protection.[61] When the National Association of Audubon Societies urged its members to write their representatives in Congress in support of the bill, American birdwatchers responded with a flood of letters and telegrams. The plea from Carl T. Keller of Boston proved typical: "As a lover of the woods, the birds and the beasts, I am much in favor of S1494 to protect the bald eagle."[62] With war looming on the horizon, the appeal to protect America's national symbol could no longer be ignored. But in a move that dismayed eagle supporters, Congress exempted Alaska—the one region of the country where eagles remained most abundant—from the new law. Thus, the Bald Eagle Protection Act represented only a partial victory.

REGARDING RAPTORS

The ongoing Audubon campaign provided only one of several factors that led to heightened protection for the eagle and other American birds of prey. Perhaps more critical was the expanding interest in these species that accompanied the growth of birdwatching as a mass leisure activity. Here was a crucial difference between the campaigns to gain sympathy for mammalian and avian predators. A strong devotion to birdwatching proved the common denominator that linked virtually every early bird-of-prey advocate in the 1920s and 1930s. In the process of regularly viewing raptors in the wild, these individuals developed a strong emotional and aesthetic bond with predatory birds, a bond that led them to speak out on their behalf. As birdwatching took hold, increasing numbers of other Americans sought to view birds of prey in their native environments. By the early 1940s, for example, thousands of birdwatchers annually flocked to the Hawk Mountain Sanctuary in the hope of catching a glimpse of the successive waves of raptors that migrated through each spring and fall; so many bird enthusiasts tried to reach the area at the peak of migration season that railroad officials had to schedule special excursion trains to handle the demand.[63] There was no analogous rush to view coyotes and wolves during this period, nor could there be, since they are secretive creatures that are extremely difficult to find in the wild.[64]

In the long run, the cultural baggage associated with birds of prey also tended to work in their favor. As Barry Lopez, Thomas Dunlap, Jon Coleman, and other scholars have shown, mammalian predators had long been feared and loathed with an intensity disproportionate with the actual threat they posed to humans or the damage they inflicted on livestock and game. Much of this reaction was informed by the deeply held cultural beliefs associated with these creatures. Wolves, for example, have often featured prominently in folk traditions and fairy tales as emblems of evil, cruelty, and rapaciousness. Hawks and eagles, on the other hand, have tended to be associated with power, nobility, majesty, and freedom. In addition to the United States, numerous other nations through the ages have adopted the eagle as their symbol. The ancient, highly ritualized practice of falconry provided another source of positive associations for birds of prey. The advent of anti-bird-of-prey propaganda in the nineteenth and early twentieth centuries only partially eclipsed this earlier, more deeply rooted tradition of viewing birds of prey favorably, while the rise of birdwatching helped to reinvigorate it.[65]

In the decades after World War II, several demographic trends led to greater public appreciation for birds of prey. The percentage of Americans who engaged in sport hunting saw an upsurge in the immediate postwar years before experiencing a gradual decline that has continued to this day.[66] Related to this trend, a long-term shift in population from rural to urban and suburban areas continued, while agricultural production became concentrated in fewer and fewer hands. At the same time, interest in birdwatching showed a marked increase. As a result of these changes, in the second half of the twentieth century Americans proved less and less likely to experience predatory birds as potential destroyers of game or livestock. If they thought about birds of prey at all, they were much more likely to do so in the context of the pleasant memories they had experienced while viewing those species in the wild.[67]

The postwar growth of birdwatching not only produced a large, increasingly vocal constituency for the preservation of raptors but also promoted greater legal protection for these species. In 1959, for example, Congress finally granted protection to Alaska's bald eagle population; three years later, an amendment to the Bald Eagle Protection Act also extended federal protection to the golden eagle, a species that is difficult to distinguish from the immature bald eagle. By 1966, twenty-one states protected all hawks and owls by law, while twenty-six additional states protected at least some of these species. That same year, Congress passed the first Endangered Species Preservation Act, which was intended to identify, publicize, and ultimately rescue native vertebrates that were teetering on the brink of extinction. Several birds of prey—including the Southern bald eagle, the Florida snail kite, and the Hawaiian hawk—were among the first seventy-eight organisms listed under this landmark federal legislation, which provided a foundation for subsequent Endangered Species Acts in 1969 and (most importantly) 1973. In extending federal protection to the nation's threatened symbol, the Bald Eagle Protection Act of 1940 served as an important model for the increasingly comprehensive and stringent federal endangered species legislation enacted three decades later, at the height of the modern environmental movement.[68]

CHAPTER NINE

SALVATION THROUGH SCIENCE?

THE FIRST LIFE-HISTORY
STUDIES OF ENDANGERED SPECIES

*I realize that there are some who feel that when a species becomes rare as is the
Ivorybill that it is useless to try to preserve it, and they quote the history of
the Heath Hen as an example. I feel, however, that had we known as much about
Grouse in general twenty years ago as we do today, the Heath Hen might
have been saved, and the same holds true for the Ivory-billed Woodpecker.
Unless we know more than at present there is no hope of saving it.*

ARTHUR ALLEN, 1936

BLIND OBSERVATIONS

On April 2, 1929, the ornithologist Alfred O. Gross entered a small observation
blind in the heart of Martha's Vineyard.[1] Both the blind—a crudely built, four-
by-six-by-six-foot wooden structure—and the surrounding landscape—the
scrub-oak barrens characteristic of the island—seemed quite familiar to him
from repeated visits over the previous six years. He had come to the site to study
a bird that the public called the heath hen and scientists knew as *Tympanuchus
cupido cupido*. One of three geographic races of the greater prairie chicken, the
heath hen had once ranged across the northeastern part of the United States, as
far south as Virginia, perhaps even South Carolina, and as far north as Maine.
Now the beleaguered bird was making a desperate final stand on a single island
off the coast of Massachusetts. A Harvard-trained biologist with a position at
Bowdoin College, Gross had been commissioned to find out as much as he could

FIGURE 51. Alfred Gross holding a heath hen, 1923. The heath hen was the first endangered species to be subject to intense scientific study. By the time Gross began his research in 1923, however, the bird had already reached critically low numbers from which it would never recover. Courtesy of the George J. Mitchell Department of Special Collections and Archives, Bowdoin College Library, Brunswick, Maine.

about the bird's life and behavior. The hope was that the scientific knowledge he gained would help save the heath hen. Instead, its numbers, which had oscillated wildly during the previous several decades, now seemed to be dangerously declining. By 1929, the world's entire heath hen population had been reduced to a single male bird. Despite this devastating setback, Gross maintained his periodic vigil on the island until the bitter end.

Accompanying him that cool spring day on the eve of the Great Depression was his friend, the best-selling children's nature writer and radio broadcaster Thornton W. Burgess.[2] A year earlier, Burgess had collaborated with one of Gross's students to obtain motion pictures and still photographs to document the remaining heath hen population, which at that point still numbered three birds. Now Gross and Burgess stood witness to an unusual, sobering display. While the heath hen normally remained close to the ground, the lone remaining male flew to the top of a nearby tree, engaged in its characteristic courtship display, and uttered its eerie mating call.[3] It seemed as if the bird was signaling a refusal to accept its fate. As Burgess revealed several months later in a letter to the ornithologist Waldron DeWitt Miller, standing in the presence of the last heath hen moved him profoundly: "What extermination means was most vividly brought home to me last year when I visited Martha's Vineyard for the purpose of making motion pictures of the last Heath Hens. A year ago in March there were but three, all males. This last March there was but one. . . . Waiting in the blind for that lone bird to appear in the early morning, I felt the full force of the tragedy."[4]

While the heath hen was certainly not the first North American organism to fall victim to humans, its final days differed from other recently extinct species in several key ways. The public learned little about where or how the final Labrador ducks, passenger pigeons, or Carolina parakeets passed their last years in the wild. Because of their seasonal movements and once extensive ranges, it proved difficult to state with any degree of certainty what the exact status of these species was at any given point in time. It was even difficult to say for certain when any of them had been pushed over the brink of oblivion, especially since there continued to be claimed sightings of these species long after they had undoubtedly been lost. For the last fifty years or so of its existence, however, the heath hen remained confined to a single, relatively small island, where it proved easier (albeit still challenging) to monitor. And monitored it was. Ornithologists, government officials, sportsmen, and birdwatchers flocked to the island to attend the slow, painful demise of the heath hen, which became a national media event, documented in newspapers, periodicals, government reports, lectures, photographs, motion pictures, and even a Radio Nature League program that Burgess broadcast weekly from WBZ in Boston. According to Christopher Cokinos, who has painstakingly researched the history of the subspecies, as the story of the last heath hen made its way across the country it "galvanized interest in conservation." "The renown of the last specimen has spread remorse," reported the *New York Tribune*, "and the resolve to rescue other vanishing species."[5]

The heath hen also became the first endangered species that trained, professional scientists studied extensively in the wild. Rare and apparently vanishing species had long enchanted naturalists, but until the early twentieth century, their interest remained quite narrow. Most of all they desired preserved specimens of these precious organisms before it was too late. Yet, in the rush to collect, preserve, and catalog the last examples of the great auk, passenger pigeon, and Carolina parakeet, naturalists failed to learn much about their life history, behavior, or ecology. By the time the heath hen was making its final stand on Martha's Vineyard, however, the scope of natural history had broadened considerably. Increasingly, naturalists sought to learn more not only about how organisms lived but also how they interacted with each other and their natural environment. The emergence of graduate training in biology at the end of the nineteenth century and in ecology in the 1920s and 1930s reinforced this trend.[6] Informed by the latest developments in wildlife research and the product of countless hours of careful, patient observation, Gross's 1928 monograph on the heath hen represented the first of its kind for any threatened species.

The loss of the final heath hen also occurred during a period of deep rancor within American wildlife circles as naturalists, sport hunters, government officials, nature lovers, and other interested parties argued about the proper scope and goals of conservation.[7] Nowhere were these divisions more pronounced than at the National Association of Audubon Societies, which faced mounting turmoil during the early 1930s as an older generation of conservationists with strong ties to the sport hunting community and the Bureau of the Biological Survey gave way to a more independent generation of leaders and staff members, who tended to be younger and have at least some training in modern natural history. Under the leadership of Audubon president John Baker, who came to power in 1934 (just two years after the last sighting of the heath hen on Martha's Vineyard), the organization gave increasing emphasis to scientific research. Consistent with this aim (and at the urging of Aldo Leopold), Baker instituted a research fellowship program designed to gather more data about the status, habitat requirements, and future prospects of endangered species in North America. Although modest in scope, that program served the needs of university programs searching for ways to fund their graduate students during the height of the Depression while helping Audubon moor its wildlife conservation policy recommendations on more solid scientific foundations.

The Audubon fellowship program resulted in a pioneering series of widely distributed research reports on several charismatic birds—the ivory-billed woodpecker, California condor, roseate spoonbill, and others—studies that provided authoritative information about the life, ecology, and population status of these vanishing species. Each of them also offered a set of specific recommendations for how to preserve individual species and the critical habitat they needed to survive. In the end, though, each revealed the limitations of science in wildlife conservation. As Baker and his colleagues discovered, scientific knowledge might be necessary to rescue endangered species, but it was hardly sufficient. Science could not save those species in the absence of the political will needed to implement the recommendations made in those pioneering reports.

THE TRIALS AND TRIBULATIONS OF THE HEATH HEN

At first glance, the heath hen seems an unlikely poster child for endangered species. About the size of a small Cornish hen, it was a relatively drab bird—streaked with various shades of chestnut brown, gray, and black—whose coloration blended in perfectly with what seemed like a drab environment—the coastal plains and shrubby barrens of the northeastern United States. By some accounts, its meat tasted too gamy to be an especially desirable source of food. Nor did sport hunters find it a challenging target, since it massed in open fields and tended to fly with great labor and in straight lines when flushed.[8] It lacked the romantic associations with the frontier West that the bison enjoyed; it did not blacken the sky in awe-inspiring flocks like the passenger pigeon; and it failed to carry nationalistic symbolism like the bald eagle. Still, the heath hen attracted a significant following, especially in its final days.

Undoubtedly part of the species' charm lay in its mating ritual, referred to variously as drumming, booming, or tooting. Beginning as early as the first of March and reaching a height between early April and the middle of May, mature heath hens engaged in what Gross characterized as an "extraordinary performance."[9] During the early morning and late afternoon, the males gathered in open fields, where they strutted about, dashed forward for a short distance, leaped and twirled in the air, stomped their feet, stretched out their necks, lifted their dangling neck feathers (called pinnea) into a dark V shape that extended above their heads, and raised their tail feathers. After completing these preliminaries, the amorous males then inflated tennis-ball-sized orange-colored air sacs on the sides of their throats and simultaneously issued their haunting call, a low pitched sound in one-and-one-half to three-second bursts that Gross compared to the cry of a tugboat sounding its horn though a distant fog.[10] Seeing dozens, even hundreds of these birds engaged in a booming frenzy proved an unforgettable spectacle, and the behavior became the basis for the generic name that Smithsonian ornithologist Robert Ridgway bestowed on the bird, *Tympanuchus* (which literally means drum neck), when he revised the genus to which it belonged in 1885. Linnaeus coined the bird's specific epithet, *cupido,* chosen because the feather tufts on its neck resembled the wings of Cupid.

Following mating, the heath hen built its nest on the ground, in a spot concealed by dense vegetation. The nest remained so well hidden that naturalists located very few of them, despite increasingly diligent searches as the their numbers became thinned.[11] Because so few nests were found, determining the average clutch size for the subspecies is difficult, but the meager records that exist show a range of between four and twelve eggs per nest. Under ideal conditions, its reproductive potential thus proved quiet high. Even when disturbed, the female rarely left the nest, a behavior that probably increased heath hen mortality during the fires that periodically swept across Martha's Vineyard.

The writings of early European settlers in North America make frequent mention of the heath hen, sometimes referred to as the heathcocke, pheysant, or grous.[12] It was once so abundant on parts of mainland Massachusetts that laborers living on the site that became the city of Boston are supposed to have implored their employers not to serve it more than a "few times a week."[13] The British naturalist and artist Mark Catesby crafted the first published illustration of the bird for his two-volume *Natural History of Carolina, Florida and the Bahamas* (1731–43).[14] By the late eighteenth and early nineteenth century the subspecies commonly appeared in the market stalls of Boston, New York, and other eastern cities. Like many other forms of North American wildlife, the heath hen faced transformation from a component of a unique ecosystem to a subsistence food item and ultimately to a valuable commodity linked to growing markets with nearly insatiable demands.

The increasing exploitation of the heath hen soon gained the attention of government officials, who enacted limited measures to ensure its perpetuation.[15] As early as 1708, the New York state legislature passed an apparently unenforced law protecting the bird on the plains of Long Island. In 1791, that same body passed a bill entitled "An Act for the preservation of the heath hen, and other game," an action that startled several representatives who wondered why officials would want to preserve Indians or other "heathen." Massachusetts passed its first in a long series of protective laws in 1831, limiting the time of hunting to November and December. Though apparently well meaning, these early laws did little to stem the decline of the heath hen population or the shrinkage of its range. Mainland Massachusetts lost the bird by 1840, though it lingered in parts of New Jersey and Pennsylvania until as late as 1869. By 1880, the final surviving members of the race were restricted to Martha's Vineyard.

At this point, naturalists and collectors began paying increasing attention to the rare bird. In 1885, the ornithologist William Brewster—owner of one of the largest private bird collections ever amassed and a founder of the American Ornithologists' Union—became the first to formally differentiate the heath hen from its western cousin, the prairie chicken. Brewster based his determination on a comparison of three specimens of the bird's eastern form borrowed from the collection of the natural history dealer F. T. Jencks to more than a hundred western specimens he had found offered for sale in Boston markets. Based on his studies, he declared the two geographical forms to be "distinct but closely related species," a judgment that his colleagues subsequently rejected when they reduced the heath hen to a subspecies (or geographical race) of the prairie chicken.

Although aware of the drastic reduction in the heath hen's original range, Brewster felt the bird remained fairly common on Martha's Vineyard and was in "no present danger of extinction."[16] In a follow-up article five years later, however, he expressed alarm about the bird's status. Based on firsthand observations and the testimony of locals, he estimated that anywhere from 120 to 200 birds remained, a number sufficiently small "to warrant grave solicitude for the continuance of this interesting remnant of a once widespread race, as well as the most strenuous efforts for its protection and encouragement."[17] Since state law already protected the birds, Brewster argued for more vigorous enforcement efforts. He also suggested that the Massachusetts Fish and Game Protective Association divert funds it had been using to introduce exotic game into the state for protection of this rapidly diminishing native species.

Brewster's articles set off a flurry of activity among collectors who clamored to possess examples of this newly described and apparently rare bird. One particularly active dealer was the wool mill operator Charles E. Hoyle of West Millbury, Massachusetts, who in the 1890s personally collected or brokered the sale of many of the two hundred or so heath hen specimens amassed in public museums and private collections before the subspecies was lost.[18] In 1895, for example, Hoyle offered Brewster a set of fourteen heath hens for $250, while claiming that in the recent past he had been able to fetch as much as $50 for a pair of the birds.[19] Three years later, Hoyle reported that his supply of heath hens seemed to be drying up. This past season his "man" on the island managed to locate only two broods, a total of a total of twenty-seven birds, all of which he collected as soon as the chicks had fully feathered. Enclosed with the package containing the specimens was a note indicating that Hoyle's unnamed collector "did not expect to ever see another one."[20] While predictions of the heath hen's demise proved premature, intense pressure from collectors represented another factor threatening the bird, and it may have contributed to a population bottleneck from which it never fully recovered.

Collectors were not the only ones to notice the decline of the heath hen. In 1905, John Howland, a deputy state game warden from Vineyard Haven, began pushing the Massachusetts Fisheries and Game Commission and the newly established National Association of Audubon Societies to do something to address the problem. He hoped an increase in the fine for killing the bird (from $20 to $100) and an active campaign to arouse greater sympathy for it on Martha's Vineyard would do the trick.[21] Both institutions responded favorably to Howland's pleas, with Audubon president William Dutcher declaring that the "experiment of trying to save a species of bird on the verge of extinction" would be of "great scientific interest."[22] The Massachusetts state legislature declared a closed season of five years for the heath hen and instituted a fine of $100 for "hunting, taking or killing, or for buying, selling or otherwise disposing of or having in possession a Heath Hen or any part thereof."[23] In 1906, a fire on the island reduced the population to around eighty birds, leading to renewed pleas for conservation. This time the state legislature responded by establishing a 1,600-acre heath hen reservation in the center of Martha's Vineyard, from land purchased with funds donated by a number of organizations and individuals, including William Brewster, John E. Thayer (whose private collection contained twenty-five heath hens), and John C. Phillips. A year later, the state enlarged the reservation through purchase of an additional six hundred acres and the rental of one thousand more.[24] By the end of that year, estimates of the number of heath hens throughout the island ran between forty-five and sixty.

The heath hen population was then subject to closer scrutiny, more intense management, and wide population swings.[25] In the hope of encouraging the bird to rebound, state officials cut firebreaks, guarded against poaching, planted winter food crops, and instituted a more aggressive predator control program on the newly established refuge. Initially, it looked like these efforts might rescue the bird. In April of 1916, Edward Forbush, the Massachusetts state ornithologist, visited the refuge and found the booming grounds teeming with strutting heath hens. In all, he counted eight hundred birds at one site, and estimates of the population throughout the island ranged as high as two thousand.[26] It looked like the subspecies might have turned the corner on the road to recovery, even to the point of fueling optimism about reintroducing it back onto the mainland. Soon after his visit to Martha's Vineyard, Forbush met with T. Gilbert Pearson (of the National Association of Audubon Societies), Winthrop Packard (of the Massachusetts State Audubon Society), George W. Field (of the Massachusetts Fisheries and Game Commission), and other ornithologists and conservationists who decided to remove surplus birds for captive breeding and transplantation experiments. Then, in May of that year, disaster struck in the form of massive fire fanned by nearly gale force winds. In all, about thirteen thousand acres burned, more than 20 percent of the island and much of the scrub-oak barren habitat the heath hens needed to survive.[27]

Despite the devastating fire, Massachusetts officials decided to continue the risky transplantation experiment.[28] In December 1916, the New York Conservation Commission, which was keen on reintroducing the bird back into its former Long Island stronghold, received eighteen captured heath hens, which it placed in a three-acre enclosure. John C. Phillips, who had previously enjoyed success raising numerous game birds (including the prairie chicken) at his estate in Wenham, Massachusetts, received eight birds, which he kept in small covered breeding pens. In his characteristically understated tone, Gross later described the results of these experiments as "uniformly unsatisfactory." Seven of the heath hens shipped to New York fell victim to predatory birds, while the remainder died of a mysterious disease. One of Phillips's females was accidentally killed, one of the males died in a fight, and the remaining birds showed absolutely no interest in mating. Frustrated, Phillips sent the surviving birds to a colleague in No Man's Land, Massachusetts, where they apparently languished. Following the fire, the capture and removal of twenty-six birds, a hard winter, and an invasion of goshawks, by the spring of 1917, the heath hen population dipped to one hundred birds.[29]

For a short while, the bird seemed to rally again. Following his 1918 count, Forbush estimated that 150 heath hens remained on the island. That same year, Massachusetts game officials asked A. K. Fisher, of the Bureau of the Biological Survey, to inspect the reservation. Not surprisingly, given the priorities of his agency at the time, Fisher's report called for a more active program of fire and vermin control, both of which were instituted.[30] By the spring of 1920, Forbush counted 314 birds, and Allan Keniston, the superintendent of the reservation, estimated that the total population had increased to 600 birds. Then, for reasons that remain unclear, the population began falling again. By 1922, Forbush counted only 117 birds. Concerned that the state of Massachusetts was spending more than $3,000 per year to maintain a reservation that seemed to be failing in its mission, the director of the Massachusetts Division of Fisheries and Game sent a questionnaire to leading ornithologists, sportsmen, and others who had expressed an interest in the plight of the heath hen. Should the state continue its extraordinary efforts to save the bird? The response proved a resounding yes, and members of the Nuttall Ornithological Club and the Brookline Bird Club urged that a life-history study be undertaken as soon as possible to discover "the conditions which govern the existence of the birds on the island." When state officials failed to sponsor such a study, Phillips spearheaded the effort to raise private funds and hire Alfred Gross to conduct it. Phillips, Barbour, Thayer, and a number of other Nuttall Club members provided the initial funding for the research, with the newly founded Federation of New England Bird Clubs picking up the tab after the first few years.[31]

ENTER ALFRED GROSS, EXIT BOOMING BEN

Alfred O. Gross seemed a natural choice to undertake a detailed scientific study of the beleaguered heath hen.[32] Born in 1883 in east-central Illinois, Gross's early interest in natural history was fostered by his father, a merchant and farmer, and his older brother, who taught him to hunt. Following high school graduation, he attended a college preparatory school, where he supported himself by working as a taxidermist at the University of Illinois Natural History Museum. During his junior year in college, he began working for the pioneer ecologist Stephen Forbes, who offered him a job directing the Illinois Statistical Ornithological Survey, the first attempt to census bird populations across an entire state.[33] After earning his B.A. degree with special honors in zoology from the University of

Illinois, Gross entered graduate school at Harvard, where opportunities for the formal study of field natural history were unavailable.[34] Though his major advisor was the physiologist George H. Parker, Gross still managed to find time to do fieldwork during his summers at Harvard's Bermuda Biological Station, where he completed research on the tropicbird, his first of many subsequent life-history studies. After earning his Ph.D., Gross accepted a position at Bowdoin College, where he would spend the next sixty years, and soon completed studies of the dickcissel, the black-crowned night heron, and the common nighthawk.

Gross first traveled to Martha's Vineyard in early April 1923 and remained on the island until July. Initially, he found the reservation, a "large area of sandy plains grown up to scrub oaks," rather bleak, "little more than an uninteresting waste." Gradually, however, he warmed to the area's "lure and attractiveness."[35] What he failed to do, though, was detect a significant population of heath hens. His initial census revealed only twenty-eight birds, an alarmingly low number that clearly did not bode well for the future of the subspecies. While aware of the limitations of his census technique, which involved actual counts of birds engaging in their "weird courtship antics" supplemented with daily reports of wardens and other reliable observers, Gross feared that many previous estimates of heath hen populations had been marred "by the personal factor" and were thus "larger than the facts warranted."[36] Concerned that a disproportionate abundance of male birds reduced the breeding potential of the small remaining population, Gross recommended the trapping of five male heath hens, a common practice among gamekeepers in England and Scotland trying to maximize the reproductive rate of bird populations. Unfortunately, all of the captured males soon died.[37] The next year's census revealed a modest but still promising increase: fifty-six heath hens, all but three of which proved male. Then the population began declining again, a development that perplexed Gross and his supporters.

Following a 1925 conference on the heath hen, the Federation of New England Bird Clubs hired an additional game warden to patrol the reservation and further ratchet up predator control efforts. Between July of 1925 and November of 1926, Keniston and the warden killed at total of 162 cats, 71 crows, 21 rats, 44 hawks, and 5 owls with little apparent effect on the heath hen.[38] Rather than rebounding, the population continued to decline. The March 1927 census revealed only thirteen heath hens, and the best estimate indicated that a total of no more than thirty birds remained on the island.

In May 1928, Gross published the results of his first five years of heath hen study in a book financed with income from a bequest that William Brewster had granted to the Boston Society of Natural History. His introduction seemed to state the obvious when it declared the heath hen was "on the verge of extinction."

Although the bird had stimulated "great human interest among sportsmen, ornithologists and bird lovers all over the world," now it was "waging what appears to be a losing fight for existence": "It is probable that under the present efficient methods of protection the birds will continue to exist for many years, alternately rising and falling in numbers but finally flickering out. From a biological point of view it would be nothing short of miraculous if the present small group of Heath Hen, chiefly males, fighting against seemingly insurmountable odds, should be able permanently to re-establish itself over any considerable part of its former range in New England."[39]

In the body of his report, Gross summarized the history of the once abundant heath hen and detailed efforts to protect it since Howland first sounded the alarm about the subspecies in 1905. Gross blamed humans as a primary factor behind the "disappearance of the Heath Hen from most of its former range." The bird offered a "striking example of a bird which has not been able to adapt itself to the changing conditions brought about by civilization."[40] Outside of his more detailed analysis of what he thought had happened to the subspecies on Martha's Vineyard, however, Gross refused to speculate about exactly how and why the heath hen declined from its formerly extensive range on the mainland. He did hint that overhunting may have played a role in its overall decline, but he remained silent about the potential impact of habitat destruction, competition from exotic species, exposure to diseases, and other factors that may have contributed to a reduction in the subspecies' range. Undoubtedly part of his reticence here was related to the fact that early records of the heath hen tended to be "uncertain and obscure."[41]

He proved less reticent when discussing the impact of humans on the Martha's Vineyard heath hen population.[42] Summer residents of the island had frequently abandoned cats upon returning to the mainland in the fall, resulting in a large feral population that preyed on all manner of birds and small mammals. Attempts at supplementing the heath hen diet with corn and other grains had backfired by attracting rats and crows to the area, animals that also preyed on the subspecies. In addition, humans had introduced domestic chickens, turkeys, and geese onto the island, birds that carried deadly poultry diseases, like blackhead. At least one of the heath hens captured in the effort to reduce the ratio of male birds in 1923 died from this disease.

Beyond these factors were others—native predators and the periodic fires—whose causes and consequences proved much more difficult to disentangle. Gross admitted that heath hen remains had been found in the stomachs of marsh hawks, red-tailed hawks, and especially goshawks, which had appeared on the island during the winters of 1916–17 and 1926–27. But he remained skeptical of the widely accepted notion that "wholesale killing" of all hawks and owls, particularly those species that fed primarily on rodents, benefited the heath hen. On the contrary, he feared that indiscriminate killing of predators might "upset the balance of nature so as to act as a boomerang to the Heath Hen." Attempts at fire suppression proved equally uncertain in their impact. Periodic fires were a friend of the heath hen, as long as the population remained reasonably large and widely dispersed, because they helped keep the bird's habitat from becoming overgrown with shrubs and trees. But as the heath hen population became tiny and confined to a relatively small area, it became much more vulnerable to fire. To complicate the matter further, fire suppression efforts on the island may have contributed to more severe burns when fire inevitably broke out.

Gross suggested two final factors that seemed to have contributed to the decline of the heath hen on Martha's Vineyard. Several observers had noted the high ratio of male to female birds in the island's heath hen population. He speculated that this imbalance might have resulted either from "excessive inbreeding" in the small remaining colony or from the fact that females proved more prone to predation and fires while incubating their eggs and caring for their young. Whatever the cause of the distorted sex ratio, Gross pointed that in closely related species, like the ruffed grouse, males were known to harass nesting females and even kill their young, and he feared something similar might be happening with the heath hens on Martha's Vineyard. In addition, Gross noted that male sterility might have played a role in the population decline.[43]

Gross's report represented the most exhaustive study of an endangered species that had ever been attempted. It provided much more detailed and reliable information than, for example, Hornaday's study of the bison, issued in 1889, or anything that was published about the passenger pigeon or the Carolina parakeet before they fell victim to extinction in the first part of the twentieth century. Still, Gross failed to produce a definitive life history and ecological study of the kind that scientists routinely began to undertake at the time he completed his research. For one thing, the rarity of the species severely limited the kind of research he could undertake. Gross, for example, proved unable to thoroughly document exactly what the heath hen ate. Rather than collecting numerous specimens of the heath hen during different times of the year and examining their stomach contents—a technique that economic ornithologists had used for decades—Gross had to rely on the "meager notes" contained in the writings of previous ornithologists, a handful of records taken from museum specimens, and his own field notes.[44] As we have already seen, his ability to learn much about the behavior of females and young was also limited by the fact that he failed to locate the nest of the heath hen, despite repeated searches.

In 1929, one year after publishing his report, Gross penned an article for *Bird-Lore* announcing that "in spite of the best efforts of conservationists, the Heath Hen has steadily decreased in numbers until, this year, apparently, but one bird remains. The death of this individual will also mean the death of its race, and then another bird will have taken its place among the endless array of extinct forms." The state of Massachusetts had spent more than $70,000 and conservation organizations, bird clubs, and individuals had donated "thousands more" to study and protect the heath hen. "For the first time in the history of ornithology," Gross pointed out in a later version of this article, "a species has been studied and photographed in its normal environment down to the very last individual."[45] In his final public statements about the bird, Gross emphasized two causes above all others for its demise: "The habit of the Heath Hen of congregating in open fields and the ease with which it was tricked and killed by the market gunners were contributing factors in its rapid decline after white man and firearms came to America."[46]

The final heath hen, nicknamed "Booming Ben," continued to live for several more years. Ironically, he almost perished in 1929 when John Phillips, who had done as much as anyone to save the species, nearly ran over him with his car.[47] Some scientists argued that the last heath hen should be captured to preserve its skin for prosperity, as had been done with the final passenger pigeon that died in the Cincinnati Zoo, but Gross and his colleagues thought it more fitting to let it "live in its natural environment among the scrub oaks of Martha's vineyard than . . . to mount it and have it collect dust on some museum shelf."[48] The last authenticated sighting of this increasingly famous lone bird occurred in March of 1932. When Booming Ben failed to make an appearance the following spring, Gross reported to Massachusetts officials that the race *Tympanuchus cupido cupido* was extinct, though naturalists never found his body.[49]

The profound sense of loss that the demise of this bird provoked was best captured in an anonymous editorial published in the *Vineyard Gazette*, which devoted an entire issue to the passing of the heath hen: "There is no survivor, there is no future, there is no life to be recreated in this form again. We are looking upon the uttermost finality which can be written, glimpsing the darkness which will not know another ray of light. We are in touch with the reality of extinction."[50] According to this unnamed writer, even scientists who had come to Martha's Vineyard "full of passion for metric measurements and Latin labels" had been inspired "to write poetry" about their experiences with the bird. The demise of the heath hen had opened up a "void in the April dawn" that would not soon be forgotten. In the end, the considerable knowledge that Gross produced

FIGURE 52. "Booming Ben," the last heath hen, 1929. As the final living heath hen, Booming Ben became a minor media celebrity that provoked broad discussion about the larger meaning of wildlife extinction. From Alfred O. Gross, "The Last Heath Hen," *Bird-Lore* 31 (1929): 253.

about the heath hen revealed much about why the species had perished, but it could not prevent this void.

ON THE TRAIL OF THE IVORYBILL

In January 1937, nearly five years after the last definitive sighting of the heath hen, James T. Tanner loaded up his car and headed south. Over the next three years, the twenty-two-year-old Cornell graduate student would devote nearly twenty-one months in the field, logging more than forty-five thousand miles on his Model A Ford and covering countless additional territory by train, foot, boat, and horse. The goal of this lengthy journey was to learn more about the elusive ivory-billed woodpecker, known to science as *Campephilus principalis,* but known to locals across its range as the "ivorybill," the "kent," the "king of the woodpeckers," or the "Lord God bird." With a wingspan of approximately thirty inches, it was the largest woodpecker in North America and among the largest of its kind in the world. This awe-inspiring species had once inhabited the moist bottomlands and swampy forests across the southeastern United States. But as its habitat had been destroyed in the years surrounding the American Civil War, its range had been dramatically reduced and its numbers dangerously thinned. Tanner hoped to discover exactly how many ivorybills remained, to identify the specific locations they inhabited, and to study their life history, behavior, and ecology. Ultimately, he and his sponsors hoped his research would inform policies that would rescue the beleaguered species.[51]

Financed with a grant from the National Association of Audubon Societies, Tanner's study came at a crucial juncture in the development of American wildlife management, when the field was experiencing a period of rapid professionalization and transformation. Within a few short years in the midst of the Great Depression, Herbert Stoddard authored the first American monograph on the biology and management of a game species (1931); Aldo Leopold published the first textbook and secured first university chair specifically devoted to the field (1933); the U.S. Bureau of the Biological Survey established a Cooperative Wildlife Research Unit Program in conjunction with eight American universities (1935); and wildlife managers and conservationists called the first North American Wildlife Conference (1936), founded their own professional organization, the Wildlife Society (1937), and began publishing the *Journal of Wildlife Management* (1937). During this same period, a handful of professionally conscious wildlife managers began calling for an increased reliance on science to learn more about the factors governing the distribution and abundance of native flora and fauna. It was time, these reformers argued, to abandon the crude "rule-of-thumb" calculations that had long characterized their field. It was also time to broaden the preoccupation with game species, which had traditionally garnered the lion's share of attention and funding.[52]

Perhaps not coincidentally, these calls for reform in wildlife management came not long after the field of natural history began experiencing a series of profound shifts of its own. Naturalists who had once learned their craft through apprenticeships and self-study were quickly giving way to university-trained biologists, while a traditional focus on collecting, naming, and describing new species was broadening to an interest in the ecology and behavior of living organisms in the field. By 1910, the Ph.D. degree had become the standard entry credential into most scientific disciplines. Within the next decade, a series of programs at places like Cornell, Michigan, and Berkeley began churning out university-trained naturalists with the scientific tools and the personal inclination to study native species in their natural environment.[53]

Modern science did offer the hope of providing authoritative information about the status of endangered species. In fact, Tanner eventually succeeded in locating a handful of surviving ivory-billed woodpeckers, in identifying the critical habitat they needed, and in offering a series of policy recommendations that might have ensured their long-term viability. But as Tanner and his sponsor, the National Association of Audubon Societies, learned, science alone could not save the species. In the early 1940s, when Tanner published the results of his ivorybill research, America seemed much more concerned with preserving private property rights, promoting economic development, and winning the war than in saving an obscure swamp-dwelling bird that ate grubs. While we will never know for certain whether Tanner's recommendations might have snatched the ivorybill from the jaws of extinction, we can be sure that the failure to implement them helped secure the species' fate.

The British naturalist Mark Catesby first described the ivory-billed woodpecker in the early eighteenth century, one of dozens of New World birds he introduced to science.[54] By the first several decades of the nineteenth century, Alexander Wilson, John James Audubon, and other naturalists had recorded the basic facts regarding the ivorybill's appearance and habitat.[55] The bird's plumage was glossy black, with a white stripe that started on each cheek and continued down the side of its neck. It also possessed a broad band of white on the trailing edge of its wings, an area that appeared as a highly visible triangular patch on its back when the bird was perched. Its bill was ivory white and approximately two and a half inches long. The male boasted a prominent scarlet crest that shone brilliantly in the sunlight, while the crest of the female was entirely black. Audubon compared the bird's call to the high false note of a clarinet, while others described it as a nasal "kent" or thought it resembled the sound of a child's tin trumpet. Although it was (and still is) often confused with its ubiquitous cousin, the pileated woodpecker (*Dryocopus pileatus*), the ivorybill was slightly larger, its bill whiter, and its large white wing patches clearly visible on its back when it rested. Early on, naturalists also recognized the ivorybill's preference for deep bottomland forests in the southeastern United States.

By the end of the nineteenth century, the ivorybill had declined to the point that it had become one of the rarest of North American birds. As such, its skins and eggs proved in great demand among bird collectors. Even naturalists who were otherwise sympathetic to avian conservation rarely passed up the opportunity to collect an example of the increasingly prized bird. For example, the ornithologists Frank Chapman and William Brewster had both been founders of Audubon bird protection movement in the late 1880s.[56] Yet, in 1890, the two were delighted to find an ivorybill during an expedition down Florida's Suwannee River and expressed no qualms about collecting the rare bird.[57] The rage for collecting remained so strong that by 1920 more than four hundred ivorybill skins were gathering dust in ornithological collections across the world, yet only than the most rudimentary facts concerning its life history and behavior were known.[58] At that point, some feared the bird might have been gone entirely.

As had been the case earlier with the heath hen, predictions of the ivorybill's demise proved premature. In 1934, Arthur A. Allen, a professor of ornithology at Cornell, got wind of a recent sighting of the ivorybill in Madison Parish, Louisiana, on an eighty-one-thousand-acre site known as the Singer Tract.[59] At the time, he was contemplating how to spend his sabbatical leave the following spring. During the previous decade, Allen had been building Cornell's ornithology and wildlife conservation program into one of the nation's largest and most successful of its kind.[60] Cornell also represented a moving force behind the American nature study movement, so not surprisingly, besides his extensive undergraduate and graduate teaching duties, Allen also remained committed to the popularization of ornithology. In the late 1920s, he began experiments to produce the first sound recording of birds in the wild. Joining him in these pioneering efforts were two of his students, Peter Paul Kellogg and Albert R. Brand, both of whom would devote much of their subsequent careers to producing bird recordings for scientific study and popular consumption.[61]

During the summer of 1934, Allen and Brand decided to mount an expedition to record the voices of birds, particularly species that seemed vanishing. Brand, a former Wall Street investment banker and now an affiliate with the American Museum of Natural History in New York, planned the itinerary and financed the venture.[62] According to Allen, one of the "first objects of the expedition was the rediscovery of the ivory-billed woodpecker, perhaps the rarest of North American birds and at one time thought to be extinct."[63] Allen and his colleagues also hoped to record the songs, take photographs, and capture motion pictures of many more common species that they would encounter along the way. Accompanying Allen throughout the fifteen-thousand-mile journey were Kellogg and another of his students, James T. Tanner. Brand and George M. Sutton, a bird artist and curator of ornithology at Cornell, also joined the expedition party at various points along the way.

The Brand–Cornell University–American Museum of Natural History Ornithological Expedition set out from Ithaca in mid-February 1935. One of the first destinations was central Florida, where Allen had found a pair of ivory-billed woodpeckers in the mid-1920s that a local taxidermist had later collected. Despite a month-long search in areas of the state the bird had once inhabited, Allen and his colleagues failed to locate a single ivorybill. Disappointed, but undaunted, in late March they headed off to Madison Parish, Louisiana, where they continued their search with the aid of two local guides. Spring rains made the muddy unpaved roads virtually impassable for their heavy trucks, so the naturalists loaded their supplies and 1,500 pounds of recording equipment onto a mule-drawn wagon and slogged their way into the heart of the Singer Tract. On the third day, seven miles from the nearest improved road, the expedition party be-

FIGURE 53. Transporting equipment into the Singer Tract, 1935. From *left* to *right:*
James T. Tanner, J. J. Kuhn, A. A. Allen, Albert, and Ike. With the help of Kuhn, a
local guide, Allen's expedition party managed to record, photograph, and film the
elusive ivorybill for the first time. The success of that expedition led to the creation of
the Audubon Graduate Fellowship program. Courtesy of James T. Tanner.

came ecstatic when they finally located a pair of nesting ivorybills. In the words
of Sutton: "The whole experience was like a dream. There we sat in the wild
swamp, miles and miles from any highway, with two ivory-billed Woodpeckers
so close to us that we could see their eyes, their long toes, even their strongly
curved claws without binoculars."[64]

The group pitched a tent less than three hundred feet from the nesting site and
dubbed their soggy new quarters "Camp Ephilus," a pun on the scientific name
of the species they had rediscovered. For the next several days, they gathered a
mountain of information on the life and behavior of the ivorybill, more data than
had ever been previously accumulated on the species in the nearly two centuries
since Catesby had first described it. The men took turns using tripod-mounted
twenty-four-power binoculars to monitor the nest from daylight until dark. In
addition, Sutton sketched the birds from life, while Allen, Tanner, Kellogg, and
their guide made the first photographs, motion pictures, and sound recordings of
the ivorybill from a makeshift blind constructed in an adjacent tree only twenty
feet from the nest opening, which was nearly fifty feet from the ground.

Allen's motion pictures and sound recordings featuring the elusive ivorybill
proved a huge hit at the numerous popular lectures he delivered following his

return to Ithaca. Perhaps no audience was more enthralled with the presentation, however, than the members assembled in New York City for the National Association of Audubon Society's annual meeting in October 1935.[65] At the time, the organization had begun devoting increasing attention to threatened species, in part because of a tremendous upheaval that the group had recently experienced. The sources of that discord were many, but much of it can be traced back to a scathing sixteen-page pamphlet issued back in 1929 and entitled *A Crisis in Conservation: Serious Danger of Extinction of Many North American Birds.*[66] As we have seen, the tract's authors charged the National Association of Audubon Societies with standing by idly while a disturbing number of avian species were rushing headlong into extinction.

It took several years for all the dust to settle, but by 1934, the association's longtime president, T. Gilbert Pearson, had been forced from office, and John Baker, a birdwatcher and no-nonsense business executive took charge of the organization's day-to-day operations. Over the next several years, Baker instituted a series of reforms aimed at restoring the credibility and effectiveness of Audubon.[67] One strategy for reinvigorating the organization was to recruit new staff and board members. An example of the latter was Aldo Leopold, whom Baker described as one of the "most original thinkers on conservation and game management subjects I know."[68] At the time of Baker's ascendancy, Leopold was making a name for himself through his thoughtful writings on conservation, his pioneering book on *Game Management* (1933), and his newly created program in that same field at the University of Wisconsin.

In the fall of 1935, Baker began discussions with Leopold on how to focus American wildlife conservation more on the needs of species facing extinction. Writing to Baker from Germany in September of that year, Leopold articulated a desire for "an inventory of all native species, especially threatened ones, giving a pretty *national work plan* as to just what is needed to perpetuate that species, and who can do it. The entry for each species should include not only protective legislation (on which NAAS has so far laid almost exclusive emphasis), but also protective actions other than legislation, and especially 'environmental control' needs. . . . Research needs should also be listed and 'assigned.'"[69] Leopold drew up a sample entry for the sand hill crane indicating exactly the kind of information required. The entry included a call for financing a research fellowship to learn more about the life history of the species.

Later that year, Leopold began circulating a "rough draft" of a paper entitled "Proposal for a Conservation Inventory of Threatened Species."[70] There he pointed out that despite the "large and sudden increase" in wildlife conservation initiatives in the United States, governmental agencies, private organizations, and individual citizens had failed to coordinate effectively.[71] Moreover, too much activity remained focused on game species, which were likely to be saved by "powerful motives of local self-interest . . . however great the blunders, delays and confusion" in getting their management properly established. The same was not necessarily true for what Leopold referred to as "wilderness game" (like grizzly bears), migratory birds, predators, and rare plant associations, or "in general all wild native forms that fly at large or have only esthetic and scientific value to man." These species represented the "threatened element in 'outdoor America'—the crux of conservation policy." What was desperately needed was an inventory of these threatened forms "in each of their respective places of survival" as well as an "inventory of the information, techniques, and devices applicable to each species in each place, and of local human agencies capable of applying them."[72] As an example of the kind of information he thought would prove useful to compile and circulate, Leopold pointed to the ivory-billed woodpecker, "a bird inextricably interwoven with our pioneer tradition—the very spirit of that 'dark and bloody ground' which has become the locus of national culture." Ornithologists had recently located a remnant population of these noble birds in a virgin forest slatted for logging. Now it was important to bring this information to the attention of the National Park Service, the Forest Service, and the Bureau of the Biological Survey so that these agencies could work together to save this remnant home of the ivorybill.[73]

"AN ANIMATED NEEDLE IN A HAYSTACK"

About the time Leopold began circulating his proposal for an inventory of threatened species, Allen wrote him. While working up his notes on the ivory-billed woodpecker, he had become "increasingly impressed" with the need to have a "well-qualified young ornithologist make an intensive study of the present status and life history of the bird, taking up where we left off last spring."[74] James Tanner, the graduate student who had accompanied the Brand–Cornell–American Museum Expedition the previous year, had expressed an interest in pursuing the project and had even approached Audubon officials to ask about the possibility of funding. Leopold replied that "nothing would delight me more than to see you follow up with a life history study of the ivory-bill. . . . I hope, too, that any research plan would include an administrative plan for the perpetuation of the species."[75]

Baker was quickly sold on the idea. He and his staff, however, began thinking in terms not simply of a single graduate fellowship to study the ivorybill, but a series of Audubon-sponsored fellowships to study other endangered species as well. Besides discussions with Allen, Baker also approached Joseph Grinnell, at the University of California at Berkeley, about a graduate fellowship to study the California condor and Charles T. Vorheis, at the University of Arizona, about another to study the desert mountain sheep. In a speech announcing the fellowship program at the 1936 Audubon Society meeting, Baker suggested that studies on the whooping crane, glossy ibis, sage grouse, grizzly bear, and mountain lion should also be initiated.[76] Baker then wrote Allen to ask whether Cornell officials would welcome such a fellowship, if Allen would be willing to recommend and supervise a suitable candidate, and when the fellowship could begin. The results, Baker hoped, would be widely circulated in special report that the Audubon Society would publish.[77] In subsequent correspondence, Baker indicated that Audubon could commit up to $1,500 a year for three years to fund the Ivory-Bill Graduate Research Fellowship.[78]

Given the difficulties in securing financing for his students, especially during the Depression, the Audubon fellowship proposal delighted Allen. During the negotiations over the specifics of the contract, he sought to reassure Baker that science might still rescue the beleaguered species. Based on the recent experience of the heath hen, some considered it futile to attempt to rescue a bird that had become as rare as the ivorybill. But "had we known as much about Grouse in general twenty years ago as we do today as we do today," Allen argued, "the Heath Hen might have been saved, and the same holds true for the Ivory-billed Woodpecker. Unless we know more than at present there is no hope of saving it."[79]

With three years of secure funding, Tanner set out on the trail of the ivorybill in January 1937. He spent his first several months in the field trying to discover how many and where ivorybills might still exist. Based on published records, specimens in collections, rumors of sightings, and interviews with local residents, he identified about fifty potential sites to survey in seven southeastern states.[80] Ironically, one of the most useful sources of information about the species turned out to be the employees affiliated with the logging industry, who tended to have a detailed knowledge of local timber stands and the wildlife they contained. When he identified what appeared to be suitable habitat in swampy forests, he searched high and low for ivorybill holes and feeding signs, extensive scaling of bark on trees that had recently died, while listening carefully for the bird's distinctive call and the characteristic loud, double rap of its beak on a tree. Tanner compared the enterprise to "searching for an animated needle in a haystack" and after his arduous investigation concluded that only two dozen ivory-bill woodpeckers still remained in about five areas in Louisiana, Florida, and possibly South Carolina.[81]

In fact, he had seen with his own eyes only six of the elusive birds, all of them in the Singer Tract in Madison Parish, Louisiana. There he spent a total of over a

FIGURE 54. A pair of ivory-billed woodpeckers exchanging places on a nest, 1935. During his studies of the Singer Tract ivorybills, Tanner discovered that the species' restricted range was related to its narrow feeding preferences. Using the knowledge he gained, the National Audubon Society pushed to have the Singer Tract preserved. Courtesy of James T. Tanner.

year documenting their ecology and life history. Among his most important discoveries, Tanner found that the species' dire status appeared linked to its narrow feeding preferences. The ivorybill's primary food source was the wood-boring larvae of several families of beetles that live between the bark and sapwood of recently dead trees. Ivorybills existed only in forests where dead and dying trees were frequent and other woodpeckers abundant, conditions normally only found in "tracts of uncut, mature timber."[82] While overzealous collecting and careless hunting might have contributed to the decline of the species in some areas, the overall diminution in range of the ivorybill coincided with the spread of the logging industry in the South: "Mature forests of large, old trees have almost disappeared, and these conditions favorable for the Ivory-bill will very probably never again prevail. Its preservation must be accomplished by saving suitable habitat or by maintaining on certain areas an adequate food supply for the birds."[83]

Ideally, areas still inhabited by the ivorybill should be preserved as "refuges and as primitive areas."[84] According to Tanner, the most promising sites included the Big Cypress Swamp and the lower Apalachicola River Swamps in Florida, possibly the Santee River Swamp in South Carolina, and most importantly, the Singer Tract in Louisiana. In these locations, he recommended securing a minimum undisturbed area of about two and one-half to three square miles for each pair of the birds. Tanner emphasized that this was a minimum requirement, and "The larger the area of the forest the better for the conservation of the birds—that is certain."[85] Where the chief obstacle to the establishment of a larger protective refuge was the market value of the standing timber, as with the Singer Tract, Tanner recommended a program of selective cutting outside the undisturbed areas that would leave dead and dying trees to supply borers to the woodpeckers. He also thought selective killing of suitable tries might encourage the growth of borer populations, thus increasing the ivorybill's food supply.[86]

Tanner's copiously illustrated account appeared in 1942 as Research Report No. 1 of the National Audubon Society.[87] In a foreword to the publication, Baker declared that the Audubon Society considered scientific research of the type found in this report as an "essential basis for wise policies governing the conservation of wildlife resources" and pronounced that his organization remained firmly committed to the proposition that "the Ivory-bill shall not, as a part of America's heritage, go the way of the Passenger Pigeon and the Great Auk."[88] Allen contributed a preface that proclaimed that "the greatest tragedy in nature is the extinction of a species. . . . Surely no intelligent human being could be indifferent to the passing of the last Ivory-billed Woodpecker."[89]

At the time of Tanner's report, the National Audubon Society had already begun efforts to protect the ivorybills on the Singer Tract. The Singer Manufacturing Company had acquired the land in 1916, and ten years later, company officials contracted with the Louisiana Department of Conservation to establish a game refuge in the area.[90] Their motivation was not to protect the ivorybill or other wildlife in the area, however, but simply to minimize the risk of accidental fire from hunters. By the time Tanner began his study in 1937, the Singer Tract consisted of about eighty-one thousand acres, four-fifths of which he considered "virgin timber."[91] That same year, the Chicago Mill and Lumber Company acquired rights to the lumber on the Singer Tract and soon began logging west of the Tensas River. By 1941, the tract had been approximately 40 percent cut over.[92] Fortunately for the ivorybill, most of the land logged by that point was not in areas where the species had recently been found.

Baker pursued numerous avenues in a desperate bid to protect the site. He appealed directly to President Roosevelt for help, and soon officials in the Forest Service, the Park Service, and the Fish and Wildlife Service became involved in the discussions.[93] However, he failed to convince any of these federal agencies to acquire the home of the ivorybill, the primary obstacle being the value of the standing timber on the Singer Tract, estimated to be worth about $2 million.[94] He pleaded with two successive governors of Louisiana to use the powers of condemnation to declare a wildlife sanctuary in the area, but both refused. With the pace of logging increasing during the war, in 1943, Baker secured a $200,000 commitment from the state of Louisiana to purchase part of the remaining uncut area in the Singer Tract as an inviolate wildlife refuge. He also convinced the governors of Louisiana, Tennessee, Arkansas, and Mississippi to sign a petition requesting the Chicago Mill and Lumber Company to waive its contractual rights to cut the remaining timber on the Singer Tract and to sell a cutover buffer area for a reasonable price.[95] At a meeting in Chicago in December of that year, however, officials of Chicago Mill and Lumber flatly refused to negotiate. They did reveal that German POWs were logging the land and that most of the lumber was being fashioned into shipping crates and tea chests for the British.[96] By 1944, only one ivorybill remained in the Singer Tract. It represented the last documented sighting of the bird in the area.[97] And still the cutting continued. (Ironically, in 1980, the federal government created the Tensas River Wildlife Refuge on part of the site).

In the end, hopes for saving the ivory-billed woodpecker were dashed on the hard rocks of political and economic reality. Clearly, science could reveal vital information about the status and ecology of this magnificent creature, but science alone could not effect the fundamental reorientation in values needed to save it. By the time American society began to show signs of such a value shift, it was too late for the ivorybill. The species was among the earliest to be listed under the first Endangered Species Act of 1966, but by then, its trail had grown increasingly cold.[98] While recent reports of sightings nurse hopes that the bird might once again grace our bottomland hardwood forests, it seems possible that the ivorybill now exists only as lifeless skins in the drawers of museums, as fading images on paper and film, and as ghosts that haunt us with what might have been.[99]

IN CONDOR COUNTRY

The California condor (*Gymnogyps californianus*) narrowly escaped a similar fate. Anyone who has seen a close-up photograph of the bird—which bears an uncanny resemblance to a Disney cartoon character—might wonder why its decline would engender concern. From the perspective of most humans, the condor is a homely creature that survives in an grisly fashion: by picking through the carcasses of dead and decaying animals, mostly large mammals. Witnessing the species in person, however, can be a different experience altogether. With its massive size, majestic flight, and wilderness haunts, it is easier to appreciate how the bird could elicit mounting concern as its numbers diminished.

Possessing a wingspan of nine to ten feet, the California condor is the largest soaring land bird in North America, and from a distance, viewers often confuse it with an aircraft.[100] Able to cover vast distances in a single day, the condor relies on thermals or winds blowing over topographic features for flight, a characteristic that generally limits food acquisition to the middle hours of the day. Its average flying speed is around thirty miles per hour, although they have been clocked at up to forty-five miles per hour. According to the conservation biologists Noel Snyder and Helen Snyder, the basic habitat needs of the California condor are relatively straightforward: an area with adequate carrion, reliable winds or thermals, and few obstructions so food can be easily discovered and safely accessed.[101] The species is monogamous, with pair bonds enduring over many years and usually terminating only with the death of one member. At the same time, it is highly sociable, gathering in large numbers at roosts, bathing areas, and feeding sites, a behavior that has facilitated census attempts, but also rendered it more vulnerable. Adult California condors face relatively few predators, save for humans, while eggs and nestlings sometimes fall victim to common ravens, golden eagles, black bears, and (formerly) grizzly bears. However, with a late date of sexual maturity (about six to eight years), an invariable clutch size of only one egg, and a tendency not to lay a second egg in a year following successful nesting, the species has an extremely slow population growth potential, so it must maintain low mortality rates to persist.

The California condor once ranged widely across North America. Remains of the species have been found as far east as New York and Florida and as far south as Nuevo León, Mexico. The extinction of much of the continent's mammalian megafauna at the end of the Pleistocene era probably contributed to a reduction in the condor's range, which was gradually confined to the southwestern portion of what would later became the United States.[102] According to some scholars, the humans who arrived on the North American continent about twelve thousand years ago may have killed off many of the large mammals that once served as the condor's food.[103] Native peoples in the western part of the continent may have also more directly limited the population of the condor by widely using it in ritual sacrifice ceremonies.[104]

Reliable information about the species proved slow in coming. Although brief accounts of the bird occasionally made their way into writings of the early Spanish settlers in North America, the California condor remained uncollected until 1792.[105] That year, the Scottish naturalist Archibald Menzies secured a specimen at Monterey that ended up at the British Museum, where it was mounted for public exhibit, formally named, and scientifically described.[106] Just over a decade later, members of the Lewis and Clark expedition encountered the bird near the mouth of the Columbia River. When a member of the exploring party brought a wounded condor back to camp, Lewis had the opportunity to sketch its head, measure its body, and describe its soft parts for the first time.[107] In the 1830s, Audubon produced a life-size drawing of the California condor for the folio edition of his *Birds of America,* although the source of the specimen that served as his model is uncertain.[108] In a short account of the species in 1890, the famed ornithologist and longtime California resident James G. Cooper became the first to sound the alarm about its status. Based on wide reading in the literature and his own firsthand observations of California avifauna over a period of nearly four decades, Cooper declared that "unless protected our great vulture is doomed to rapid extinction."[109]

Cooper's dire prediction coincided with and probably contributed to a growing interest in the condor within the ranks of bird collectors. Between 1881 and 1910, they took a minimum of 111 skins and 49 eggs of the species from the wild; at least 20 birds and 7 eggs were collected in 1897 and 1898 alone.[110] As had been the case with the heath hen and the ivory-billed woodpecker, the increasingly rare condor became a highly valued commodity within natural history circles, and that fact, coupled with its low reproductive potential, placed the species in jeopardy. In 1895, H. R. Taylor, a California collector and natural history dealer, offered to pay $250 each for up to three eggs of the bird.[111] While it remains unclear whether anyone took Taylor up on his offer, a decade later he reported that his hired collector had gathered nine condor eggs. In 1908, another California collector, Kelly Truesdale, established a dubious record when the Massachusetts ornithologist, sportsman, and collector John E. Thayer (the very same gentleman whose museum contained no less than two dozen heath hen specimens) paid him $300 for a single condor egg.[112] In his detailed study of the California condor published in 1978, the biologist Sanford Wilbur concluded that "the disappearance of condors from various parts of their range coincided with peaks of collecting activity in those areas."[113] In their recent monograph on the species, Snyder and Snyder are equally blunt in their analysis of the impact of collecting on the California condor: the large take of birds at the hands of collectors "must be considered one of the significant historical stresses on the species."[114]

Even as condor collecting reached a peak, however, a modest movement to protect the bird began. Initially, the center of that movement was a group of California ornithologists and bird collectors who expressed not only a growing pride in but also a deep concern about the native fauna of their ecologically diverse home state. In 1893, four such individuals founded the Cooper Ornithological Club.[115] The organization published brief accounts of its activities in a series of short-lived natural history journals before launching its own *Bulletin* in January 1899. A year later, the journal was rechristened *The Condor,* a nod to the growing attachment its members felt to this increasingly beleaguered species, whose range now appeared to be confined largely to California. In addition to the observation and promotion of bird study, a stated goal of the Cooper Club was "conservation of Birds and Wild-Life, in general, for the sake of the future."[116] Members of this organization undoubtedly played a role in lobbying for the passage of a rarely enforced state law, passed in 1905, that made it illegal to kill condors or collect their eggs.

A year later, the naturalist and pioneering bird photographer William L. Finley began taking pictures of a nesting pair of California condors in Eaton Canyon, located in the San Gabriel Mountains, just north of Los Angeles. Guiding him through unfamiliar territory on the day that he first found the condor nest were two Cooper Club mainstays, Walter Taylor and Joseph Grinnell, the latter of whom had begun editing *The Condor* only one year previously. Together with his long-time partner, Herman Bohlman, Finley repeatedly returned to the nest site. They encountered adult birds so unfazed by their presence that they did not have to conceal themselves in a blind to photograph them, behavior at odds with later claims that the species was exceedingly wary of humans. In all, Finley and Bohlman produced more than 250 condor images, a remarkable accomplishment considering the heavy equipment available to them at the time and the nesting site's challenging terrain. Finley also issued a series of articles based on his observations of the condor pair and their nestling, which he captured and transported back to his home along the Willamette River in Oregon. He later donated his pet condor, named "General," to the New York Zoological Society, where the bird lived for eight years before dying from ingesting a rubber band that had been carelessly thrown into its cage.

Finley was decidedly pessimistic concerning the fate of the species. In the second of four articles published in *The Condor,* he confirmed Cooper's earlier suspicion that the species "will soon become extinct." Since the time when white settlers had first arrived in the western United States, its breeding range and overall population had greatly diminished. Although the bird survived in the "wilder mountainous" regions of central and southern California, it was "nowhere common." Finley highlighted two principle causes for the decline of the condor: overhunting and predator control campaigns. Most species would be able to withstand these onslaughts, but the condor had the misfortune to possess an usually slow reproductive rate. Given these conditions, Finley concluded, it was "not surprising that the condor numbers are decreasing, and unless the needed protection is given, this bird will undoubtedly follow the Great Auk."[117]

Finley's writings and photographs initially failed to ignite broader concern about the fate of the California condor. For the next several decades, members of the Cooper Club continued to gather scattered information about the species,

FIGURE 55. Pair of California condors in Eaton Canyon, 1906. Photograph by William Finley and Herman Bohlman. Finley, who with his partner Bohlman took hundreds of photographs of the California condor, predicted that without more vigorous protection, the species would soon go extinct. Reproduced by permission of the Museum of Vertebrate Zoology, University of California, Berkeley, catalog no. 9763.

while East Coast naturalists occasionally offered grim reports about its status. In 1913, for example, New York Zoo director and radical conservationist William Temple Hornaday declared that the "existence of this species hangs on a very slender thread." In his inimitable style, Hornaday issued a nationalistic appeal to save the species: "The California Condor is one of only two species of condor now living, and it is the only one found in North America. As a matter of national pride, and a duty to posterity, the people of the United States can far better afford to lose a million dollars from their national treasury than to allow that bird to become extinct. Its preservation for all time is distinctly a white man's burden upon the state of California."[118] John C. Phillips remained more circumspect in his 1926 list of extinct and vanishing birds of the Western Hemisphere.[119] There he reported an estimate from Joseph Grinnell, who thought that about fifty condor pairs survived in southern California and perhaps as many as twenty-five pairs in lower California. Although acknowledging that the bird had recently become highly sought after by collectors, he believed that state protection was helping the species hold its own. Three years later, Miller, Van Name, and Quinn highlighted the dire plight of the condor in their vituperative pamphlet, *A Crisis in Conservation: Serious Danger of Extinction of Many North American Birds.* They included the species among a long list of "birds beyond saving," doomed creatures that would soon "join the brotherhood of ghosts."[120] In 1933, Alex Wetmore confirmed that discouraging prognosis when he claimed that the remaining population of California condors had fallen to only ten birds, though he failed to reveal the source of his estimate in his article for *National Geographic* magazine.[121]

SECURING THE SISQUOC SANCTUARY

A year after Wetmore declared the California condor all but extinct, the wealthy Santa Barbara County rancher, businessman, and budding conservationist Robert E. Easton invited seven guests to view the species on a remote forty-acre parcel of meadow and sandstone outcropping known as Montgomery Potrero.[122] The land was owned by the Sisquoc Investment Company, of which Easton served as secretary and general manager. Included among the party was the California ornithologist, Cooper Club member, and retired Union Oil chemical engineer Ernst Dyer, who produced the first motion pictures of the condor at the site. Initially, Easton and Dyer tried to keep word of their find secret, for fear that egg collectors, hunters, and others would descend on the area and harm the birds. When sensational accounts of their find began to appear in California newspapers, including one story that depicted a condor carrying off a baby fawn in its talons, Dyer decided to set the record straight in an issue of *The Condor* published in early 1935.

Dyer's account of the adventure bears a strong resemblance to Thorton Burgess's response to viewing the last heath hens on Martha's Vineyard only four years earlier. After witnessing a group of about ten condors feasting on the carcass of an old ranch horse, Dyer reported that he and other members of the expedition party "realized that we were, at last, beholding a portion—too large a portion—of the pitiful remnant of that great race that once ranged, even in prehistoric times, from Baja California in the south, northward through the full length of Alta California up into Oregon and Washington, perhaps even further." "As to their future prospects," Dyer continued, "the trend of their past history undeniably points to their ultimate extinction unless conservation measures are promptly put into effect."[123]

Dyer's article alerted the leadership of the National Association of Audubon Societies, which had become increasingly sensitive to the issue of endangered species. Shortly after the article appeared (and not long after Arthur Allen set off on his cross-country expedition to photograph and record rare American birds), Joseph Grinnell wrote to Audubon president John Baker asking if his organization would sponsor a field study of the condor to learn more about its life history and current status.[124] Grinnell also recommended that Baker contact Easton, who had recently approached the California Fish and Game Commission and local Forest Service officials about improving protection for the condor. Apparently, he had found a generally sympathetic ear in the person of Stephen A. Nash-Boulden, who was supervisor of the Santa Barbara National Forest, and his deputy, Cyril Robinson, who had been systematically collecting data on condor sightings for several years. In response to Grinnell's request, Baker pointed out that he had already lobbied Ferdinand A. Silcox, chief of the Forest Service, on behalf of the condor, and that the National Association of Audubon Societies was willing to offer $50 a month for a period of two months to study the condor.[125] Grinnell found Baker's initial offer of support insufficient and recommended that he approach Easton about additional funding for a proper study.[126]

In April 1936, Baker ventured to the Santa Barbara National Forest to meet with Nash-Boulden and Robinson.[127] Prior to his visit, he had corresponded with Aldo Leopold, who, as we have already seen, had begun advocating for systematic studies of endangered species throughout North America. In addition, Audubon officials had been in contact with Easton and Dyer about the plight of the Sisquoc condors. Easton expressed particular concerned about Nash-Boulden's effort to criss-cross the Santa Barbara National Forest with a dense network of roads, part of a campaign to control fires in the area.[128] One of the proposed roads would penetrate directly through the Sisquoc Falls area, the site of prime condor habitat, thereby allowing public access to this previously remote site. In an effort to stop construction of the road, Easton offered Nash-Boulden right-of-way through land owned by the Sisquoc Ranch as an alternative route. Nash-Boulden then agreed to a moratorium on the projected road until the situation could be studied further. After corresponding with Silcox, observing conditions firsthand, and speaking with members of the Cooper Ornithological Club, Baker agreed to raise funds to place two Forest Service employees in the field for thirty days to get a better handle on how many condors inhabited the Sisquoc Falls area and a second site near the White-Acre Peak-Hopper Mountain area.[129] The National Audubon Society and the Cooper Ornithological Club then raised $270 to fund this limited study, which Robinson supervised. In his unpublished report, Robinson estimated that about fifty condors survived within the Santa Barbara National Forest (which had recently been renamed the Los Padres National Forest).[130] By October, Robinson informed Baker that the Forest Service had not only decided not to build the Sisquoc road but also to limit access to the area to protect the condor and maintain its habitat.[131]

Unfortunately, no clear administrative procedure existed for taking this action. Since its founding in 1905, the Forest Service's primary focus had been on timber management within the boundaries of national forests.[132] When it came to managing wildlife on these federally owned areas, however, the agency had almost invariably deferred to state authority. On the rare occasions when it had done otherwise, it had provoked resentment, controversy, and even litigation in the western states where the vast majority of the national forests were located. For example, in the 1920s, the agency's efforts to control the overpopulation of deer in the Kaibab National Forest in Arizona led to a Supreme Court case when state game officials arrested Forest Service employees attempting to cull herds in the area.[133] Although the court ultimately affirmed the right of the U.S. government to protect its property by removing excess wildlife, uncertainty about the exact limits of federal authority remained.

Silcox had gone on record in support of greater protection for endangered species, and in 1936, he hired the botanist and ecologist Homer Shantz to become the agency's first director of wildlife management. Early the next year, his agency issued regulation T-9-1, a rarely invoked rule that prohibited trespassing of any area closed by the chief of the Forest Service, for the "perpetuation and protection of (1) rare and vanishing species of plants or animals, (2) special biological communities, or (3) historical or archaeological places or objects of interest." It also prohibited entry to any area closed for scientific experiments or other similar purposes.[134] The first site specifically set aside under this new regulation was the 1,200-acre Sisquoc Condor Sanctuary. Thanks to the untiring efforts of Easton, Robinson, Baker, and others, the Forest Service had finally taken an important first step in protecting the condor and preserving endangered species more generally.

KOFORD'S CONDORS

Even as he was struggling to establish the Sisquoc Condor Sanctuary, Baker was also laying plans for an ambitious Audubon fellowship program that would fund the study of endangered species. A confluence of events—the loss of the final heath hen, Allen's search for the elusive ivory-billed woodpecker, Dyer's article on the threatened California condor, Leopold's call for a systematic inventory of endangered species, and the administrative shake-up within Audubon—led to the creation of the fellowship proposal that Baker announced at the fall 1936 meetings of the American Ornithologists' Union and the National Association of Audubon Societies.[135] Shortly after those meetings, Baker sent Grinnell a letter declaring that his organization was interested in offering a graduate fellowship "in cooperation with the Museum of Vertebrate Zoology, and under your supervision."[136] In subsequent correspondence, he outlined a tentative proposal that called for a three-year program to study the California condor that carried a stipend of $1,500 per year.

Grinnell, the longtime editor of *The Condor* and one of the most respected naturalists in the United States, seemed a logical choice to approach about the fellowship project.[137] A bird collector from his youth, Grinnell earned his master's degree (1901) and his Ph.D. (1913) from Stanford, where he studied under the famed evolutionary biologist David Starr Jordan. Along with the philanthropist and naturalist Annie M. Alexander, Grinnell proved instrumental in founding the Museum of Vertebrate Zoology in 1908. The institution's two organizers both hoped to establish a different kind of natural history museum at Berkeley, one where research took precedence over exhibition and where representative specimens of the Pacific coast's unique terrestrial vertebrates might be preserved before those species were forever lost to development. A meticulous researcher eager to raise the standards of natural history practice, Grinnell instituted a detailed set of procedures for fieldwork, specimen documentation, and research presentation that became the Museum of Vertebrate Zoology's hallmark. He also developed fundamental concepts in ecology—like the "niche" and "competitive exclusion"—while making important contributions to other areas of evolutionary biology, particularly the relationship between geographical distribution, climate, and physical environment.

In addition to his considerable scientific achievement, Grinnell became deeply concerned about the many threats facing the wildlife of his home state. Two years before Baker approached him in 1937, he had written an article, "Why We Need Birds and Mammals," that provided a succinct expression of his basic conservation philosophy. While humans had once deeply respected the natural world and imbued it with a spiritual dimension, in recent times science had rendered them "more familiar with the workings of nature." Unfortunately, this familiarity had led to the destruction of forests, the denuding of hillsides and valleys, and the extinction or near extinction of many animals. But because of the close interdependence of all living things, in the end the careless devastation of any one part of nature would rebound back to humanity. In language that echoed the notion of an ecological ethic that Aldo Leopold was groping toward during this period, Grinnell called for a fundamental reorientation of values:

> We need to return to an attitude of deep respect toward nature. The relationships which have been set up through the ages between wild birds, mammals, and plants, in fact among all forms of life, can not be disturbed unless we are willing to accept the consequences—and these may be exceedingly serious for us. Where change in existing conditions appears to be unavoidable we should proceed with a caution

befitting our inadequate knowledge of the forces with which we deal; and above all we should avoid the capacious heedlessness that has characterized our exploitation of our surroundings so often in the recent past.[138]

While fearful that his position at a state-supported university hamstrung his ability to speak out freely in public, behind the scenes Grinnell worked tirelessly to end large-scale predator poisoning campaigns and to provide greater protection for all forms of wildlife, particularly birds and mammals. He also became mentor to numerous students who shared his devotion to scientific research and wildlife conservation. By the time Baker wrote him in late 1937, Grinnell's graduate program in ornithology and mammalogy was among the most active in the nation.

While delighted at the prospect of funding for one of his students, Grinnell realized that numerous issues still needed ironing out. In his review of Baker's proposal, Alden Miller, a Grinnell protégé and colleague, worried about the problem of maintaining high research standards in a sponsored study, particularly since the Audubon Society was ultimately interested in "conservation publicity," an agenda that might lead the organization to unduly "pressure for results of a certain type."[139] Moreover, there were turf issues to consider, since the ornithologist Harry Harris and the Los Angeles petroleum geologist and wildlife photographer J. R. "Bill" Pemberton, both prominent Cooper Club members, had been studying the condor for many years. Then there was challenge of finding funding for the fellowship. Initially, Baker hoped to convince a single wealthy donor to endow the entire Audubon fellowship program with sufficient funds to keep it going for many years, perhaps indefinitely.[140] When he failed to convince anyone to pledge the necessary $250,000 he estimated would be needed, Baker began soliciting piecemeal donations for each of the studies. One particularly effective fund-raising technique involved screening motion pictures of the condor at local Audubon Societies and sportsmen's clubs in the region surrounding its habitat. Pemberton led this campaign, which featured movies he had made to garner support for the project.[141]

Finding just the right person to conduct the study also presented a challenge. Carl Koford proved an inspired choice.[142] Born in Oakland, California, as an undergraduate he trained in forestry and zoology at the University of Washington before transferring to Berkeley during his senior year. After graduation in 1937, he served a term as a Civilian Conservation Corps officer in Montana and then spent a summer studying ground squirrels at the Hastings Reservation in Carmel Valley, a biological research facility that had been donated to the university. The fact that he had spent forty-two consecutive days perched in a tree observing his

FIGURE 56. Carl Koford with "Oscar," a California condor nestling, 1939. A careful, dedicated researcher, Koford spent thousands of hours in the field studying the California condor. Based on his extensive experience with the species, he concluded that the condor was a wary bird that would fare poorly in captivity, a claim that would later be challenged. Courtesy of the Western Foundation of Vertebrate Zoology.

subject impressed Grinnell, who felt Koford possessed the requisite "perseverance, ruggedness, and natural instincts for observation which the field study of the California condor absolutely required."[143] One of his close associates later characterized him as "a very unusual fella" with keen eyesight, a probing mind that "was analyzing things twenty-four hours a day," and a temperament that allowed him to endure countless hardships to learn more about the life and behavior of his research subject.[144]

In February 1939, just over a year after Baker had first proposed the fellowship, twenty-four-year-old Koford boarded a bus to Los Angeles to begin his study. With him he carried a rucksack, a pair of binoculars, a notebook, and a letter of introduction from Grinnell.[145] J. R. Pemberton took the enthusiastic young naturalist under his wing, freely sharing his extensive notes on the condor and inviting Koford to join him on an upcoming trip to the Hopper Mountain area, where he had recently seen eighteen of the rare birds.[146] Because he did not own an automobile, during the first year of his study Koford had to repeatedly hike long distances across difficult terrain to restock supplies while in the field.

By the time he finally finished his study, Koford had spent nearly five hundred days viewing living condors in the wild, recording his observations on about 3,500 pages of narrowly scrawled notes, and producing 350 still photographs of the birds and their habitat. In addition to his extensive fieldwork, he interviewed many "personnel of government forest and agricultural agencies, ranchers, hunters, old-timers, and others" with knowledge of the condor and spent considerable time in museums, zoos, and libraries tracking down information on the species.[147]

In addition to gathering data on an increasingly rare bird that resided in remote areas and covered considerable territory while foraging, Koford faced numerous additional challenges in his attempt to bring the condor study to successful completion. One early complication was the death of Grinnell in May 1939. Alden Miller, Grinnell's right-hand man at the Museum of Vertebrate Zoology, then assumed supervision of the project. Fortunately, Miller not only had a personal interest in Koford's research but he also maintained close contact with him from the beginning, so the transition proved relatively smooth.[148] Nearly two and one half years into the three-year study Koford faced an even greater challenge when, in June 1941, he was called into active duty in the navy. Concerned that he might be forced to remain in the military for many years and that his data might soon become obsolete, Baker floated the idea of asking someone else to write up Koford's copious notes.[149] Miller immediately shot down the idea, fearing that allowing someone else to complete Koford's research would result in a "badly diluted second-hand product." Moreover, it would represent a real disservice to Koford who had "given all his energy in complete devotion to a scientific problem and ideal" and who rightfully expected that his work would lead to a doctoral degree.[150] In the end, Baker acquiesced on the issue; for the moment he was forced to settle for an interim report from Koford, though on several subsequent occasions, he again raised the possibility of hiring someone else to write up the final report.[151] As it turned out, Koford remained in the navy until after the war ended and did not get back into the field to finish his research until February 1946.

A year later, California ornithologists and National Audubon Society staff successfully lobbied the Forest Service for the establishment of a second, much larger condor sanctuary in the Los Padres National Forest. Miller and Baker had first seriously discussed the idea of pushing for additional protective areas for the species sometime in late 1940 or early 1941. Initially, Baker approached Ira Gabrielson, chief of the Bureau of the Biological Survey, about the possibility of having his agency take over the Sespe State Game Refuge, an area within Los Padres National Forest that Koford had identified as prime condor habitat. The hope was that the Biological Survey would close the refuge to hunters and others who might harm the condor, something the state of California had refused to do. Apparently Gabrielson "listened sympathetically" to the idea but failed to act on it.[152]

Once Koford had been called up for active duty, Baker began using his interim report as a tool to push for the creation of a second condor sanctuary administered by the Forest Service. His initial sense of urgency diminished, however, when Cyril Robinson, who had by then been promoted to acting supervisor of the Los Padres National Forest, closed the Sespe State Game Refuge to public access on June 1942.[153] Events following World War II—the reopening of closed sites in the national forest, the end of gas rationing, and recent reports from Koford that condors were still occupying the proposed protected area—led to a renewed campaign for the establishment of an additional sanctuary. Setting boundaries based on recommendations from Koford, the Forest Service created the nearly thirty-five-thousand-acre Sespe Condor Sanctuary in 1947 (and enlarged the protected area to more than fifty thousand acres four years later).[154] The Forest Service and the National Audubon Society split the cost of a special patrolman to guard the site during the eight most critical months during the year.[155] The move proved popular among condor enthusiasts but had its detractors among local residents who resented the closure of areas previously open to hunting and other forms of recreation. Responding to the earlier creation of the Sisquoc Sanctuary, H. H. Sheldon argued in *Field and Stream* that the species had "outlived its time and is on the trail of the dodo." By establishing a sanctuary to save an obviously relict organism, humans were "sacrificing the recreational value of a large area in exchange for the extremely doubtful preservation of a bird of no value, esthetically or otherwise."[156]

Meanwhile, Baker was keenly disappointed to learn in 1947 that Koford needed at least two more years to finish his doctoral dissertation, a document that would require additional editing to make it suitable for a broader audience.[157] With support from a fellowship granted by the University of California, Koford successfully defended his dissertation in 1950. By this point, the National Audubon Society had issued two formal reports on endangered species and had one more in progress. In addition to Tanner's monograph on the ivory-billed woodpecker (Research Report No. 1, 1942), Audubon staff member Robert Porter Allen had completed his study on the roseate spoonbill (No. 2, 1942) and was working on a second report on the whooping crane (No. 3, 1952).[158]

According to Koford, Baker insisted on "readable and breezy" prose like that found in the previous Audubon reports.[159] Tanner had written his book under the watchful eye of A. A. Allen, who had considerable experience communicating to a popular audience, while Robert Porter Allen proved an accomplished writer with a gift for making his work come alive for the reader. Koford and Miller, on the other hand, were clearly more comfortable expressing themselves in the more turgid style typical of most academic science writing. In part as a result of their differences, Koford failed to complete the numerous revisions on his report that Baker had requested before he left for Peru early in 1951 to begin studying the vicuña. Further delays occurred when the airline on which Koford shipped the edited version from Lima lost the manuscript for seven months.[160] It was not finally published until 1953, fourteen years after it was begun.

KOFORD'S CONCLUSIONS

Koford's landmark report offered the first detailed study of living condors in the field. In his forward to the book, Baker once again reiterated the importance of obtaining solid facts through scientific research as "the essential basis for wise policies governing the conservation of wildlife resources." Humans had so reduced the California condor population that it had become the "third rarest living species of North American bird" and on "the verge of the extinction." Now it was up to humans to promote its "survival and restoration."[161] Alden Miller's preface vehemently denied the claim that the condor was a relict species beyond saving, a "passing remnant of the Pleistocene." To understand the true nature of the California condor (and how it might be rescued from the brink of extinction) a scientist needed "to live with them, think about them daily, gather facts in abundance, analyze and reflect," just as Carl Koford had done. Koford's exhaustive report offered a solid foundation for the conservation of the condor, Miller argued; now his sound program of protection and education needed implementation.[162]

In the body of his copiously illustrated study, Koford summarized what he had learned about the distribution, life, and behavior of this rare bird. A species that had once ranged widely across much of North American had now been reduced to a modest final stronghold, a wishbone-shaped swath of land just north of Los Angeles that cut through seven counties.[163] Based on an unexplained extrapolation from the maximum counts of condors seen by several reliable witnesses over a number of years, Koford estimated that about sixty condors remained and that the population size was currently stable, neither increasing nor decreasing.[164] Because of the population's small size and low breeding potential, however, the failure of even a single nest or the prevention of the death of a single condor could have important implications for the ultimate survival of the species.[165]

Unlike Finley, who presented the condor as tolerant of humans, Koford argued that it was a particularly sensitive species that would readily desert its nest if disturbed.[166]

In the conservation section of his book, Koford outlined his conclusions about the threats the condor faced and his recommendations about how to preserve the species. He began by refuting the largely pessimistic attitude taken toward the fate of the species since Cooper had first lamented its decline in 1890. The major single obstacle to perpetuating the species, he asserted, was the lack of commitment on the part of the many individuals who had the power to affect its fate.[167] Among the most important perils threatening the continued existence of the condor were wanton shooting by thoughtless hunters and ranchers; collecting, which he believed had "contributed significantly to the decline of the condor in some areas"; inadvertent poisoning and trapping by those engaged in campaigns to reduce mammalian pest and predator populations; various kinds of accidents; road building, trail construction, and oil development that not only directly disturbed habitat but also made condor country more accessible to the public; and photographers who bothered the birds in an effort to obtain still or motion pictures.

Koford dismissed as impractical several proposals to resurrect the species.[168] Artificial feeding, by regularly placing carcasses near roosts, would be costly to implement, attract carnivores that sometimes preyed on condors, and cause the remaining birds to concentrate in relatively small areas, where they proved more vulnerable to disaster or disease. The trapping and transplanting of condors to unoccupied territories would also fail since the birds were able to cover such a large territory and a pair that had been moved from its original nest site would likely return back to that site. Koford was particularly adamant in his opposition to artificial breeding of the species in captivity, which he said should only be attempted as a last resort, "after all efforts to maintain the natural population have failed."

The San Diego Zoo had successfully raised the Andean condor (*Vultur gryphus*) in a large cage, but for several years leading up to the publication of the condor report, Koford, Miller, and Baker had strongly opposed a proposal to grant that institution the necessary permit to capture a pair of wild California condors.[169] Their reasons were ostensibly scientific. In his report, Koford questioned whether condors raised in captivity would possess the basic skills and experiences necessary to survive in the wild. He also worried that the release of captive condors might introduce diseases into the wild population. But in the end, the arguments against captive breeding were as much about aesthetics as science. Koford, for example, spoke of the grace and majesty of the species in its native habitat, an aesthetic treasure that greatly diminished when the species was forced to live in artificial circumstances:

> The beauty of a California condor is in the magnificence of its soaring flight. A condor in a cage is uninspiring, pitiful, and ugly to one who has seen them soaring over the mountains. Condors are so few that their recreational value is one of quality rather than of quantity. As Leopold . . . points out, the recreational value of wildlife is inverse in proportion to its artificiality. The thrill of seeing a condor is greatly diminished when the birds are being raised in captivity. Our objective should be to maintain and perhaps to increase the natural population of condors.[170]

To achieve that goal, Koford offered a series of recommendations, beginning with greater legal protection. While state law had protected the condor at least since 1905, apparently only one person had ever been prosecuted for violating that law; he received a $50 fine after being convicted trying to sell a condor he shot near Pasadena in 1908. "As a striking and unique species, threatened with extinction," Koford declared, "the condor is of national interest and is deserving of strict federal protection." A tentative start in that direction had been made in 1942, with the ratification of the Convention on Nature Protection and Wild Life Protection in the Western Hemisphere. The annex of this convention listed the condor as one of several species whose protection was "declared to be of special urgency and importance." Moreover, the federal government agreed to safeguard it "as completely as possible," though as yet nothing had come of this pledge.[171]

Beyond federal legislation, Koford offered a several other suggestions for rescuing the condor. In keeping with his notion about the sensitivity of the species to human disturbance, Koford called for the continued closure of its nesting and roosting grounds. All the recently used nesting sites that he knew about were either inside the boundaries of national forests or, in one case, within a federal Indian reservation. While several of these areas had been closed to public entry with the creation of the Sisquoc and Sespe Condor Sanctuaries, oil drilling was still being permitted within a large portion of the Sespe, several prime condor roosting sites were located on private property, and hunters still used some trails within sanctuaries that were supposed to be closed. In addition to addressing these problems, Koford thought the Forest Service should stop maintaining certain trails and roads that crisscrossed condor country. Unfortunately, most of the condor feeding areas remained on private property, where the birds faced harassment by the public. Here Koford called for an educational campaign to help those who came in contact with the bird to "understand something of the relation of the condor to its environment." And finally, Koford thought that ranchers, stockmen, and trappers should be encouraged to leave animal carcasses in the wild rather than destroying them. At the same time, Koford reiterated, "carcasses of poisoned animals must be regarded as dangerous to condors until they are proven otherwise." To shed further light on this issue, he called for studies using turkey vultures as surrogates for the condor.[172]

Koford's report received accolades in the handful of publications that reviewed it.[173] But it seems to have had little effect on slowing the decline of the condor. The National Audubon Society cosponsored a second condor report in 1965, which was also completed under the supervision of Alden Miller. There Eben and Ian McMillan—brothers, ranchers, and amateur naturalists—estimated that only forty-two condors remained, a number that had been calculated by relying on Koford's earlier estimate.[174] Like the ivory-billed woodpecker, the condor was among the first species to receive protection under the first Endangered Species Act (1966), but listing initially did little to stem the species' ongoing population decline. Although strongly opposed by Koford, Miller, McMillan, and others, calls for a more aggressive captive breeding program for the species became increasingly hard to ignore as the condor population continued to plummet toward disaster.

THE LIMITS OF SCIENCE

The 1920s and 1930s marked a watershed in the evolution of policies and scientific practices related to endangered species in the United States. While naturalists had long shown a keen interest in the plight of vanishing wildlife, for the first time they possessed the training, conceptual tools, techniques, financial backing, and desire to begin more thorough investigations of those species in the field. The three studies chronicled in this chapter were born of an optimistic belief that if enough could be learned about the life history, behavior, and ecology of vanishing animals, they might be snatched from the jaws of extinction. Consistent with that belief, all three studies identified numerous hazards that seemed to be pushing their respective subjects toward the brink of annihilation, and all three offered concrete suggestions about how to reverse that decline. Although each pointed to human actions as the primary source of endangerment, their recommendations proved quite modest, at least in the context of later, more substantial and interventionist efforts to save species threatened with extinction. In the cases of Tanner and Koford, the single most important proposal was to restrict human access to and modifications of areas known to be prime ivorybill and condor habitat. Yet, given the larger political, social, and economic climate of the time, even those modest recommendations faced stiff resistance from individuals who had prior claims on the landscapes on which these species depended to survive. While Koford and his colleagues overcame much of that resistance, the National Audubon Society failed to stop logging on the Singer Tract. Clearly, science alone could achieve only so much without a larger change in values.

Beyond the lack of political will to fully implement the recommendations that Tanner and Koford made were limitations in the science they pursued. Their pioneering studies clearly went far beyond previous research on endangered species, but they remained restricted in scope and scale. Each was undertaken by a single individual, with a modest budget, and over a fairly narrow span of time. While Tanner visited several locations where recent sightings of the species had been claimed, he managed to actually find and observe only five ivorybills at a single site. Exactly how many of these charismatic birds still roamed the earth, and where were they? How representative were the behaviors he witnessed at the Singer Tract? How much variability did the species exhibit in terms of its behavior, food requirements, and habitat needs? Despite Tanner's knowledge and dedication, these critical questions remained unanswered following his study— and probably unanswerable for a single naturalist, no matter how capable and devoted.

Similarly, while Koford made numerous important discoveries during his four years of intensive research on the California condor, at the end of the day he too lacked accurate information about the size of the remaining condor population, the number of total number of breeding pairs, and mortality rates, all critical pieces of information for developing an effective plan to rescue the species.[175] Despite evidence to the contrary, he also argued that the California condor was an extremely wary, sensitive species, and thus not a good candidate for captive breeding experiments. The conservation biologists Noel Snyder and Helen Snyder argue that this perspective on the condor was not only inaccurate but also dangerous for the long-term well-being of the beleaguered species, which continued to decline long after the creation of the Sespe Sanctuary.[176]

The years following World War II witnessed a dramatic increase in public support for endangered species preservation and an impressive growth in the size of American universities, both of which lead to more funding for research on the issue. Growing public concern about extinction also fostered a political atmosphere in which stronger protective legislation could prevail. Just as they had done since the nineteenth century, naturalists would continue to play a key role in shaping that legislation.

CHAPTER TEN

"THE NATION'S FIRST RESPONSIBILITY"

SAVING ENDANGERED SPECIES IN THE AGE OF ECOLOGY

*Can society, whether through sheer wantonness or callous neglect,
permit the extinction of something beautiful and grand in nature without risking
extinction of something beautiful and grand in its own character?*

EDITORIAL IN THE *CHRISTIAN SCIENCE MONITOR*, 1954

*Nothing is more priceless and worthy of preservation than the rich array
of animal life which our country has been blessed.*

RICHARD NIXON, 1973

RESCUING THE WHOOPING CRANE

Here, finally, was a chance to snatch the whooping crane from the jaws of extinction. Or so the Fish and Wildlife Service biologist Ray C. Erickson hoped after seeing a memo calling for innovative ideas that newly inaugurated President John F. Kennedy circulated to federal agencies shortly after coming to office in early 1961.[1] The outspoken Erickson responded with a proposal to begin captive breeding experiments with the sandhill crane (*Grus canadensis*), a species that was closely related to the critically endangered whooping crane (*Grus americana*) but not nearly so rare. At the time, only about three dozen whooping cranes remained in the wild, and recommendations to capture any of them had faced stiff opposition. But, given its critically low numbers and sluggish reproductive rate, Erickson had concluded that the species was probably doomed unless some

FIGURE 57. Ray Erickson with whooping crane chick, ca. late 1960s. Erickson played a key role in pushing the U.S. Fish and Wildlife Service to expand its captive breeding initiatives with endangered species like the whooping crane. Courtesy of the U.S. Fish and Wildlife Service.

of the remaining wild birds were taken into captivity. Initial experiments with sandhill cranes offered a safe way to develop the techniques needed to begin a risky breeding program with its much rarer cousin, if and when the Fish and Wildlife Service mustered the political will to grant its approval. Indeed, within a few years after authoring his letter, the forty-three-year-old Erickson would play a central role in establishing a new captive breeding facility for whooping cranes and other endangered species at the agency's research station and refuge on the banks of Maryland's Patuxent River.

Other than a willingness to speak his mind freely, little in Erickson's background suggested he would become a central player in pushing the federal government to become more active, interventionist, and systematic in its approach to endangered wildlife.[2] Born in St. Peter, Minnesota, during the final year of the First World War, he earned his A.B. in biology from Gustavus Adolphus College several months before the United States entered the Second World War. In 1942, he completed his master's degree at Iowa State University, one of a handful of American universities where the Fish and Wildlife Service had recently established a Cooperative Wildlife Research Unit. At the time, wildlife management training and practice remained firmly focused on game species, and Erickson's graduate work and early career proved no exception. He completed his master's thesis on the breeding habits of the canvasback duck at the Malheur Refuge in southeastern Oregon in 1942, and following a lengthy stint as a naval officer, he wrote a doctoral dissertation on the life history and ecology of that same species.[3] Following graduation in 1948, he signed on as a wildlife management biologist at Malheur, where he conducted game inventories, explored the impact of livestock grazing on waterfowl, and completed other studies for the Fish and Wildlife Service. In 1955, Erickson moved east to Washington, D.C., to become head of habitat management in the Fish and Wildlife Service's Division of Wildlife Refuges. There he quickly found himself embroiled in the controversy over the whooping crane.

As the tallest North American bird, the conspicuous, predominantly white whooping crane was a quintessentially charismatic species that attracted growing public attention as its numbers dwindled.[4] Although probably never as abundant as its sandhill cousin, during the Pleistocene, the whooping crane had ranged across much of the continent's once soggy interior. Climatic changes at the end of the ice age dried up much of its marsh habitat, however, and its range and numbers declined dramatically. While a few early European observers claimed to have witnessed large masses of the birds blackening the skies for hours, more recent analysis suggests a much smaller population during the last several centuries; by the 1870s, it probably numbered less than a thousand individuals.[5] Overhunting of the large, conspicuous bird combined with habitat destruction—as plains farmers and ranchers drained the wetlands the species needed for nesting—to push the whooping crane into a dangerous downward spiral.

The federal government took the first step to protect the species when it declared a permanent closed season on the whooping crane under the terms of the Migratory Bird Act of 1918, an action that failed to halt its decline. The next major step came in 1937, when President Franklin Roosevelt signed an executive order establishing the Aransas Migratory Wildlife Refuge, a 47,261-acre site known as Blackjack Peninsula on the southern Gulf coast of Texas. The area provided a home for the largest migratory whooping crane population on earth, but by then, the bird was perched perilously on the brink of extinction.[6] An initial census at Aransas in 1938 revealed only fourteen migratory whoopers inhabiting the newly established refuge. But even there, the generally shy bird was not completely insulated from the intrusions of civilization; the ink had barely dried on Roosevelt's executive order when the Army Corps of Engineers began dredging a channel for an intercoastal waterway that skirted one edge of the site, military officials started bombing and machine-gun practice on a nearby island, and oil companies initiated drilling in the bay that bordered the refuge.[7]

To make matters worse, precious little was known about the life history and behavior of the endangered whooping crane. Even the location of its northern breeding site remained a deep mystery. Hoping to gather the scientific knowledge needed to save the whooping crane, in 1945 the National Audubon Society and the Fish and Wildlife Service joined in a loose-knit partnership dubbed the "Whooping Crane Cooperative Project." The Audubon Society assigned a plucky staff member, Robert Porter Allen—the author of a previous report on the roseate spoonbill—to study the whooping crane.[8] Between 1946 and 1950, Allen spent more than two years in the field, searching for the bird across much of its known range and extensively observing its behavior at the Aransas Refuge. Beyond fieldwork along the bird's migration route between Texas and Saskatchewan, he also completed extensive aerial surveys in Alaska and western Canada. While Allen failed to locate the northern home of the whooping crane, he compiled much vital information about its life history, behavior, and habitat needs that he incorporated into Research Report No. 3 of the National Audubon Society, published in 1952.

The accidental discovery of the whooping crane's nesting site in 1954, continued oscillation in the Aransas population, and a proposal from the Fish and Wildlife biologist John Lynch to initiate a captive breeding program prompted a gathering of whooping crane enthusiasts and experts in Washington.[9] By the time the Fish and Wildlife Service called the conference in 1956, only two dozen whooping cranes survived the annual trek to Aransas, despite nearly two decades of protection at the site. For several years before the meeting, a small but increasingly vocal group of ornithologists, aviculturalists, and zoo officials had been pushing for captive breeding experiments using the species.[10] They argued that it was high time for some of the remaining wild birds to be bred under the care of an experienced aviculturalist. In addition to Lynch, S. Dillon Ripley (an avid aviculturalist, curator of ornithology at Yale, and later president of the International Committee for Bird Preservation) and Fred Bard (director of the Saskatchewan Museum of Natural History, Regina) strongly advocated rounding up at least some portion of the remaining wild whooping crane population. In the late 1950s, a group that would dub itself the Whooping Crane Conservation Association coalesced around the proposition that an organized captive breeding program was essential to rescue the species.[11]

Allen and his colleagues at the National Audubon Society, however, strongly opposed the idea on both practical and ideological grounds. It was too early to take such a drastic, dangerous step, they argued, especially when success remained uncertain. Each bird taken into captivity meant one less bird in the perilously small wild population. And even if some way could be found to breed the species in captivity, there was no guarantee that it could be successfully reintroduced into the wild. After all, the birds that game breeders produced through artificial propagation methods were notorious for faring poorly when they were released.[12] Moreover, the prospect of reducing the species to captivity disturbed Allen and his colleagues on aesthetic and philosophical grounds. "It is not our wish to see this noble species preserved behind wire, a faded, flightless, unhappy imitation of his wild, free-flying brethren," Allen wrote. "We are dedicated to preserving the whooping crane in a wild state."[13] Facing an impasse between the pro- and anticaptivity factions at the meeting, the Fish and Wildlife Service responded as bureaucracies often do: it appointed a committee, the Whooping Crane Advisory Group, to study the problem.

That committee approved Erickson's 1961 proposal for captive breeding experiments with the sandhill crane. Later that summer, the Fish and Wildlife Service opened a small makeshift facility at the Monte Vista National Wildlife Refuge in southern Colorado, a site along the migratory route of the sandhill crane. There staff biologists developed procedures for successfully maintaining and breeding the birds. They found that if they collected the eggs of wild sandhills rather than chicks, for some reason the birds were less likely to contract diseases at the facility.[14] Based on this experience and an analysis of mortality trends among the whooping crane, Erickson began arguing that since sandhills (and whoopers) tended to lay two eggs, but only one chick typically survived, the removal of a single egg presented little risk to the overall wild population of either species.

Emboldened by the results of these initial captive breeding experiments and aviculturalists' success with raising other rare birds around the world, Erickson proposed a new initiative devoted to the propagation, laboratory research, and field ecology of species threatened with extinction.[15] What he envisioned was the creation of an Endangered Wildlife Research Station at Patuxent, an old 355-acre estate in Maryland where the service had maintained a modest research complex since the 1930s. Erickson put together the initial proposal, helped secure a $350,000 appropriation from Congress, and obtained a position as director of wildlife research and propagation at the new facility. Soon after the opening of what was touted as a "modern-day Noah's ark" in 1965, Erickson and his staff were experimenting not only with the whooping and sandhill cranes but also with several other endangered species (or their surrogates), including the masked bobwhite quail, the Andean condor (a stand-in for the California condor), and the black-footed ferret.[16]

The Endangered Wildlife Research Station represented one of a series of increasingly interventionist federal initiatives aimed at addressing the problem of vanishing wildlife during this period. This chapter shows how a variety of social, scientific, and political forces pushed the federal government to identify and preserve species threatened with extinction. The emergence of a vigorous postwar environmental movement signaled a public that was not only more aware of basic ecological concepts but also increasingly concerned about the fate of rare and endangered species. At the same time, scientists in the United States and abroad continued to play a leading role not only in producing authoritative knowledge about endangered species but also in prodding governmental officials and private organizations to act on that knowledge. Responding to this pressure, the administrations of Lyndon B. Johnson and Richard M. Nixon joined with several key members of Congress to offer leadership on this and other environmental issues. The Endangered Species Act of 1973 represented the culmination of the impressive growth in concern about declining wildlife. Through that landmark legislation, the federal government pledged to use its power to ensure that no species would ever again go extinct. Rejecting the previous piecemeal and haphazard approach to the problem of endangered species, it placed itself at the center of a systematic, comprehensive program to rescue the whooping crane and hundreds of other endangered species, both at home and abroad.

THE AGE OF ECOLOGY

"The Age of Ecology opened on the New Mexican desert, near the town of Alamagordo, on July 16, 1945, with a dazzling fireball of light and a swelling mushroom cloud of radioactive gases."[17] So does historian Donald Worster assert in the vivid description of the Trinity Test—the first atomic bomb detonation—that appeared in his pioneering history of ecology. Worster argues that postwar concern about emerging nuclear technology helped transform the previously obscure biological subdiscipline of ecology into a pervasive popular term, one that by the late 1960s became synonymous with concern about environmental degradation more broadly. Following up on this lead, scholars have subsequently found that the links between the bomb, ecology, and the emergence of the postwar environmental movement are even deeper than Worster realized.

The bomb fostered a fundamental questioning about progress, while fueling apocalyptic fears.[18] By the time of the Trinity Test, many Manhattan Project scientists already understood that they had unleashed a powerful new force capable of inflicting unprecedented destruction. But because that first atomic test was shrouded in secrecy, the public failed to learn of the bomb's existence until it leveled the Japanese cities of Hiroshima and Nagasaki at the end of World War II. With military officials tightly controlling the flow of information from those devastated cities, at first Americans seemed both grateful for the new technology and relieved to see an end to the long war.[19] However, several public intellectuals quickly appreciated that with the release of the nuclear genie, civilization had reached a fundamental turning point. In a remarkable editorial completed only a few hours after President Truman's terse announcement that an atomic bomb had been detonated over Hiroshima, *Saturday Evening Post* editor Norman Cousins warned that "Modern Man is Obsolete," while raising the specter of possible human extinction.[20]

The bomb's apocalyptic potential received further elaboration when the editors of the *Bulletin of Atomic Scientists* began prominently featuring a "Clock of Doom" on the periodical's cover with the June 1947 issue.[21] After initially setting the symbolic clock at seven minutes until midnight to signal the precarious position that humans now found themselves, the editors soon decided to move the minute hand forward or backward as the threat of nuclear war (and potential nuclear annihilation) appeared to wax and wane. That threat grew ominously with the development of thermonuclear weapons, so-called hydrogen bombs, in the early 1950s. The first of this new generation of weapons, which Americans tested in the Marshall Islands in 1952, was roughly a thousand times more powerful than the bombs that had leveled Hiroshima and Nagasaki, and their destructive potential was limited only by the amount of hydrogen fuel that could be detonated. Only nine months later, the Soviet's successfully tested their own primitive thermonuclear device, leading the *Bulletin* editors to push the clock's hands to just two minutes before midnight. Nuclear fear reached epidemic proportions by the end of the 1950s, as Cold War tensions heightened, nuclear weapons proliferated, and new delivery systems went into production.

Even as it provoked terror in the hearts of humans around the world, nuclear technology proved a godsend for academic ecology in the postwar period.[22] As the Cold War began heating up in the late 1940s and early 1950s, the federal government decided to play a much larger role in funding basic research of all sorts, and with federal leavening, the American scientific enterprise expanded like dough in a warm oven. Beginning in 1948, the Atomic Energy Commission (AEC) created a Division of Biology and Medicine that supported a wide range of studies in the life sciences and became a major sponsor of ecological research. The sudden influx of grant money proved particularly significant for a field like ecology that had previously operated on shoestring budgets. For example, Eugene Odum, who pioneered the ecosystem approach that became a central paradigm for postwar ecological research in America, parlayed a long series of AEC grants into one of the nation's premier ecological research centers at the University of Georgia.[23]

While the number of academic ecologists dramatically expanded in the postwar era, the circulation of ecological knowledge remained confined to a relatively small circle of scientists and policymakers until the late 1950s, when growing public concern about fallout from atmospheric testing of nuclear weapons led to the popularization of basic ecological concepts.[24] Several fallout products posed a potential threat to human health, but attention focused on the dangers of strontium 90, a beta emitter with a half-life of twenty-eight years.[25] The primary fear here was not acute radiation poisoning but long-term exposure. Because its molecular structure resembled calcium, this toxic substance traveled up food chains much like that essential nutrient. Beginning in 1958, the St. Louis Committee for Nuclear Information—the brainchild of the biologist-turned-political-activist Barry Commoner and his physicist colleague John Fowler—began publicizing how the strontium 90 that regular testing of nuclear weapons was spewing into the atmosphere eventually settled on the ground. From there, it traveled up the food chain, from the soil to plants to cows, and ultimately to the humans who ingested their milk. The March 1959 issue of *Consumer Reports* warned that although strontium 90 levels in milk supplies remained below limits that federal officials deemed safe, they were continuing to rise with each new atomic test. No one could determine the exact risk that increasing human exposure to strontium 90 represented, but many Americans found the very fact that it was happening quite disturbing. A series of popular science fiction movies—*Them!* (1954), *Godzilla* (1956), *Attack of the Crab Monsters* (1957), and others—featured giant mutants produced by exposure to radioactive fallout, further fueling anxiety about the issue.[26]

Although public concern about fallout greatly diminished in early 1963, when the world's major nuclear powers signed a treaty banning the atmospheric testing of nuclear weapons, a widespread fear of chemical contamination from pesticides soon came to supplant it. The impetus behind this shift was Rachel Carson's blockbuster book, *Silent Spring* (1962), a powerful exposé on the dangers that newly introduced synthetic pesticides, especially DDT, posed to humans and the natural world.[27] Carson drew from myriad scientific studies to document her bold claims, including extensive research on the effects of pesticides on wildlife that Fish and Wildlife Service biologists had conducted at Patuxent since World War II.

FIGURE 58. Rachel Carson at Hawk Mountain, ca. 1950s. Photo by Shirley A. Briggs. Carson's best-selling book, *Silent Spring*, warned of the dangers of modern pesticides and helped ignite the modern environmental movement. Courtesy of the Lear/Carson Collection, Connecticut College.

She also relied on the research of several field naturalists. Charles Broley, a retired banker and amateur naturalist from Winnipeg, banded more than one thousand fledgling bald eagles between 1939 and 1949.[28] By the mid-1950s, Broley noted that about 80 percent of the nests along Florida's west coast were no longer producing young, and even adult eagles had become increasingly rare. Broley speculated that newly introduced synthetic pesticides might be responsible for the population crash he had witnessed. Similarly, the naturalist and Hawk Mountain curator Maurice Broun noted a marked decline in the number of juvenile bald eagles that migrated annually along the Kittatinny Ridge. From 1935 to 1939, 40 percent of the eagles observed at Hawk Mountain Sanctuary were yearlings, which are easily identified by their dark plumage. Between 1955 and 1959, however, they declined to only 20 percent of the total count.[29]

Carson's book offered a plausible scenario for both sets of observations. The newly introduced pesticides were broad-spectrum toxins—Carson called them "biocides"—that not only failed to break down in the environment but also concentrated as they moved up the food chain. Top predators, like hawks and eagles, were thus ingesting alarming amounts of these poisons as they were increasingly deployed to kill insect pests. Reproductive failures like these offered just one of many disturbing symptoms of pesticide poisoning in wildlife that Carson documented in her shocking book. Because humans were also part of food chains,

Carson argued, they too faced exposure to these dangerous new substances being carelessly broadcast across the landscape. Thus *Silent Spring* not only introduced the potential hazards of pesticide contamination to a broad reading audience but also served as a basic primer in ecology.

Part of Carson's effectiveness related to her rhetorical strategy of drawing a link between the dangers of strontium 90 and the new synthetic pesticides introduced on the market in the postwar era.[30] She peppered *Silent Spring* with references to fallout and radiation more generally and strontium 90 in particular, all of which would have been familiar to her readers. Like strontium 90, she argued that new synthetic pesticides like DDT were man-made, persistent, indiscriminate, invisible, and pervasive toxins that concentrated as they moved up food chains to threaten humans and the environment. Nor did Carson remain immune from the apocalyptic currents that swept through generations living under the shadow of the bomb. She dedicated her book to Albert Schweitzer and chose a pessimistic epigraph from his hand: "Man has lost the capacity to foresee and forestall. He will end by destroying the earth."[31] It would be a mistake to credit Carson with single-handedly launching the modern environmental movement; but in popularizing ecological concepts, raising concern about chemical contamination, and motivating the public to act, she clearly changed how Americans thought about their relationship to the natural world. Following on the heels of public concern about the dangers of atomic weapons and radioactive fallout, *Silent Spring* helped put ecology on the nation's cultural and political map.

INTERNATIONAL INITIATIVES: THE SURVIVAL SERVICE COMMISSION

Along with the diffusion of ecological concepts, a second postwar development further prodded the U.S. government to act more decisively on the issue of endangered species: the creation of the International Union for the Conservation of Nature and Natural Resources (IUCN), which was originally established in 1948 as the International Union for the Protection of Nature (IUPN).[32] A few months before its creation the British biologist, conservationist, and statesman of science Julian Huxley had written to his brother, the novelist and cultural critic Aldous, to express his deep fear about the future of life on the planet: "Meanwhile I come to feel more and more that no system of morals is adequate which does not include within the sphere of moral relationships, not only other human beings, but animals, plants, even things." "We have done quite monstrously badly by the earth we live in," Huxley continued. "If we don't do something about it pretty soon, we shall find that, even if we escape atomic warfare, we shall destroy our civilization by destroying the cosmic capital on which we live."[33] Huxley hoped that a newly created international institution—the United Nations Educational, Scientific, and Cultural Organization (UNESCO)—would promote ethical extension, encourage the deployment of scientific expertise to better manage natural resources, and raise public consciousness about how humanity was damaging the natural environment. As the first director-general of UNESCO, he played a key role both in founding the IUPN at Fontainebleau later that same year and in appropriating the funds to keep the fledgling organization afloat during its formative years.

Unlike existing organizations devoted to international conservation, the new IUPN was a hybrid institution, composed of representatives from national governments and nongovernmental organizations. Among the individuals who signed the Final Act creating the International Union were delegates from eighteen governments and ninety-four nongovernmental organizations (the vast majority of which were western European).[34] The organization sought both to build a geographically expansive membership and to achieve a global impact, although initially its membership was primarily based in the Northern Hemisphere while its activities focused on the Southern. This extensive network of conservationists exchanged experiences, pooled information, and shared a commitment to using the power of science, especially the science of ecology, to shield the natural world from the onslaught of civilization.

Rescuing endangered species provided a primary focus for the IUCN from its earliest days. The union's constitution, for example, declared that the organization would take a special interest in the "preservation of species threatened with extinction," and at its first conference in 1949, held at Lake Success, New York, members approved a resolution highlighting a modest list of thirteen birds and fourteen mammals deemed of particular concern.[35] At that same meeting the prominent American naturalist and conservationist Harold J. Coolidge proposed the creation of an "International Survival Office," a branch of the IUPN that would be charged with "assembling, evaluating, and disseminating information on all living species believed to be threatened with extinction."[36] Conference delegates endorsed Coolidge's recommendation by voting to found the Survival Service.

Although IUPN officials initially placed the Belgium naturalist J. M. Vrydagh in charge of the Survival Service, from the beginning Coolidge played a central role in expanding its activities. Soon he formally assumed chairmanship of what would be renamed the Survival Service Committee (and later the Survival Service Commission).[37] A prime mover behind the American Committee for International Wild Life Protection, Coolidge also became deeply involved in the newly established IUPN, serving as what one insider later called a "President in Exile" even when not formally serving in that office. He constantly prodded the board and secretariat to act on a long stream of proposals while working tirelessly to raise the funds that kept the organization afloat during its first decade. Although dedicated and energetic, the loud, gravelly voiced New Englander also proved a "demanding taskmaster" who sometimes rubbed his European colleagues the wrong way.[38]

The Survival Service's protective mandate potentially encompassed all flora and fauna facing extinction, but initially it focused on endangered mammals. Coolidge was anxious to update Glover Allen's *Extinct and Vanishing Mammals of the Western Hemisphere* (1942) and Francis Harper's *Extinct and Vanishing Mammals of the Old World* (1945), compilations that the American Committee had sponsored, he played a major role in producing, and the passage of time was quickly rendering obsolete.[39] Beyond this work, the first fruit of the Survival Service's activities was an illustrated book, *Les Fossiles de demain* (1954), which detailed what was known about the life history, ecology, and conservation status of thirteen of the fourteen mammals on the original Lake Success endangered list.[40] A planned English edition failed to materialize, but the committee did manage to place a short illustrated article on the topic in *Life* magazine.[41] Soon after its formation the committee turned the study of endangered birds over to the International Committee for Bird Preservation, thus heading off a potential turf battle with that more established organization. The decision also smoothed over relations with the prickly James C. Greenway, the ornithologist who was still struggling to complete a catalog of rare and vanishing birds for the American Committee, which bankrolled the Survival Service for several years.[42] Not until the 1960s did the committee begin paying much attention to other forms of wildlife beyond birds and mammals.

Committee members quickly discovered that endangered species also tended to be understudied.[43] To address this problem, Coolidge raised the funds to hire a young ecologist, Lee Merriam Talbot, to work on IUPN's behalf.[44] Talbot possessed an impressive pedigree: his grandfather was C. Hart Merriam (the first director of the Bureau of the Biological Survey) and his father was an associate director of the Research Branch of the Forest Service. He also boasted impressive credentials, including a bachelor's degree in wildlife conservation from Berkeley, a working knowledge of several languages, and extensive field experience.[45] After gaining board approval for his mission, in 1955 Talbot spent six months traveling forty-two thousand miles and visiting thirty countries in the Middle East and South and Southeast Asia. In addition to meeting with numerous government officials, scientists, and conservationists, he also surveyed the status of several endangered species, including three kinds of Asian rhinoceros, the Asian lion, the Arabian oryx, and the Syrian wild ass. Talbot's mission resulted in a published book presenting his findings (*A Look at Threatened Species*), a series of popular magazine articles, and numerous proposals for follow-up studies.[46] Over the next fifteen years, Talbot would not only continue pioneering research on endangered species but also help author an international treaty and several major federal laws to address the problem.

One project inspired by Talbot's survey was an attempt to rescue the Arabian oryx (*Oryx leucoryx*), a beautiful cream-colored antelope that once ranged across the Syrian, Iraqi, and Arabian deserts.[47] Its two long, straight horns appear as one when viewed from the side, a fact that has led scholars to claim the species as a possible source for the mythical unicorn. Admired for its speed, strength, and toughness, desert tribesmen had long hunted the oryx. As a result, by the late eighteenth century, the species was already becoming rare, and by the middle of the twentieth century, it was confined to a large uninhabited sand desert in the southern third of the Arabian peninsula, an area known as Rubʿ al-Khali, or "empty quarter." The species continued to decline even here, one of the most forbidding environments on the earth, but it was probably not in serious jeopardy until after World War II, when wealthy Arab potentates began using jeeps and machine guns to mow down the remaining herds. At its fourth biennial assembly in Copenhagen, held in August 1954, the IUPN passed a resolution warning that the "most beautiful and rare" Arabian oryx was "in great danger of extinction owing to its having been, and still being, hunted from motor cars."[48] Following his return from the Middle East, Talbot called for a captive breeding program to save the beleaguered species, which he estimated had dwindled to no more than a few hundred animals.

At the time, the Survival Service was struggling to formulate a consistent policy on the captive breeding of vanishing species. The specific episode that prompted discussion of the issue was a 1958 report that Burmese authorities had granted a Swiss animal dealer permission to capture six young Sumatran rhinoceroses for several European and Asian zoos. The species had become critically endangered due to overhunting (its horn was thought to possess medicinal properties) and habitat destruction. Soon after assuming chair of the newly renamed Survival Service Commission in 1958, the British conservationist and Fauna Preservation Society secretary C. L. Boyle alerted his colleagues about the situation and asked for input on a set of guidelines for capturing rare species. Clearly, situations would arise when such action might be appropriate, but Boyle warned that "if everybody joins in the merry hunt for endangered animals the[ir] position becomes even worse than before."[49]

The IUCN voiced strong opposition to the proposed capture of Sumatran rhinoceros, an action that generated protests from members of the International Union of Directors of Zoological Gardens at their next annual meeting.[50] Boyle thought that response highlighted the need for clear guidelines on the capture of endangered species from the wild, so he crafted a draft for discussion. There Boyle warned that demand from zoos, circuses, and private individuals placed increasing pressure on critically low species, thereby "threaten[ing] their existence in the wild." Remaining survivors were often scattered at various zoos around the world, where they could not successfully perpetuate themselves over the long haul. Ideally, endangered species should be preserved in their native habitat, but when their situation was otherwise "hopeless," capture from the wild might serve as "insurance against the extermination of the species." Even so, competition between zoos to obtain endangered species from the wild should be avoided at all costs.[51]

The Survival Service Commission discussions about the capture of rare species led to a symposium on "Zoos and Conservation" held at the London Zoo in 1964. Participants agreed that captive breeding might reduce the demand on wild stocks of endangered species. The removal of rare animals from their natural habitat should be carefully monitored, however, and appropriate standards for their care and perpetuation established. Moreover, individual zoos should specialize in one or more endangered species that they might be especially suited to breed rather than competing to possess the most rarities, thereby allowing solitary animals to languish in widely scattered locations. To facilitate the long-term breeding of vanishing animals in captivity, the Survival Service agreed to begin coordinating the maintenance of stud books for endangered species.[52]

Before completing negotiations on these guidelines, the Survival Service Commission became involved in an attempt to remove some of the last Arabian oryxes from the wild as part of what they hoped would serve as a model captive breeding program. In the spring of 1961, Boyle learned that a hunting party from Qatar had recently journeyed into Rubʿ al-Khali desert, where they "ran down and shot no less than 28 Oryx." Fearful that this slaughter signaled the end of the oryx "as a free-living and breeding species," Boyle began planning an expedition to capture any remaining specimens he might locate.[53] Clearly, in Boyle's mind, the oryx provided an example of a species whose future seemed hopeless without this kind of drastic intervention. The expedition managed to capture only three oryxes—two males and a female—which following a lengthy quarantine period, were housed at the Phoenix Zoo. They were soon joined by several other specimens, including a lone female loaned by the Zoological Society of London, another from the sultan of Kuwait, and two pairs from the Riyhad Zoo. The birth of two calves brought the so-called World Herd to eleven oryx by early 1964. The animals bred well in captivity, and within fourteen years, officials began releasing their descendants back into the wild.[54]

During this period, the IUCN also became a major player in a campaign to establish a biological research station in the Galápagos Islands.[55] As we have seen, since the 1930s, naturalist and conservationists on both sides of the Atlantic had struggled to provide greater protection for the archipelago's unique biota along with improved facilities for researchers who sought to follow in Darwin's footsteps.[56] In 1954, the young Austrian zoologist Iranäus Eibl-Eibesfeldt visited the islands and was horrified by the devastation he witnessed, despite the protective decree the Ecuadorian government had issued two decades earlier. Eibl-Eibesfeldt approached the IUCN with his concerns, and in 1957, the organization joined with UNESCO to sponsor a five-month expedition to systematically survey the island's wildlife and select an appropriate site for a research station. Julian Huxley (who had long sought protection of the Galápagos tortoise), S. Dillon Ripley (who was now secretary of the Smithsonian Institution), and Harold J. Coolidge (who was just assuming additional duties as chair of the IUCN Commission on National Parks and Protected Areas) vigorously lobbied fellow naturalists, government officials, and the public on behalf of the research station idea. The Ecuadorian government responded favorably in 1959, on the centennial anniversary of Darwin's *Origin of Species,* by formally setting aside the islands' undeveloped regions as a national park. A year later, the first buildings of the Charles Darwin Research Station began rising over Academy Bay. As historian of science Edward Larson has noted, "For the first time, the Galápagos Islands had a permanent home for science and conservation."[57]

SEEING RED: THE IUCN RED DATA BOOKS

The Survival Service Commission, like the IUCN as a whole, struggled financially until 1961, when the World Wildlife Fund (WWF) began.[58] The basic idea had been bandied about at least since the mid-1950s, but it failed to get off the ground until 1960, when Julian Huxley published a series of three articles for *The Observer* in which he called attention to the desperate plight of Africa's fauna. Areas that only fifty years ago had been "swarming with game," he warned, were "now bare of all large wild life."[59] Huxley's impassioned plea generated a series of discussions that included the British conservationist and ornithologist Max Nicholson and the wildlife artist, former Olympic yachtsman, and IUCN board member Peter Scott. Beyond his many other accomplishments, Scott was also an avid aviculturalist who established a center for the captive breeding and

FIGURE 59. Peter Scott with nenes, ca. early 1950s. Photo by Philippa Scott. An aviculturalist, artist, conservationist, and sportsman, Scott played a central role in founding the World Wildlife Fund and expanding the activities of the International Union for the Conservation of Nature and Natural Resources' Species Survival Commission. Here he is pictured with several nenes, a species of endangered goose endemic to Hawaii, that he successfully bred at his estate. © Philippa Scott.

ecological study of rare waterfowl at his estate in western England. Together with several colleagues, Nicholson and Scott drew up the blueprint for the World Wildlife Fund, which was officially launched on September 26, 1961. This remarkably successful fund-raising organization embraced the practice of using sentiment to tap into the growing public interest in ecology and the environment in a way that IUCN, which was largely staffed by scientists, had failed to do. During its first five years, the fledgling WWF raised $1.9 million, which it used to sponsor numerous projects that not only helped rescue endangered wildlife but also the chronically cash-strapped IUCN.[60]

Scott took over formal leadership of the Survival Service Commission beginning with the 1963 IUCN meeting in Nairobi. One of his first reforms was to create a series of specialist groups—networks of experts on a particular group of endangered species—to gather information on their status and to develop recovery plans. Scott's inspiration probably came from the rhinoceros group he had established in 1962 as part of a WWF campaign to highlight the desperate predicament of these large charismatic animals, which were not only in increasingly dire straights but also "uniquely capable of arousing interest and sympathy in the cause of coordination."[61] Immediately after assuming chair of the Survival Service Commission, Scott proposed the creation of nineteen specialist groups to focus on endangered species. He hoped the new organizational structure would not only produce more accurate information on endangered species but also push the commission beyond its narrow focus on birds and mammals.[62]

Scott also implemented the Red Data Book project that he had proposed two years earlier. In December 1961, only a few months after helping found the World Wildlife Fund, Scott developed a system of color-coded notebooks to help WWF and IUCN officials set priorities and manage projects efficiently. One of these, the "Red Book," contained information from the Survival Service Commission's card index of endangered species that Boyle had begun compiling two years earlier, together with a similar ICBP project covering birds and "additions in other classes of animals, including invertebrates, and also plants."[63] The first Red Data Books, loose-leaf sheets held in plastic binders, were intended for internal use by commission members and IUCN staff. They included concise information on the present and former distribution of endangered species, status and population estimates, breeding rate, reasons for decline, proposed and actual protection measures, numbers and breeding potential in captivity, and a brief list of references.[64] Noel Simon compiled the mammal volume, while Jack Vincent, an ICBP official who was working with the Survival Service Commission, compiled the bird volume.

Increasing demand for both works led to the idea of publication for a wider audience.[65] Initially, the commission decided to publish a specialist edition of the Red Book with a format similar to the original edition and a separate, highly illustrated popular edition that contained accessible essays on species on the Survival Service Commission endangered list. The specialist volumes on mammals and birds appeared in 1966, and the popular edition came out three years later.[66] Specialist volumes on reptiles and amphibians, freshwater fishes, and angiosperms followed in 1968, 1969, and 1970.[67] The Red Data Books, widely accepted as authoritative, represent one of the most important legacies of the Survival Service Commission and continue to be updated regularly.

UDALL'S AGENDA

Secretary of Interior Stewart L. Udall was one of many American conservationists who were closely following the work of the Survival Service Commission. In September 1963, he traveled to Nairobi to attend that nation's independence ceremonies and deliver an address at the IUCN biennial conference.[68] At the time, the young, outspoken forty-three-year-old had been pushing the Kennedy administration to take a more activist stance on conservation issues but with only limited success.[69] An old-school conservationist when he first came to Washington in 1954 to serve in the House of Representatives, by the early 1960s Udall found himself swept up in the emerging environmental movement, with its broader ecological perspective.[70] While he never fully relinquished his enthusiasm for dam building in the West, as secretary of interior, he also advocated policies that reflected a growing sympathy with the values of what was initially termed the "new conservation" and soon became known as "environmentalism."

Two events proved pivotal in his transformation: the publication of Rachel Carson's *Silent Spring,* which he defended following its appearance in 1962, and research for his own book on the evolution of American conservation thinking, *The Quiet Crisis,* published a year later.[71] In the final chapter of that book, entitled "Notes on a Land Ethic for Tomorrow," Udall began with a famous epigraph from Leopold's *Sand Country Almanac:* "We abuse land because we regard it as a commodity belonging to us. When we see land as a community to which we belong, we may begin to use it with love and respect." He continued by echoing Leopold's call for the development of new ethic governing our relationship with nature, an ethic "as honest as Thoreau's *Walden,* and as comprehensive as the sensitive science of ecology. It should stress the oneness of our resources and the live-and-help-live logic of the great chain of life."[72] In a guide to further reading at the end of his book, Udall acknowledged his intellectual debt to Leopold and claimed that "if asked to select a single volume which contains a noble elegy for the American earth and a plea for a new land ethic, most of us at Interior would vote for Aldo Leopold's *A Sand County Almanac.*"[73]

Udall's claim about his staff's sympathy with Leopold's ideas notwithstanding, there were few tangible signs that the Fish and Wildlife Service had adopted either a more ecological perspective or a systematic interest in protecting species facing the threat of extinction. The agency's report for fiscal year 1961, for example, trumpeted its accomplishments in managing game species and administering federal wildlife law but contained not a word about endangered species.[74] And two years later, when Udall published *The Quiet Crisis,* the federal government was still leading a more than five-decade-long campaign to eradicate predators in the West. In 1963, that program resulted in the destruction of nearly two hundred thousand mammals, including nearly three thousand gray wolves, a species that had been all but extinguished in the lower forty-eight states. Other endangered species—the black-footed ferret in the northern Great Plains and the California condor—also suffered collateral damage from the massive federal predator poisoning campaign.[75]

That is not to say that the Fish and Wildlife Service entirely ignored the issue of endangered species. We have already seen how in the first decades of the twentieth century that agency's predecessor, the Bureau of the Biological Survey, worked closely with the American Bison Society to established a series of western refuges that nursed the buffalo back from the brink of extinction. Under the leadership of the plainspoken naturalist and conservationist Ira Gabrielson, who headed the Biological Survey and its successor, the Fish and Wildlife Service, from 1935 to 1946, the problem of "rare and vanishing wildlife" received more sustained attention.[76] In the mid-1930s, for example, the agency founded the Red Rock Lakes National Waterfowl Refuge in southwestern Montana to protect the remnant trumpeter swan population (estimated at only thirty-three birds in the wild at the time), and it created the Aransas National Waterfowl Refuge in southeastern Texas to save the whooping crane (although it failed to respond to calls to establish an ivorybill refuge on the Singer Tract in Louisiana).[77] During Gabrielson's watch, Fish and Wildlife Service naturalists also served as informal consultants on the series of endangered species research projects the National Audubon Society commissioned, helped secure passage of the Bald Eagle Protection Act (1940), and contributed to *Fading Trails: The Story of Endangered American Wildlife* (1942), a popular book highlighting the many forms of native wildlife that were struggling to endure.[78]

During the 1950s, the agency continued to address the issue of endangered species on multiple fronts. In 1957, for example, it gained Congressional approval to purchase a 4,038-acre refuge in south Florida designed to protect the key deer, a diminutive subspecies that had dwindled down to twenty-five individuals, and to fund Hawaii's captive breeding program for the nene, an endangered relative of the Canada goose that had evolved into a sedentary, largely flightless species. In addition to its continued work with the whooping crane, the agency published a brief pamphlet, *Protecting Our Endangered Birds* (1959), that highlighted that and six other avian species facing grave danger, including the Eskimo curlew, ivory-billed woodpecker, Attwater's prairie chicken, Everglade kite, California condor, and nene. While the Fish and Wildlife Service pursued these and other scattered projects related to endangered species, the agency lacked a systematic program to deal with the growing problem of extinction, which remained relatively low on the overall list of agency priorities.

Two international conferences held on American soil during the summer of 1962 highlighted the problem of endangered species and helped place the issue more squarely on the agency's agenda. The first was a meeting of the International Committee for Bird Preservation, which was held in New York City, presided over by S. Dillon Ripley, and attended by delegates from more than thirty nations. Not surprisingly, a prominent theme of the conference was the growing list of birds threatened with extinction around the world.[79] The first World Conference on National Parks, convened in Seattle only a few weeks later. Although organized by the IUCN and hosted by the National Park Service, the primary moving force behind the conference was the indefatigable Harold J. Coolidge, who was also the founding chair of the IUCN's recently established Commission on National Parks. One resolution approved by the 250 delegates called for protecting each plant or animal threatened with extinction by including an area of appropriate habitat large enough to maintain a reproducing population of the species within a national park or nature reserve, an idea that would soon find expression in the first federal Endangered Species Act.[80]

Although Secretary Udall pursued multiple conservation initiatives, by this point, the problem of endangered species was firmly on his radar screen. In May 1963, for example, he delivered an address to the Fund for Preservation of Wildlife and Natural Areas in Boston that warned about the fate that apparently awaited a disturbing number of species that had taken "millennia to evolve." In language that echoed Leopold, Udall asked a series of rhetorical questions linking ethics, ecology, and endangered species: "How much longer can man—who patently controls nature now—justify the unthinking destruction of wildlife habitats which causes countless life forms to disappear forever? Does not man's control over nature imply a responsibility to his fellow-creatures on this planet? Must not the entire scope of life on earth be considered as a unity—a complex tangle of interrelationships with each member of the host of living things entitled to his place at the planetary table?" Udall emphasized that in his view "the Nation's first responsibility surely is the preservation of American wildlife in the wild."[81]

To provide the funds needed to expand habitat protection for endangered species, Udall stressed the importance of the Land and Water Conservation Fund Act, which passed Congress in 1964. It authorized the Fish and Wildlife Service to utilize recreational user fees, surplus land sales, and a motorboat fuel tax for many purposes, including "the preservation of species of fish and wildlife that are threatened with extinction."[82] Congress, balked, however, when the agency proposed allocating $3.1 million in fiscal year 1965 from the new fund to buy habitat for the whooping crane, Mexican duck, Florida sandhill crane, and several vanishing Hawaiian birds. When Fish and Wildlife Service Director Gottschalk testified before a House subcommittee about the planned expenditures, he found himself skewered by its chair, Rep. Winfield Denton (D-Indiana). Since only about thirty-two whooping cranes remained, Denton calculated that spending a $1.1 million dollars to buy five thousand additional acres of habitat for the endangered species meant that the agency was wasting about $35,000 per bird. Was this a "wise use of the taxpayer's money?" he asked rhetorically. Gottschalk responded that the issue of endangered wildlife could not be addressed solely in economic terms; what was ultimately at stake was "the prevention of one entire species from disappearing from the face of the earth."[83] Unmoved by Gottschalk's reasoning, the subcommittee rejected the plan and informed the Fish and Wildlife Service that it needed Congressional approval for a new law specifically authorizing habitat purchase.

With Udall's encouragement, several naturalists in the Fish and Wildlife Service simultaneously pursued a second strategy for addressing the growing threat to the nation's biodiversity. While attending the IUCN meeting in Nairobi in late 1963, Udall had learned firsthand about the work of Survival Service Commission and its ambitious Red Book. Not long after his return to Washington, he announced that the Department of Interior was initiating a new project "for the preservation of species of native wildlife."[84] A key part of that initiative was creation of a new committee modeled on the Survival Service Commission— the Committee on Rare and Endangered Wildlife Species (CREWS)—that was charged with creating its own official list of native vertebrates threatened with extinction. The committee, which began compiling its inventory on January 1964, chose a presentation format that was virtually identical to the one used for the data sheets in the IUCN Red Book.[85] Beyond Erickson, eight other Fish and Wildlife staff members initially served on CREWS, which began circulating a preliminary listing of endangered species in July of that year. Compiled after consultation with more than three hundred specialists and released with much fanfare, the first official federal endangered species list contained entries on about sixty vertebrate species or subspecies, including sixteen mammals, thirty-five birds, three reptiles, and six fish.[86]

Captive breeding represented another component of the Fish and Wildlife Service's emerging strategy for engaging more deeply with the problem of vanishing species. Success with raising the Hawaiian nene and sandhill crane in captivity—along with continued prodding from Ripley, Scott, and other aviculturalists—convinced Erickson that this risky approach held great potential in certain cases. In 1963, he authored a detailed proposal for establishing a new Endangered Wildlife Research Station at Patuxent that would be devoted to raising rare animals while studying their behavior, physiology, and habitat needs in the wild. "I saw it as a three-legged stool," Erickson later recalled, "Propagation, laboratory research, and field ecology—with no leg more important than any other."[87] Nothing came of the proposal until a year later, when he ventured to Capital Hill to present a photograph of the whooping crane to Karl Mundt, a senator from South Dakota who served as the ranking minority member of the Senate Appropriations Committee. One of Mundt's aides, Rod Kreger, had seen the cranes migrating through the state and was worried that his children would never experience the same thrill. Was there anything the senator could do to help preserve this species? Kreger asked him. "I could think of a few things," Erickson replied drolly. The following spring Mundt attached an amendment to an appropriation bill that allocated $350,000 for staff salaries, animal pens, buildings, and related costs at the new center, which formally opened its doors in 1965. Officials appointed Erickson to head wildlife research and propagation and moved the experiments begun with sandhill cranes at Monte Vista to Patuxent.[88]

Thwarted from using newly available Land and Water Conservation Funds to purchase additional habitat for the species that CREWS had identified as vulnerable to extinction, Fish and Wildlife Service staff also authored legislation specifically authorizing the agency to acquire new refuge land for endangered species.[89] In June 1965, Udall submitted their handiwork to both houses of Congress, identical bills that directed the secretary of interior to "initiate and carry out a comprehensive program to conserve, protect, restore, and where necessary to establish wild populations, propagate selected species of native fish and wildlife, including game and nongame migratory birds, that are found to be threatened with extinction."[90] The two bills, which proved uncontroversial, sailed through with few amendments. Only a handful of individuals (mostly representing national conservation organizations) appeared at the House and Senate hearings to consider the legislation, and their testimony proved overwhelmingly favorable. The only substantive debate was over the expanded federal role over wildlife regulation, an area that had primarily been the purview of states.

The final version to emerge from the Conference Committee, known as the Endangered Species Preservation Act of 1966, required the Department of Interior to continue compiling an official list of endangered species.[91] It also directed various departments in the federal government to protect listed species but only wherever "practicable" and consistent with the primary mandates of those departments. At the same time, the new law consolidated the confusing array of federal refuges and preserves into the National Wildlife Refuge System and authorized the expenditure of modest funds, up to $5 million per year and up to $750,000 for any one area, to purchase new refuges specifically for endangered species.

While the Endangered Species Preservation Act of 1966 represented a significant milestone in the federal government's engagement with the issue of endangered species, it proved far from the "comprehensive program" that its promoters claimed. For one thing, it only addressed the issue of vertebrate species threatened with extinction, not invertebrates, plants, or subspecies. At the same time, it covered only native species, not the growing list of foreign species declining around the globe. Moreover, it remained largely voluntary in approach; it did not require anything of federal agencies whose activities might threaten endangered species nor did it limit the actions of local or state governments, corporations, or private individuals. Perhaps most importantly, the act prohibited the destruction of those species only on federal refuges; it offered no protection for the many endangered species residing outside those areas. Aware of these limitations and anxious to begin addressing them, Fish and Wildlife staff began formulating new legislation even before the 1966 act was signed into law.[92]

The situation with the American alligator (*Alligator mississippiensis*) highlighted the limitations of the first Endangered Species Act and led to increasingly vocal calls for its reform. This giant, water-loving reptile once inhabited swamps, marshes, and lakes across much of the southeastern United States, but it was especially abundant in the states of Florida and Louisiana. Like other carnivorous predators, it had long been perceived as an evil, threatening presence, but it seems to have been largely spared from the systematic campaigns of destruction that targeted wolves, mountain lions, and bears beginning with the arrival of European settlers in the New World.[93] The alligator was, however, commodified on a massive scale. Beginning in the second half of the nineteenth century, a brisk trade in hides was supplemented by a strong demand for alligator souvenirs of various sorts, including live babies, adult teeth, and stuffed individuals.[94] According to a U.S. Department of Agriculture report published in 1893, Florida represented the center of commerce in the species, with as many as 2.5 million hides having been shipped out of the state since 1880. The report's author predicted that it was "only a question of time when this valuable fishery resource . . . will become exhausted."[95] Despite that warning, however, the alligator failed to garner support during the turn-of-the-century wildlife conservation movement that rescued the bison and many plume birds from impending extinction. Compared with birds and mammals, few scientists studied alligators or other reptiles, and relatively few Americans seemed concerned about their plight.[96]

Two factors further accelerated the alligator's decline in the post–World

War II era. The first was increased habitat destruction. After the war, Florida's population grew at an unprecedented rate as the mass adoption of air conditioning and the initiation of large-scale DDT spray campaigns helped make the state's otherwise muggy, buggy landscape more appealing.[97] Developers drained vast expanses of wetlands to provide the housing tracts, roads, and strip malls that supported the influx of new residents. At the same time, the alligator faced renewed pressure from hunters responding to a robust postwar demand for alligator leather. In response, naturalists, game officials, and wildlife conservationists joined forces to secure more stringent legislation in Florida (which closed the hunting season on alligators in 1961) and Louisiana (which did the same in 1963). In 1964, the alligator was also among the sixty species that CREWS included in its first federal inventory of endangered species, and it received limited protection under the terms of the Endangered Species Preservation Act of 1966. But this legislation failed to stop poachers from carrying illegally obtained alligator hides across state lines, where they could sell them without fear of arrest.

Anxious about the future of the species, the world-renowned herpetologist Archie Carr, who had already been leading a campaign to rescue the endangered green turtle, entered the fray.[98] As part of that initiative, in 1967, he published an article in *National Geographic* entitled "Alligators: Dragons in Distress." Using lyrical prose reminiscent of Aldo Leopold, Carr told of his first encounter with voice of an alligator in a cypress swamp on the banks of the St. Johns River. There, on a clear spring night, he experienced three alligators bellowing in a "ponderous, pulsing chorus, half sound, half shaking of the earth" that "seemed to rock the whole swamp." Now, several decades later, the same site was hushed, the cypresses having long since been reduced to lumber and a song that was "200 million years old, an echo of the Age of Reptiles, when cold-blooded creatures ruled the earth," had fallen entirely silent.[99]

Two things were forcing the alligator's decline here and across its entire range, Carr argued: illegal overhunting fueled by a "fantastic rise in the price of hides" and the "destruction of habitat by drainage and development projects."[100] He stressed the central role that the species played in the ecology of its wetland habitat and warned of persistent illegal hunters who cruised previously inaccessible gator habitat in newly developed airboats and glades buggies. To save the beleaguered species, Carr called for a change in fashion that would eliminate the demand for alligator hides and advocated extending the protection of the Lacey Act—the turn-of-the-century federal legislation that forbid taking illegally caught birds and mammals across state lines—to the alligator. Protecting the species would be a challenge, Carr argued, but in doing so, Americans would prove that "we have the sense and soul to cherish a wild creature that was here

FIGURE 60. Archie Carr and sea turtle hatchlings, 1961. Carr was a skilled naturalist, gifted writer, and passionate advocate for wildlife. Already well known for his work on sea turtle conservation, in the 1960s he also led a campaign to rescue the endangered American alligator. Courtesy of the Archie Carr, Jr. Papers, Department of Special and Area Studies Collections, George A. Smathers Library, University of Florida.

before any warm-blooded animal walked the earth, and that, given only a little room, would live on with us and help keep up the fading color of the land."[101] In the Age of Ecology, even a predatory species like the alligator was finding growing support.

While Carr and other reptile enthusiasts pushed for greater protection for the American alligator, the international dimensions of the extinction crisis received increasing attention. Popular television shows—like Marlin Perkins's *Wild Kingdom* (begun in 1961), a series of National Geographic Society specials (begun the same year), and *The Undersea World of Jacques Cousteau* (begun in 1968)—brought vivid, compelling images of endangered wildlife into America's living rooms while warning of their increasingly desperate plight.[102] A string of full-length motion pictures, including the *Serengeti Shall Not Die* (1960), *Wild Gold* (1961), *Born Free* (1966), and *Living Free* (1972), offered larger-than-life views of African wildlife and habitat.[103] Newspaper and magazine headlines raised troubling questions about the future of wildlife by asking "Are the Days of the Arctic's King Running Out?" and "Can Africa's Wildlife Be Saved?" They also offered coverage of the global trade in large, charismatic species like polar bears, elephants, leopards, and rhinos. And a series of popular books—*Vanishing*

Wildlife (1963), *The Auk, the Dodo, and the Oryx* (1967), and *Extinct and Vanishing Animals* (1967)—provided detailed accounts of the myriad threats endangered species faced around the world.[104] In the preface to the popular edition of the IUCN Red Data Book, published in 1969, the American nature writer Joseph Wood Krutch declared that "never in the history of the world has there been so much concern over man's violation of the natural environment and the disappearance of what was once an abundant wildlife."[105]

Sensing growing public support and emboldened by the easy passage of the first Endangered Species Act, Fish and Wildlife staff prepared to strengthen the law.[106] On February 8, 1967, they submitted draft legislation to Rep. John Dingell (D-Michigan), the chair of the subcommittee on Fisheries and Wildlife Conservation of the House Committee on Merchant Marine and Fisheries. Dingell, an avid outdoorsman who was to play an important role in shepherding this and subsequent endangered species legislation through Congress, introduced one version of the bill in the House. No one testified against the proposed law at committee hearings Dingell held in October 1967, and the full House approved it in early August 1968. It looked like similar legislation might sail through the Senate as well, but when fur, leather, and pet industry representatives got wind of it, they pressured Fish and Wildlife officials and the Senate Commerce Committee for changes in the bill. Furriers, tanners, and exotic pet importers argued that extinction was an international issue and that passing unilateral legislation would unduly harm U.S. companies without halting the destruction of endangered species. Resigned that a new endangered species bill was likely to pass in some form, however, they proposed a series of amendments, including several that the Department of Interior endorsed. In October 1968, the Senate Commerce Committee reported favorably on an amended version of the bill that adopted the Interior Department's suggestions. Differences between the House and Senate versions meant that a conference committee would have to convene, and the legislation died when Congress adjourned later in October.

Several endangered species bills were reintroduced the next year, following a series of meetings between representatives of conservation groups, the two relevant congressional committees, the Interior Department, and the fur and leather industries. The final version of the law represented an incremental expansion of the 1966 act. To address growing concerns about the plight of the alligator, the Endangered Species Conservation Act of 1969 officially broadened the definition of "wildlife" beyond the more traditional notion of mammals and birds. Now endangered reptiles, amphibians, and even invertebrates were also eligible for federal protection.[107] Consistent with this expansion, the new law expanded the Lacey Act to prohibit the interstate sale or transportation of endangered reptiles, amphibians, mollusks, and crustaceans taken in violation of state or foreign laws. To meet concerns about the international trade in species threatened with extinction, the legislation also directed the Fish and Wildlife Service to expand its official list of endangered species to include organisms not found in the United States and banned the importation of any such species or its products. It also called for an international convention on trade in endangered species, which the IUCN had been working toward for several years by that time.

THE FUR FLIES OVER BIG CATS

On the first of January 1970, less than a month after signing the Endangered Species Conservation Act of 1969, President Nixon called reporters to the signing ceremony for another landmark piece of legislation, the National Environmental Policy Act.[108] With much fanfare, he predicted that the 1970s would be remembered as the decade of the environment, a time "when this country regained productive harmony between man and nature."[109] While the nation failed to achieve that lofty goal, during the next few years environmentalism did wash across the United States like a tidal wave. A long series of public opinion polls revealed that environmental issues enjoyed unprecedented traction with the nation's citizens.[110] So did the celebration of the first Earth Day, on April 22, 1970, when more than twenty million Americans participated in activities ranging from teach-ins and community cleanup drives to guerilla actions aimed at corporate polluters. Reflecting the growing concern with the state of the environment, traditional conservation groups—like the National Audubon Society, the National Wildlife Federation, and the Sierra Club—broadened their scope while experiencing impressive membership growth. At the same time, newer groups—like the Environmental Defense Fund, the Fund for Animals, Friends of the Earth, and the National Resources Defense Council—proliferated. The green tenor of the times also produced an outpouring of federal environmental legislation in the early 1970s, including not only the National Environmental Policy Act, but also the Clean Air Act Amendments, the Federal Water Pollution Control Act Amendments, the Federal Environmental Pesticide Control Act, the Marine Mammal Protection Act, and many others.

Ecology, in its newer, more expansive sense, loomed increasingly large in public discourse at the time. Peter Jannsen, the education editor at *Newsweek* magazine declared that the mass outpouring of protest during Earth Day signaled the dawning of the "Age of Ecology," a time when "tens of thousands of young people have turned their love toward the earth in an effort to repair ravished landscapes, oil-filled harbors and over-crowded cities."[111] In a cover story featured that same year, the editors of *Time* magazine proclaimed the biologist Barry Commoner the "Paul Revere of Ecology." Similarly, the April 21, 1970 issue of *Look* magazine presented a picture of the new ecology flag, a green and white standard commonly hoisted at environmental gatherings and worn on patches and pin-on buttons.[112]

It was in this politically supercharged atmosphere that the Endangered Species Act of 1973 gained passage. Increasingly, Americans pushed the federal government to do more to protect organisms threatened with extinction and to act before those organisms reached critically low numbers from which they might never recover. Federal bills to achieve these goals were introduced in 1970 and 1971, but they failed to go very far until early 1972, when, jockeying to burnish his image as an environmental president, President Nixon delivered a message to Congress calling for a "stronger law to protect endangered species of wildlife." Nixon claimed that "even the most recent act to protect endangered species, simply does not provide the kind of management tools needed to act early enough to save a vanishing species."[113] The same day that Nixon delivered this address, Representative Dingell introduced an administration bill that had been authored by staff in the Fish and Wildlife Service, the Department of the Interior, and others, and ten days later, Senator Mark Hatfield introduced identical legislation in the Senate. While the legislation failed to pass in 1972, a similar bill was reintroduced the next year. This time the bill received broad support from environmental groups and an aroused public; it passed by overwhelming margins in both houses of Congress.

The highly publicized, precipitous decline of many of the world's spotted cats—tigers, leopards, cheetahs, jaguars, ocelots, and others—proved a key factor demonstrating the need for stronger endangered species legislation.[114] In 1962, the American First Lady Jacqueline Kennedy sparked a fashion trend when she began sporting a leopard skin coat at various public events.[115] A year later, as part of a campaign aiming to counter that trend, the National Audubon Society convinced the IUCN to pass a resolution decrying the growing fad for wearing spotted cat fur as a "threat to the continued existence of these kinds of animals."[116] Despite that warning, by the late 1960s, the United States and Europe were annually importing the pelts of an estimated ten thousand leopards, fifteen thousand jaguars, three to five thousand cheetahs, and two hundred thousand "ocelots" (a trade name covering not only ocelots proper but several similar species, like the margay and oncilla). Although obtaining precise data on the population status of these generally secretive and often nocturnal animals proved notoriously difficult, cat supporters worried that the unprecedented rate of exploitation vastly exceeded the ability of these top-level predators—which also faced habitat destruction and in some cases systematic hunting from local residents fearful of their attacks—to reproduce.

Concerned that numerous spotted cat species were rapidly being hunted into oblivion, a coalition of naturalists, wildlife conservationists, and humanitarians pursued several strategies. They not only publicized the issue in a series of newspaper and magazine articles—with titles like "The Big Cats in Trouble" and "The Sad Tale of the Tiger"—but, in keeping with the times, they also organized protests and boycotts.[117] In early 1970, for example, a newly established environmental group, Friends of the Earth, launched a campaign to convince wealthy Americans to stop purchasing furs made from wildlife, especially of endangered species like spotted cats. That campaign featured pickets at stores offering furs and a full-page advertisement in the fall fashion edition of a New York trade publication, *Women's Wear Daily,* that included signatures from a hundred well-known personalities who signed a pledge not to buy products made from "wild and endangered animals." One New York furrier responded by announcing that he would no longer advertise in the *Women's Wear Daily* while decrying those responsible for the protest: "Ecology and its ramifications is suddenly very fashionable, and the little old ladies that used to help the pigeons crap up our city are now minor heroines."[118]

Evidence that publicity about the plight of spotted cats was making an impact came quickly. By the fall of that same year, the editors of *Vogue,* a leading American fashion monthly, declared that they were "deeply concerned about preserving animals threatened with extinction" and would henceforth refuse to publish photographs showing the skins of such animals.[119] In New York, naturalists, cat supporters, and others concerned about the continued trade in endangered species successfully lobbied for a state law prohibiting the sale of goods made from several species of spotted cat, the red wolf, the polar bear, the vicuña, and all alligators, caimans, and crocodiles.[120] Furriers complained that the legislation would sound the death knell of their industry, in one breath, and in the next, strongly denied that they carried goods made with endangered species. After surviving a series of legal challenges in state and federal courts, New York's Mason Act went into effect on September 1970. Within the next several months, California passed a similar bill, and five other states considered comparable legislation.

Buoyed by success at the state level, those concerned about the fate of the spotted cats also lobbied for stronger federal protection. Fish and Wildlife Service staff had listed several spotted cats, but they had been slow to add additional species or subspecies, claiming that there was insufficient evidence that they faced "worldwide extinction," the listing standard required under the terms of the Endangered Species Act of 1969.[121] Eley P. Denson, a staff assistant at the

Fish and Wildlife Service's small Office of Endangered Species and International Activities, infuriated supporters of expanded listing for spotted cats when he publicly branded the Friends of the Earth anti–wild fur campaign as "overemotional" and motivated by a desire for publicity and money. He added further fuel to the fire by admitting that he would buy a leopard coat for his wife if he could afford one.[122]

Cat experts responded to the cacophony of competing claims about the status of the world's cats and the failure of the Fish and Wildlife Service to act by calling a symposium on their ecology and conservation in March 1971. There the Berkeley ecologist Norman Myers presented compelling evidence that several species of spotted cat were in significant trouble.[123] At the same meeting, Harold J. Coolidge, then president of the IUCN, and Paul Leyhausan, chair of the Survival Service Commission's Cat Specialist group, led a roundtable discussion in which they called for legal protection of threatened species, like the spotted cats, before they became critically endangered. "If we wait until a species is low enough in numbers to qualify for the Red Book," Coolidge warned, "it is often too late to save it."[124] During the ensuing deliberation, Coolidge offered a resolution that the assembled experts voted unanimously to pass: "The Federal (U.S.) Government should back up the efforts of state governments in banning the importation and sale of skins and products made from the wild cats of any country of the world."[125] The final pressure for listing came in January 1972, when Cleveland Amory, the president and founder of the Fund for Animals, briefed Dick Cavett, a member of the organization's board, on the situation after learning that Interior Secretary Rogers Morton was to appear on Cavett's popular television show. Morton, who was clearly caught off guard by Cavett's question about why the spotted cats were not on the endangered species list, replied that his department was in the process of adding some of them.[126]

Faced with increasing pressure not only from conservation organizations but also from cat experts, the Fish and Wildlife Service finally responded. In late 1971, a staff mammalogist, John L. Paradiso, prepared a status report on the wild cats of the world that recommended listing all species and subspecies of the cheetah, leopard, tiger, snow leopard, jaguar, ocelot, margay, and tiger cat.[127] Several days later, following his appearance on the *Dick Cavett Show*, Secretary Morton issued a press release announcing the formal proposal to list these forms. There Morton was quoted as saying, "We must act now, because a world without great cats is unthinkable."[128] The final listing came a month later, but much to the dismay of some large cat supporters, for the next year the Interior Department officials routinely granted "economic hardship" exemptions to safari hunters so they could bring their feline trophies back to the United States.[129] For the growing number of Americans worried about the plight of the world's wildlife, the episode highlighted the need for stronger federal legislation that would grant the federal government the power to list species before they reached critically low numbers from which they may not be able to return. It also suggested that future legislation should include provisions to force the agency to act when it was dragging its feet on listing a particular species or refusing to carry out the mandates of the Endangered Species Act.

SAVING MARINE MAMMALS

Concern about the plight of marine mammals provided additional pressure for strengthening federal endangered species legislation. That concern flowed not only from a growing environmental sensibility but also from the increasing exposure Americans had to these charismatic creatures, both directly and through the media. Zoos and aquariums had long featured displays of polar bears, seals, and dolphins, while whale skeletons and models proved popular in natural history museums.[130] But it was the development of so-called oceanariums, with their performing dolphins and (later) whales that captivated the public's imagination while highlighting the intelligence, social nature, and beauty of these marine creatures. The first oceanarium was Marine Studios (soon renamed Marineland), which opened just south of St. Augustine, Florida, in 1938.[131] By 1953, about seven hundred thousand visitors annually flocked to the popular tourist attraction, which spawned numerous imitators. One of the most successful of these was Sea World, which opened in San Diego in 1964.[132] A year later, that facility began thrilling audiences with performances of "Shamu," the first in a series of orca (or killer) whales with the same name that became Sea World's principal draw and official mascot.

The growth of whale watching as a recreational activity also promoted concern about the species. In the mid-1940s, the naturalist Carl Hubbs began leading his students in annual gray whale counts around San Diego, and by the mid-1950s, commercial operators took tourists to view migrating gray whales along the coast of southern California.[133] In 1967, two hunger activists, Bemi DeBus and Clark Cameron, created the American Cetacean Society, a California-based organization devoted to whale conservation.[134] From the beginning, the society sought to educate the public about the increasingly dire status of whales with the hope of inspiring action to rescue them from the impending extinction, and a primary means for achieving that goal was sponsoring whale watching trips. By the 1970s, the business of whale watching had spread to Canada, New England, and Hawaii .[135]

Besides viewing marine mammals in oceans both real and simulated, Americans also experienced them vicariously in books, magazines, records, television shows, and movies. In 1966, the whale enthusiast Scott McVay published an article for *Scientific American* entitled "The Last of the Great Whales," the first in a series of publications that drew attention to the increasingly desperate plight of the largest mammals on the earth.[136] Four years later, the biologist Roger Payne produced a popular LP record released as *Songs of the Humpback Whale*. Those eerie, haunting vocalizations, which to some listeners sounded liked funeral dirges, soon began appearing in other recordings.[137] Following its debut in 1968, the celebrated television show *The Undersea World of Jacques Cousteau* regularly featured mesmerizing images of marine mammals in their natural surroundings. Also influential in shaping public perceptions of these creatures was the movie *Flipper*, released by MGM in 1963. The original movie, sequel, and television spin-off, which featured the antics of a playful Lassie-like cetacean and his boyhood companion, Sandy, helped transform the dolphin into an endearing glamour species and enduring cultural icon.[138] So did the many publications of the neurophysiologist-turned-dolphin-guru John C. Lilly, whose books, *Man and Dolphin* (1961) and *The Mind of a Dolphin* (1967), celebrated the keen intelligence and sophisticated communication skills of these charismatic cetaceans.[139] Lilly's research inspired Robert Merles to write a bestselling novel, *The Day of the Dolphin* (1967, English translation 1969), which in turn spawned a popular motion picture of the same name in 1973. In both the book and the movie the basic plot was identical: dolphins trained to speak English saved the world from nuclear annihilation.

Americans, were shocked, then, to learn that hundreds of thousands of what they had come to appreciate as bright, fun-loving, sociable creatures died each year at the hands of large-scale tuna fishers.[140] For reasons that remain unclear, yellowfin tuna (*Thunnus albacores*) in the eastern tropical Pacific Ocean often swim directly underneath groups of spinner and spotted dolphins (*Stenella longirostris* and *S. attenuata*), making them relatively easy to spot from a distance. In the late 1950s and early 1960s, tuna fishers who plied those warm waters developed a method known as purse seining that involved setting massive nylon nets, as long as a mile and up to six hundred feet deep, on groups of dolphins visible from the surface. Confident that a large school of yellowfins was likely to be swimming underneath, the fishers then hauled their enormous net on board, sorted out the tuna, and threw the captured dolphins and any other unwanted fish overboard. This gruesome process resulted in significant mortality and death for the as many as two thousand dolphins that might be captured in a single set. Although fishers soon developed a "back down" procedure that involved reversing the ship's motors once the net was partially hauled and herding the dolphins toward its submerged rim, the procedure proved both time consuming and often unsuccessful.

The first publicity about the immense cetacean destruction at the hand of tuna fishers came from William F. Perrin, a graduate teaching assistant at UCLA who was researching the biology of spinner and spotted dolphins and who began publishing on the issue in 1968.[141] Although it was difficult to get a handle on the precise scope of the problem, estimates of the number of dolphins killed between 1959 and 1970 ranged to four million or more, an obviously unsustainable number that presented a clear and present danger to spinner and spotted dolphin populations. The fact that the public increasingly perceived dolphins as intelligent, sociable creatures further heightened the sense of the alarm.

At about the same time the story about the tuna industry's wholesale destruction of dolphins began breaking, Americans also learned grisly details about the harvest of another marine mammal with broad public appeal: the harp seal (*Phoca groenlandica*) in Canada. Harp seals, which gather in large aggregations to whelp on North Atlantic and Arctic pack ice, had long been pursued for their oil.[142] Beginning in 1951, sealers aggressively hunted them for their hide as well. The shift came after Norwegian tanners discovered a way to preserve the long snow-colored fur of newborn pups, so-called whitecoats, whose fur is actually fetal hair that mammals usually shed while still in the womb. That year alone sealers took a total of over 312,000 baby and adult harp seals. Hunters clubbed the defenseless whitecoats, only a few day's old, using a long wooden pole with a curved iron spike on one end, and skinned their hide, which was primarily used for trim and trinkets. As early as 1960, the biologist David Sergeant warned the without stronger regulation of the annual harvest, the harp seal was in "grave danger of catastrophic decline in numbers within a very few years."[143] But it was not until haunting pictures of the brutal slaughter of seal pups began circulating widely that a horrified public responded. The story first broke in 1964 when a film produced by the Artek Studios of Montreal showed a seal pup being skinned alive on the ice off eastern Canada.[144] Brian Davies, who founded the International Fund for Animal Welfare in 1969, proved an early leader in the crusade to stop the practice. Over the next several years, a series of articles, books, photographs, and film clips raised consciousness about the annual harp seal kill and led to growing calls for reform.

A third stream also fed the rising tide of concern about the fate of marine mammals: the fear that overhunting was driving whales to the point of extinction. Whaling has had a long, sordid history consisting of various seafaring nations aggressively exploiting one species after another in succession, mainly for their oil and more recently for their meat. Once the most valuable and easily caught species became depleted—beginning with rights, bowheads, and sperm whales in the mid-nineteenth century—whalers moved on to less attractive ones.[145] The development of harpoon guns, explosive-tipped harpoons, and steam-driven vessels in the second half of the nineteenth century permitted pursuit of swifter species, while the appearance of modern factory ships in 1925 freed up whalers from having to haul their catch on shore for processing. Blue whales, the largest animal ever to inhabit the earth, dominated the catches through the 1930s, but the take of that awe-inspiring species dropped precipitously by the 1950s. Fin whales represented the largest part of the catch from the 1950s until the early 1960s, when their numbers also began dramatically declining. Sei whales were generally ignored until the late 1950s, but they too had diminished greatly by the 1960s. Stocks of humpback whales, never very numerous, also collapsed during this period.

This wasteful pattern of exploitation had not gone entirely unnoticed or entirely unregulated.[146] In 1931, twenty-six nations signed a League of Nations–sponsored convention that established basic rules for harvesting whales, though this initial agreement appears to have been only partially observed. Six years later, nine nations, including every one that was active in whaling at the time except Japan, signed the First International Whaling Convention, which tightened up those initial regulations. Finally, following the end of World War II, sixteen nations negotiated the International Convention for the Regulation of Whaling (1946), an agreement that created the International Whaling Commission (IWC) to regulate whaling among its member nations. However, for much of its history, the IWC remained largely unsuccessful in achieving its basic goal of "safeguarding for future generations the great natural resources represented by the whale stocks."[147] The commission struggled with many obstacles: a lack of hard scientific data about the status of whales, overly optimistic harvest quotas, varying levels of governmental support, and spotty reporting of catches, to name just a few. One particularly active figure in the commission's deliberations was the mammalogist and U.S. National Museum curator Remington Kellogg, who, from 1930 to 1964, served as an official U.S. delegate to every major international meeting about whales.[148]

Faced with IWC foot-dragging and mounting public pressure, seven months after the passage of the Endangered Preservation Species Act of 1969, the Interior Department proposed adding eight whales to its official list of imperiled animals.[149] The July 1970 proposal included not only the threatened gray whale and the practically extinct blue, right, bow, and humpback whales but also several species—the finback, sei, and sperm whales—that were still being pursued commercially. Fearful that listing the sperm whale would halt the lucrative market in sperm-whale oil for submarines, the Pentagon and the Commerce Department opposed the proposal on the grounds that the finback, sei, and sperm whale did not face immediate extinction and therefore did not qualify for listing under existing endangered species legislation. Fish and Wildlife Service Director Gottschalk backed down, but Udall's successor, the flamboyant Interior Secretary Walter J. Hickel, decided to list them anyway and submitted the final rule to the printer on the Wednesday before Thanksgiving in 1970. It turned out to be one of his last official acts, for two days later Nixon fired Hickel for openly criticizing the administration's handling of the Vietnam War. Pentagon and Commerce Department officials pressured Hickel's second in command at Interior, E. U. Curtis Bohlen, to withdraw the listing before it was published, but he resisted, and the ruling became final.

Heightened concern about the fate of dolphins, seals, whales, and other mammals that spent much or all of their lives underwater also led to the passage of the Marine Mammal Protection Act. As one measure of the issue's saliency, in 1971, legislators introduced no less than thirty different federal bills relating to marine mammal protection. One group of bills called for a total ban on almost all taking of these creatures, a second mandated humane harvesting techniques, a third provided funding for additional research, and a fourth combined elements of the other three.[150] The bill that was finally signed into law on October 21, 1972, represented a compromise between preservationists who had called for a complete ban on killing and harassing marine mammals and conservationists who argued for a more management-oriented approach that protected species from overhunting to the point of population collapse but still promoted their utilization.[151]

The Marine Mammal Protection Act declared that certain species and populations were facing extinction as a result of overexploitation and that they "should not be permitted to diminish beyond the point at which they cease to be a significant functioning element in the ecosystem of which they are a part."[152] The goal was not simply to stem the tide of extinction, then, but to achieve the "optimum sustainable population" for each species, which was defined as "the number of animals which will result in the maximum productivity of the population of the species keeping in mind the carrying capacity of the habitat and the health of the ecosystem of which they form a constituent element."[153] To achieve this ambitious goal, the act declared a moratorium on the taking of marine mammals throughout areas subject to U.S. jurisdiction and by any person or vessel subject to that jurisdiction on the high seas. It also defined "taking" quite broadly, including to "harass, hunt, capture, or kill" or to attempt to do these things. Waivers to this general moratorium might be granted in certain instances, but not for depleted species or stock.[154] The Marine Mammal Protection Act proved important not only for extending the protection it afforded, but also because several of its key provisions, including its broad definition of "taking," were soon incorporated into yet another round of revised endangered species legislation.

ENDANGERED SPECIES REDUX

In an editorial published less than a year after the passage of the Endangered Species Act of 1969, New York Zoological Society Director William C. Conway offered a pointed critique of the legislation and its implementation.[155] While the official U.S. endangered species list at the time had grown to 131 mammals, 118 birds, 22 reptiles and amphibians, several fish, and one mollusk, it generally failed to include species, like the spotted cats and whales, that were exploited commercially. Even more troubling, for the most part, the new law covered only "those species already on the very brink of extinction," forms for which recovery often proved difficult, if not impossible. To do a more effective job of protecting vanishing species subjected to trade, Conway argued, the federal list should include all species and subspecies being hunted commercially but not being harvested on a "sustain-yield" basis; all forms of wildlife that are difficult to distinguish from protected forms (like the various subspecies of any species and the many crocodilians); and those "whose ranges or populations are shrinking or already small." The goal of endangered species legislation should not be simply to preserve small, scattered, remnant populations of rare species but to maintain "wildlife populations large and widespread enough to fulfill their ecological roles, and to withstand natural disasters such as drought or disease, and unpredictable man-made catastrophes, such as oil slicks." While subsequent decisions to list many spotted cats and whales addressed several of Conway's specific concerns, they failed to come to terms with his most basic complaint: the Endangered Species Acts of 1966 and 1969 failed to achieve their lofty goal because they did not require the government to take action until it was too late, when a species or subspecies had already reached such critically low numbers that it could not usually recover. It was a fear that many of Conway's colleagues had come to share.

Their concern eased somewhat with the successful negotiation of the Convention on International Trade in Endangered Species (CITES) in February 1973, an agreement that had been in the works for more than a decade.[156] At the 1961 IUCN meeting in Arusha, Tanzania (then Tanganyika), delegates from several wildlife exporting nations argued that they could never control poaching as long as the international market in wildlife products, like elephant ivory, rhinoceros horns, and cats skins, remained unregulated.[157] One session at the meeting, chaired by Lee Talbot, tackled the difficult wildlife trade question, which was gaining increasing attention at the time. The discussion at that session, in turn, led to a proposal for a convention on international wildlife trade, and two years later, at the IUCN meeting in Nairobi, delegates adopted a formal resolution calling for such a convention. Within a year, Talbot and Wolfgang Burhenne, chair of the IUCN's Commission on Legislation, hammered out an initial draft. A long, arduous process of revision and rewriting followed—with input from the Survival Service Commission, the Commission on Legislation, the IUCN's Executive Board, and following the 1966 Assembly in Lucerne, thirty-nine governments and eighteen international organizations.

The IUCN began circulating a second draft of the proposed convention in 1969, but several fundamental challenges hindered progress.[158] First, nations involved in the discussions pursued a bewildering variety of approaches to foreign trade regulation and to the protection of wild species, approaches that proved difficult to reconcile. Second, the framers of the convention struggled with the problem of defining exactly what constituted an endangered species, a thorny issue that had also plagued the Survival Service Commission and the Committee on Rare and Endangered Wildlife Species as they drew up their own authoritative lists of vanishing species. And third, the interests of "exporter" nations, which tended to favor stricter trade controls, often differed from those of "importer" nations, which wanted flexible regulations that allowed them to continue the lucrative trade in animal products. Pressure for successful resolution of these challenges increased in 1969, when the United States passed its second Endangered Species Act, legislation that not only banned the import of listed species but also called for an international convention on trade in those species to be held two years later. Russell Train (an IUCN board member and chair of the newly created Council on Environmental Quality [CEQ]), Lee Talbot (then a senior scientist at the CEQ and an IUCN board member), and Harold Coolidge (who was president of the IUCN) pressed Washington to pursue successful completion of the negotiations. Working closely with the Kenyan wildlife official Peter Olindo, Americans also recommended stricter provisions in the draft convention. A campaign to conclude the negotiations during the 1972 United Nations Conference on the Human Environment, the so-called Stockholm Conference, failed.

While the third draft of the convention was circulating in 1971, U.S. officials spread word that they were interested in convening a conference to iron out any remaining differences between the participating nations and to open the convention for signature. That meeting, held in February and March 1973, attracted delegates from eighty-eight nations, including every major player except the People's Republic of China. Not surprisingly, given the previous history of the convention, discussions quickly mired down in a series of conflicting amendments and seemingly intractable disputes.[159] As prospects for a strong agreement dimmed, Department of Interior officials turned the tide with a surprise press conference. There Assistant Secretary of Interior Nathaniel Reed announced that federal agents in New York had cracked a major international fur smuggling ring. During an eighteen-month period, Reed charged, thirty-three defendants (the prestigious New York furriers Vesely-Forte, Inc., among them) had handled one hundred thousand smuggled pelts of spotted cats, including hides equaling one-tenth of the world's surviving cheetah population.[160] The announcement not only undermined furriers' long-standing claims that they did not deal in endangered wildlife but also brought home the point that the delegates in Washington were not engaged in a purely theoretical exercise.

The final agreement, which the United States and seventy-nine other nations signed on March 2, 1973, was a long, complex document that consisted of twenty-five articles and ran into thirty-six single-spaced pages. CITES established a worldwide system of import and export permits for species threatened with extinction.[161] The convention placed 375 species and genera thought to be in grave danger in Appendix 1 and required export permits from the country of origin and import permits from the country of destination for the international movement of any of these. While the agreement did not ban all trade in species listed in this appendix, import and export permits for these species were only to be granted under "exceptional circumstances." Initially included in this category were most of the spotted cats, most crocodilians (including the American alligator), numerous birds and primates, and five species of whales. Less endangered organisms, some 250 animals and plants, were placed in Appendix 2. Trade in these species would still be allowed, but importing countries would require each specimen or product to have an export permit. Appendix 3 consisted of those species endangered in a limited area but not globally. The convention granted individual countries the right to list here those species they deemed threatened within their own boundaries, and only they could withdraw the species from that appendix. The agreement entered force in July 1975, after ten nations, including the United States, formally ratified it. While CITES lacked any mechanism to punish nations that failed to enforce its provisions, it was nonetheless a broad, substantive agreement that recognized individual nations as the primary protectors of their own flora and fauna while promoting international cooperation to guard against overexploitation of endangered organisms through trade across national boundaries. The formal recognition of various degrees of vulnerability to extinction represents one of the most significant innovations of the convention.[162]

While they were pushing to complete negotiations for CITES, scientists, environmentalists, and policy makers also sought to strengthen federal laws related to endangered species.[163] Among many shortcomings, the bills passed in 1966 and 1969 failed to outlaw the taking of listed species except on federal refuges, or when they were already protected by state law, they failed to require federal agencies to halt activities that might destroy endangered species, and they failed to provide a mechanism for identifying and protecting species before they reached the brink of extinction. One version of a bill that sought to address these defects came from the hand of the Undersecretary of Interior E. U. Curtis Bohlen, a passionate environmentalist and former state department official. Nixon presented Bohlen's bill to Congress in February 1972, with the argument that existing legislation did not "provide the kind of management tools needed to act early enough to save a vanishing species."[164] (John Dingell introduced the administration bill in the House, and Mark Hatfield did the same in the Senate.) Distracted by the impending elections, however, the Ninety-Second Congress failed to act on either measure.

Several individuals worked behind the scenes to strengthen Bohlen's original bill and steer it through Congress. Frank Potter, counsel for the House Merchant Marine and Fisheries Committee, and the ubiquitous Lee Talbot were among those. Talbot, working within the administration, and Potter, working on Capital Hill, pursued the bill as it slowly worked its way through the legislative process both in 1972 and in 1973, when four new endangered species bills gained introduction into Congress. Also involved in the ongoing revision process were the biologist Earl Baysinger (who was Goodwin's replacement at the Office of Endangered Species), Clark Bavin (who headed the Fish and Wildlife Service's law enforcement division), and Representative Dingell. Little controversy emerged during the consideration of these bills following their reintroduction in 1973, and they enjoyed broad support in the popular media.[165] The Senate voted unanimously, 92–0, to pass its version in July 1973. The House was also strongly united in support of its version, which passed by a vote of 390–12 in September 1973. Not surprisingly, the final bill that emerged from the conference committee enjoyed broad support as well. In December, the Senate voted, again unanimously, to accept the conference report. The next day, when the House considered the report, Representative Dingell noted, "It would be no exaggeration to say that scarcely a voice has been heard in dissent."[166] The House accepted the conference report by a vote of 345–4.

A week later Nixon signed the popular, comprehensive bill, which superseded the two previous acts (except the provisions in the 1966 bill consolidating the National Refuge System). During the signing ceremony, the president declared that "nothing is more priceless and more worthy of preservation than the rich array of

FIGURE 61. Confiscated spotted cat pelts, 1984. The Endangered Species Act of 1973 and the Convention on International Trade in Endangered Species offered powerful weapons in the campaign to reduce the traffic in species threatened with extinction. The U.S. Fish and Wildlife Service seized these pelts as part of Operation Trophy Kill. Courtesy of the U.S. Fish and Wildlife Service.

life with which our country has been blessed," and indeed, the bill granted the federal government unprecedented power to pursue the goal of preventing extinction not only of native species but also those that resided abroad.[167] Borrowing from the Marine Mammal Protection Act and responding to those who argued that earlier legislation failed to mandate action until it was too late, the new act required the secretary of the interior to maintain a list of species, subspecies, and/or isolated populations thought to be either "endangered" or almost so ("threatened").[168] All animals (except insect pests, bacteria, and viruses) were to be eligible for listing, including those species that merely resembled endangered or threatened species and thus were difficult to distinguish from those in actual danger. Plants were also eligible for protection once the Smithsonian Institution had completed a review of their status.[169] Biological criteria alone were to be used in selecting organisms for the list, and anyone with appropriate evidence could propose additions or deletions. To address the recurring problem of bureaucratic foot dragging on the issue of listing, private citizens and organizations could file suit to force the secretary to act on a proposed listing. Once listed, organisms could not be legally imported, exported, or "taken" within the United States, its territorial waters, or the high seas. Again, following the Marine Mammal Protection Act, taking was defined extremely broadly to include "harassing, harming, pursuing, hunting, shooting, wounding, killing, trapping, or capturing" protected forms, or even attempting to do the same. Those who violated the terms of the act faced stiff penalties of up to $20,000 and one year of imprisonment for each count. The bill also required federal agencies and departments to review any actions they undertook directly, funded, or permitted to make sure they did not jeopardize listed species or modify any habitat designated as critical to their survival.

NATURE ON A LEASH

In his pioneering study of the Endangered Species Act, historian Steven Yaffee argued that the strong, prohibitive endangered species legislation that Congress finally passed in late 1973 represented the "peak" of the "wave of environmentalism" that washed across American society in the late 1960s and early 1970s. To be sure, strong public support for the idea of preventing extinction played a key role in the act's passage by overwhelming margins. But as this and previous chapters have shown, it would be a mistake to conclude that the Endangered Species Act was merely an artifact of the Age of Ecology. Indeed, for more than a century preceding its passage, a small, dedicated group of enthusiasts had been sounding the alarm about the decline of wildlife, organizing like-minded citizens, researching vanishing species, developing arguments for preservation, cultivating public support, and lobbying for protective policies at the state and federal level. Naturalists haunted by the specter of extinction—individuals like Joel A. Allen, William T. Hornaday, Frank M. Chapman, Victor Shelford, and Harold J. Coolidge—played a central role in the earliest efforts to define and address the problem, and naturalists continued to be central to the passage and implementation of the series of federal endangered species acts that culminated in the 1973 legislation.

While that Endangered Species Act of 1973 represented the culmination of this long dialogue about the problem of extinction, it also presented discontinuities with the past in at least two key respects. One of these was the comprehensiveness of its coverage. As we have seen, before 1973 concern about endangered species had been largely focused on so-called charismatic megafauna, relatively large birds and mammals (and to a much lesser extinct, fish, reptiles, and mollusks), species of wildlife that had attracted the most scientific research and public concern. As historian Shannon Petersen has recently shown, the media coverage of and Congressional hearings considering the Endangered Species Act of 1973 also focused almost entirely on a handful of familiar organisms—bald eagles, gray wolves, grizzly bears, cheetahs, whales, and elephants—that had enjoyed increasing public support in the Age of Ecology. Except for a few actual or potential pests, however, the legislation itself offered the same strong protection to *all* vanishing animals—regardless of size or class—and (after further study) even to plants.

Clearly, the authors of this landmark legislation understood the broad extent of its protective mandate, but Congress and the public failed to appreciate its full scope until five years after the passage of the Endangered Species Act. When opponents of the Tellico Dam proposed for the lower Tennessee River successfully pushed for the listing of the snail darter, a small nondescript fish that threatened to bring the more than $200 million federal project to a screeching halt, the broad protection the ESA provided was brought home dramatically. In a landmark decision handed down in 1978, the U.S. Supreme Court upheld the sweeping act, which it called "the most comprehensive legislation for the preservation of endangered species ever enacted by any nation."[170]

Beyond is comprehensiveness, the 1973 act was also much more systematic than previous legislation. Before the first Endangered Species Act of 1966, many state and federal agencies, conservation organizations, and private individuals had attempted to rescue species that seemed threatened with extinction. But their actions remained uncoordinated, piecemeal, and ad hoc. The 1966 act had taken an initial step in creating a more methodical approach to the problem of vanishing species by requiring the Fish and Wildlife Service to begin maintaining a listing of native species endangered with extinction, while the 1969 act extended that official listing to foreign species as well. But it was not until the 1973 legislation, with its broad notion of "taking" and harsh penalties for those who harassed or harmed any listed species, that the federal government placed itself at the center of a far-reaching program to identify, monitor, and rescue species that seemed to be struggling to survive. Once the Fish and Wildlife Service made a decision to declare a species as endangered or threatened, that decision affected not only other state and federal agencies but also private landowners. All were barred from any activity that might harm a listed species or its critical habitat, wherever it might be found. And once a vanishing species had been listed, the federal government was required to commission a detailed recovery plan designed to nurse that species back to self-sustaining population levels. For some critically endangered species—like the California condor and the red wolf—those recovery plans would include highly controversial campaigns to capture every remaining example of a species as part of a last-ditch captive breeding program.

The charismatic whooping crane, which had long served as a poster creature for the federal government's efforts to respond to the growing problem of impending extinction, offers a prime example of the increasingly intensive

FIGURE 62. Captive whooping cranes at Patuxent. Photograph by John Gottschalk. To nurse the endangered whooping crane back from the brink of extinction, the U.S. Fish and Wildlife Service has engaged in intensive management efforts, including a captive breeding program based at Patuxent. While the population of whooping cranes has slowly but steadily increased as a result of these efforts, intensive management must continue for the foreseeable future. Courtesy of the National Archives and Records Administration, Record Group 22-G, Box 6, Folder 85, no. 5732.

management that rare species faced once they were formally declared endangered. In 1967, Fish and Wildlife Service biologist Ray Erickson accompanied a party of U.S. and Canadian officials to retrieve six whooping crane eggs from the perilously small population nesting at Wood Buffalo National Park in western Canada. Over the next seven years, fifty eggs, nearly half of the total number produced at the site, endured the arduous journey to Patuxent to build up a captive flock of the birds.[71] But whooping cranes proved more difficult to foster than the sandhill surrogates that Erickson and his staff had been working with until that point. Bacterial infections, parasite infestations, and leg disorders led to unusually high mortality rates in the birds, while the chicks that managed to survive into adulthood exhibited behavioral abnormalities that delayed their initial breeding attempts. Because none of the breeding pairs could copulate on their own, officials had to artificially inseminate them three times per week during the breeding season.

By 1975, the Patuxent flock began producing its first eggs, but few of the eighty or so laid over the next five years resulted in viable young. As a result, in 1975, the Fish and Wildlife Service began transferring whooping crane eggs from the Woods Buffalo–Aransas population and the captive Patuxent flock to the nests of sandhill cranes in the Grays Lake National Wildlife Refuge in southwestern Idaho. The agency abandoned that experiment more than a decade later when it realized that the improperly imprinted male and female whoopers were failing to breed at the site.[172] A new initiative to establish a nonmigratory breeding population in south Florida began in 1993; the jury remains out on whether that project will prove effective. Meanwhile, officials have established two additional captive flocks to serve as an insurance policy against disease or other catastrophic loss of the species: one at the headquarters of the International Crane Foundation in Wisconsin and the other at the Calgary Zoo.[173]

In addition to enduring the removal of more than four hundred eggs between 1967 and 1994, the Woods Buffalo–Aransas population has also seen its habitat intensely managed. Wildlife officials have regularly instituted controlled burns, planted crops, and manipulated water levels at the many stopover points along the crane's annual migration route. At Aransas, the Fish and Wildlife Service has constructed ponds for fresh drinking water, burned several thousand acres annually, created new marshlands, and built concrete mats to control erosion along the edge of the refuge bordering the Gulf Intercoastal Waterway. Since 1985, biologists have routinely floated the eggs at the nesting site in Woods Buffalo National Park to confirm their viability, and when they detect an infertile egg, they replace it with one thought to be viable. As a result of these and other interventions, the number of wild whoopers had reached as many 267 by the year 2000. But the population is still not safe from the threat of extinction, and intense management will likely continue into the foreseeable future. As the nature writer Peter Matthiessen has recently noted about the whooping crane, "If man wants the last wild land and life to illuminate his world, he will have to pay dearly to undo his damage. . . . The fierce nature of the species offers hope that it may yet prevail, but it cannot survive without strong assistance and commitment from the creature that drove it to the abyss of extinction in the first place."[174]

Ironically, the kind of "strenuous intervention" that Matthiessen seems to endorse here means that endangered species must in effect become partially domesticated, subjected to continued human surveillance, manipulation, and control to ensure their continued perpetuation. It is one of the many environmental ironies we have learned to live with in the twenty-first century. While the plethora of environmental laws passed during the Age of Ecology have blunted our impact on the earth, the economic system in which we find ourselves enmeshed and the choices we make on a daily basis continue to destroy species at an unprecedented rate, even as we profess a strong desire to prevent their extinction.

CONCLUSION

For in order to care deeply about something important it is first necessary to know
about it. So let us resume old-fashioned expeditions at a quickened pace, solicit
money for permanent field stations, and expand the support of young scientists—
call them "naturalists" with pride—who by inclination and the impress of early
experience commit themselves to deep knowledge of particular groups of organisms.

EDWARD O. WILSON, 2000

For more than two centuries now, naturalists have been grappling with the problem of extinction. As we have seen, however, how they framed and responded to that problem have both changed fundamentally during that time. For Thomas Jefferson and most of his late eighteenth-century colleagues, the basic challenge was how to reconcile the burgeoning collection of fossil specimens that they were so eagerly amassing with the prevailing notion of a stable, harmonious natural environment. Initially, that static worldview portrayed extinction as anathema, an intractable obstacle to the admission that fossils might represent the remains of lost species. For far longer than might seem reasonable to later generations of naturalists, Jefferson and his contemporaries continued to nurse hopes that living examples of these organisms might still be roaming the uncharted regions of the globe.

During the decades following the publication of Jefferson's *Notes on the State of Virginia,* those hopes fell victim to a series of seismic scientific shifts that not only established that the natural world had experienced profound transformations in the past but also catapulted the study of extinction to front and center within natural history. Geologists pushed the age of the earth back from a mere thousands to millions, even billions, of years, thus embracing a notion that would later be termed "deep time"; anatomists developed new comparative techniques allowing them to more precisely differentiate between closely allied species; and naturalists found a raft of exotic new plants and animals as they ventured to the far corners of the world but failed to encounter any living mastodons, mammoths, or other giant creatures whose fossil bones they were unearthing. These developments culminated in the research of the French comparative anatomist and paleontologist Georges Cuvier, whose studies of living and fossil elephants presented compelling evidence that extinction had in fact taken place in the past. Cuvier's subsequent claim proved even bolder: periodic species loss had been a recurrent theme over the earth's long history.

Coming hard on the heels of the discovery of extinction was recognition that island species seemed particularly vulnerable to disappearance. During the 1830s and the 1840s, naturalists reconstructed the stories of the moa of New Zealand and the dodo of Mauritius, two large flightless birds that had both fallen victim to human predation and habitat modification centuries earlier. They also documented the plight of the more recently exterminated great auk, a third flightless bird that until a few years earlier had congregated on rocky islands in the North Atlantic. As naturalists grappled with the anthropogenic loss of these insular species, Charles Darwin and Alfred Russell Wallace groped toward the theory of evolution by natural selection. Firsthand experience with island flora and fauna proved crucial to both men's evolutionary ruminations, which not only captivated the imaginations of their contemporaries once they were finally put into print in the mid-nineteenth century but also portrayed species loss as essential to the teeming diversity of life on earth. As Darwin himself proclaimed, "We need not marvel at extinction . . . [for] the extinction of old forms and the production of new forms are intimately linked."[1] Indeed, rather than lamenting contemporary extinctions, even when humans were clearly culpable, naturalists initially seemed to shrug them off as a routine part of nature's operations.

That indifference gave way to deep concern in the second half of the nineteenth century, as apparently clear-cut examples of human-caused species decline proliferated, romanticism diffused throughout the broader culture, and well-publicized cases of contemporary continent-wide extinctions (or near extinctions) unfolded. Paradigmatic cases of the latter took place in North America, where the nearly simultaneous decline of the bison and the passenger pigeon—economically valuable species whose prodigious masses had once provoked expressions of awe and wonder—triggered alarm about the future of the continent's once abundant wildlife. Naturalists responded by issuing the first scientific publications on native animals threatened with extinction, widely publicizing their dire findings, and establishing conservation committees within scientific organizations. They also founded Audubon Societies, pushed for the creation of wildlife preserves, lobbied for the passage of protective legislation, and engaged in limited experiments with captive breeding of endangered species. The naturalists active in this early phase of the wildlife conservation movement in the United States worried not only about the aesthetic and economic loss that extinction represented but also about its symbolic dimensions. Nationalism and nostalgia loomed large in the pioneering campaigns to rescue the bison and other native fauna that seemed on the verge of vanishing.

As the United States began assuming a larger role in world affairs during the early decades of the twentieth century, so did American naturalists begin halting engagement with the issue of extinction on a more global scale. Initially, they fixed their conservation gaze on Africa, a continent that was not only experiencing rapid transformation at the hands of European colonial powers but also the first stirrings of a protectionist consciousness. At the height of the Great Depression, though, American naturalists shifted their sights southward, toward Latin America, a region with a much longer tradition of U.S. interaction and intervention. There they struggled to secure protection for the Galápagos Islands (the beleaguered archipelago that helped spark Darwin's evolutionary thinking), negotiated a migratory bird treaty with Mexico, and, on the eve of America's entry into World War II, drafted the Convention on Nature Protection and Wildlife Preservation in the Western Hemisphere, which nineteen American republics would ultimately ratify. The organization most responsible for this landmark treaty, the American Committee for International Wild Life Protection, also commissioned a groundbreaking, multivolume inventory of lost and vanishing mammals and birds around the world.

Until the 1920s, American wildlife conservation focused almost exclusively on the plight of vulnerable *individual* species, particularly those birds, mammals, and fish considered economically valuable. With limited success, ecologists pushed to broaden this concern to threatened *associations* of organisms, an approach more in keeping with their interest in the relationship between living things and their environment. Working through the Ecology Society of America's Committee for the Preservation of Natural Conditions, in 1927, ecologists published an inventory of more than a thousand U.S. sites that were either preserved or thought worthy of saving. At the same time, they developed a new rationale for nature protection—the need to document the environmental conditions at relatively undisturbed sites to provide a baseline for ecological research—while lobbying to protect a diversity of vulnerable locations across North America. Ecologists and ecological arguments also played an important role in contentious debates about predator control that raged in the 1930s and 1940s.

By that time, American naturalists had begun calling for an expanded role for science in wildlife management as well as more attention to the specific needs of threatened species. Responding to those calls, along with charges that it had unduly neglected the needs of wildlife facing extinction, in the late 1930s, the National Association of Audubon Societies established a graduate fellowship program to research the life history and ecology of endangered species. The first two Audubon-sponsored projects—James Tanner's ivory-billed woodpecker study and Carl Koford's California condor study—uncovered a wealth of information about the habitat needs of and the threats facing their respective subjects. Both naturalists also offered a series of conservation recommendations consistent with those findings. In the end, though, neither succeeded in reversing the ongoing decline of their study subjects. Science might be necessary to rescue endangered species, they discovered, but it was hardly sufficient. What was ultimately needed was a broader political, social, and cultural shift to support additional research and to implement the often costly management recommendations that scientists made.

The modern environmental movement represented just such a shift. In the post–World War II era, and especially in the 1960s and 1970s—a period sometimes heralded as the Age of Ecology—Americans grew increasingly uneasy about the myriad threats to their quality of life, the dangers of modern pesticides and industrial pollution, and the destruction of wilderness and wildlife that had accompanied the robust economic growth of the previous decades. This emerging consciousness proved fertile ground for one of the most stringent and comprehensive pieces of environmental legislation ever enacted: the Endangered Species Act of 1973. That act offered a powerful mandate for the preservation of organisms facing extinction, while placing the federal government at the center of a systematic, comprehensive program to rescue those species. In short, the Endangered Species Act restored extinction to a position of anathema, incorporating into federal law the abiding concern about preserving wild species that naturalists had been articulating for more than a century.

According to the terms of that bold legislation, listing decisions, critical habitat designations, and recovery plans were all to be based on the best available scientific evidence.[2] Not surprisingly then, given the many threats wildlife faced by the end of the twentieth century, the scope and scale of endangered species research expanded dramatically in the years following its passage. Studies that had once been undertaken by lone individuals laboring in the field over relatively limited periods increasingly gave way to large, multidisciplinary teams over much longer intervals. Those teams not only struggled to disentangle the multiple, complex causes of species decline but also developed increasingly interventionist responses to those threats.

While the committee hearings that preceded the Endangered Species Act and the publicity surrounding its passage focused on only a handful of vanishing wildlife—so-called charismatic megafauna like the bald eagle, whooping crane, and grizzly bear—the act itself embraced an immense range of animals and (later) plants. The full breadth of that coverage came home dramatically in an environmental controversy that attracted wide media attention shortly after its passage.[3] While snorkeling on the Little Tennessee River in August 1973, the University of Tennessee ichthyologist David Etnier encountered a new species of fish that he later named the snail darter (*Percina tanasi*).[4] Two years later, opponents of the Tellico Dam, a Tennessee Valley Authority (TVA)–sponsored project being constructed just downstream from the discovery site, successfully petitioned the secretary of the interior to list the three-inch-long fish under the terms of the Endangered Species Act. Not surprisingly, TVA officials strongly objected to the proposed listing, arguing that the project was not bound by the prohibitive legislation because it had commenced before the law had passed, that Congress continued to fund the project after that date, and that the dam was essentially complete. After forcing the Department of Interior to list the snail darter, dam opponents took the TVA to court to enjoin the Tellico Dam, a case that eventually made it to the U.S. Supreme Court.

The language of the act was plain and brooked no exceptions, the nation's highest court declared in a landmark decision handed down in 1978.[5] To protect the endangered but otherwise unremarkable snail darter, the TVA would have to abandon the more than $100 million project, despite the fact that it was nearly finished. During a House committee hearing a year previously, one representative estimated that if the Tellico Dam were to remain incomplete, the cost of protecting this formerly obscure fish would be unacceptable: "$50,000 to $100,000 for each snail darter."[6] Following the Court's decision, Congress established a special cabinet-level review committee, nicknamed the "God Squad," with the authority to grant exemptions to the act for federal projects. When that newly formed committee refused to let the TVA off the hook, Congress itself specifically exempted the Tellico Dam from the Endangered Species Act. Construction was completed, the floodgates closed, and the reservoir filled. Ironically, not long afterward, Etnier and his colleagues found the snail darter at several additional sites, including small colonies in other rivers in the region and a large population in the Hiwassee River, where the TVA had transplanted the species in an early bid to keep the Tellico Dam project alive. Based on these new discoveries, in 1984 federal officials down-listed the species from "endangered" to "threatened."

The Endangered Species Act was not only ambitious in scope, but it also granted the authority for extraordinary interventions to rescue species teetering on the brink of extinction. A particularly striking example involved the California condor, the large, charismatic vulture that Carl Koford began studying in the 1930s and 1940s.[7] As we have already seen, Koford, his advisor Alden Miller, and the National Audubon Society vehemently opposed proposals to take any remaining condors into captivity for experiments to rebuild the declining wild population, which Koford conservatively estimated at about sixty birds in 1953. Condors were wary creatures that would not do well in captivity, they argued, and even if the birds could be persuaded to reproduce, the resulting offspring would lack the essential social and survival skills needed for successful reintroduction into their native habitat. They also worried about the aesthetic implications of reducing this huge soaring bird, a magnificent symbol of wilderness if ever there was one, to a mean captive existence. To save the condor, the Audubon coalition consistently stressed the importance of preserving its wilderness habitat on the outskirts of Los Angeles.

Yet, despite protection of nearly five hundred thousand acres of prime condor habitat by 1985, condor loss seemed to be accelerating rather than declining. Using a newly developed photographic census technique, researchers estimated that only twenty-one condors remained in the wild by 1982, nineteen by 1983, fifteen by 1984, and nine by 1985.[8] The precipitous decline proved troubling to all parties in the controversy over how to protect the beleaguered bird. Fearful that a recovery strategy based on habitat protection alone was utterly failing, officials at the Condor Research Center, created in 1980 as a joint venture of the Fish and Wildlife Service and the National Audubon Society, sought permission to begin a limited captive breeding program. They proposed stocking that program with birds hatched from eggs removed from the nests of wild condors, which would almost invariably lay a second egg if the first one were lost. After finally securing the necessary permission from state and federal officials in 1983, over the next three years condor researchers removed sixteen eggs from five breeding pairs for artificial incubation and also captured a small number of nestling birds. By 1985, with the number of egg-laying condors in the wild crashing to a single pair, members of the federally commissioned Condor Recovery Team sought permission to trap all nine of the remaining wild condors for incorporation into the captive breeding program. When the Fish and Wildlife Service and the California Fish and Game Commission signed off on the controversial proposal, the National Audubon Society filed suit to block its implementation. That legal challenge failed, however, and by the end of 1987, biologists captured the last known free-flying condor in existence. At that point all the remaining condors in the world—a meager total of only twenty-seven birds—were crowded into specially designed cages at the Los Angeles Zoo and the San Diego Wild Animal Park.

Although the decision to capture every last free-flying condor proved contentious, researchers enjoyed great success with breeding the species. By 1992, when the total population reached fifty-two birds, the Condor Recovery Team felt confident enough to release eight captive-bred condors at the Sespe Condor Sanctuary in the Los Padres National Forest. Just as critics had long warned, however, the released birds exhibited a range of behavioral problems, including a potentially deadly attraction to humans and their structures. While researchers discovered ways to reduce the severity of those problems through behavioral aversion programs and other changes to the breeding protocol, they have persisted to some degree. Even more troubling, many released birds developed lead poisoning as a result of ingesting shot from the carcasses of game animals. As of May 2008, there were 332 California condors, including 152 surviving in the wild.[9] But it remains to be seen whether they will ever recover as an entirely self-sustaining population. The real world no longer seems entirely safe for this awe-inspiring species.

Just as the final wild California condors were being taken into captivity in the mid-1980s, the mother of all endangered species conflicts was not only grabbing media headlines but also serving as a lightening rod for criticism of the federal endangered species program. In the early 1970s, Eric Forsman, a wildlife graduate student at Oregon State University, began the first studies on the ecology of the northern spotted owl (*Strix occidentalis caurina*), one of three spotted owl subspecies native to western North America. He and his advisor, Howard Wight, soon became convinced that the bird was in trouble. The northern spotted owl needed old-growth habitat to survive, they concluded, and most nesting pairs resided on federal land slated for timber sales.[10] In the face of bureaucratic foot-dragging and scientific uncertainty both about what constituted a minimum viable population of the subspecies as well as the amount of old growth habitat each pair needed to survive, their repeated warnings about the plight of the bird remained largely unheeded. In the mid-1980s, however, wilderness enthusiasts and old-growth forest advocates placed increasing pressure on the federal government to act, while scientists moved toward a consensus about the ecological requirements of the subspecies. In 1986, an expert panel convened by the American Ornithologists' Union and the National Audubon Society called for a spotted owl management program that would support a minimum of 1,500 pairs of birds across an extensive network of old-growth habitat.[11]

The Fish and Wildlife Service sought to remain out of the dispute. Indeed, during the first term of the Reagan presidency, a notoriously antienvironmental administration, the agency listed only a handful of new species, while its endangered species program experienced debilitating budget cuts.[12] A series of court orders and an embarrassing General Accounting Office report—which concluded that politics, not science, had been responsible for the continued failure to list the northern spotted owl—finally forced the agency's hand.[13] In June 1990, officials reluctantly declared the subspecies "threatened," and facing yet another court order a year later, they proposed designating a staggering amount of critical habitat necessary to perpetuate the northern spotted owl: a total of 11.6 million acres, including 3 million acres in private hands.[14] That same year, a federal judge issued an injunction against the resumption of timber sales in federally owned old-growth forests in the West until the Forest Service completed an acceptable spotted owl management plan.

That series of legal and administrative maneuvers brought the simmering controversy to a full rolling boil. Timber interests warned that logging restrictions associated with protecting the spotted owl would eliminate as many as 168,000 jobs in the Pacific Northwest, a devastating blow to the region's already shaky economy. In framing the controversy narrowly in terms of "jobs" versus "environmental protection," they found allies not only in angry loggers and sawmill operators (some of whom sported bumper stickers that read "Kill a Spotted Owl—Save a Logger") but also among political conservatives who denounced the Endangered Species Act as a prime example of overbearing environmental regulation. On the other side of the issue were those who viewed the spotted owl not only as a unique organism worthy of protection in its own right but also an important symbol of wilderness. For some scientists and environmentalists, the bird served as a so-called umbrella species whose preservation offered the hope of rescuing an entire vulnerable, venerable ecosystem: the ancient forest of the Pacific Northwest. Spotted owl advocates also challenged industry claims about the projected level of job loss associated with rescuing the declining subspecies, pointing out that the region's timber industry was already experiencing significant downsizing due to increased mechanization, overcutting for short-term profits, and the export of unprocessed logs for milling overseas.

In 1994, following a highly publicized summit that brought together representatives from both sides of the controversy, President Bill Clinton announced a plan that he hoped would not only rescue the spotted owl but also the Pacific Northwest's ailing timber industry. Not surprisingly, both sides expressed disappointment with the compromise plan, which gained judicial approval shortly after its release.[15]

The snail darter, California condor, and spotted owl controversies played out in an atmosphere of growing public and scientific interest in the issue of wildlife extinction. In 1979, the ecologist Norman Myers published *The Sinking Ark*, a book warning that the current extinction rate of up to one species per day could increase to one species per hour within a decade. "By the time human communities establish ecologically sound life-styles, the fallout of species could total several million," Myers continued. "This would amount to a biological debacle greater than all mass extinctions of the geological past put together."[16] Two years later, Paul and Anne Ehrlich offered their own sobering contribution to the genre: *Extinction: The Causes and Consequences of the Disappearance of Species.* Pushing organisms to the point of annihilation was like removing rivets from the body of an airplane, they cautioned; at some point, the craft would no longer remain airworthy. It will be a challenge, the Ehrlichs acknowledged, but the accelerating rate of extinction could and must be "stopped before the living structure of our spacecraft [i.e., Spaceship Earth] is so weakened that at a moment of stress it fails and civilization is destroyed."[17] As one measure of the mounting attention devoted to the subject, during the 1950s a total of only 13 English-language books contained the word "extinction" in their title. That figure rose to 79 by the 1970s, 104 by the 1980s, and 237 by the 1990s.[18]

Not long after the Ehrlichs issued their alarming book, officials from the National Academy of Sciences and the Smithsonian Institution began planning for a national forum on the issue of wild species loss.[19] While searching for an attention-grabbing name for the gathering, the biologist and social activist Walter G. Rosen coined the term "biodiversity," a contraction of the words "biological diversity."[20] The Pulitzer Prize–winning naturalist Edward O. Wilson, who was positioning himself as a leading voice in the growing chorus of scientists warning about the dangers of species loss, initially rejected Rosen's neologism as too flashy, but he soon came to embrace the term. The National Forum on BioDiversity, held in Washington, D.C., convened to much fanfare in September 1986. Organizers downlinked the final evening's proceedings to an estimated audience of ten thousand viewers at over one hundred sites, while numerous reports of the historic gathering appeared in the media. Dubbing themselves the Club of Earth, one group of preeminent biologists in attendance (including Wilson, Ehrlich, Peter Raven, Thomas Eisner, Ernst Mayr, and others) called a press conference to publicize the grave risk that biodiversity loss posed. "The species extinction crisis," they warned ominously, "is a threat to civilization second only to the threat of thermonuclear war."[21] Following the meeting, the new term "biodiversity" quickly caught on in both scientific and popular circles.[22]

In an attempt to address the biodiversity crisis at home while avoiding future "train wrecks" like the spotted owl controversy, Clinton's secretary of the interior, Bruce Babbitt, announced a series of endangered species initiatives shortly after coming to office in 1993.[23] His "first priority" was creation of the National Biological Survey.[24] Michael Kosztarab, an entomologist at Virginia Polytechnic Institute and State University, had floated the idea in the mid-1970s, after discovering that unlike his native Hungary and many other European countries, the United States lacked basic descriptions and identification manuals for a wide variety of organisms. During the previous two centuries, naturalists had amassed mounds of information on the nation's flora and fauna, he found, but the data was widely scattered, frustratingly spotty, and inconsistently reported. Not until the mid-1980s, though, with the heightened concern about biological diversity and the development of more powerful computer technology, did the idea of a coordinated, large-scale survey of the nation's biological wealth finally gain traction. In 1985, the Association of Systematics Collections—a national organization concerned with collections and research in systematic biology—devoted a symposium to the issue.[25] Although participants offered a ringing endorsement of the national survey proposal as well as a various suggestions about how to implement it, the idea failed to garner the necessary funding or sponsorship until Babbitt's appointment as secretary of interior.

After receiving input from the National Research Council, Babbitt launched a scaled-down version of the original survey idea. He transferred 1,200 scientists and 600 other employees from seven Department of Interior agencies into the newly created National Biological Survey.[26] The hope was that in the course of compiling a more comprehensive inventory of the nation's biota, the new agency might identify and initiate appropriate remedial action before declining species reached a point where they would have to be listed under the terms of the relatively inflexible Endangered Species Act. Just as its proponents had envisioned, the establishment of the National Biological Survey heightened visibility while improving coordination of federal research on vanishing species. But being in the spotlight also exacted a potentially heavy toll when conservatives who opposed the Endangered Species Act—and the risk the law presented to private property rights—threatened to yank funding from the new agency after gaining control of Congress during the 1994 elections.[27] Hoping to blunt mounting criticism of the agency he had just established, Babbitt changed its name to the National Biological Service in 1995. A year later, he changed its name again, to the Biological Resources Division, reduced its staffing level, and tucked it away in the U.S. Geological Survey, where it remains to this day.

Meanwhile, naturalists pushed for a much more ambitious project to inventory the earth's biota. A series of developments—declining levels of federal funding for non-defense-related research, fears about the diminishing status of systematics relative to other biological disciplines, concerns that the global rate of species loss had reached crisis proportions, and a sharp increase in estimates of how many species inhabit the earth—led to an far-reaching research initiative dubbed Systematics Agenda 2000.[28] Undoubtedly, the success of several other massive federally funded science projects begun in the 1980s, including the $12 billion Superconducting Super Collider and the $3 billion Human Genome Project, also prodded systematic researchers into action. In 1991, the same year that ecologists articulated their own sweeping research agenda known as the Sustainable Biosphere Initiative, biological systematists scheduled a series of meetings to begin strategizing about the future of their discipline.[29] Three years later, they issued a detailed report calling for a twenty-five-year-long international program to "discover, describe and inventory global species diversity," to synthesize the resulting inventory into a "predictive classification system that reflects the history of life," and to develop the information systems necessary to make the results easily retrievable so that they could benefit both science and society. If the Systematics 2000 agenda were fully implemented, one entomologist noted, "systematic biology would have to move from almost the bottom of the league of science funding to the top of 'big science' requiring a budget at least comparable, for example, with astronomy or nuclear physics."[30] While funding for systematics research did increase, much more remains to be done to meet the ambitious goals of this report.[31]

At the same time they pursued increased support for biological systematics, naturalists and conservationists worried that existing treaties failed to adequately address the growing biodiversity crisis. As early as 1982, the International Union for the Conservation of Nature began calling for a new treaty that would strengthen the existing patchwork of more than three hundred international environmental agreements then in place, thereby providing the tools needed to "conserve biodiversity at the genetic, species, and ecosystem levels."[32] Five years later, the Governing Council of the United Nations Environment Program also expressed strong interest in the idea. Developing nations, however, voiced suspicion that a new global biodiversity treaty would promote a "Northern agenda," preventing full utilization of their own natural resources while forcing them to shoulder the costs of implementing the agreement. They also expressed concern about the growing practice of "bioprospecting," the appropriation of biological compounds for commercial development without adequate compensation to the nations from which those compounds originated. Negotiators managed to allay those fears by the time of the Earth Summit, a United Nations Conference on Environment and Development held in Rio de Janeiro in June 1992, where the Convention on Biological Diversity was formally opened for signature. The final version, which went into force in 1993, articulated three main goals— the conservation, sustainable use, and equitable sharing of benefits derived from biodiversity—and was informed by three broad political principles: countries have the right to exploit their own natural resources, wealthy countries have an obligation to help developing nations abide by the agreement, and "species-rich but cash-poor nations" should share in the profits from their resources.[33]

The United States proved notably absent among the 150 nations that signed the historic agreement during the Earth Summit and the only industrialized nation that refused to do so. The Bush administration opposed the convention on the grounds that it might harm American biotechnology companies by restricting their ability to protect inventions and patents abroad.[34] The Clinton administration seemed more receptive to the idea and signed the agreement in 1993, but to this day, the Senate has refused to ratify it.[35] Thus the United States effectively abandoned its long-established role as a world leader in endangered species protection. Ironically, in a debate that echoed controversies about collecting rare specimens a century earlier, biologists who had sought a biodiversity treaty later decried the ensuing red tape that hindered research in nations that had implemented it.[36]

While calling for increased funding for biological systematics and helping to craft the document that became known as the Convention on Biological Diversity, conservation-minded scientists were also coalescing into a new mission-oriented biological discipline specifically focused on countering the growing threat of extinction.[37] As this book has shown, the field that came to be known as "conservation biology" has deep roots that stretch back well over a century. But its immediate origins can be traced to the tireless advocacy of Michael Soulé, who earned his doctorate at Stanford for research on the genetic fitness of wild plant and animal populations. In 1978, Soulé, whose major advisor was the quintessential biologist-activist Paul Ehrlich, organized the first International Conference on Conservation Biology. According to a later account, "an odd assortment of academics, zoo-keepers, and wildlife conservationists" showed up for that initial gathering, during which Soulé issued an impassioned plea for a bolder response to the growing extinction crisis: "The world was on the verge of the worst biological extinction in 65 million years" he implored, "and it was high time academics and conservationists overcame the barriers between their field to work together and save plants and animals."[38]

One product of the conference was an influential edited volume that Soulé and his doctoral student Bruce Wilcox published in 1980: *Conservation Biology: An Evolutionary-Ecological Perspective*. There they hailed conservation biology as an emerging "mission-oriented discipline comprising both pure and applied science" that focused "the knowledge and tools of all biological disciplines, from molecular biology to population, on one issue—nature conservation."[39] The foreword to the book, contributed by the tropical biologist and World Wildlife Fund staff member Thomas Lovejoy, reflected the generally alarmist tone of the new venture: "Hundreds of thousands of species will perish, and this reduction of 10 to 20 percent of the earth's biota will occur in about half a human life span." Lovejoy decried the "reduction of the biological diversity of the planet" as "the most basic issue of our time."[40]

A series of meetings and publications followed, leading to the Second International Conference on Conference on Conservation Biology in 1985. On the final day of the gathering, conference participants voted to formally organize the Society for Conservation Biology.[41] After the meeting, Soulé offered an expanded definition of the field he had been striving to create. The goal of conservation biology, he wrote, "is to provide principles and tools for preserving biological diversity." Given the increasingly desperate plight of the world's biota, it was by necessity a "crisis discipline," one that required its practitioners to act through a "synthetic, eclectic, and multidisciplinary structure" and "before knowing all the facts." Moreover, conservation biology was normative rather than merely descriptive: its practitioners viewed the "untimely extinction of populations and species as bad" and sought to deploy modern science to avoid that undesirable outcome whenever and wherever possible.[42] When the society's board convened in March 1986, it named David Ehrenfeld as inaugural editor of its new journal, *Conservation Biology.*

After facing limited opposition, the field of conservation biology prospered. Its creators experienced initial resistance from a handful of wildlife managers and ecologists who dismissed the fledgling enterprise as a passing fad that offered little or nothing beyond what they had already been doing for years.[43] Still, by 1992, more than five thousand scientists had signed the membership rolls of the Society for Conservation Biology, at least sixteen graduate programs offered advanced training in the field, and federal agencies and private foundations established significant funding opportunities for biodiversity research. "The speed with which this idea has caught on has been nothing short of incredible," noted the University of Wisconsin wildlife ecologist Stanley A. Temple, who was the president of the Society of Conservation Biology at the time.[44]

Despite this success, some practitioners openly worried about the direction of the apparently thriving new field. In 1996, Reed F. Noss, the second editor of *Conservation Biology,* published a thoughtful editorial lamenting the fact that, like many of his colleagues, he had not been venturing out into the field much any more.[45] Increasingly, he and his fellow conservation biologists were turning to theory building, GIS software, and computer modeling to gain the knowledge they hoped would rescue the world's endangered biota. These were important, powerful tools, Noss admitted, but they were also apt to lead scientists astray if used "in the absence of a firm foundation in field experience, void of the 'naturalist's intuition' that is gained only by many years of immersion in raw Nature." "Scientific abstractions and fancy technologies are no substitutes for the wisdom that springs from knowing the world and its creatures in intimate, loving detail," he continued. Not only did field studies provide invaluable insights into the natural world and the raw data needed to build theories and models, but they also fostered a strong bond between a naturalist and his or her research subjects: "Empathy for living things comes from many years of observing them in their natural environments, which is why field biologists have always been among the most adamant defenders of wild Nature." It was thus critical to resist the widespread trend toward "indoor biology" and to provide opportunities for field experience and field study at all educational levels, Noss concluded.[46]

Four years later, Edward O. Wilson echoed many of Noss's concerns while offering his own call for biological reform. In an editorial entitled "The Future of Conservation Biology," he argued that for the young field to "mature into an effective science," a "renaissance of systematics and natural history" was desperately needed. Estimates of the number of extant species ranged as high as over 100 million, while the number of named and described species was only between 1.5 and 8 million. Detailed knowledge of the earth's biota lagged behind even further still, with less than 1 percent of the named species having been studied beyond the bare bones of habitat preference and the basic anatomy needed to initially name and describe them. "The full exploration of the living part of this planet, will be an adventure of megascience," Wilson predicted, "summoning the energy and imagination of our best minds." Writing not long after molecular biologists crowed about sequencing the first human chromosome, an important milestone in the massive Human Genome Project, Wilson called on fellow conservation biologists to support and encourage young "naturalists" pursuing the equally challenging task of cataloging the earth's full biota.[47] The knowledge they produced would be critical not only to conservation biology but also to cultivating empathy for the living world, Wilson noted, "For in order to care deeply about something it is first necessary to know about it."[48]

Surely, Noss and Wilson are on to something in highlighting the critical role that establishing a connection with living organisms plays in the development of a conservation commitment. Indeed, this book has shown that for well more than a century now the specter of extinction has moved naturalists of all sorts to become passionate, engaged advocates for rescuing vanishing species and threatened landscapes. Undoubtedly, a desire to maintain the opportunity for research accounts for part of their persistent desire to act on behalf of nature. But naturalists' dedication to conservation runs much deeper than this simple utilitarian calculation would suggest. Noss, for example, waxes eloquently about the powerful emotional response he experiences whenever he returns to the field and encounters "old, long-forgotten friends—the wildflowers, ferns, trees, salamanders, fungi, and beetles." His response is by no means unique. Noss, Wilson, and Soulé are simply the most recent torchbearers of a venerable naturalist tradition begun by the likes of Henry David Thoreau, Victor Shelford, and Aldo Leopold. In the process of researching the natural world, these and other naturalists have developed strong emotional bonds with the creatures they have come to know and ultimately love. Their deep sense of connection with the organisms and places they study has driven them to become leading champions for their preservation. We should not only be grateful for their ongoing commitment to protecting biodiversity but also join them in their noble and worthy struggle.

NOTES

INTRODUCTION

1. On the history of this landmark legislation, see Steven Lewis Yaffee, *Prohibitive Policy: Implementing the Federal Endangered Species Act* (Cambridge, Mass.: MIT Press, 1982); Shannon Petersen, *Acting for Endangered Species: The Statutory Ark* (Lawrence: University Press of Kansas, 2002); Brian Czech and Paul R. Krausman, *The Endangered Species Act: History, Conservation Biology, and Public Policy* (Baltimore: Johns Hopkins University Press, 2001); Charles C. Mann and Mark L. Plummer, *Noah's Choice: The Future of Endangered Species* (New York: Alfred A. Knopf, 1995); Dale D. Goble, J. Michael Scott, and Frank W. Davis, eds., *The Endangered Species Act: Renewing the Conservation Promise*, vol. 1 (Washington, D.C.: Island Press, 2006); Congressional Research Service, *A Legislative History of the Endangered Species Act of 1973, as Amended in 1976, 1977, 1978, 1979, and 1990* (Washington, D.C.: Government Printing Office, 1982).

2. House Committee on Merchant Marine and Fisheries, *Endangered and Threatened Species Conservation Act of 1973*, Report no. 93–412, republished in Congressional Research Service, *Legislative History*, 143.

3. Most discussion at the time centered around concerns that the proposed legislation might usurp state authority on wildlife issues. Petersen, *Acting for Endangered Species*, 28.

4. Quoted in Congressional Research Service, *Legislative History*, 196.

5. Richard M. Nixon, "Statement on Signing of the Endangered Species Act of 1973, December 28, 1973," *Public Papers of the Presidents of the United States: Richard Nixon* (Washington, D.C.: Government Printing Office, 1975), 1027–28.

6. On the history of the modern environmental movement, see Samuel P. Hays, *Beauty, Health, and Permanence: Environmental Politics in the United States, 1955–1985* (Cambridge: Cambridge University Press, 1989); Hal Rothman, *The Greening of a Nation? Environmentalism in the United States since 1945* (Fort Worth, Tex.: Harcourt Brace, 1998); Donald Fleming, "Roots of the New Conservation Movement," *Perspectives in American History* 6 (1972): 7–91; Riley E. Dunlap and Angela G. Mertig, eds., *American Environmentalism: The U.S. Environmental Movement, 1970–1990* (Washington, D.C.: Taylor and Francis, 1992); Kirkpatrick Sale, *The Green Revolution: The American Environmental Movement, 1962–1992* (New York: Hill and Wang, 1993); and Philip Shabecoff, *A Fierce Green Fire: The American Environmental Movement* (New York: Hill and Wang, 1993).

7. Quoted in Shabecoff, *Fierce Green Fire*, 113.

8. On the history of natural history, see Paul Farber, *Finding Order in Nature: The Naturalist Tradition from Linnaeus to E. O. Wilson* (Baltimore: Johns Hopkins University Press, 2000); David Elliston Allen, *The Naturalist in Britain: A Social History*, 2nd ed.

(Princeton, N.J.: Princeton University Press, 1994); Nicholas Jardine, James A. Secord, and Emma C. Spary, eds., *Cultures of Natural History* (Cambridge: Cambridge University Press, 1996); Peter J. Bowler, *The Norton History of the Environmental Sciences* (New York: W. W. Norton, 1993); and Janet Browne, "Natural History," in *The Oxford Companion to the History of Modern Science*, ed. J. L Heilbron (Oxford: Oxford University Press, 2003), 553–59.

9. Paula Findlen, *Possessing Nature: Museums, Collecting, and Scientific Culture in Early Modern Italy* (Berkeley: University of California Press, 1994); and Brian Ogilvie, *The Science of Describing: Natural History in Renaissance Europe* (Chicago: University of Chicago Press, 2006).

10. Richard Drayton, *Nature's Government: Science, Imperial Britain, and the "Improvement" of the World* (New Haven, Conn.: Yale University Press, 2000); Lucile Brockway, *Science and Colonial Expansion: The Role of the British Botanic Gardens*, reprint ed. (New Haven, Conn.: Yale University Press, 2002); and Londa Schiebinger, *Plants and Empire: Colonial Bioprospecting in the Atlantic World* (Cambridge, Mass.: Harvard University Press, 2004).

11. Allen, *Naturalist in Britain*.

12. Farber, *Finding Order*, 22–36; Paul Farber, *Discovering Birds: The Emergence of Ornithology as a Scientific Discipline* (Baltimore: Johns Hopkins University Press, 1996).

13. See the example of the American Ornithologists' Union, which was established in 1883, discussed in Mark V. Barrow, Jr., *A Passion for Birds: American Ornithology after Audubon* (Princeton, N.J.: Princeton University Press, 1998).

14. Quoted in Farber, *Finding Order*, 9.

15. See, for example, Nigel Rothfels, *Savages and Beasts: The Birth of the Modern Zoo* (Baltimore: Johns Hopkins University Press, 2002); Elizabeth Hanson, *Animal Attractions: Nature on Display in American Zoos*

(Princeton, N.J.: Princeton University Press, 2002); Gregg Mitman, *Reel Nature: America's Romance with Wildlife on Film* (Cambridge, Mass.: Harvard University Press, 1999); and Cynthia Chris, *Watching Wildlife* (Minneapolis: University of Minnesota Press, 2006).

16. On the growing popularity of bird-watching, see Barrow, *Passion for Birds*, chap. 7; Stephen Moss, *A Bird in the Bush: A Social History of Birdwatching* (London: Aurum Press, 2004); Felton Gibbons and Deborah Strom, *Neighbors to the Birds: A History of Birdwatching in America* (New York: W. W. Norton, 1988); Joseph Kastner, *A World of Watchers: An Informal History of the American Passion for Birds—from Its Scientific Beginnings to the Great Birding Boom of Today* (New York: Alfred A. Knopf, 1986); and Scott Weidensaul, *Of a Feather: A Brief History of American Birding* (Orlando, Fla.: Harcourt, 2007).

17. Marston Bates decried this tendency in his classic *The Nature of Natural History* (New York: Charles Scribner's Sons, 1950).

18. On graduate education and its impact on ornithology, see Barrow, *Passion for Birds*, 184–90.

19. On the relationship between traditional natural history and biology, see Lynn Nyhart, "Natural History and the 'New' Biology," in *Cultures of Natural History*, ed. Nicholas Jardine, James A. Secord, and Emma C. Spary (Cambridge: Cambridge University Press, 1996), 426–43.

20. On the emergence of ecology and its relationship to laboratory biology, see Robert Kohler, *Landscapes and Labscapes: Exploring the Lab-Field Border in Biology* (Chicago: University of Chicago Press, 2002); and Joel B. Hagen, *An Entangled Bank: The Origins of Ecosystem Ecology* (New Brunswick, N.J.: Rutgers University Press, 1992). On the rise of ethology, see Richard W. Burkhardt, *Patterns of Behavior: Konrad Lorenz, Niko Tinbergen, and the Founding of Ethology* (Chicago: University of Chicago Press, 2005).

21. Charles Elton, *Animal Ecology* (London: Sigdwick and Jackson, 1927), 1.

22. David S. Wilcove and Thomas Eisner, "The Impending Extinction of Natural History," *Chronicle Review* 47, no. 3 (September 15, 2000): B24. See also Robert Michael Pyle, "The Rise and Fall of Natural History," *Orion* (Autumn 2001): 17–23; and David J. Schmidly, "What It Means to Be a Naturalist and the Future of Natural History," *Journal of Mammalogy* 86, no. 3 (2005): 449–56.

23. See, for example, Roderick Nash, *Wilderness and the American Mind*, 4th ed. (New Haven, Conn.: Yale University Press, 2001); and William C. Cronon, "The Trouble with Nature; or, Getting Back to the Wrong Nature," *Environmental History* 1 (1996): 7–28.

24. Henry David Thoreau, *The Journal of Henry David Thoreau*, ed. Bradford Torrey and Francis H. Allen (Boston: Houghton Mifflin, 1949), 5:135.

25. C. William Beebe, *The Bird: Its Form and Function* (New York: Henry Holt, 1906), 18.

26. On the commodification of nature in America, see William Cronon, *Changes in the Land: Indians, Colonists, and the Ecology of New England* (New York: Hill and Wang, 1983), and Ted Steinberg, *Down to Earth: Nature's Role in American History* (New York: Oxford University Press, 2002).

27. William Temple Hornaday, *Our Vanishing Wildlife: Its Extermination and Preservation* (New York: New York Zoological Society, 1913), 14, see also p. 63 and his statement that "*no species can withstand systematic slaughter for commercial purposes*" in William Temple Hornaday, *Wild Life Conservation in Theory and Practice* (New Haven, Conn.: Yale University Press, 1914), 9.

28. Quoted in Sharon Kingsland, *The Evolution of American Ecology, 1890–2000* (Baltimore: Johns Hopkins University Press, 2005), 93.

29. See, for example, the discussion in Hornaday, *Our Vanishing Wildlife*, 213–33.

30. Harold J. Coolidge, Jr., "International Nature Protection Article for the Bulletin of the Pan American Union," p. 5, Box 2, Folder: Coolidge, H. J.—4. Photographs for Pan American Union International Nature Protection Article, American Committee for International Wild Life Protection (Secretary's Office), Wildlife Conservation Society, New York.

31. See, for example, Paul and Anne Ehrlich, *Extinction: The Causes and Consequences of the Disappearance of Species* (New York: Random House, 1981); and Norman Myers, *A Wealth of Wild Species: Storehouse for Human Welfare* (Boulder, Colo.: Westview Press, 1983).

32. Aldo Leopold, *A Sand County Almanac and Sketches Here and There* (New York: Oxford University Press, 1949), 210.

33. Aldo Leopold, *Round River* (New York: Oxford University Press, 1993), 145–46.

34. See, for example, Henry Fairfield Osborn, *Preservation of Wild Animals of North America* (Washington, D.C.: Boone and Crockett Club, 1904), which was also incorporated into George B. Grinnell, ed., *Native Game in Its Haunts: The Book of the Boone and Crockett Club* (New York: Forest and Stream Publishing, 1904), 349–73; Henry Fairfield Osborn, "The Preservation of the World's Animal Life," *American Museum Journal* 7, no. 4 (1912): 123–24; Henry Fairfield Osborn and Harold E. Anthony, "Close of the Age of Mammals," *Journal of Mammalogy* 3, no. 4 (November 1922): 219–37; Henry Fairfield Osborn and Harold E. Anthony, "Can We Save the Mammals?" *Natural History* 22, no. 5 (September–October 1922): 389–405.

35. See, for example, Nash, *Wilderness and the American Mind*, 67–83; and Charlotte M. Porter, *The Eagle's Nest: Natural History and American Ideas, 1812–1842* (University: University of Alabama Press, 1986).

36. Hugh E. Strickland and Alexander G. Melville, *The Dodo and Its Kindred; or the History, Affinities, and Osteology of the Dodo,*

Solitaire, and Other Extinct Birds of the Islands Mauritius, Rodriguez, and Bourbon (London: Reve, Benham, and Reeve, 1848), 5.

37. Francis B. Sumner, "The Need for a More Serious Effort to Rescue a Few Fragments of Vanishing Nature," *Scientific Monthly* 10, no. 3 (1920): 236–48; and Francis B. Sumner, "The Responsibility of the Biologist in the Matter of Preserving Natural Conditions," *Science* 54, no. 1385 (1921): 39–43.

38. Charles C. Adams, *Guide to the Study of Animal Ecology* (New York: Macmillan, 1913), 25, 30.

39. John Muir, *A Thousand-Mile Walk to the Gulf*, ed. William F. Badé (Boston: Houghton Mifflin, 1916), 98.

40. Roderick Frazier Nash, *The Rights of Nature: A History of Environmental Ethics* (Madison: University of Wisconsin Press, 1989), 39.

41. Leopold, *Sand County Almanac*, 204.

42. Leopold, *Sand County Almanac*, 211.

43. Hornaday, *Our Vanishing Wildlife*, 7.

44. See, for example, Edward O. Wilson, *Naturalist* (Washington, D.C.: Island Press, 1994); and the accounts of the many biologists interviewed in David Takacs, *The Idea of Biodiversity: Philosophies of Paradise* (Baltimore: Johns Hopkins University Press, 1996), 236–47.

45. Frank M. Chapman, "Bird Clubs in America," *Bird-Lore* 17 (1915): 348.

46. Quoted in Takacs, *The Idea of Biodiversity*, 139.

47. T. Gilbert Pearson, *Adventures in Bird Protection: An Autobiography* (New York: D. Appleton-Century, 1937), 71.

48. See the discussion in Daniel McKinley, *The Carolina Parakeet in Florida* (Gainesville: Florida Ornithological Society, 1985), 50–53.

49. A number of recent studies document the transnational character of conservation discourse or offer comparative perspectives. See, for example, Thomas Dunlap, *Nature and the English Diaspora: Environment and History in the United States, Canada, Australia,* and New Zealand (Cambridge: Cambridge University Press, 1999); and Ian Tyrell, *True Gardens of the Gods: Californian-Australian Environmental Reform, 1860–1930* (Berkeley: University of California Press, 1999).

50. On America's leadership role in defining and promoting national parks, for example, see William M. Adams, *Against Extinction: The Story of Conservation* (London: Earthscan, 2004), 77–79.

51. See the discussion in Michael J. Bean, *The Evolution of National Wildlife Law*, rev. ed. (New York: Praeger Publishers, 1983), 9–7. The quote from *Geer v. Connecticut*, below, is on p. 16.

52. R. C. Benedict, "Game Laws for the Conservation of Wild Plants," *Science* 58, no. 1490 (July 20, 1923), 39–41.

53. Stephen Kellert, "Social and Perceptual Factors in the Preservation of Animal Species," in *The Preservation of Species: The Value of Biological Diversity*, ed. Bryan G. Norton (Princeton, N.J.: Princeton University Press, 1986), 50–73, on p. 62, argues that the public is most concerned about the possible extinction of "creatures that are large, aesthetically attractive, phylogenetically similar to human beings, and regarded as possessing capacities for feeling, thought, and pain."

54. See the statement in Edward S. Ayensu and Robert A. DeFilipps, *Endangered and Threatened Plants of the United States* (Washington, D.C.: Smithsonian Institution and the World Wildlife Fund, 1978), 1. The possible exception was forests, which enjoyed much attention and regulation by this time, especially those on publicly owned lands.

CHAPTER 1

1. Quoted in Silvio A. Bedini, *Thomas Jefferson: Statesman of Science* (New York: Macmillan, 1990), 1. On Jefferson's multifarious scientific interests, see also Edwin T. Martin, *Thomas Jefferson: Scientist* (New York: Collier Books, 1961); John C. Greene, *Ameri-*

can *Science in the Age of Jefferson* (Ames: Iowa State University Press, 1984); and I. Bernard Cohen, ed., *Thomas Jefferson and the Sciences* (New York: Arno Press, 1980).

2. The events surrounding the preparation and publication of *Notes on the State of Virginia* are recounted in Bedini, *Thomas Jefferson*, 91–108, 112–14, and passim; and Thomas Jefferson, *Notes on the State of Virginia*, ed. William Peden (Chapel Hill: University of North Carolina Press, 1955), xi–xxv, quoted on p. xxv. All citations to Jefferson's *Notes* refer to this edition.

3. All of the books in note 1 above deal extensively with Jefferson's interest in fossils. See also Silvio A. Bedini, *Thomas Jefferson and American Vertebrate Paleontology*, Virginia Division of Mineral Resources Publication 61 (Charlottesville, Va.: Department of Mines, Minerals and Energy, 1985); Julian Boyd, "The Megalonyx, the Megatherium, and Thomas Jefferson's Memory Lapse," *Proceedings of the American Philosophical Society* 102, no. 5 (1958): 420–35; Thomas Horrocks, "Thomas Jefferson and the Great Claw," *Virginia Cavalcade* 35, no. 2 (1985): 70–79; Henry Fairfield Osborn, "Thomas Jefferson as a Paleontologist," *Science* 82, no. 2136 (1935): 533–38; Howard C. Rice, Jr., "Jefferson's Gift of Fossils to the Museum of Natural History in Paris," *Proceedings of the American Philosophical Society* 95, no. 6 (1951): 597–627; George Gaylord Simpson, "The Discovery of Fossil Vertebrates in North America," *Journal of Paleontology* 17, no. 1 (1943): 26–38; and W. J. T. Mitchell, *The Last Dinosaur Book: The Life and Times of a Cultural Icon* (Chicago: University of Chicago Press, 1998), 110–23.

4. Jefferson, *Notes*, 45.

5. The best single source on the American incognitum is Paul Semonin, *American Monster: How the Nation's First Prehistoric Creature Became a Symbol of National Identity* (New York: New York University Press, 2000), which I have relied on extensively in the chapter that follows.

6. On Jefferson and Buffon, see especially, Bedini, *Thomas Jefferson*, 32, 95–96, 98–99, 148–51, 162–63, 166, 182–83, 185–86, and passim; Greene, *American Science*, 10, 28–30, 163. American responses to Buffon's theory are detailed in Antonello Gerbi, *The Dispute of the New World: The History of a Polemic, 1750–1900*, trans. Jeremy Moyle (Pittsburgh: University of Pittsburgh Press, 1973). The best biography of Buffon is Jacques Roger, *Buffon: A Life in Natural History*, ed. L. Pearce Williams, trans. Sara Lucille Bonnefoi (Ithaca, N.Y.: Cornell University Press, 1997).

7. Jefferson, *Notes*, p. 47.

8. The episode is described in Semonin, *American Monster*, 185.

9. Rice, "Jefferson's Gift."

10. Thomas Jefferson, "A Memoir on the Discovery of Certain Bones of a Clawed Kind in the Western Parts of Virginia," *Transactions of the American Philosophical Society* 4 (1799): 246–60, reproduced in Keir B. Sterling, ed., *Selected Works in Nineteenth-Century North American Paleontology* (New York: Arno Press, 1974); Boyd, "Megalonyx"; and Horrocks, "Thomas Jefferson and the Great Claw."

11. Jefferson, *Notes*, 53–54.

12. Jefferson, *Notes*, 54.

13. Thomas Jefferson, "Instructions to Lewis," in *Letters of the Lewis and Clark Expedition, with Related Documents, 1783–1854*, 2nd ed., ed. Donald Jackson (Urbana: University of Illinois Press, 1978), 61–66, on p. 63. On the scientific results of the Lewis and Clark Expedition, see Greene, *American Science;* Paul Russell Cutright, *Lewis and Clark, Pioneering Naturalists* (Urbana: University of Illinois Press, 1969).

14. See Thomas Jefferson to John Adams, April 11, 1823, in *The Adams-Jefferson Correspondence: The Complete Correspondence between Thomas Jefferson and Abigail and John Adams*, ed. Lester J. Cappon (Chapel Hill: University of North Carolina Press, 1959), 2:592.

15. The worldview of Jefferson and his contemporaries, especially as it relates to the issues of the order and age of the earth, is ably summarized in Peter J. Bowler, *Evolution: The History of an Idea*, rev. ed. (Berkeley: University of California Press, 1989); and Peter J. Bowler, *The Norton History of the Environmental Sciences*, 1st American ed. (New York: W. W. Norton, 1993).

16. On the development of natural theology, see Bowler, *Evolution*, 53–55, 123–25; Bowler, *Environmental Sciences*, 44, 97–98, 150–53, 169, 172, 177–78; and Neal C. Gillespie, "Natural History, Natural Theology, and Social Order: John Ray and the 'Newtonian Ideology,'" *Journal of the History of Biology* 20 (1987): 1–49.

17. On Ray and his influence, see C. E. Raven, *John Ray, Naturalist: His Life and Works* (Cambridge: Cambridge University Press, 1950).

18. See the lucid discussion in Bowler, *Evolution*, 53–55, and John Ray, *The Wisdom of God Manifested in the Works of the Creation* (London: Samuel Smith, 1691). The quote that follows is from page 12 of that influential book.

19. The classic study of this seminal concept is Arthur O. Lovejoy, *The Great Chain of Being: A Study in the History of an Idea* (Cambridge, Mass.: Harvard University Press, 1966). For a handy review of scholarship since the first edition of Lovejoy (originally published in 1936), see William F. Bynum, "The Great Chain of Being after Forty Years: An Appraisal," *History of Science* 13 (1975): 1–28; and Bowler, *Evolution*, 55, 59–63, 66, 83–85, 93, 113, and 123.

20. Bynum, "Great Chain," 20.

21. On Linnaeus's life, ideas, and influence, see Lisbet Koerner, *Linnaeus: Nature and Nation* (Cambridge, Mass.: Harvard University Press, 1999); Wilfred Blunt, *The Compleat Naturalist: A Life of Linnaeus* (New York: Viking, 1971); James L. Larson, *Reason and Experience: The Representation of Natural*

Order in the Work of Carl Von Linné (Berkeley: University of California Press, 1971); and Tore Frängsmyr, ed., *Linnaeus: The Man and His Work* (Berkeley: University of California Press, 1983).

22. Sten Lindroth, "The Two Faces of Linnaeus," in *Linnaeus: The Man and His Work*, ed. Tore Frängsmyr (Berkeley: University of California Press, 1983), 1–62, on p. 16.

23. Quoted in Roger, *Buffon*, 88. See also Lovejoy, *Great Chain of Being*, 229–30.

24. Bynum, "Great Chain of Being," 4, points out that Lovejoy devoted over a third of his book to eighteenth-century literary, philosophical, and biological uses of the idea.

25. Quoted in Lovejoy, *Great Chain of Being*, 60.

26. One of the few historical treatments of the concept is Frank N. Egerton, "Changing Concepts of the Balance of Nature," *Quarterly Review of Biology* 48 (1973): 322–50. See also Clarence J. Glacken, *Traces on the Rhodian Shore: Nature and Culture in Western Thought from Ancient Times to the End of the Eighteenth Century* (Berkeley: University of California Press, 1967) and A. J. Jansen, "An Analysis of 'Balance in Nature' as an Ecological Concept," *Acta Biotheoretica* 21 (1972): 86–114, which deals more with theoretical issues.

27. Egerton, "Changing Concepts," 324.

28. Carl Linnaeus, "The Oeconomy of Nature," in *Miscellaneous Tracts Relating to Natural History, Husbandry, and Physick*, ed. Benjamin Stillingfleet (London: J. Dodsley, Baker and Leigh, and T. Payne, 1791), 39–149. This was originally published as a thesis in 1749 by one of Linnaeus's students, Isaac Biberg. It was common practice at the time for a student to defend a thesis written by his mentor.

29. Linnaeus, "Oeconomy of Nature," 114.

30. Linnaeus, "Oeconomy of Nature," 119. Linnaeus returned to some of these same general themes a few years later in Carl Linnaeus, "On the Police of Nature," in *Select Dissertations from the Amœnitates Academicæ,* ed. F. J. Brand (London: G. Robinson and J. Robson, 1781), 126–66.

31. Quoted in John C. Greene, *The Death of Adam: Evolution and Its Impact on Western Thought* (Ames: Iowa State University Press, 1959), 131.

32. Bowler, *Evolution*, 67–68, 72–77.

33. Francis C. Haber, *The Age of the World: Moses to Darwin* (Baltimore: Johns Hopkins University Press, 1959); Bowler, *Evolution*, 4–5, 39, 46–49; Martin J. S. Rudwick, *Bursting the Limits of Time: The Reconstruction of Geohistory in the Age of Revolution* (Chicago: University of Chicago Press, 2005); and Claudine Cohen, *The Fate of the Mammoth: Fossils, Myth, and History* (Chicago: University of Chicago Press, 2002).

34. For a detailed exploration of the use of biblical chronology to calculate the age of the earth, see Jack Repcheck, *The Man Who Found Time: James Hutton and the Discovery of Earth's Antiquity* (New York: Perseus, 2003), 25–43.

35. Adrienne Mayor, *The First Fossil Hunters: Paleontology in Greek and Roman Times* (Princeton, N.J.: Princeton University Press, 2000), on p. 9.

36. Or so argues Martin Rudwick in his important study, Martin J. S. Rudwick, *The Meaning of Fossils: Episodes in the History of Paleontology*, 2nd ed. (Chicago: University of Chicago Press, 1985), which I have relied on heavily in the brief account that follows. See also Claudine Cohen's, *The Fate of the Mammoth*.

37. Rudwick, *Meaning of Fossils*, 1–48.

38. Rudwick, *Meaning of Fossils*, 49–100.

39. Rudwick, *Meaning of Fossils*, 53–56, 61–65, 73–76. For a more popular account of Steno's contributions, see Alan Cutler, *The Seashell on the Mountaintop: A Story of Science, Sainthood, and the Humble Genius Who Discovered a New History of the Earth* (New York: Dutton, 2003).

40. John Ray, *Observations Topographical, Moral and Physiological, Made in a Journey through Part of the Low Countries, Germany, Italy, and France* (London: J. Martyn, 1673), 113–31.

41. Rudwick, *Meaning of Fossils*, 64–65.

42. Rudwick, *Meaning of Fossils*, 65.

43. The details of this hoax were first revealed in Melvin E. Jahn and Daniel J. Woolf, *The Lying Stones of Dr. Johann Bartholomew Adam Beringer, Being His Lithographiæ Wirceburgensis* (Berkeley: University of California Press, 1963).

44. Rudwick, *Meaning of Fossils*, 89–90.

45. Quote in Greene, *Death of Adam*, 136.

46. Haber, *Age of the World*, 115–36; Roger, *Buffon*, 409–13; and Rudwick, *Bursting the Limits of Time*, 139–50.

47. Martin J. S. Rudwick, *Scenes from Deep Time: Early Pictorial Representations of the Prehistoric World* (Chicago: University of Chicago Press, 1992), 17.

48. Quoted in Haber, *Age of the World*, 168. On Hutton and his importance, see Repcheck, *The Man Who Found Time*.

49. Rudwick, *Meaning of Fossils*, 105–7.

50. Adrienne Mayor, *Fossil Legends of the First Americans* (Princeton, N.J.: Princeton University Press, 2005).

51. George Gaylord Simpson, "The Beginnings of Vertebrate Paleontology," *Proceedings of the American Philosophical Society* 86 (1942): 130–88, on pp. 130, 135.

52. The best account of this initial find is Semonin, *American Monster*, 15–40. On the widely accepted idea that these bones represented a lost race of giants, see Cohen, *The Fate of the Mammoth*, 23–40.

53. On the fossil excavations in the Hudson, see Henry Fairfield Osborn, "Mastodons of the Hudson Highlands," *Natural History* 23 (1923): 3–34; and John M. Clarke, "Mastodons of New York: A List of Discoveries of Their Remains, 1705–1902," *Bulletin of the New York State Museum*, no. 69 (1902): 921–33.

54. Thomas D. Matijasic, "Science, Religion, and the Fossils at Big Bone Lick," *Journal of the History of Biology* 20 (1987):

413–21; and Willard Rouse Jillson, *Big Bone Lick: An Outline of Its History, Geology, and Paleontology* (Louisville, Ky.: Big Bone Lick Association, 1936). See also Stanley Hedeen, *Big Bone Lick: The Cradle of American Paleontology* (Lexington: University Press of Kentucky, 2008), which appeared after this chapter was completed.

55. The episode is recounted in Simpson, "The Beginnings of Vertebrate Paleontology," 135–39; Semonin, *American Monster*, 87; and Mayer, *Fossil Legends*, 1–8.

56. The Crohgan expedition is discussed in Semonin, *American Monster*, 104–10, quoted on p. 105; and Whitfield J. Bell, Jr., "A Box of Old Bones: A Note on the Identification of the Mastodon, 1766–1806," *Proceedings of the American Philosophical Society* 93, no. 2 (1949): 169–77.

57. Simpson, "The Beginnings of Vertebrate Paleontology," 132.

58. My account of Buffon and Daubenton's ideas about the American incognitum and related species is from Semonin, *American Monster*, 111–35.

59. On the Siberian mammoth, see Cohen, *The Fate of the Mammoth*, 61–81; and Semonin, *American Monster*, 62–83.

60. William Hunter, "Observations on the Bones, Commonly Supposed to Be Elephants Bones, Which Have Been Found near the Ohio River," *Philosophical Transactions of the Royal Society of London* 58 (1769), 34–45, on pp. 37, 38, and 43.

61. Hunter, "Observations on the Bones," 45; emphasis added.

62. Quoted in Semonin, *American Monster*, 158.

63. John Stuart to Thomas Jefferson, April 11, 1796, in *The Papers of Thomas Jefferson*, ed. Barbara B. Oberg (Princeton, N.J.: Princeton University Press, 2002), 29:64–65 (hereafter cited as *Papers of Thomas Jefferson*). The episode is detailed in Bedini, *Thomas Jefferson and American Vertebrate Paleontology*; Bedini, *Thomas Jefferson*, 268–74; Horrocks,

"Thomas Jefferson and the Great Claw"; and Boyd, "Megalonyx."

64. Thomas Jefferson to John Stuart, November 10, 1796, *Papers of Thomas Jefferson*, 29:205–6.

65. Thomas Jefferson to Archibald Stuart, May 26, 1796, *Papers of Thomas Jefferson*, 29:113.

66. Thomas Jefferson to David Rittenhouse, July 3, 1796, *Papers of Thomas Jefferson*, 29:138–39.

67. Quoted in Boyd, "Megalonyx," 424.

68. Boyd, "Megalonyx," 424–26.

69. An English translation of Cuvier's paper on the megatherium as well as the circumstances surrounding its creation are detailed in Martin J. S. Rudwick, *Georges Cuvier, Fossil Bones, and Geological Catastrophes: New Translations and Interpretations of the Primary Texts* (Chicago: University of Chicago Press, 1997), 25–32. See also José M. López Piñero, "Juan Buatista Bru (1740–1799) and the Description of the Genus *Megatherium*," *Journal of the History of Biology* 21, no. 1 (1988): 147–63.

70. The manuscript copy of the original memoir, along with notations about the many changes that Jefferson made to it, is found in *Papers of Thomas Jefferson* 29:291–304. His memoir was published two years later as Thomas Jefferson, "A Memoir on the Discovery of Certain Bones of a Clawed Kind in the Western Parts of Virginia," *Transactions of the American Philosophical Society* 4 (1799): 246–60.

71. Jefferson, "Memoir," 258.

72. Jefferson, "Memoir," 252.

73. Jefferson, "Memoir," 256.

74. Jefferson, "Memoir," 256.

75. George Turner, "Memoir on the Extraneous Fossils, Denominated Mammoth Bones: Principally Designed to Shew, That They Are the Remains of More Than One Species of Non-Descript Animal," *Transactions of the American Philosophical Society* 4 (1799): 510–18, on p. 512.

76. Turner, "Memoir," 516.

77. Turner, "Memoir," 518.

78. For a fuller discussion of this issue, see chap. 3.

79. On nationalism and natural history in the early republic, see Charlotte M. Porter, *The Eagle's Nest: Natural History and American Ideas, 1812–1842* (University: University of Alabama Press, 1986), 7–11, 30, 48–56; Philip J. Pauly, *Biologists and the Promise of American Life: From Meriwether Lewis to Alfred Kinsey* (Princeton, N.J.: Princeton University Press, 2000), 18–19; Greene, *American Science*, 7, 10, 24, 128, 102, 188, 416–19; and Paul Semonin, "'Nature's Nation': Natural History as Nationalism in the New Republic," *Northwest Review* 30 (1992): 6–41.

80. Semonin, *American Monster*, 186.

81. On the origins and implications of Peale's Museum, see Charles Coleman Sellers, *Mr. Peale's Museum: Charles Willson Peale and the First Popular Science Museum of Natural Science and Art* (New York: W. W. Norton, 1980); Edgar P. Richardson, Brooke Hindle, and Lillian B. Miller, *Charles Willson Peale and His World* (New York: Harry N. Abrams, 1982); David R. Brigham, *Public Culture in the Early Republic: Peale's Museum and Its Audience* (Washington, D.C.: Smithsonian Institution Press, 1995); and Lillian B. Miller, ed., *New Perspectives on Charles Willson Peale: A 250th Anniversary Celebration* (Pittsburgh: University of Pittsburgh Press, 1991). See also Lillian B. Miller, Sidney Hart, and David C. Ward, eds., *The Selected Papers of Charles Willson Peale and His Family,* 5 vols. (New Haven, Conn.: Yale University Press, 1983–2000). The account that follows is drawn primarily from vol. 5 of that series, *The Autobiography of Charles Willson Peale,* 111–30; Sellers, *Mr. Peale's Museum*, 9–48; Semonin, *American Monster*, 174–78, 186–92, 195; and Bell, "Box of Old Bones."

82. A copy of the solicitation is reproduced in Sellers, *Mr. Peale's Museum*, 23.

83. Turner, "Memoir," 515.

84. Semonin, *American Monster*, 313–14.

85. Because it eventually led to the creation of one of his most famous paintings, *The Exhumation of the Mastodon,* the circumstances surrounding Peale's expedition to retrieve a complete fossil skeleton of the American incognitum has been told repeatedly. See, for example, Sellers, *Mr. Peale's Museum*, 113–58; Charles Coleman Sellers, "Unearthing the Mastodon: Peale's Greatest Triumph," *American Heritage* 30, no. 5 (1979): 16–23; George Gaylord Simpson and H. Tobien, "The Rediscovery of Peale's Mastodon," *Proceedings of the American Philosophical Society* 98, no. 4 (1954): 279–81; Lillian B. Miller, "Charles Willson Peale as History Painter: The Exhumation of the Mastodon," *American Art Journal* 13 (1981): 47–68; and Laura Rigal, "Peale's Mammoth," in *American Iconology: New Approaches to Nineteenth-Century Art and Literature,* ed. David C. Miller (New Haven, Conn.: Yale University Press, 1993), 18–38. Most of these accounts lean heavily on Peale's autobiography, reproduced as vol. 5 of Miller, Hart, and Ward, eds., *The Selected Papers of Charles Willson Peale and His Family.*

86. Miller et al., eds., *Selected Papers of Charles Willson Peale,* 5:287.

87. Quoted in Miller et al., eds., *Selected Papers of Charles Willson Peale,* 5:286.

88. The painting accurately captures the excitement and drama of Peale's expedition, but it also depicts many individuals who were not present at the time. See the discussion in Miller, "Charles Willson Peale as History Painter."

89. Sellers, *Mr. Peale's Museum*, 142.

90. The handbill is reproduced in Sellers, *Mr. Peale's Museum*, 146.

91. Quoted in Porter, *The Eagle's Nest*, 31.

92. Sellers, *Mr. Peale's Museum*, 143.

93. Rembrandt Peale, *Historical Disquisition on the Mammoth, or, Great American Incognitum, an Extinct, Immense, Carnivorous Animal Whose Fossil Remains Have Been Found in North America* (London: E. Lawrence, 1803), iv.

94. Peale, *Historical Disquisition,* 9.

95. Quoted in Semonin, *American Monster,* 344. Goforth's travails, Clark's success, and Jefferson's response is recounted on pp. 344–56 of that work; and Rice, "Jefferson's Gift."

96. On Cuvier's life and work, see William Coleman, *Georges Cuvier, Zoologist: A Study in the History of Evolution Theory* (Cambridge, Mass.: Harvard University Press, 1964); Dorinda Outram, *Georges Cuvier: Vocation, Science and Authority in Post-Revolutionary France* (Manchester: Manchester University Press, 1984); and Rudwick, *Meaning of Fossils,* 101–45.

97. On the German historicist tradition and it influence on paleontology more generally and Cuvier specifically, see Nicholas A. Rupke, "The Study of Fossils in the Romantic Philosophy of History and Nature," *History of Science* 21 (1983): 389–413; David R. Oldroyd, "Historicism and the Rise of Historical Geography," *History of Science* 17 (1979): 191–213, 227–57; and especially Philip F. Rehbock, "The Discovery of Species Extinction at the Beginning of the 19th Century and Its Acceptance in Britain" (paper presented at the History of Paleontology Conference, Cambridge, 1997).

98. Quoted in Rehbock, "Discovery of Species Extinction," 8.

99. Coleman, *Georges Cuvier,* 107–8; and Rudwick, *Meaning of Fossils,* 113.

100. Rudwick, *Meaning of Fossils,* 105–6.

101. Originally published in abstract form as Georges Cuvier, "Mémoire sur les espèces d'éléphans tant vivantes que fossils," *Magasin encyclopédique* 3 (1796): 440–45, which is translated with commentary in Rudwick, *Georges Cuvier,* 13–24. The full publication is Georges Cuvier, "Mémoire sur les espèces d'éléphans tant vivantes que fossiles," *Académie des sciences, Mémoires* 2 (1799): 1–22.

102. Rudwick, *Meaning of Fossils,* 101.

103. Georges Cuvier, "Sur les éléphans vivans et fossiles," *Annales du Muséum national d'histoire naturelle* 8 (1806): 1–58, 93–155, 249–69; Georges Cuvier, "Sur le grand mastodonte," *Annales du Muséum national d'histoire naturelle* 8 (1806): 270–312; Georges Cuvier, "Sur différentes dents du genre des mastodontes," *Annales du Muséum national d'histoire naturelle* 8 (1806): 401–24. All three were reprinted in Georges Cuvier, *Recherches sur les ossemens fossiles de quarupèdes* (Paris: Deterville, 1812). See also Georges Cuvier, "Memoir upon Living and Fossil Elephants," *Philosophical Magazine* 26 (1806–7): 158–69, 204–11, 302–12; and Georges Cuvier, "Additional Memoir upon Living and Fossil Elephants," *Philosophical Magazine* 28 (1807–8): 258–64, 359–66; 29:52–65, 244–54; 30:15–25.

104. Cuvier's most important publications on fossil quadrupeds were reprinted in Cuvier, *Recherches sur les ossemens fossiles de quarupèdes* (1812).

105. Rudwick, *Meaning of Fossils,* 127. The British geologist and engineer William Smith, who is now generally credited as one of the founders of stratigraphy, was developing the idea of fossils as markers of different geological epochs in England at about the same time. See Simon Winchester, *The Map That Changed the World* (New York: HarperCollins, 2001).

106. A prominent exception was Cuvier's colleague, Jean-Baptiste Lamarck, who posited the notion that species were constantly undergoing progressive development and so could never go extinct. See Bowler, *Evolution,* 82–88; and Richard W. Burkhardt, *Spirit of System: Lamarck and Evolutionary Biology* (Cambridge, Mass.: Harvard University Press, 1977).

107. Benjamin Smith Barton, *A Discourse on Some of the Principal Desiderata in Natural History, and on the Best Means of Promoting the Study of This Science in the United States* (Philadelphia: Printed by Denham and Town, 1807), 19. On Barton's life and contributions, see Joseph Ewan and Nesta Dunn Ewan, *Benjamin Smith Barton, Naturalist and Physi-*

cian in *Jeffersonian America*, ed. Victoria C. Hollowell, Eileen P. Duggan, and Marshall Crosby (St. Louis: Missouri Botanical Garden Press, 2007).

108. Barton, *Discourse,* 20.

109. Benjamin Smith Barton, *Archaeologiae Americanae Telluris Collectanea et Specimena; or, Collections, with Specimens, for a Series of Memoirs on Certain Extinct Animals and Vegetables of North-America* (Philadelphia: Printed for the Author, 1814), 33.

110. Donald K. Grayson, "Nineteenth-Century Explanations of Pleistocene Extinctions: A Review and Analysis," in *Quaternary Extinctions: A Prehistoric Revolution*, ed. Paul Martin and Richard G. Klein (Tucson: University of Arizona Press, 1984), 5–93, on p. 6.

111. Variations of this idea would become common in the nineteenth century. See Peter J. Bowler, *Fossils and Progress: Paleontology and the Idea of Progressive Evolution in the Nineteenth Century* (New York: Science History Publications, 1976).

112. Richard Harlan, *Fauna Americana: Being a Description of the Mammiforous Animals Inhabiting North America* (Philadelphia: Anthony Finley, 1825), 1, reprinted in Keir B. Sterling, ed., *Fauna Americana* (New York: Arno, 1974). This book was based on the work of the French mammalogist Anselme Desmarest. On Harlan and his rocky relationship with John Godman, see Greene, *American Science*, 312–18.

113. John Godman, *American Natural History*, 3 vols. (Philadelphia: H. C. Carey and I. Lea, 1826–28), 2:208–9

114. On Fleming and his ideas about extinction, see Philip F. Rehbock, "John Fleming (1785–1857) and the Economy of Nature," in *From Linnaeus to Darwin: Commentaries on the History of Biology and Geology*, ed. Alywne Wheeler and James H. Price (London: Society for the History of Natural History, 1985), 129–40.

115. John Fleming, "Remarks Illustrative of the Influence of Society on the Distribution of British Animals," *Edinburgh Philosophical Journal* 11 (1824), 287–305, on p. 288.

116. Fleming, "Remarks," 290.

117. Fleming, "Remarks," 291.

118. Fleming, "Remarks," 303.

119. Fleming, "Remarks," 304.

120. Charles Lyell, *Principles of Geology* (London: John Murray, 1832), 2:141.

121. On Lyell and his influence, see Worster, *Nature's Economy*, 138–44; Bowler, *Evolution*, 134–41, 155–58.

122. Worster, *Nature's Economy*, 143.

123. Lyell, *Principles of Geology*, 2:144.

124. Lyell, *Principles of Geology*, 2:145.

125. Lyell, *Principles of Geology*, 2:150.

126. Lyell, *Principles of Geology*, 2:156.

CHAPTER 2

1. On the growing interest in creating an inventory of world's species and the development of natural history more broadly, see Nicholas Jardine, James Secord, and Emma C. Spary, eds., *Cultures of Natural History* (Cambridge: Cambridge University Press, 1996); and Paul Lawrence Farber, *Finding Order in Nature: The Naturalist Tradition from Linnaeus to E. O. Wilson* (Baltimore: Johns Hopkins University Press, 2000).

2. A convenient starting place for the history of exploration and its relationship to the development of natural history is Tony Rice, *Voyages of Discovery: Three Centuries of Natural History Exploration* (New York: Clarkson N. Potter, 1999).

3. On botanical gardens as biological clearinghouses, see Richard H. Drayton, *Nature's Government: Science, Imperial Britain, and the "Improvement" of the World* (New Haven, Conn.: Yale University Press, 2000); Lucile Brockway, *Science and Colonial Expansion: The Role of the British Royal Botanic Gardens* (New York: Academic Press, 1979); and Donald P. McCracken, *Gardens of Empire: Botanical Institutions of the Victorian British Empire* (London: Leicester University Press,

1997). On zoological gardens and acclima-
tization societies, see Michael A. Osborne,
*Nature, the Exotic, and the Science of French
Colonialism* (Bloomington: Indiana University
Press, 1994); and Eric Baratay and Elisabeth
Hardouin-Fugier, *Zoo: A History of Zoological
Gardens in the West* (London: Reaktion Books,
2002).

4. See, for example, Paula Findlen,
*Possessing Nature: Museums, Collecting, and
Scientific Culture in Early Modern Italy* (Berke-
ley: University of California Press, 1994);
and Oliver Impey and Arthur MacGregor,
*The Origins of Museums: Cabinets of Curiosities
in Sixteenth- and Seventeenth-Century Europe*
(Oxford: Oxford University Press, 1985).

5. On Linnaeus's life and work, see
Lisbet Koerner, *Linnaeus: Nature and Nation*
(Cambridge, Mass.: Harvard University Press,
1999); James L. Larson, *Reason and Experi-
ence: The Representation of Natural Order in the
Work of Carl Von Linné* (Berkeley: University
of California Press, 1971); and Wilfred Blunt,
The Compleat Naturalist: A Life of Linnaeus
(New York: Viking, 1971).

6. Janet Browne's wonderful book
remains the starting point for exploring the
early history of biogeography and my main
source for the paragraph that follows: Janet
Browne, *The Secular Ark: Studies in the History of Biogeography* (New Haven, Conn.: Yale
University Press, 1983).

7. See Browne, *Secular Ark*, 25–26, 33–38.

8. Charles Lyell, *Principles of Geology,
Being an Attempt to Explain Former Changes of
the Earth's Surface by Reference to Causes Now
in Operation* (London: John Murray, 1832),
2:70, 90. Lyell relied heavily on the work of
Augustin de Candolle.

9. Richard H. Grove, *Green Imperialism:
Colonial Expansion, Tropical Island Edens and
the Origins of Environmentalism, 1600–1860*
(Cambridge: Cambridge University Press,
1995).

10. The species has been the subject of
several recent monographs, including Errol

Fuller, *Dodo: A Brief History* (New York:
Universe Publishing, 2002); Clara Pinto-
Correia, *Return of the Crazy Bird: The Sad,
Strange Tale of the Dodo* (New York: Coper-
nicus Books, 2003); and Ben Van Wissen,
ed., *Dodo:* Raphus cucullatus [Didus ineptus]
(Amsterdam: ISP/Zoölogisch Museum–
Universiteit van Amsterdam, 1995). See also
Stephen Jay Gould, "The Dodo in the Caucus
Race," *Natural History* 105, no. 11 (1996):22–
33; Masauji Hachisuka, *The Dodo and Kindred
Birds, or the Extinct Birds of the Mascarene
Islands* (London: H. F. & G. Witherby, 1953);
and David Quammen, *The Song of the Dodo:
Island Biogeography in an Age of Extinctions*
(New York: Scribner, 1996).

11. Hugh E. Strickland and Alexander G.
Melville, *The Dodo and Its Kindred; or the
History, Affinities, and Osteology of the Dodo,
Solitaire, and Other Extinct Birds of the Islands
Mauritius, Rodriguez, and Bourbon* (London:
Reeve, Benham, and Reeve, 1848), on p. 4,
was the first to note that the dodo's general
appearance suggested "gigantic immaturity."
The dodo is an example of a species that main-
tains juvenile characteristics even after the
species has reached sexual maturity, a process
known as neoteny.

12. Pinto-Correia, *Return of the Crazy
Bird*, 17–46, provides a brief history of Arab,
Portuguese, and Dutch encounters with
Mauritius.

13. Pinto-Correia, *Return of the Crazy
Bird*, 136–37.

14. On the various names used to describe
the species, see Strickland and Melville, *The
Dodo and Its Kindred*, 9–27, and Hachisuka,
The Dodo and Kindred Birds, 35–37.

15. In 1752, Paul H. G. Moehring provided
the original name for the dodo's genus,
Raphus, a Latin form of the Dutch word *reet*,
meaning rump. In the tenth edition of his
Systema naturae (1758), Linnaeus dubbed the
species *Struthio cucullatus*, meaning "cuckoo-
like ostrich." Eight years later, Linnaeus
rechristened the bird *Didus ineptus*, a name

that was subsequently rejected for *Raphus cucullatus*.

16. Quoted in Pinto-Correia, *Return of the Crazy Bird*, 144.

17. Gould, "The Dodo in the Caucus Race," was to first to make this point.

18. Even less is known about the white dodo and the Rodrigues solitaire than the dodo of Mauritius. In the absence of surviving physical remains, Fuller, *Dodo*, 168–72, questions whether the white dodo ever existed. The few pictures that purport to show this species, he argues, may simply be albinisitic examples of the Mauritius dodo or they may have been painted that way for artistic effect.

19. Quammen, *Song of the Dodo*, 267.

20. Or so claims Strickland and Melville, *The Dodo and Its Kindred*, 4–5.

21. Strickland and Melville, *The Dodo and Its Kindred*, 37–38.

22. Biographical information on Strickland comes from G.S.B., "Hugh Edwin Strickland," in *Dictionary of National Biography* (London: Oxford University Press, 1921–22), 19:50–52.

23. Strickland and Melville, *The Dodo and Its Kindred*, 5.

24. Strickland and Melville, *The Dodo and Its Kindred*, 3.

25. Strickland and Melville, *The Dodo and Its Kindred*, 5.

26. Strickland and Melville, *The Dodo and Its Kindred*, 7, 22.

27. The best discussion of these paintings is Pinto-Correia, *The Return of the Crazy Bird*, 49–88.

28. See the account in R. F. Ovenell, "The Tradescant Dodo," *Archives of Natural History* 19, no. 2 (1992): 145–52.

29. Lyell, *Principles of Geology*, 2:151.

30. Strickland and Melville, *The Dodo and Its Kindred*, 6.

31. Strickland and Melville, *The Dodo and Its Kindred*, 33–35.

32. Strickland and Melville, *The Dodo and Its Kindred*, 39–43. Recent DNA evidence

seems to confirm Strickland and Melville's hunch. See Anonymous, "DNA Suggests Dodo Bird a Relative of the Pigeon," *Roanoke Times*, March 3, 2002, A-17.

33. See Fuller, *Dodo*, 123–28; Pinto-Correia, *Return of the Crazy Bird*, 170–73; and George Clark, "Account of the Late Discovery of Dodo's Remains in the Island of Mauritius," *Ibis* 2 (1865): 141–46.

34. Richard Owen, *Memoir on the Dodo (Didus ineptus, Linn.)* (London: Taylor and Francis, 1866), on p. 39.

35. Owen, *Memoir on the Dodo*, 49.

36. Biographical sketches of the Newton brothers are found in Barbara Mearns and Richard Mearns, *The Bird Collectors* (San Diego, Calif.: Academic Press, 1998), 174–76; and Errol Fuller, *The Great Auk* (New York: Harry N. Abrams, 1999), 379–84.

37. Mearns and Mearns, *Bird Collectors*, 175.

38. Morton N. Cohen, *Lewis Carroll: A Biography* (New York: Alfred A. Knopf, 1995).

39. Pinto-Correia, *Return of the Crazy Bird*, 194–95.

40. Andrew C. Kitchener, "On the External Appearance of the Dodo, *Raphus cucullatus* (L., 1758)," *Archives of Natural History* 20, no. 2 (1993): 279–301.

41. The definitive modern treatment of Owen's life and work and the source of the biographical sketch here is Nicolaas Rupke, *Richard Owen: Victorian Naturalist* (New Haven, Conn.: Yale University Press, 1994).

42. The story was first pieced together in T. Lindsay Buick, *The Discovery of Dinornis* (New Plymouth, New Zealand: Thomas Avery and Sons, 1936), a book that is part of a trilogy that also includes T. Lindsay Buick, *The Mystery of the Moa: New Zealand's Avian Giant* (New Plymouth, New Zealand: Thomas Avery and Sons, 1931); and T. Lindsay Buick, *The Moa-Hunters of New Zealand* (New Plymouth, New Zealand: Thomas Avery and Sons, 1937). See also Atholl Anderson, *Prodigious Birds: Moas and Moa-Hunting in Prehistoric*

New Zealand (Cambridge: Cambridge University Press, 1989); and Trevor H. Worthy and Richard N. Holdaway, *The Lost World of the Moa: Prehistoric Life of New Zealand* (Bloomington: Indiana University Press, 2002).

43. Buick, *The Discovery of Dinornis*, 107–8, reproduces the letter. According to Anderson, *Prodigious Birds*, 92, the rendering of the bird's name as "movie" was apparently a misunderstood reference to the Maori name for the North Island, *Ika na Maui* (or "movie"), which means the Fish of Maui.

44. Buick, *The Discovery of Dinornis*, 111–12.

45. Buick, *The Discovery of Dinornis*, 115.

46. See the discussion in Rupke, *Richard Owen*, 117–28.

47. Quoted in Anderson, *Prodigious Bird*, 18.

48. Richard Owen, "On the Bone of an Unknown Struthius Bird from New Zealand," *Proceedings of the Zoological Society of London* 7 (1839): 169–71; Anderson, *Prodigious Bird*, 15.

49. Anderson, *Prodigious Birds*, 1.

50. Anderson, *Prodigious Birds*, 13. Later in his book (p. 91), Anderson points out that the term "moa" is a very common Polynesian root and word with a bewildering variety of meanings.

51. This priority dispute is detailed in Anderson, *Prodigious Birds*, 11–14.

52. Anderson, *Prodigious Birds*, 14–16.

53. On Owens's ideas about dinosaurs, see the discussion in Rupke, *Richard Owen*, 130–37; and Adrian Desmond, *Hot-Blooded Dinosaurs: A Revolution in Paleontology* (New York: Dial Press, 1976).

54. Richard Owen, *Memoirs on the Extinct Wingless Birds of New Zealand*, 2 vols. (London: John Van Voorst, 1879).

55. Rupke, *Richard Owen*, 127.

56. See the discussions in Anderson, *Prodigious Birds*, 97–187.

57. Quoted in Worthy and Holdaway, *The Lost World of the Moa*, 56.

58. Worthy and Holdaway, *The Lost Worlds of the Moa*, 14–15; Susan Sheets-Pyenson, *Cathedrals of Science: The Development of Colonial Natural History Museums during the Late Nineteenth Century* (Kingston: McGill-Queen's University Press, 1988), 26–28, 49–51, 70, 80–83.

59. Sheets-Pyenson, *Cathedrals of Science*, 81.

60. Quoted in Anderson, *Prodigious Bird*, 3.

61. The episode is detailed in Fuller, *The Great Auk*, 80–87.

62. In addition to Fuller's heavily illustrated monograph, *The Great Auk*, see also Christopher Cokinos, *Hope Is the Thing with Feathers: A Personal Chronicle of Vanished Birds* (New York: Jeremy P. Tarcher/Putnam, 2000); Ron Freethy, *Auks: An Ornithologist's Guide* (New York: Facts on File, 1987); Jeremy Gaskell, *Who Killed the Great Auk?* (Oxford: Oxford University Press, 2000); and Anthony Gaston and Ian L. Jones, *The Auks: Alcidae* (Oxford: Oxford University Press, 1998). A convenient overview of the species may be found in William A. Montevecchi and David A. Kirk, "Great Auk (*Pinguinus impennis*)," in *The Birds of North America*, ed. A. Poole and F. Gill (Washington, D.C.: American Ornithologists' Union, 1996), no. 260.

63. Gaskell, *Who Killed the Great Auk*, 49–59.

64. Quoted in Freethy, *Auks*, 38.

65. Quoted in Gaskell, *Who Killed the Great Auk*, 109.

66. Fuller, *The Great Auk*, 68.

67. Accounts of Audubon's visit to Labrador are found in Gaskell, *Who Killed the Great Auk*, 31–40; Francis Hobart Herrick, *Audubon the Naturalist: A History of His Life and Time* (New York: D. Appleton-Century, 1938), 2:26–66; and Maria R. Audubon, ed., *Audubon and His Journals*, 2 vols. (New York: Charles Scribner's Sons, 1897), 343–45.

68. Scott Russell Sanders, ed., *Audubon*

Reader: The Best Writings of John James Audubon (Bloomington: Indiana University Press, 1986), 79–82, on p. 82.

69. Herrick, *Audubon the Naturalist*, 50.

70. The history of Audubon's auk is detailed in Fuller, *The Great Auk*, 154–60.

71. Fuller, *The Great Auk*, 107, 259, and 332.

72. Fuller, *The Great Auk*, 120–242, details the history of each of the known surviving specimens.

73. See the biographical sketch of Wolley in Fuller, *The Great Auk*, 377–79.

74. See the discussion in Fuller, *The Great Auk*, 416–19.

75. Quoted in Gaskell, *Who Killed the Great Auk*, 134.

76. Fuller, *The Great Auk*, 96.

77. Fuller, *The Great Auk*, contains dozens of examples of great auk images that appeared in advertising and art.

78. Janet Browne, *Charles Darwin: The Power of Place* (New York: Alfred A. Knopf, 2002); Janet Browne, *Charles Darwin: Voyaging* (New York: Alfred A. Knopf, 1995); Adrian Desmond and James Moore, *Darwin: The Life of a Tormented Evolutionist* (New York: Warner Books, 1991), are modern biographies that combine findings from the raft of scholars who have studied Darwin with the authors' own compelling insights. On Darwin's life and work, see also Gavin De Beer, "Charles Robert Darwin," in *Dictionary of Scientific Biography*, 3:565–76; David Kohn, ed., *The Darwinian Heritage* (Princeton, N.J.: Princeton University Press, 1985); Dov Ospovat, *The Development of Darwin's Theory: Natural History, Natural Theology, and Natural Selection, 1838–1859* (Cambridge: Cambridge University Press, 1981); and Peter J. Bowler, *Evolution: The History of an Idea*, rev. ed. (Berkeley: University of California Press, 1989).

79. De Beer, "Charles Robert Darwin," 3:566. On Darwin's *Beagle* voyage and its importance to the development of his ideas, see Richard Darwin Keynes, *Fossils, Finches and*

Fuegians: Darwin's Adventures and Discoveries on the Beagle (Oxford: Oxford University Press, 2003); Browne, *Charles Darwin: Voyaging*, 167–340; and Desmond and Moore, *Darwin*, 101–95.

80. Browne, *Charles Darwin: Voyaging*, 186–90.

81. Bowler, *Evolution*, 140.

82. Quoted in Browne, *Charles Darwin: Voyaging*, 223.

83. Desmond and Moore, *Darwin*, 159–60.

84. Charles Darwin, *Journal of Researches into the Geology and Natural History of the Various Countries Visited by the HMS Beagle, under the Command of Captain Fitzroy, R.N., from 1832 to 1836* (London: Henry Colburn, 1839), 211–12.

85. On the importance of the Galápagos Islands to Darwin, see Edward J. Larson, *Evolution's Workshop: God and Science in the Galápagos Islands* (New York: Basic Books, 2001), 61–85; Frank J. Sulloway, "Darwin and His Finches: The Evolution of a Legend," *Journal of the History of Biology* 15 (1982): 1–53; Frank J. Sulloway, "Darwin's Conversion: The *Beagle* Voyage and Its Aftermath," *Journal of the History of Biology* 15 (1982): 325–96; and Frank J. Sulloway, "Darwin's Early Intellectual Development: An Overview of the *Beagle* Voyage (1831–1836)," in *The Darwinian Heritage*, ed. David Kohn (Princeton, N.J.: Princeton University Press, 1985), 121–49.

86. Quoted in Larson, *Evolution's Workshop*, 74–75.

87. Larson, *Evolution's Workshop*, 76–77.

88. Larson, *Evolution's Workshop*, 77.

89. On Darwin's initial insight into natural selection as a possible mechanism for evolution, see Ospovat, *The Development of Darwin's Theory*, 60–86; and M. J. S. Hodge and David Kohn, "The Immediate Origins of Natural Selection," in *The Darwinian Heritage*, ed. David Kohn (Princeton, N.J.: Princeton University Press, 1985), 185–206.

90. From Darwin's D Notebook, quoted

in Ospovat, *The Development of Darwin's Theory*, 61–62.

91. Francis Darwin, ed., *The Foundations of the Origin of Species: Two Essays Written in 1842 and 1844* (Cambridge: Cambridge University Press, 1909), 113.

92. Darwin, *Foundations*, 112.

93. Charles Darwin, *On the Origin of Species by Means of Natural Selection, or the Preservation of Favoured Races in the Struggle for Life* (London: John Murray, 1859), 317.

94. Darwin, *Origin*, 109–10.

95. Darwin, *Origin*, 322.

96. Darwin, *Origin*, 388–406.

97. Darwin, *Origin*, 397.

98. Darwin, *Origin*, 398.

99. Relative to Darwin, Wallace has been unduly neglected by historians, though that lacuna has recently been addressed by two biographies: Peter Raby, *Alfred Russel Wallace: A Life* (Princeton, N.J.: Princeton University Press, 2001); and Michael Shermer, *In Darwin's Shadow: The Life and Science of Alfred Russel Wallace: A Biographical Study on the Psychology of History* (Oxford: Oxford University Press, 2002).

100. Alfred Russel Wallace, *A Narrative of Travels on the Amazon and the Rio Negro* (London: Reeve, 1853); and Alfred Russel Wallace, *Palm Trees of the Amazon and Their Uses* (London: J. Van Voorst, 1853).

101. Quoted in Raby, *Alfred Russel Wallace*, 122; Jane R. Camerini, ed., *The Alfred Russel Wallace Reader* (Baltimore: Johns Hopkins University Press, 2002), 115–16.

102. The essay is reproduced in Charles H. Smith, ed., *Alfred Russel Wallace: An Anthology of His Shorter Writings* (Oxford: Oxford University Press, 1991), 220–31, on p. 222.

103. Smith, *Alfred Russel Wallace*, 224.

104. Quoted in Raby, *Alfred Russel Wallace*, 126.

105. Raby, *Alfred Russel Wallace*, 131.

106. Alfred W. Crosby, *Ecological Imperialism: The Biological Expansion of Europe,*

900–1900 (Cambridge: Cambridge University Press, 1986).

107. Bowler, *Evolution*, 246–81, 307–32.

108. Darwin, *Origin*, 490.

109. Darwin, *Origin*, 488.

110. Quoted in Donald Worster, *Nature's Economy: A History of Ecological Ideas*, rev. ed. (Cambridge: Cambridge University Press, 1994), 180.

CHAPTER 3

1. Benson John Lossing, *A Centennial Edition of the History of the United States: From the Discovery of America, to the End of the First One Hundred Years of American Independence* (Hartford, Conn.: T. Belknap; Chicago: F.A. Hutchinson, 1876), xxi.

2. Theodore D. Woolsey et al., *The First Century of the Republic: A Review of American Progress* (New York: Harper and Brothers, 1876), 9.

3. Dee Alexander Brown, *The Year of the Century: 1876* (New York: Charles Scribner's Sons, 1966), and Robert Rydell, *All the World's a Fair: Visions of Empire at American International Expositions, 1876–1916* (Chicago: University of Chicago Press, 1984) both explore the historical background and cultural meanings of America's centennial celebrations. See also Robert C. Post, ed., *1876: A Centennial Exhibition* (Washington, D.C.: Smithsonian Institution, 1976).

4. Recent exceptions include Frank Sulloway, "Joel Asaph Allen," in *Dictionary of Scientific Biography* 17:20–23; and Mark V. Barrow, Jr., *A Passion for Birds: American Ornithology after Audubon* (Princeton, N.J.: Princeton University Press, 1998). The account of Allen and his early work on extinction that follows draws heavily on the latter publication. Other important sources of information on Allen include J. A. Allen, *Autobiographical Notes and a Bibliography of the Scientific Publications of Joel Asaph Allen* (New York: American Museum of Natural History,

1916); and Frank M. Chapman, "Joel Asaph Allen, 1838–1921," *Memoirs of the National Academy of Sciences* 21 (1927): 1–20.

5. Allen's western expeditions are chronicled in his *Autobiographical Notes*, 20–33.

6. J. A. Allen, *The American Bisons, Living and Extinct, Memoirs of the Museum of Comparative Zoology* 4, no. 10 (Cambridge, Mass.: Welch, Bigelow, 1876).

7. Allen, *The American Bisons*, 180; see also pp. 55, 70; emphasis added.

8. J. A. Allen, "The North American Bison and Its Extermination," *The Penn Monthly* 7 (1876): 214–24.

9. Allen, "The North American Bison," 215, 216,

10. J. A. Allen, "The Extirpation of the Larger Indigenous Mammals of the United States," *The Penn Monthly* 7 (1876): 794–806.

11. See, for example, Allen, "Extirpation of the Larger Indigenous Mammals," 795, 798.

12. J. A. Allen, "The Extinction of the Great Auk at Funk Islands," *American Naturalist* 10 (1876): 48.

13. J. A. Allen, "Decrease of Birds in Massachusetts," *Bulletin of the Nuttall Ornithological Club* 1 (1876): 53–60.

14. J. A. Allen, "On the Decrease of Birds in the United States," *The Penn Monthly* 7 (1876): 931–44.

15. Allen, "On the Decrease of Birds," 932.

16. See, for example, Allen, "Extirpation of the Larger Indigenous Mammals," 805–6; Allen, "On the Decrease of Birds," 943–44; and Allen, "The North American Bison," 223–24.

17. On the contributions of sportsmen to wildlife protection, see John F. Reiger, *American Sportsmen and the Origins of Conservation* (New York: Winchester Press, 1975).

18. Allen, "On the Decrease of Birds," 944.

19. Allen, "The North American Bison," 215, 224.

20. On the expedition to Yellowstone that Allen accompanied, see John M. Carroll, ed.,

The Yellowstone Expedition of 1873 (Mattituck, N.Y.: J. M. Caroll, 1986). On the circumstances surrounding the creation of Yellowstone National Park, see Roderick Nash, *Wilderness and the American Mind*, 3rd ed. (New Haven, Conn.: Yale University Press, 1982), 108–16; and Alfred Runte, *National Parks: The American Experience*, 2nd ed. (Lincoln: University of Nebraska Press, 1987), 33–47. On early wildlife policies and controversies at Yellowstone, see James A. Pritchard, *Preserving Yellowstone's Natural Conditions: Science and the Perception of Nature* (Lincoln: University of Nebraska Press, 1999); and Karl Jacoby, *Crimes against Nature: Squatters, Poachers, Thieves, and the Hidden History of American Conservation* (Berkeley: University of California Press, 2001), 81–146.

21. Allen, "On the Decrease of Birds," 944.

22. James Turner, *Reckoning with the Beast: Animals, Pain, and Humanity in the Victorian Mind* (Baltimore: Johns Hopkins University Press, 1980), provides a concise, insightful history of the humane movement in the United States and Great Britain, while Lisa Mighetto, *Wild Animals and American Environmental Ethics* (Tucson: University of Arizona Press, 1991); and Diane L. Beers, *For the Prevention of Cruelty: The History and Legacy of Animal Rights Activism in the United States* (Athens: Ohio University Press, 2006), document humanitarians' role in wildlife conservation.

23. The account of Bergh's buffalo protection campaign that follows comes from Andrew C. Isenberg, *The Destruction of the Bison: An Environmental History, 1750–1920* (Cambridge: Cambridge University Press, 2000), 145–56.

24. A. W. Schorger, *The Passenger Pigeon: Its Natural History and Extinction* (Madison: University of Wisconsin Press, 1955), 165.

25. On the perception of superabundance of New World resources, see William Cronon, *Changes in the Land: Indians, Colonists, and the*

Ecology of New England (New York: Hill and Wang, 1983), 22–25; and Stewart L. Udall, *The Quiet Crisis* (New York: Holt, Rinehart and Winston, 1963). For an early formulation of the "myth of inexhaustibility," see Hugh Hammond Bennett, *Soil Conservation* (New York: McGraw-Hill, 1939).

26. William Wood, *New England's Prospect*, ed. Alden Vaughan (Amherst: University of Massachusetts Press, 1977 [1634]), on pp. 56, 52, and 50.

27. Nash, *Wilderness and the American Mind*. See also David Nye, *America as Second Creation: Technology and Narratives of New Beginnings* (Cambridge, Mass.: MIT Press, 2003).

28. Cronon, *Changes in the Land*, examines differing native American and European conceptions of property. A recent work that challenges the pervasive notion of Native Americans as wise environmental stewards is Shepard Krech III, *The Ecological Indian: Myth and History* (New York: W. W. Norton, 1999).

29. For a recent synthesis of American environmental history stressing the importance of the commodification of nature, see Ted Steinberg, *Down to Earth: Nature's Role in American History* (New York: Oxford University Press, 2002).

30. The list of landscape transformations that followed colonization of North America comes from Cronon, *Changes in the Land*.

31. On early American wildlife law, see James A. Tober, *Who Owns the Wildlife?: The Political Economy of Conservation in Nineteenth-Century America* (Westport, Conn.: Greenwood Press, 1981), 3–40; and Thomas A. Lund, *American Wildlife Law* (Berkeley: University of California Press, 1980), 19–34.

32. On wildlife regulation as a form of social control, see Daniel Justin Herman, *Hunting and the American Imagination* (Washington, D.C.: Smithsonian Institution Press, 2001); and Louis S. Warren, *The Hunter's Game: Poachers and Conservationists in Twentieth-*

Century America (New Haven, Conn.: Yale University Press, 1997).

33. Tober, *Who Owns the Wildlife?*, 24–25.

34. The treatment of romanticism that follows owes much to Nash, *Wilderness and the American Mind*, 44–66; and Barbara Novak, *Nature and Culture: American Landscape Painting, 1825–1875* (New York: Oxford University Press, 1980).

35. Here I am also in debt to Nash, *Wilderness and the American Mind*, 84–95.

36. On Thoreau as a naturalist, see Laura D. Walls, *Seeing New Worlds: Henry David Thoreau and Nineteenth-Century Natural Science* (Madison: University of Wisconsin Press, 1995); and Robert D. Richardson, *Henry Thoreau: A Life of the Mind* (Berkeley: University of California Press, 1986). A convenient source of his views on nature is Robert L. Dorman, *A Word for Nature: Four Pioneering Environmental Advocates, 1845–1913* (Chapel Hill: University of North Carolina Press, 1998).

37. Henry David Thoreau, *The Journal of Henry David Thoreau*, ed. Bradford Torrey and Francis H. Allen (Boston: Houghton Mifflin, 1949), 5:135. In 8:221–22, he uses the analogy of a concert to talk about nature.

38. Thoreau, *Journal*, 12:387.

39. On American nationalism and natural history during this period, see Charlotte M. Porter, *The Eagle's Nest: Natural History and American Ideas, 1812–1842* (University: University of Alabama Press, 1986).

40. Perry Miller, *Nature's Nation* (Cambridge, Mass.: Belknap Press of Harvard University Press, 1967).

41. Audubon's life and work have been subject to extensive study. See, for example, Alice Ford, *John James Audubon: A Biography* (New York: Abbeville Press, 1988); Shirley Streshinsky, *Audubon: Life and Art in the American Wilderness* (New York: Villard Books, 1993); William Souder, *Under a Wild Sky: John James Audubon and the Making of the Birds of America* (New York: North Point

Press, 2004); and Richard Rhodes, *John James Audubon: The Making of an American* (New York: Alfred A. Knopf, 2004).

42. Ford, *John James Audubon*, 243.

43. On the diffusion of Audubon's and other natural history images, see Margaret Welch, *The Book of Nature: Natural History in the United States, 1825–1875* (Boston: Northeastern University Press, 1998).

44. Marsh's life and work are chronicled in David Lowenthal, *George Perkins Marsh: Prophet of Conservation* (Seattle: University of Washington Press, 2000), and more succinctly treated in Dorman, *A Word for Nature*, 3–45.

45. Lowenthal, *George Perkins Marsh*, xv.

46. Quoted in Lowenthal, *George Perkins Marsh*, 184–85.

47. Quoted in Lowenthal, *George Perkins Marsh*, xvi. Lowenthal has also edited a recent edition of George Perkins Marsh, *Man and Nature: Or, Physical Geography as Modified by Human Action*, ed. David Lowenthal (Seattle: University of Washington Press, 2003 [1864]).

48. Marsh, *Man and Nature*, 36.

49. Marsh, *Man and Nature*, 87.

50. Quoted in Lowenthal, *George Perkins Marsh*, 283.

51. Herman, *Hunting and the American Imagination*, 126; on the values and rituals associated with hunting in the South, see Nicolas W. Proctor, *Bathed in Blood: Hunting and Mastery in the Old South* (Charlottesville: University Press of Virginia, 2002).

52. Herman, *Hunting and the American Imagination*, 114–58.

53. Reiger, *American Sportsmen and the Origins of Conservation*, 25–49.

54. See the summary in Tober, *Who Owns the Wildlife?*

55. On the fish culture movement, see Joseph Taylor, *Making Salmon: An Environmental History of the Northwest Fisheries Crisis* (Seattle: University of Washington Press, 1999), 68–98; John C. Cumbler, "The Early Making of an Environmental Consciousness: Fish, Fisheries Commissions, and the Con-

necticut River," *Environmental History Review* 15 (1991): 73–91; Darin Kinsey, "'Seeding the Water as the Earth': The Center and Periphery of a Western Aquaculture Revolution," *Environmental History* 11 (2006): 527–66; J. T. Bowen, "The History of Fish Culture as Related to the Development of Fishery Programs," in *A Century of Fisheries in North America*, ed. Norman G. Benson (Washington, D.C.: American Fisheries Society, 1970), 71–93; Richard W. Judd, *Common Lands, Common People: The Origins of Conservation in New England* (Cambridge, Mass.: Harvard University Press, 1997), 146–59; and G. Brown Goode, "Epochs in the History of Fish Culture," *Transactions of the American Fish Culturalists' Association* 10 (1881): 34–59.

56. On river damming and its impact in the mid-nineteenth century, see Ted Steinberg, *Nature Incorporated: Industrialization and the Waters of New England* (Cambridge: Cambridge University Press, 1991); and Frederick Mather, "Poisoning and Obstructing the Waters," *Transactions of the American Fish Culturists' Association* 4 (1875): 14–19.

57. Robert B. Roosevelt, "Centennial Meeting," *Transactions of the American Fish Culturists' Association* 6 (1877): 10–17.

58. Taylor, *Making Salmon*, 69–70.

59. George P. Marsh, *Report Made under Authority of the Legislature of Vermont, on Artificial Propagation of Fish* (Burlington, Vt.: Free Press, 1857); the body of this report is reproduced in Stephen C. Trumblack, ed., *So Great a Vision: The Conservation Writings of George Perkins Marsh* (Hanover, N.H.: Middlebury College Press, 2001), 62–72. I relied on that reprint.

60. Marsh, *Artificial Propagation of Fish*, 67.

61. Marsh, *Artificial Propagation of Fish*, 68–69.

62. Quote in Taylor, *Making Salmon*, 71.

63. The creation of these state commissions is chronicled in Goode, "Epochs"; on the creation of the American Fish Culturists'

Society, see Livingston Stone, "The Origin of the American Fisheries Society," *Transactions of the American Fisheries Society* 27 (1898): 56–61; Frederick Mather, "Recollections of the Early Days of the American Fish Cultural Association," *Transactions of the American Fish Culturalists' Society* 8 (1879): 55–60; and Ward T. Bower, "History of the American Fisheries Society," *Transactions of the American Fisheries Society* 40 (1911): 323–58.

64. Quoted in Bower, "History of the American Fisheries Society," 325.

65. Dean C. Allard, *Spencer Fullerton Baird and the United States Fish Commission* (New York: Arno Press, 1978).

66. Taylor, *Making Salmon*, 77.

67. As an enduring icon of the American West, the bison has been the subject of much scholarly exploration. Some convenient entry points to this vast body of literature include David A. Dary, *The Buffalo Book: The Full Saga of the American Animal* (Chicago: Sage Books, 1974); Harold P. Danz, *Of Bison and Man* (Niwot: University of Colorado Press, 1997); Larry Barsness, *Heads, Hides, and Horns: The Compleat Buffalo Book* (Fort Worth: Texas Christian University Press, 1985); and especially, Isenberg, *Destruction of the Bison*. I have relied on all four of these accounts in the section that follows.

68. Isenberg, *Destruction of the Bison*, 23–30, has recently calculated through a variety of means that the maximum sustainable bison population on the entire Great Plains was somewhere in the range of 27–30 million.

69. Meriwether Lewis and William Clark, *The History of the Lewis and Clark Expedition*, ed. Elliott Coues (New York: Harper, 1893), 3:148–49.

70. Quoted in Danz, *Of Bison and Men*, 17.

71. Virtually every account of the bison contains some version of this incident. See, for example, Isenberg, *Destruction of the Bison*, 103–8; Barsness, *Heads, Hides, and Horns*, 133; and Dary, *The Buffalo Book*, 73.

72. George Catlin, *North American Indians: Being Letters and Notes on Their Manners, Customs, and Conditions*, 2 vols. (London: George Catlin, 1880), 1:288–95.

73. Mark Spence, *Dispossessing the Wilderness: Indian Removal and the Making of National Parks* (New York: Oxford University Press, 1999).

74. Tom McHugh, quoted in Krech, *The Ecological Indian*, 128.

75. Isenberg, *Destruction of the Bison*, 110.

76. Quoted in Dary, *The Buffalo Book*, 129.

77. David Smits, "The Frontier Army and the Destruction of the Buffalo, 1865–1883," *Western Historical Quarterly* 25 (1994) : 312–38. Isenberg, *Destruction of the Bison*, 128–29, argues that western soldiers in the 1860s and 1870s were too few in number and otherwise preoccupied to make much of a difference.

78. See the discussion in Isenberg, *Destruction of the Bison*, 130–32.

79. See the various estimates in Isenberg, *Destruction of the Bison*, 136–37.

80. The literature on the passenger pigeon remains more modest than that on the bison. See, for example, David E. Blockstein, "Passenger Pigeon: *Ectopistes migratorius*," in *The Birds of North America*, ed. A. Poole and F. Gill (Philadelphia: Birds of North America, 2002), no. 611; Jennifer Price, *Flight Maps: Adventures with Nature in Modern America* (New York: Basic Books, 1999); Christopher Cokinos, *Hope Is the Thing with Feathers: A Personal Chronicle of Vanished Birds* (New York: Jeremey P. Tarcher/Putnam, 2000); and Schorger, *Passenger Pigeon*.

81. Schorger, *Passenger Pigeon*, 204–5. Schorger admitted that his estimate was simply a "guess," but it has become the standard figure used by most scholars who write about the species.

82. In addition to utilizing the passenger pigeon for food, Native Americans also consumed vast quantities of the nuts that were the species' preferred food source. See Thomas W.

Neumann, "Human-Wildlife Competition and the Passenger Pigeon: Population Growth from System Destabilization," *Human Ecology* 13 (1985): 389–410. I thank William Woods for bringing this article and this general argument to my attention.

83. Quoted in Price, *Flight Maps*, 1.

84. Quoted in Schorger, *Passenger Pigeon*, 201.

85. Quoted in Cokinos, *Hope Is the Thing with Feathers*, 201.

86. John James Audubon, "Passenger Pigeon," in *Audubon Reader: The Best Writings of John James Audubon*, ed. Scott Russell Sanders (Bloomington: Indiana University Press, 1986), 117–18.

87. Audubon, "Passenger Pigeon," 120.

88. See the discussion in Price, *Flight Maps*, 8–18.

89. Audubon, "Passenger Pigeon," 121.

90. Schorger, *Passenger Pigeon*, 144–57, offers a compelling overview of the commercial market for the species.

91. On the various methods of obtaining the species, see Schorger, *Passenger Pigeon*, 167–98.

92. On passenger pigeons as targets for trap shooters, see Schorger, *Passenger Pigeon*, 157–66.

93. William Brewster, "The Present Status of the Wild Pigeon (*Ectopistes migratorius*) as a Bird of the United States, with Some Notes on Its Habits," *The Auk* 6 (1889): 285–91, on pp. 286 and 291.

94. On legislation aimed at protecting the species, see Schorger, *Passenger Pigeon*, 226–27.

95. See the lengthy discussion in Cokinos, *Hope Is the Thing with Feathers*, 228–57.

96. Useful historical overviews of this movement include William Dutcher, "History of the Audubon Movement," *Bird-Lore* 7 (1905): 45–57; Robin W. Doughty, *Feather Fashions and Bird Preservation: A Study in Nature Protection* (Berkeley: University of California Press, 1975); Frank Graham, Jr.,

The Audubon Ark: A History of the National Audubon Society (New York: Alfred A. Knopf, 1990); and Barrow, *Passion for Birds*, 102–53. The account of the movement that follows is largely reproduced from that latter publication.

97. Details surrounding the formation and early activities of the AOU bird protection committee are included in Barrow, *Passion for Birds*, 111–17.

98. J. A. Allen, "The Present Wholesale Destruction of Bird-Life in the United States," *Science* 7 (1886): 191–95. In addition to this essay, Allen was apparently primarily responsible for writing several of the other entries in this collection, including "The Relation of Birds to Agriculture" (pp. 201–2), "Bird-Laws" (pp. 202–4), "Destruction of Birds for Millinery Purposes" (pp. 196–97), and "An Appeal to the Women of the Country on Behalf of the Birds" (pp. 204–5). See, Allen, *Autobiographical Notes*, 138.

99. Allen, "Present Wholesale Destruction," 191, 192, and 194.

100. Allen, "Present Wholesale Destruction," 194, 195.

101. Allen, "Present Wholesale Destruction," 195.

102. On economic ornithology, see Matthew D. Evenden, "The Laborers of Nature: Economic Ornithology and the Role of Birds as Agents of Biological Pest Control in North American Agriculture, ca. 1880–1930," *Forest and Conservation History* 39 (1995): 172–83.

103. The AOU played a crucial role in the creation of the Division of Economic Ornithology and Mammalogy. See Barrow, *Passion for Birds*, 59–61.

104. The first version of this law was published in the *Science* supplement, 204.

105. Convenient biographical sketches of Grinnell's life may be found in John H. Mitchell, "A Man Called Bird," *Audubon Magazine* 89 (March 1987): 81–104; and A. K. Fisher, "In Memoriam: George Bird Grinnell," *The Auk* 56 (1939): 1–12. See also the recent biography, Michael Punke, *Last Stand: George Bird Grinnell, the Battle to Save the Buffalo, and the Birth of the New West* (New York: Smithsonian Books/Collins, 2007), which appeared after this chapter was drafted.

106. Quoted in Mitchell, "A Man Called Bird," 91

107. On *Forest and Stream*'s role in wildlife conservation, see Reiger, *American Sportsmen and the Origins of Conservation*.

108. Dutcher, "History of the Audubon Movement," 45.

109. Anonymous, "The Audubon Society," *Forest and Stream* 26, no. 3 (1886): 41.

110. See, for example, Anonymous, "The Audubon Society," *Audubon Magazine* 1 (1887): 20.

111. Anonymous, "Membership of the Audubon Society," *Audubon Magazine* 1 (1887): 19.

112. Anonymous, "The Great Auk," *Audubon Magazine* 1 (1887): 29–30.

113. Membership data is reproduced in Doughty, *Feather Fashions*, 102.

114. Anonymous, "Discontinuance of the Audubon Magazine," *Audubon Magazine* 2 (1889): 262.

115. Hemenway's role in the creation of the Massachusetts Audubon Society is detailed in Graham, *Audubon Ark*, 14–18; Joseph Kastner, "Long before Furs, It Was Feathers That Stirred Reformist Ire," *Smithsonian Magazine* 25 (July 1994); and John H. Mitchell, "The Mothers of Conservation," *Sanctuary* (January–February 1996): 1–20.

116. Quoted in William Dutcher, "Report of the A.O.U. Committee on Protection of North American Birds," *The Auk* 14 (1897): 21–32, on p. 31.

117. Price, *Flight Maps*, 57–109, stresses the gender dynamics of the revived Audubon movement.

118. Biographical information on Dutcher comes from Graham, *Audubon Ark*, 19–22, 26–33, 42–47, and 68–73; and T. S. Palmer, "In Memoriam: William Dutcher," *The Auk*

38 (1921): 501–13. On Dutcher's work for the AOU bird protection committee and the Audubon movement, see also Barrow, *Passion for Birds*, 128–34.

119. Theodore Whaley Cart, "The Lacey Act: America's First Nationwide Wildlife Statute," *Forest History* 17 (October 1973): 4–13.

120. That campaign is detailed in Barrow, *Passion for Birds*, 133.

121. The details that follow are from Barrow, *Passion for Birds*, 133–34.

122. On Chapman's life and career, see Mark V. Barrow, Jr., "Naturalists as Conservationists: American Scientists, Social Responsibility, and Political Activism before the Bomb," in *Science, History and Social Activism: A Tribute to Everett Mendelsohn*, ed. Garland E. Allen and Roy M. MacLeod (Dordrecht: Kluwer Academic Publishers, 2001), 217–33; François Vuilleumier, "Dean of American Ornithologists: The Multiple Legacies of Frank M. Chapman of the American Museum of Natural History," *The Auk* 122 (2005): 389–402; and Ernst Mayr, "Frank Michler Chapman," *Dictionary of Scientific Biography* 17 (1990): 152–53.

CHAPTER 4

1. The account of Hornaday's bison expedition that follows is drawn from several secondary sources—including James Andrew Dolph, "Bringing Wildlife to Millions: William Temple Hornaday, the Early Years; 1854–1896" (Ph.D. diss., University of Massachusetts, 1975); Hanna Rose Shell, "Last of the Wild Buffalo," *Smithsonian Magazine* 30, no. 11 (February 2000): 26–30; Hanna Rose Shell, "Introduction: Finding the Soul in the Skin," in William Temple Hornaday, *The Extermination of the American Bison* (Washington, D.C.: Smithsonian Institution Press, 2002), viii–xxiii; and Gregory J. Dehler, "An American Crusader: William Temple Hornaday and Wildlife Protection in America, 1840–1940" (Ph.D. diss., Lehigh University, 2001)—and two primary accounts: William Temple Hornaday, *Extermination of the American Bison*, and "The Passing of the Buffalo," *The Cosmopolitan* 4 (1887): 85–98, 231–43, on p. 88.

2. Hornaday, "Passing of the Buffalo," 85, and Shell, "Introduction," xix, both stress Hornaday's sense that preserving properly mounted examples of vanishing species offered the possibility of atonement for driving those species to extinction.

3. On Hornaday's early life and accomplishments, see Dolph, "Bringing Wildlife to Millions;" and Mary Anne Andrei, *Nature's Mirror: The Taxidermists Who Shaped America's Natural History Museums and Saved Our Endangered Species* (Chicago: University of Chicago Press, forthcoming). Other convenient sources of biographical information include Dehler, "An American Crusader"; Pamela M. Henson, "William Temple Hornaday," in *Biographical Dictionary of American and Canadian Naturalists and Environmentalists,* ed. Keir B. Sterling et al. (Westport, Conn.: Greenwood Press, 1997), 378–81; Phillip Drennon Thomas, "William Temple Hornaday," in *American National Biography*, ed. John A. Garraty and Mark C. Carnes (New York: Oxford University Press, 1999), 11:210–11.

4. On Ward and his central position in the American natural history community during the second half of the nineteenth century, see Mark V. Barrow, Jr., "The Specimen Dealer: Entrepreneurial Natural History in America's Gilded Age," *Journal of the History of Biology* 33 (2000): 493–534; and Sally G. Kohlstedt, "Henry A. Ward: The Merchant Naturalist and American Museum Development," *Journal of the Society for the Bibliography of Natural History* 9 (1980): 647–61.

5. Karen Wonders, *Habitat Dioramas: Illusions of Wilderness in Museums of Natural History,* Acta Universitatis Upsaliensis, Figura Nova Series 25 (Uppsala: Universitatis Upsaliensis, 1993), 116.

6. The figures are from Hornaday, *Extermination*, 542 and 545.

7. Quoted in Wonders, *Habitat Dioramas*, 121. Wonders and Andrei detail the techniques Hornaday used to mount the specimens.

8. Dolph, "Bringing Wildlife to Millions," 466–68.

9. Quoted in Hornaday, *Extermination*, 546.

10. Quoted in Wonders, *Habitat Diorama*, 122.

11. The claim comes from Shell, "Last of the Wild Buffalo," xviii. Charles R. Knight, the artist for the $10 note issued in 1901 later claimed that he worked from a specimen in the National Zoo, not Hornaday's bison group. See David A. Dary, *The Buffalo Book: The Full Saga of the American Animal* (Chicago: Sage Books, 1974), 282–83.

12. Hornaday, *Extermination*, 464.

13. On the history and cultural meanings of American zoos, see R. J. Hoage and William A. Deiss, eds., *New Worlds, New Animals: From Menagerie to Zoological Park in the Nineteenth Century* (Baltimore: Johns Hopkins University Press, 1996); Elizabeth Hanson, *Animal Attractions: Nature on Display in American Zoos* (Princeton, N.J.: Princeton University Press, 2002); and Vernon Kisling, Jr., "Zoological Gardens in the United States," in *Zoo and Aquarium History: Ancient Animal Collections to Zoological Gardens*, ed. Vernon Kisling, Jr. (Boca Raton, Fla.: CRC Press, 2001), 147–80. For zoos outside the United States, see Harriet Ritvo, *The Animal Estate: The English and Other Creatures in the Victorian Age* (Cambridge, Mass.: Harvard University Press, 1987), 205–42; Nigel Rothfels, *Savages and Beasts: The Birth of the Modern Zoo* (Baltimore: Johns Hopkins University Press, 2002); and Eric Baratay and Elisabeth Hardouin-Fugier, *Zoo: A History of Zoological Gardens in the West* (London: Reaktion Books, 2002).

14. The story of the founding of the U.S. National Zoo that follows comes from

Dolph's meticulously researched dissertation, "Bringing Wildlife to the Millions," 566–648; and Helen L. Horowitz, "The National Zoological Park: 'City of Refuge' or Zoo?" *Records of the Columbia Historical Society* 49 (1973–74): 405–29.

15. Quoted in Dolph, "Bringing Wildlife to Millions," 572.

16. Quoted in Dolph, "Bringing Wildlife to Millions," 575.

17. Quoted in Dolph, "Bringing Wildlife to Millions," 582.

18. The account of the founding of the Boone and Crockett Club that follows is from John Reiger, *American Sportsmen and the Origins of Conservation* (New York: Winchester Press, 1975), 114–41.

19. Quoted in Reiger, *American Sportsmen*, 119.

20. Reiger, *American Sportsmen*, 123–33; on the enforcement of antipoaching legislation at Yellowstone, see Karl Jacoby, *Crimes against Nature: Squatters, Poachers, Thieves, and the Hidden History of American Conservation* (Berkeley: University of California Press, 2001), 99–146.

21. On the ideology that informed sport hunting at the end of the nineteenth and the beginning of the twentieth centuries, see Daniel J. Herman, *Hunting and the American Imagination* (Washington, D.C.: Smithsonian Institution Press, 2001), 218–36.

22. See, for example, Richard Slotkin, "Nostalgia and Progress: Theodore Roosevelt's Myth of the Frontier," *American Quarterly* 33 (1981): 608–35.

23. Gray Brechin, "Conserving the Race: Natural Aristocracies, Eugenics, and the U.S. Conservation Movement," *Antipode* 28 (1996): 229–45, makes this point most strongly. See also Brian Regal, *Henry Fairfield Osborn: Race and the Search for the Origins of Man* (Aldershot, England: Ashgate, 2002), 102–35; Ronald Rainger, *An Agenda for Antiquity: Henry Fairfield Osborn and Vertebrate Paleontology at the American Museum of Natural History, 1890–1935* (Tuscaloosa: University of Alabama Press, 1991), 105–22; and Minna Alexandra Stern, *Eugenic Nation: The Faults and Frontiers of Better Breeding in Modern America* (Berkeley: University of California Press, 2005), 115–49.

24. Grant has received more attention for his views on eugenics than his role in wildlife conservation. Exceptions include Bridge, *Gathering of Animals;* Susan R. Schrepfer, *The Fight to Save the Redwoods: A History of Environmental Reform, 1917–1978* (Madison: University of Wisconsin Press, 1983), 3–17; Helen L. Horowitz, "Animal and Man in the New York Zoological Park," *New York History* 56 (1975): 426–55; and Thomas J. Curran, "Madison Grant," in *Biographical Dictionary of American and Canadian Naturalists and Environmentalists*, ed. Keir B. Sterling et al. (Westport, Conn.: Greenwood Press, 1997).

25. Quoted in William Bridges, *Gathering of Animals: An Unconventional History of the New York Zoological Society* (New York: Harper and Row, 1974), 4.

26. The history of the New York Zoological Society that follows comes from Bridges, *Gathering of Animals;* and Horowitz, "Animal and Man."

27. Osborn served as president of the New York Zoological Society from 1909 to 1924; Grant served as the institution's secretary (1895–1924), chairman of the Executive Committee (1908–36), and president (1925–37).

28. Quoted in Bridges, *Gathering of Animals*, 16. This was several years before the German animal entrepreneur Carl Hagenbeck pioneered the use of more naturalistic animal exhibits at his establishment on the outskirts of Hamburg.

29. Quoted in Bridges, *Gathering of Animals*, 19.

30. Quoted in Bridges, *Gathering of Animals*, 31, 32.

31. Hornaday's struggles to maintain the bison and other endangered species are

chronicled in Bridges, *Gathering of Animals*, 71–74, 265–66.

32. Raymond Gorges, *Ernest Harold Baynes: Naturalist and Crusader* (Boston: Houghton Mifflin, 1928) offers a celebratory biography of Baynes, who has generally been ignored by historians. On the American Bison Society, see pp. 70–91 of that book; Bridges, *Gathering of Animals*, 257–70; Andrew C. Isenberg, *Destruction of the Bison: An Environmental History, 1750–1920* (Cambridge: Cambridge University Press, 2000), 166–89; Martin S. Garretson, *The American Bison: The Story of Its Extermination as a Wild Species and Its Restoration under Federal Protection* (New York: New York Zoological Society, 1938), 205–14; and Larry Barsness, *Heads, Hides, and Horns: The Compleat Buffalo Book* (Fort Worth: Texas Christian University Press, 1985), 160–89. Hornaday, *Extermination*, 525, estimates that in 1889 there were 635 bison in the wild, 256 in captivity, and 200 under protection in Yellowstone National Park. In 1903, Frank Baker, of the U.S. National Zoo, estimated that there were a total of 1,644 bison still in existence (as recorded in Garretson, *The American Bison*, 212).

33. Quoted in Barsness, *Heads, Hides, and Horns*, 161.

34. See, for example, Ernest Harold Baynes, "In the Name of the American Bison," *Harper's Weekly* 24 (1906): 404–6.

35. Bridges, *Gathering of Animals*, 258–61.

36. Clara Ruth, "Preserves and Ranges Maintained for Buffalo and Other Big Game," *Wildlife Research and Management Leaflet*, no. BS-95 (1937), 2. The federal government had first set the land aside as a forest reserve in 1901.

37. See, for example, William Temple Hornaday, "The Founding of the Wichita National Bison Herd," *Annual Report of the American Bison Society* 1 (1908): 55–69, on pp. 55–56; William Temple Hornaday, "Report of the President on the Founding of the Montana National Bison Herd," *Annual Report*

of the *American Bison Society* 3 (1910): 1–18, on pp. 17–18; Anonymous, "The National Buffalo Herd: A Gift from the New York Zoological Society to Start a Great Southwestern Herd" (unpublished ms., ca. 1907); and William T. Hornaday to Charles E. Hughes, June 28, 1907; American Bison Society Papers, President's Office, 1905–12, Box 1, Wildlife Conservation Society, New York (hereafter cited as ABSP/WCS). In this same collection, Henry F. Osborn to William Temple Hornaday, January 24, 1906, argues that inbreeding was one of the greatest dangers facing the species.

38. The quote comes from Edmund Seymoure, "National Buffalo Herd," Box 2, ABSP/WCS.

39. See, Isenberg, *Destruction of the Bison*, 170.

40. American Bison Society, *Appeal for the Buffalo* (New York: American Bison Society, ca. 1906), Box 4, ABSP/WCS.

41. Franklin W. Hooper, "President's Annual Report," *Annual Report of the American Bison Society* 7 (1914): 9–15.

42. Hornaday, "Founding of the National Bison Herd," provides an official account of the project.

43. J. Alden Loring, "The Wichita Buffalo Range," *Annual Report of the New York Zoological Society* 10 (1906): 3–22.

44. See, for example, "Zoos Bison Herd Accepted," *New York Times*, July 9, 1906, 14.

45. T. S. Palmer, "Our National Herds of Buffalo," *Annual Report of the American Bison Society* 10 (1916): 40–62. See also Barsness, *Heads, Hides, and Horns*, 158–72, which offers a detailed account of the fate of the Pablo herd.

46. William Temple Hornaday to Gilbert N. Haugen, May 4, 1908, Box 2, Folder 12, ABSP/WCS.

47. See, for example, William Temple Hornaday to J. C. Cannon, March 30, 1908, Box 1, Folder 15; William Temple Hornaday to Hon. F. W. Mondell, April 21, 1908, Box 1, Folder 25; ABSP/WCS. The note questioning

the legitimacy of the solicitation appears in
J. K. Moore to Edward Seymour, July 10,
1908, Box 3, Folder 3 in that same collection.

48. Ruth, "Preserves and Ranges," 4–6.

49. Quoted in Isenberg, *Destruction of the Bison*, 183.

50. Hornaday, "Report of the President," 17–18.

51. Palmer, "Our National Herds," 49–50.

52. See, for example, Anonymous, "Seventh Annual Meeting," *Annual Report of the American Bison Society* 6 (1913): 16–21; and Anonymous, "Eighth Annual Meeting," *Annual Report of the American Bison Society* 7 (1914): 17–24.

53. Anonymous, "Ninth Annual Meeting of the Board of Managers of the Society," *Annual Report of the American Bison Society* 10 (1916): 30.

54. Details of the search are chronicled in Anonymous, "A Last Attempt to Locate and Save from Extinction the Passenger Pigeon," *The Auk* 27 (1910): 112; Anonymous, "Notes and News," *The Auk* 27 (1910): 243; C. F. Hodge, "A Last Effort to Find and Save from Extinction the Passenger Pigeon," *Bird-Lore* 12 (1910): 52; C. F. Hodge, "The Passenger Pigeon Investigation," *The Auk* 28 (1911): 49–53; C. F. Hodge, "A Last Word on the Passenger Pigeon," *The Auk* 29 (1912): 169–75; and Anonymous, "Notes and News," *The Auk* 31 (1914): 366–67.

55. Ruthven Deane, "Some Notes on the Passenger Pigeon (*Ectopistes migratorius*) in Confinement," *The Auk* 13 (1896): 234–37.

56. Ruthven Deane, "The Passenger Pigeon (*Ectopistes migratorius*) in Confinement," *The Auk* 25 (1908): 181–83; Ruthven Deane, "The Passenger Pigeon—Only One Pair Left," *The Auk* 26 (1909): 429.

57. The story of Whitman and his flock that follows comes from Philip J. Pauly, *Biologists and the Promise of American Life: From Meriwether Lewis to Alfred Kinsey* (Princeton, N.J.: Princeton University Press, 2000), 162–64; Willard F. Hollander, "Charles O.

Whitman's Flock," *National Pigeon Association News* (1961): 24–29; C. H. Ames, "Breeding the Wild Pigeon," *Forest and Stream* 56 (1901): 464; W. Wade, "Breeding the Wild Pigeon," *Forest and Stream* 56 (1901): 485; and Deane, "The Passenger Pigeon in Confinement," 181–82.

58. Deane, "The Passenger Pigeon in Confinement," 182.

59. Kisling, "Zoological Gardens in the United States," 157; Schorger, *The Passenger Pigeon*, 27–29.

60. Deane, "The Passenger Pigeon— Only One Pair Left," 429.

61. Christopher Cokinos, *Hope Is the Thing with Feathers: A Personal Chronicle of Vanished Birds* (New York: Jeremey P. Tarcher/Putnam, 2000), 258–78, does a wonderful job of reconstructing the story of Martha. See also D. H. Eaton, "The Last Surviving Passenger Pigeon," *Forest and Stream* 81, no. 6 (1914): 165–66.

62. Quoted in Cokinos, *Hope Is the Thing with Feathers*, 270; see R. W. Shufeldt, "Anatomical and Other Notes on the Passenger Pigeon (*Ectopistes migratorius*) Lately Living in the Cincinnati Zoological Gardens," *The Auk* 32 (1915): 28–41.

63. The account of the Carolina parakeet that follows comes from Mikko Saikku, "The Extinction of the Carolina Parakeet," *Environmental History Review* 14 (1990): 1–18; Daniel McKinley, *The Carolina Parakeet in Florida*, ed. John William Hardy, Special Publication No. 2 (Gainesville: Florida Ornithological Society, 1985); Noel F. R. Snyder and Keith Russell, "Carolina Parakeet (*Cornuropsis carolinensis*)," in *The Birds of North America*, ed. A. Poole and F. Gill (Philadelphia: Birds of North America, 2002), no. 667; Cokinos, *Hope Is the Thing with Feathers*, 7–58; and Noel F. R. Snyder, *The Carolina Parakeet: Glimpses of a Vanished Bird* (Princeton, N.J.: Princeton University Press, 2004).

64. Quoted in Snyder and Russell, "Carolina Parakeet," 3.

65. J. A. Allen, "On the Mammals and Winter Birds of East Florida," *Bulletin of the Museum of Comparative Zoology* 2 (1871): 161–450, on pp. 308–9.

66. Charles E. Bendire, *Life Histories of North American Birds*, U.S. National Museum, Special Bulletin No. 3 (Washington, D.C.: Government Printing Office, 1895), 4.

67. McKinley, *Carolina Parakeet*, 4, has located 806 specimens in museums around the world. See also Paul Hahn, *Where Is That Vanished Bird? An Index to the Known Specimens of Extinct North American Species* (Toronto: Royal Ontario Museum, University of Toronto, 1963), 291–339.

68. Daniel McKinley, "The Last Days of the Carolina Parrakeet: Life in the Zoos," *Avicultural Magazine* 83 (1977): 42–49; see also Arthur A. Prestwich, *Records of Birds Bred in Captivity* (London: Arthur A. Prestwich, 1952), 79–80.

69. Most of what we know about the source of Ridgway's parakeets comes from a brief passage in Amos William Butler, "Some Bird Records from Florida," *The Auk* 48 (1031): 436–39; see also McKinley, "The Last Days," 45–46; Cokinos, *Hope Is the Thing with Feathers*, 48–50.

70. Paul Bartsch, "A Pet Carolina Parakeet," in *The Bird Watcher's Anthology*, ed. Roger Tory Peterson (New York: Harcourt, Brace, 1957), 148–51; Bartsch also translated an article, Dr. Nowotny, "The Breeding of the Carolina Paroquet in Captivity," *The Auk* 25 (1898): 28–32.

71. Quoted in Bridges, *Gathering of Animals*, 215–16.

72. Lee S. Crandall, "The Carolina Parakeet," *Zoological Society Bulletin* 49 (1912): 834–36; Bridges, *Gathering of Animals*, 216.

73. George Laycock, "The Last Parakeet," *Audubon* (March 1969): 21–25.

74. See, for example, Laycock, "The Last Parakeet"; Snyder and Russell, "Carolina Parakeet," 3–7.

75. Casey A. Wood, "Lessons in Aviculture from English Aviaries," *The Condor* 28 (1926): 1–30, on p. 25. See also Lord Tavistock, "Saving Rare Parrots in Captivity," *The Condor* 38 (1926): 184–86. For a contrary view of the prospect, see W. L. McAtee, "Preservation of Species in Aviaries," *The Auk* 52 (1935): 128–29.

76. Snyder, *Carolina Parakeet*, 12, analyzes a large number of late reports of the parakeet.

77. See the many examples in Dary, *Buffalo Book*, 279–85.

78. William Temple Hornaday, *Our Vanishing Wildlife: Its History and Preservation* (New York: New York Zoological Society, 1913).

79. Hornaday, *Our Vanishing Wildlife*, 10. Hornaday's list included several Caribbean species that he had gleaned from Lionel Walter Rothschild, *Extinct Birds: An Attempt to Unite in One Volume a Short Account of Those Birds Which Have Become Extinct in Historical Times—That Is, within the Last Six or Seven Hundred Years: To Which Are Added a Few Which Still Exist, but Are on the Verge of Extinction* (London: Hutchinson, 1907).

80. Hornaday, *Our Vanishing Wildlife*, 18.

81. Hornaday, *Our Vanishing Wildlife*, 7.

82. Hornaday, *Our Vanishing Wildlife*, 54, 56.

83. Hornaday, *Our Vanishing Wildlife*, 101.

84. The chapter in which he does this is entitled "Destruction of Song-Birds by Southern Negroes and Poor Whites," but almost all of the discussion centers on the actions of African Americans.

85. Hornaday, *Our Vanishing Wildlife*, 376.

86. Hornaday, *Our Vanishing Wildlife*, 387, 322, 323.

87. Hornaday, *Our Vanishing Wildlife*, 386; William Temple Hornaday, "The Destruction of Our Birds and Mammals," *Second Annual Report of the New York Zoological Society* (1898): 77–126.

88. Hornaday, *Our Vanishing Wildlife*, 390, 392.

89. Joseph Grinnell, "Conserve the Collector," *Science*, 49, no. 1050 (February 15, 1915): 229–32.

CHAPTER 5

1. Harold J. Coolidge, Jr., "Wanted—One Giant Gorilla," *Sportsman* 4, no. 5 (1928). 41–44, 98, on p. 44. Additional details may be found Harold J. Coolidge, Jr., "Notes on the Gorilla," in *The African Republic of Liberia and the Belgium Congo, Based on the Observation Made and the Material Collected during the Harvard African Expedition, 1926–1927,* ed. Richard P. Strong (Cambridge, Mass.: Harvard University Press, 1930), 2:623–25. On the symbolic dimensions of gorilla hunting during this period, see Donna Haraway, *Primate Visions: Gender, Race, and Nature in the World of Modern Science* (New York: Routledge, 1989).

2. Coolidge, "Wanted—One Giant Gorilla," 44.

3. Coolidge, "Wanted—One Giant Gorilla," 98.

4. Biographical details on Coolidge, who has yet to gain the historical attention he deserves, may be found in Victoria Croninshield Drake, "The Pioneering International Vision of Harold Jefferson Coolidge: American Conservationist" (B.A. thesis, Harvard University, 1983); Harvard Class of 1927, *Fiftieth Anniversary Report* (Cambridge, Mass.: Printed for the University, 1977), 154–57; and Walter Sullivan, "Harold Coolidge, Expert on Exotic Mammals," *New York Times*, February 16, 1985, 24. See also his curriculum vitae in the Harold Jefferson Coolidge Papers, HUGFP 78.10, Harvard University Archives, Cambridge, Mass. (hereafter cited as HJCP/HUA), and the interview of him published as "Profile of Harold Jefferson Coolidge," *Environmentalist* 1 (1981): 65–74.

5. The episode is recounted in Victor H. Cahalane, "History of the American Committee for International Wildlife Protection,"

8–9, ms. in Victor Cahalane Collection (no. 1020), American Heritage Center, Laramie, Wyoming.

6. Phillips's life and conservation accomplishments are documented in numerous biographical sketches written by his admiring colleagues: Glover M. Allen, "In Memoriam: John Charles Phillips, M.D.," *The Auk* 56, no. 3 (1939): 221–26; Anonymous, "John C. Phillips, Scientist, Dies," *Boston Herald*, November 14, 1938, 1, 23; and Thomas Barbour, *John Charles Phillips* (Cambridge, Mass: Privately Printed, 1939).

7. The history of this organization is recounted in Cahalane, "History of the American Committee"; John P. Droege, "Specimen Collectors to Conservationists: Scientists and International Conservation before the Second World War" (paper presented at the History of Science Society Annual Meeting, New Orleans, 1994); and the minutes of the annual meetings and reports of the secretaries found in HJCP/HUA, and the Alexander Wetmore Papers, RU 7006, Smithsonian Institution Archives, Washington, D.C. (hereafter cited as AWP/SIA). See also John C. Phillips, "The Work of the American Committee for International Wild Life Protection," *Proceedings, North American Wildlife Conference, February 3–7, 1936* (1936): 51–55; John C. Phillips and Harold J. Coolidge, Jr., *The First Five Years: The American Committee for International Wild Life Protection* (Cambridge, Mass.: American Committee for International Wild Life Protection, 1934); Anonymous, *Brief History of the Formation of the American Committee for International Wild Life Protection* (Cambridge, Mass.: American Committee for International Wild Life Protection, 1930); and Robert Boardman, *International Organization and the Conservation of Nature* (Bloomington: Indiana University Press, 1981), 33–34, 36, 50, 58, 182.

8. Phillips, "The Work of the American Committee," 54.

9. The best account of the Inland Fisher-

ies Treaty is Kurkpatrick Dorsey, *The Dawn of Conservation Diplomacy: U.S.-Canadian Wildlife Protection Treaties in the Progressive Era* (Seattle: University of Washington Press, 1998), 19–107, which I relied on heavily in this section. See also the précis of his argument in Kurkpatrick Dorsey, "Scientists, Citizens, and Statesmen: U.S.-Canadian Wildlife Protection Treaties in the Progressive Era," *Diplomatic History* 19, no. 3 (1995): 407–29; and Margaret Beattie Bogue, "To Save the Fish: Canada, the United States, the Great Lakes, and the Joint Commission of 1892," *Journal of American History* 79, no. 4 (1993): 1429–54.

10. See Dorsey, *Dawn of Conservation Diplomacy*, 105–64. See also Kurkpatrick Dorsey, "Putting a Ceiling on Sealing: Conservation and Cooperation in the International Arena, 1909–1911," *Environmental History Review* 15, no. 3 (1991): 27–45; and Briton Busch, *The War against the Seals: A History of the North American Seal Fishery* (Montreal: McGill-Queen's University Press, 1985). For an account that stresses the contribution of Henry Elliott, see James Thomas Gay, "Henry W. Elliott: Crusading Conservationist," *Alaska Journal* 3, no. 4 (1973): 211–16.

11. Dorsey, *Dawn of Conservation Diplomacy*, 163.

12. On the events that led to the passage of the Migratory Bird Treaty, see Dorsey, *Dawn of Conservation Diplomacy*, 165–237; and John C. Phillips, *Migratory Bird Protection in North America: The History of Control by the United States Federal Government and a Sketch of the Treaty with Great Britain* (Cambridge, Mass.: American Committee for International Wild Life Protection, 1934).

13. The account of the formation and early years of the ICBP that follows is from Thomas Gilbert Pearson, *Adventures in Bird Protection: An Autobiography* (New York: D. Appleton-Century, 1937), 370–419; the *Bulletin of the International Committee for*

Bird Preservation 1–5 (1927–39); and Boxes A-110–11, C-48, National Audubon Society Records, New York Public Library, New York (hereafter cited as NASR/NYPL).

14. On Delacour and his multifarious ornithological, avicultural, and conservation activities, see Jean Delacour, *The Living Air: Memoirs of an Ornithologist* (London: Country Life Limited, 1966); Ernst Mayr, "In Memoriam: Jean (Theodore) Delacour," *The Auk* 103 (1986): 603–5; and Roger F. Pasquier, "In Memoriam: Jean Delacour," *American Birds* 40, no. 1 (1986): 46–48.

15. *Bulletin of the International Committee for Bird Protection* 1 (1927): 3. In 1928, the group adopted the name the International Committee for Bird Preservation, and in 1960, the International Council for Bird Preservation.

16. T. Gilbert Pearson to Percy Lowe, February 8, 1923, Box A-110, NASR/NYPL.

17. *Bulletin of the International Committee for Bird Preservation* 2 (1929): 7.

18. On the Paris Convention, which was designed to protect migratory birds, see Sherman Strong Hayden, *The International Protection of Wild Life: An Examination of Treaties and Other Agreements for the Preservation of Birds and Mammals* (New York: Columbia University Press, 1942), 89–106; and Boardman, *International Organization and the Conservation of Nature*, 26–31.

19. *Bulletin of the International Committee for Bird Preservation* 4 (1935): 14.

20. Vernon N. Kisling, Jr., "Zoological Gardens of the United States," in *Zoo and Aquarium History: Ancient Animal Collections to Zoological Gardens*, ed. Vernon N. Kisling, Jr. (Boca Raton, Fla.: CRC Press, 2001), 147–80.

21. The best account of Roosevelt's African exploits and the one I have relied on most heavily is Bartle Bull, *Safari: A Chronicle of Adventure* (New York: Viking, 1988), 157–83. See also Joseph L. Gardner, *Departing Glory: Theodore Roosevelt as Ex-President* (New

York: Charles Scribner's Sons, 1973); Gary Rice, "Trailing a Celebrity: Press Coverage of Theodore Roosevelt's African Safari, 1909–1910," *Theodore Roosevelt Association Journal* 21 (1996): 4–16; Theodore Roosevelt, *African Game Trails: An Account of the African Wanderings of an American Hunter-Naturalist* (New York: Charles Scribner's Sons, 1920); and Emily Hahn, "My Dear Selous . . . ," *American Heritage* 14, no. 3 (1963): 40–42, 92–99.

22. Bull, *Safari,* 169.

23. Bull, *Safari,* 166.

24. The Roosevelt expedition brought back more than twenty-three thousand specimens, leading Smithsonian officials to boast that their East African collections represented "probably the most complete and systematic of any in the world." Anonymous, "The Smithsonian African Expedition," *Science* 37, no. 949 (1913): 364–65.

25. Rice, "Trailing a Celebrity," 13.

26. See, for example, Anonymous, "The Rainey African Collection," *Science* 36, no. 914 (1912): 11–12.

27. Biographical information on Akeley comes primarily from Penelope Bodry-Sanders, *Carl Akeley: Africa's Collector, Africa's Savior* (New York: Paragon House, 1991). See also Mary L. Jobe Akeley, *The Wilderness Lives Again: Carl Akeley and the Great Adventure* (New York: Dodd, Mead, 1940); Karen Wonders, *Habitat Dioramas: Illusions of Wilderness in Museums of Natural History,* Figura Nova Series 25 (Uppsala: Acta Universitatis Upsaliensis, 1993), 170–77; Mary Anne Andrei, *Nature's Mirror: The Taxidermists Who Shaped America's Natural History Museums and Saved Our Endangered Species* (Chicago: University of Chicago Press, forthcoming); and Haraway, *Primate Visions,* 26–58.

28. Quoted in Wonders, *Habitat Dioramas,* 172.

29. Bodry-Saunders, *Carl Akeley,* 141.

30. Wonders, *Habitat Dioramas,* 173.

31. The expedition is chronicled in Bodry-Sanders, *Carl Akeley,* 176–200; and Haraway, *Primate Visions,* 31–34.

32. On Merriam and his conservation work, see Stephen R. Mark, *Preserving the Living Past: John C. Merriam's Legacy in the State and National Parks* (Berkeley: University of California Press, 2005).

33. The quote is from Bodry-Sanders, *Carl Akeley,* 197. Carl Akeley's wife, Mary L. Jobe Akeley, authored several additional accounts of the creation of Parc Albert National, including "Africa's Great National Park," *Natural History* 29 (1929): 638–50; "Belgian Congo Sanctuaries," *Scientific Monthly* 33, no. 4 (1931): 289–300; "King Albert Inaugurates the Parc National Albert," *Natural History* 30 (1930): 193–95; and "National Parks in Africa," *Science* 74, no. 1928 (1931): 584–88.

34. Various versions of the Johnson's affiliation with the American Museum and the resulting film, *Simba,* are found in Bodry-Sanders, *Carl Akeley,* 208–13; Osa Johnson, *I Married Adventure: The Lives and Adventures of Martin and Osa Johnson* (New York: William Morrow and Company, 1989); Pascal James Imperato and Eleanor M. Imperato, *They Married Adventure: The Wandering Lives of Martin and Osa Johnson* (New Brunswick, N.J.: Rutgers University Press, 1992), especially 114–18. Gregg Mitman, *Reel Nature: America's Romance with Wildlife on Film* (Cambridge, Mass.: Harvard University Press, 1999), 26–35, places the Johnsons' work into broader context.

35. Henry R. Carey, "Saving the Animal Life of Africa—a New Method and a Last Chance," *Journal of Mammalogy* 7, no. 2 (1926): 73–85.

36. Carey, "Saving the Animal Life of Africa," 76.

37. Carey, "Saving the Animal Life of Africa," 77.

38. Hobley's visit is covered in Madison Grant, "An Appeal for Conservation of Wild Life in the British Empire," *Natural History* 30 (1930): 216–17.

39. A brief, breezy insider history of the organization is presented in Richard Fitter and Sir Peter Scott, *The Penitent Butchers: The Fauna Preservation Society, 1903–1978* (London: Fauna Preservation Society, 1978), while a more sophisticated account is provided in John M. MacKenzie, *The Empire of Nature: Hunting, Conservation, and British Imperialism* (Manchester: Manchester University Press, 1988). See also David K. Prendergast and William M. Adams, "Colonial Wildlife Conservation and the Origins of the Society for the Preservation of Wild Fauna of the Empire (1903–1914)," *Oryx* 37 (April 2003): 251–60.

40. Burham's speech is discussed in Grant, "An Appeal for the Conservation," 216. On the life and career of Burnham, see Philip De Vencentes, "Frederick Russell Burnham," in *Dictionary of American Biography, Supplement 4 (1946–1950)*, ed. John A. Garraty and Edward T. James (New York: Charles Scribner's Sons, 1974), 126–28.

41. Grant, "An Appeal for Conservation," 216.

42. See, for example, Harold J. Coolidge to Kermit Roosevelt, February 7, 1930; Kermit Roosevelt to Harold J. Coolidge, February 10, 1930; and Harold J. Coolidge to Kermit Roosevelt, February 11, 1930, Box 2, Folder: Boone and Crockett Club, HJCP/HUA.

43. Anonymous, *Brief History of the Formation of the American Committee*, on p. 3.

44. The activities of this meeting are recorded in "Minutes for 1930" Box 28, Folder: Minutes of Committee Meetings, HJCP/HUA.

45. "Minutes for 1930," 2.

46. Harold J. Coolidge to C. W. Hobley, ca. March 1930, Box 4, Folder: Hobley, C. W., American Committee for International Wild Life Protection (Secretary's Office), 1930–61, Wildlife Conservation Society, New York (hereafter cited as ACIWLP/WCS).

47. John C. Phillips to Alexander Wetmore, July 22, 1930, RU 7006, Box 78, AWP/SIA.

48. "Minutes for 1930," 2.

49. John C. Phillips to C. W. Hobley, October 15, 1935, Box 4, Folder: Hobley, C. W., ACIWLP/WCS.

50. MacKenzie, *The Empire of Nature*, 234–44, provides a convenient summary of controversies and policies surrounding tsetse fly control.

51. C. W. Hobley to Harold J. Coolidge, June 26, 1930, Box 4, Folder: Hobley, C. W., ACIWLP/WCS.

52. "Minutes for 1930," 5–6. Richard P. Strong, Joseph C. Bequaert, and L. R. Cleveland, *Report on the Available Evidence Showing the Relation of Game to the Spread of the Tsetse Fly Borne Diseases in Africa*, Special Publication of the American Committee for International Wild Life Protection, Vol. 1, No. 1 (Cambridge, Mass.: American Committee for International Wild Life Protection, 1931), on p. 46.

53. The quotes are from Harold J. Coolidge to C. W. Hobley, July 8, 1933, Box 4, Folder: Hobley, C. W., ACIWLP/NYZS.

54. The history of the London Convention that follows comes from Hayden, *The International Protection of Wildlife*, 21–61; and MacKenzie, *The Empire of Nature*, 201–24.

55. This description of the provisions of the 1900 Convention is from Hayden, *The International Protection of Wildlife*, 37–38.

56. C. W. Hobley to Harold J. Coolidge, December 15, 1932, Box 4, Folder: Hobley, C. W., ACIWLP/WCS.

57. See, for example, Harold J. Coolidge to C. W. Hobley, June 30, 1933, Box 4, Folder: Hobley, C. W.; and Harold J. Coolidge to Madison Grant, September 1, 1933; and Harold J. Coolidge to Madison Grant, October 19, 1933, Box 3, Folder: Grant, Madison, ACIWLP/WCS.

58. Elisabeth Hone, *African Game Protection: An Outline of Existing Game Reserves and National Parks of Africa with Notes on Certain Species of Big Game Nearing Extinction, or Needing Additional Protection*, Special

Publication of the American Committee for International Wild Life Protection, Vol. 1, No. 3 (Cambridge, Mass.: American Committee for International Wild Life Protection, 1933), which was prepared by the assistant secretary of the American Committee.

59. Press Release, October 29, 1933, Box 1, Folder: Cadwalader, Charles, ACIWLP/WCS.

60. The account of the provisions of the 1933 Convention that follows is largely drawn from MacKenzie, *The Empire of Nature*, 216–33.

61. Anonymous, *The London Convention for the Protection of Fauna and Flora with Maps and Notes on Existing African Parks and Reserves*, Special Publication of the American Committee for International Wild Life Protection, No. 6 (Cambridge, Mass.: American Committee for International Wild Life Protection, 1935), on p. 6. See also the follow-up report, Anonymous, *Ratifications and Applications of the London Convention for the Protection of African Flora and Fauna*, Special Publication of the American Committee for International Wild Life Protection, No. 10 (New York: American Committee for International Wild Life Protection, 1940).

62. Marquess of Tavistock, "Alleged Excessive Collecting," *The Auk* 51 (1934): 428–29.

63. Frank M. Chapman, "Alleged Excessive Collecting," *The Auk* 51 (1934): 429–30. Several months after this initial exchange, Tavistock published a partial retraction of his charges, but he continued to insist that "the number of Masked Parrakeets taken was unnecessary and excessive." A copy of Tavistock's partial retraction is found in Frank M. Chapman, "Alleged Excessive Collecting," *The Auk* 52 (1935): 128.

64. On the committee's effort to get Chapman to respond, see Harold J. Coolidge to Thomas Barbour, March 1, 1935, Box 1, Folder: Barbour, Thomas, ACIWLP/WCS.

65. Frank M. Chapman, "The Whitney South Sea Expedition," *Science* 81, no. 2091 (1935): 95–97.

66. Details of the episode are found in Harold J. Coolidge to William Gregory, February 25, 1931, Box 3, Folder: Gregory, William, ACIWLP/WCS; and Harold J. Coolidge to Mr. Sherwood, December 1, 1931, copy appended to Harold J. Coolidge to Alexander Wetmore, December 2, 1931, RU 7006, Box 78, Folder 1, AWP/SIA. See also Imperato and Imperato, *They Married Adventure*, 160–64.

67. The first version, written by Leonard Stejneger, Herbert Friedmann, and Gerrit Smith, is found in Leonard Stejneger to Alexander Wetmore, June 7, 1934, RU 7006, Box 78, Folder 1, AWP/SIA.

68. Minutes, American Committee for International Wild Life Protection, Fifth Annual Meeting, December 13, 1934, p. 4, in RU 7006, Box 79, Folder 3, AWP/SIA.

69. See the discussions in Harold J. Coolidge to Alexander Wetmore, January 9, 1935, January 14, 1935, January 21, 1935, March 8, 1935, and May 21, 1935; Alexander Wetmore to Harold J. Coolidge, January 11, 1935, January 17, 1935, March 21, 1935, and May 17, 1935, RU 7006, Box 78, Folder 2, AWP/SIA.

70. See Joseph Grinnell, "Conserve the Collector," *Science* 41 (1915): 229–32.

71. Joseph Grinnell to Harold J. Coolidge, February 4, 1935, Box 3, Folder: Grinnell, Joseph, ACIWLP/WCS.

72. Cadwalader's concerns are summarized in Harold J. Coolidge to Alexander Wetmore, January 14, 1935, RU 7006, Box 78, Folder 2, AWP/SIA; see also Charles M. B. Cadwalader to Alexander Wetmore, March 9, 1935, in that same collection.

73. Secretary's Report, American Committee for International Wild Life Protection, Activities during 1935, December 12, 1935, ms. in Box 7, Folder 5, ACIWLP/WCS.

74. The episode is recounted in Mark V. Barrow, Jr., *A Passion for Birds: American*

Ornithology after Audubon (Princeton, N.J.: Princeton University Press, 1998), 150–53.

75. A brief account of the expedition may be found in Anonymous, "The Asiatic Primate Expedition," *Science* 85, no. 2192 (1937): 11–12.

76. Harold J. Coolidge to Henry Maurice, February 17, 1936, copy appended to Harold J. Coolidge to W. Reid Blair, February 25, 1939, Box 2, Folder: Coolidge, Harold J., January–May 1939, ACIWLP/WCS.

77. Harold J. Coolidge to P. J. Van Tienhoven, February 25, 1939, Box 27, Folder: ACIC Colleagues and Friends: P. J. Van Tienhoven, HJCP/HUA.

78. See Alexander Wetmore to Harold J. Coolidge, April 29, 1929, RU 7006, Box 78, Folder 2, AWP/SIA.

79. The biographical information that follows comes from Richard Mearns and Barbara Mearns, *The Bird Collectors* (San Diego, Calif.: Academic Press, 1998), 296–301; and Miriam Rothschild, *Dear Lord Rothschild: Birds, Butterflies and History* (Glennside, Pa.: Balaban Publishers, 1983).

80. Mearns and Mearns, *The Bird Collectors*, 297.

81. Lionel Walter Rothschild, *Extinct Birds: An Attempt to Unite in One Volume a Short Account of Those Birds Which Have Become Extinct in Historical Times—That Is, within the Last Six or Seven Hundred Years: To Which Are Added a Few Which Still Exist, but Are on the Verge of Extinction* (London: Hutchinson, 1907), which contains accounts of 166 birds.

82. The activities of Bangs and the MCZ Bird Department during this period are recorded in Mark V. Barrow, Jr., "Gentlemanly Specialists in the Age of Professionalization: The First Century of Ornithology at Harvard's Museum of Comparative Zoology," in *Contributions to the History of North American Ornithology*, ed. William E. Davis and Jerome A. Jackson (Cambridge, Mass.: Nuttall Ornithological Club, 1995), 55–94.

83. A detailed history of the project may be found in Walter J. Bock, "A Special Review: Peters' 'Check-List of Birds of the World' and a History of Avian Checklists," *The Auk* 107 (1990): 629–48.

84. James C. Greenway, *Extinct and Vanishing Birds of the World*, Special Publication No. 13 (New York: American Committee for International Wild Life Protection, 1958), v.

85. John C. Phillips, "An Attempt to List the Extinct and Vanishing Birds of the Western Hemisphere, with Some Notes on Recent Status, Location of Specimens, Etc.," in *Verhandlungen des VI. Internationalen Ornithologen-Kongresses in Kopenhagen 1926*, ed. F. Steinbacher (Berlin: Dornblüth, 1929), 503–34.

86. "Report on History of Project of American Committee for International Wild Life Protection for a Study of Vanishing and Recently Extinct Species of Mammals and Birds," ms. in Box 3, Folder: Harper Progress Reports, ACIWLP/WCS.

87. "Report of the American Committee for International Wild Life Protection for the Boone and Crockett Club Meeting, December 19, 1932," ms. in Box 7, Folder 5, ACIWLP/WCS.

88. "Report on History of Project," 3.

89. Writing to Hobley in 1935, Coolidge complained that "recovery in this country is very slow indeed. You made your trip at a very fortunate time indeed. In all five years since then with constant effort we have barely raised as much as you did on your single visit." Harold J. Coolidge to C. W. Hobley, January 23, 1935, Box 4, Folder: Hobley, C. W., ACIWLP/WCS.

90. Christopher J. Norment, "Francis Harper (1886–1972)," *Arctic* 53, no. 1 (2000): 72–75; Ralph S. Palmer, "Francis Harper," *The Auk* 90 (1973): 737–38; and Ralph S. Palmer, "Obituary: Francis Harper, 1886–1972," *Journal of Mammalogy* 54 (1973): 800–801.

91. Francis Harper to John C. Phillips,

July 29, 1935, August 6, 1935, Box 3, Folder: Harper, Francis, ACIWLP/WCS.

92. Record in "Minutes, American Committee for International Wild Life Protection, Sixth Annual Meeting, December 12, 1935," ms. in RU 7006, Box 79, Folder 3, AWP/SIA.

93. John C. Phillips to Francis Harper, March 27, 1936, Box 3: Folder: Harper, Francis, ACIWLP/WCS.

94. "Report on History of Project," 4.

95. See Harold J. Coolidge to Alexander Wetmore, May 18, 1936, RU 7006, Box 78, Folder 2, AWP/SIA. For more details about Leopold's vision of his conservation inventory, see chapter 6 below.

96. A copy of the form letter Harper used to solicit information from correspondence is found with John C. Phillips to Tordis Graim, August 3, 1936, Box 3, Folder: Graim, Tordis, ACIWLP/WCS.

97. Francis Harper, "An Inventory of Recently Extinct Animals," unpublished address for Annual Meeting of the American Society of Mammalogists, May 7, 1937, Box 3, Folder: Harper Progress Reports, ACIWLP/WCS.

98. John C. Phillips to Francis Harper, August 26, 1937, Box 3, Folder: Harper, Francis, ACIWLP/WCS.

99. Harold J. Coolidge to E. L. Gill, February 18, 1936, Box 1, Folder: Barbour, Thomas, ACIWLP/WCS.

100. John C. Phillips to Tordis Graim, May 4, 1937, Box 2, Folder: Graim, Tordis, ACIWLP/WCS.

101. Francis Harper to John C. Phillips, August 25, 1937, Box 3, Folder: Harper, Francis, ACIWLP/WCS.

102. Thomas Barbour to Harold J. Coolidge, December 7, 1937, Box 3, Folder: Colleagues and Friends: Barbour, Thomas, HJCP/HUA.

103. Fairfield Osborn, "The Work of the New York Zoological Society," *Science* 86, no. 2218 (1937): 6–7.

104. On the negotiations between the American Committee and the New York Zoological Society, see Harold J. Coolidge to Alexander Wetmore, December 3, 1937, RU 7006, Box 78, Folder 2, AWP/SIA.

105. Francis Harper, "Progress Report on Investigation of Extinct and Vanishing Mammals," ca. December 1936, ms. in Box 3, Folder: Harper Progress Reports, ACIWLP/WCS.

106. Harold J. Coolidge to Francis Harper, November 21, 1938, Box 27, Folder: ACIC Harper Report, June 1938–June 1939, HJCP/HUA.

107. Harper, "Progress Report"; Harold J. Coolidge to Francis Harper, December 28, 1938; Francis Harper to Harold J. Coolidge, January 3, 1939, Box 27, Folder: ACIC Harper Report, June 1938–June 1939, HJCP/HUA.

108. Francis Harper to C. M. B. Cadwalader, April 17, 1939, Box 27, Folder: ACIC Harper Report, June 1938–June 1939, HJCP/HUA; he would eventually complete all but seventeen of the Old World forms.

109. Allen breezed through the project. See his "Progress Report on American Mammals Recently Extinct or Nearing Extinction," May 19, 1939, Box 3, Folder: Colleagues and Friends: Allen, Glover, HJCP/HUA; Glover Allen to Harold J. Coolidge, August 23, 1939, Box 27, Folder: ACIC Harper Report, July 1939–42, HJCP/HUA; Glover Allen, "Report to the Committee on Wildlife Protection," December 7, 1939, Box 3, Folder: Harper-Allen Report Recipients, ACIWLP/WCS; Harold J. Coolidge to W. R. Blair, January 23, 1940, Box 27, Folder: ACIC: Reports, Contributions, Solicitations, 1935–42, HJCP/HUA.

110. C. M. B. Cadwalader to Harold J. Coolidge, October 9, 1939, October 17, 1939, November 14, 1939, February 16, 1940, October 1, 1940; Harold J. Coolidge to C. M. B. Cadwalader, July 3, 1940; Harold J. Coolidge, "Statement of Harper-Allen Report," October 18, 1940; Box 27, Folder: ACIC Harper Report, July 1939–42, HJCP/HUA; "Note Concerning the Harper Report," October 17, 1939, Box 3, Folder: Harper-

Allen Report Recipients, ACIWLP/WCS; C. M. B. Cadwalader to Childs Frick, November 7, 1940, Box 3, Folder: Francis Harper, ACIWLP/WCS and Box 27, Folder: ACIC Harper Report, July 1939–42, HJCP/HUA; Alexander Wetmore to Harold J. Coolidge, January 27, 1941, Box 3, Folder: ACIC Colleagues and Friends: Alexander Wetmore, HJCP/HUA.

111. Glover M. Allen, *Extinct and Vanishing Mammals of the Western Hemisphere, with the Marine Species of All the Oceans,* Special Publication No. 11 (New York: American Committee for International Wild Life Protection, 1942).

112. Allen, *Extinct and Vanishing Mammals,* viii–ix.

113. Allen, *Extinct and Vanishing Mammals,* 2.

114. Francis Harper, *Extinct and Vanishing Mammals of the Old World,* Special Publication No. 12 (New York: American Committee for International Wild Life Protection, 1945).

115. Harper, *Extinct and Vanishing Mammals,* v–vi.

116. Harper, *Extinct and Vanishing Mammals,* 22.

117. W. L. McAtee, "Review of Francis Harper, Extinct and Vanishing Mammals of the Old World," *Science* 102, no. 2646 (1945): 287–88.

118. Charles Elton, "Inquest of the World's Mammal Fauna," *Journal of Animal Ecology* 14, no. 2 (1945): 155–56. Additional reviews include Anonymous, "Review of Glover M. Allen, Extinct and Vanishing Mammals of the Western Hemisphere," *Wildlife Review,* no. 36 (1943): 6; Anonymous, "Review of Francis Harper, Extinct and Vanishing Mammals of the Old World," *Wildlife Review,* no. 44 (1945): 9–10; Lewis A. Follansbee, "Review of Glover M. Allen, Extinct and Vanishing Mammals of the Western Hemisphere," *Journal of Mammalogy* 24, no. 3 (1943): 405–6; A. H. Schultz, "Review of Francis Harper, Extinct and Vanishing Mammals of the Old World," *Quarterly Review of Biology* 21, no. 1 (1946): 93–94; and Karl P. Schmidt, "Review of Glover M. Allen, Extinct and Vanishing Mammals of the Western Hemisphere," *American Midland Naturalist* 30, no. 1 (1943): 269.

119. G. S. Meyers to the American Committee for International Wild Life Protection, July 6, 1945, RU 7006, Box 78, Folder: Extinct and Vanishing Fish Volume, AWP/SIA.

120. Alexander Wetmore to Harold J. Coolidge, August 17, 1945, RU 7006, Box 78, Folder: Extinct and Vanishing Fish Volume, AWP/SIA.

121. Childs Frick to Alexander Wetmore, October 31, 1945, RU 7006, Box 78, Folder: Extinct and Vanishing Fishes Volume, AWP/SIA.

122. Greenway, *Extinct and Vanishing Birds of the World.*

CHAPTER 6

1. In an act of incredible generosity, Kurk Dorsey shared his archival notes and two unpublished papers dealing with the history of the Migratory Bird Treaty with Mexico, both of which have been key to my account of this important agreement. See Kurkpatrick Dorsey, "Diplomats and Vertebrates: International Wildlife Protection Treaties in the Americas" (paper presented at the New Hampshire International Seminar, Durham, N.H., February 7, 1997); and Kurkpatrick Dorsey, "Hawks and Doves: The 1936 Migratory Bird Treaty with Mexico" (paper presented at the 1995 Annual Meeting of the Society for Historians of American Foreign Relations, Annapolis, Md., June 21–24, 1995).

2. The biographical sketch that follows comes from Victoria Cooper and William E. Cox, eds., *Guide to the Papers of Alexander Wetmore, ca. 1848–1979, and Undated* (Washington, D.C.: Archives and Special Collections of the Smithsonian Institution, 1990); Paul H. Oehser, "In Memoriam: Alexander Wetmore," *The Auk* 97 (1980): 608–15;

S. Dillon Ripley and James A. Steed, "Alexander Wetmore: June 18, 1886–December 7, 1978," *Biographical Memoirs of the National Academy of Sciences* 56 (1987): 597–626; and John K. Terres, "Smithsonian 'Bird Man': A Biographical Sketch of Alexander Wetmore," *Audubon Magazine* 50 (1948): 161–67.

3. Terres, "Smithsonian 'Bird Man,'" 160.

4. Oehser, "In Memoriam," 612.

5. A copy of this report may be found in Alexander Wetmore Papers, RU 7006, Box 140, Folder 4 Smithsonian Institution Archives, Washington, D.C. (hereafter cited as AWP/SIA).

6. On U.S. relations with Latin America during this period, see Lars Schoultz, *Beneath the United States: A History of U.S. Policy toward Latin America* (Cambridge, Mass.: Harvard University Press, 1998); Peter H. Smith, *Talons of the Eagle: Dynamics of U.S.–Latin American Relations* (New York: Oxford University Press, 1996).

7. Quoted in Dorsey, "Diplomats and Vertebrates," 10.

8. See Frank Sulloway, "Darwin and His Finches: The Evolution of a Legend," *Journal of the History of Biology* 15 (1982): 1–53; Jerome Weiner, *The Beak of the Finch: A Story of Evolution in Our Time* (New York: Alfred A. Knopf, 1994); and especially Edward J. Larson, *Evolution's Workshop: God and Science in the Galápagos Islands* (New York: Basic Books, 2001).

9. These are discussed in detail in Larson, *Evolution's Workshop*, 114–43.

10. William Beebe, *Galápagos: World's End* (New York: G. P. Putnam's Sons, 1924), on pp. xi–xii. On Beebe's life and work, see Robert Henry Welker, *Natural Man: The Life of William Beebe* (Bloomington: Indiana University Press, 1975); and Carol Grant Gould, *The Remarkable Life of William Beebe: Explorer and Naturalist* (Washington, D.C.: Island Press, 2004).

11. Larson, *Evolution's Workshop*, 55.

12. Charles Haskins Townsend, "Impending Extinction of the Galápagos Tortoises," *Zoological Society Bulletin* 27, no. 2 (1924): 55–56. Two years earlier, Barton Warren Evermann had called attention to the threatened status of the Galápagos tortoise in "The Conservation of Mammals and Other Vanishing Animals of the Pacific," *Scientific Monthly* 14, no. 3 (1922): 261–67, on p. 265.

13. Biographical information comes from Charles Haskins Townsend, "Old Times with the Birds: Autobiographical," *The Condor* 29 (1927): 224–32; and William Bridges, *Gathering of Animals: An Unconventional History of the New York Zoological Society* (New York: Harper and Row, 1974), 161–64, 192–211.

14. Townsend, "Old Times with the Birds," 225.

15. Townsend, "Impending Extinction," 55–56.

16. Charles Haskins Townsend, "The Galápagos Islands Revisited," *Bulletin of the New York Zoological Society* 31 (1928): 148–69, on p. 156. See also Charles Haskins Townsend, "The Galápagos Tortoises in Their Relation to the Whaling Industry: A Study of Old Logbooks," *Zoologica* 4, no. 3 (1925). 55–135; and Charles Haskins Townsend, "The Whaler and the Tortoise," *Scientific Monthly* 21, no. 2 (1925): 166–72.

17. Townsend, "The Whaler and the Tortoise," 169.

18. Townsend, "The Galápagos Islands Revisited," 151.

19. Bridges, *Gathering of Animals*, 431. See also Charles Haskins Townsend, "Giant Tortoises: Nearing Extinction in the Galápagos, They Are Being Propagated in the United States," *Scientific American* 144 (1931): 42–44.

20. Larson, *Evolution's Workshop*, 166.

21. Charles Haskins Townsend, "The Astor Expedition to the Galápagos Islands," *Bulletin of the New York Zoological Society* 33, no. 4 (1930): 133–55, on pp. 135 and 138; see also Bridges, *Gathering of Animals*, 432.

22. Townsend, "The Astor Expedition," 153.

23. Biographical information on Swarth comes from M. E. Davidson, "Harry Schelwaldt Swarth," *Science* 82, no. 2137 (1935): 562–63; and Larson, *Evolution's Workshop*, 162–67.

24. Larson, *Evolution's Workshop*, 163; Harry S. Swarth, "The Avifauna of the Galápagos Islands," *Occasional Papers of the California Academy of Sciences* 18 (1931): 1–299, on pp. 29–30.

25. See G. Dallas Hanna, "The Templeton Crocker Expedition of the California Academy of Sciences," *Science* 76, no. 1974 (1932): 375–77.

26. See H. S. Swarth, "Statement Regarding the Fauna and Flora Refuge upon the Galápagos Islands," ms. appended to Betty Hone to Alexander Wetmore, December 26, 1933, RU 7006, Box 79, Folder 2, AWP/SIA. Swarth also submitted a copy of his proposal to the Ecuadorian government. See Anonymous, "Urges Use of Galápagos: Scientist Would Set Aside Islands as Wild Life Sanctuary," *New York Times*, March 21, 1933, 15.

27. Herbert Friedmann, "In Memoriam: Robert Thomas Moore," *The Auk* 81 (1964): 326–31; and Anonymous, "Robert T. Moore Dead," *New York Times*, November 3, 1958, 37.

28. See the announcement in "Minutes, American Committee for International Wild Life Protection, Fourth Annual Meeting," December 18, 1933, copy in RU 7006, Box 79, Folder 3, AWP/SIA.

29. See, for example, Alexander Wetmore to Harold J. Coolidge, April 27, 1933; and Harold J. Coolidge to Alexander Wetmore, May 16, 1933, RU 7006, Box 79, Folder 2, AWP/SIA. Thomas Barbour played a leading role in founding this newly created biological station, which opened in 1924, and several of the institutions that were represented on the American Committee helped to finance it. See Joel B. Hagen, "Problems in the

Institutionalization of Tropical Biology: The Case of Barro Colorado Island Biological Laboratory," *History and Philosophy of the Life Sciences* 12, no. 2 (1990): 225–47.

30. In September 1933, Coolidge asked Madison Grant not to back the idea until the committee could check it out more fully: "We have reason to believe that there is an economic money-making proposition behind it all, about which Dr. Swarth may know nothing." Harold J. Coolidge to Madison Grant, September 1, 1933, Box 3, American Committee for International Wild Life Protection (Secretary's Office), World Conservation Society, New York (hereafter cited as ACI-WLP/WCS). See also Alexander Wetmore, untitled note, December 28, 1933, RU 7006, Box 78, Folder 1, AWP/SIA.

31. Copies of the Moore/Egas proposal, the American Committee's resolution, and the correspondence it used to promote the proposal are attached to Betty Hone to Alexander Wetmore, May 29, 1934, RU 7006, Box 79, Folder 2, AWP/SIA.

32. Alexander Wetmore to Harold J. Coolidge, June 1, 1934, RU 7006, Box 79, Folder 2, AWP/SIA.

33. The basic terms of the decree are outlined in Robert T. Moore, "The Protection and Conservation of the Zoological Life of the Galápagos Archipelago," *Science* 82, no. 2135 (1935): 519–21, which is also the source of all the direct quotations in this paragraph. A copy of the actual decree itself is appended to Harold J. Coolidge to Alexander Wetmore, June 28, 1935, RU 7006, Box 79, Folder 2, AWP/SIA.

34. See Anonymous, "Victor Wolfgang von Hagen," in *Current Biography 1942*, ed. Maxine Block (New York: H. W. Wilson, 1942), 859–60.

35. Details of the original plans for expedition come from Anonymous, "Scientific Notes and News," *Science* 78, no. 2034 (1933): 575–78; Anonymous, "Memorial to Charles Darwin," *Science* 81, no. 2112 (1935): 608;

Wolfgang von Hagen to Robert Moore, April 2, 1935, appended to Harold J. Coolidge to Alexander Wetmore, June 28, 1935; Robert T. Moore to BAAS Galápagos Committee, April 19, 1937 (draft), appended to Robert T. Moore to Alexander Wetmore, April 15, 1937; and Harold J. Coolidge to Julian Huxley, August 6, 1935, appended to Harold J. Coolidge to Alexander Wetmore, August 12, 1935, RU 7006, Box 79, Folder 2, AWP/SIA.

36. Wolfgang von Hagen to Harry S. Swarth, December 28, 1935, appended to Harold J. Coolidge to Alexander Wetmore, February 24, 1936, RU 7006, Box 79, Folder 2, AWP/SIA.

37. The episode is mentioned in Anonymous, "Victor Wolfgang von Hagen," 859, and Anonymous, "Colorful Honduran Birds Arrive for Zoo; Shipment of Nine Quetzals Is First to U.S.," *New York Times,* October 30, 1937, 2. See also Victor Wolfgang von Hagen, "Capturing the Royal Bird of the Aztecs," *Travel* 72 (December 1938): 38–41, 55, 57.

38. Their adventures are chronicled in a long series of articles and two books: Victor Wolfgang von Hagen, *Ecuador the Unknown: Two and a Half Years' Travels in the Republic of Ecuador and the Galápagos Islands* (London: Jarrods Publishers, 1939); and Victor Wolfgang von Hagen, *Ecuador and the Galápagos Islands* (Norman: University of Oklahoma Press, 1949).

39. See the announcement in Anonymous, "Galápagos Wild Life: A Permanent Research Station Proposed for Its Protection," *New York Times,* June 14, 1936, XX9; and Anonymous, *Decreto declarando parques nacionales de reserva, las islas de Galápagos: no. 31, de 14 de mayo de 1936* (Quito, Ecuador: Comité Provisional para la Protección de la Fauna del Archipiélago de Colón, 1936), 21.

40. See Anonymous, "Ecuador Claims Islands," *New York Times,* December 26, 1936, 15; Anonymous, "31 Canal Zone Planes Will Drill in Pacific," *New York Times,* February 3, 1936, 11; George M. Lauderbaugh, "Galápa-

gos Islands," in *Encyclopedia of Latin American History and Culture,* ed. Barbara A. Tenenbaum (New York: Charles Scribner's Sons, 1996), 5; and Larson, *Evolution's Workshop,* 175.

41. On Ecuador's political and economic turmoil during this period, see Linda Alexander Rodríguez, "Ecuador since 1830," in *Encyclopedia of Latin American History and Culture,* ed. Barbara A. Tenenbaum (New York: Scribner's, 1996), 450–55.

42. Harold J. Coolidge to Alexander Wetmore, December 18, 1935, RU 7006, Box 79, Folder 2, AWP/SIA.

43. Von Hagen's visit to Páez is described in von Hagen, *Ecuador the Unknown,* 214–21.

44. A copy of the legislation is found in Anonymous, *Decreto Declarando Parques Nacionales de Reserva,* 3–6 (in Spanish), 15–18 (in English).

45. Alexander Wetmore to Harold J. Coolidge, June 27, 1935 and August 9, 1935, RU 7006, Box 79, Folder 2, AWP/SIA.

46. Harold J. Coolidge to Julian Huxley, August 6, 1935, appended to Harold J. Coolidge to Alexander Wetmore, August 12, 1935, RU 7006, Box 79, Folder 2, AWP/SIA. Coolidge noted that von Hagen claimed not to have known that Moore was involved with Galápagos conservation even though Coolidge had informed him of such.

47. Alexander Wetmore to Robert T. Moore, February 5, 1937, RU 7006, Box 79, Folder 1, AWP/SIA.

48. Robert T. Moore to Alexander Wetmore, January 28, 1937, and Frederico Páez to V. M. Egas, May 6, 1937, appended to Robert T. Moore to Alexander Wetmore, June 28, 1937, RU 7006, Box 78, Folder 1, AWP/SIA.

49. Details of these and other charges are found in Robert T. Moore to Alexander Wetmore, May 5, 1937, RU 7006, Box 79, Folder 1, AWP/SIA.

50. A copy of the article is found with Robert T. Moore to Alexander Wetmore, May 5, 1937, while the idea of widely circulating it is mentioned in a document attached to Robert T. Moore to Alexander Wetmore, RU 7006, Box 79, Folder 1, AWP/SIA.

51. O. J. R. Howarth to Secretary, ACI-WLP, March 12, 1937, and Robert T. Moore to BAAS Galápagos Committee, April 19, 1937 (draft), attached to Robert T. Moore to Alexander Wetmore, April 15, 1937, RU 7006, Box 79, Folder 1, AWP/SIA.

52. O. J. R. Howarth to Robert T. Moore, April 13, 1937, and "Galápagos Committee Memorandum," appended to Robert T. Moore to Alexander Wetmore, May 5, 1937; Robert T. Moore to Alexander Wetmore, September 4, 1937; RU 7006, Box 79, Folder 1, AWP/SIA.

53. Robert Moore to Alexander Wetmore, September 4, 1937, RU 7006, Box 79, Folder 1, AWP/SIA.

54. John C. Phillips to Alexander Wetmore, December 22, 1937, and Robert T. Moore to Members of the Galápagos Committee, May 5, 1938; RU 7006, Box 79, Folder 1, AWP/SIA.

55. On the founding of the Darwin Research Station, see Larson, *Evolution's Workshop*, 175–96.

56. Skottsberg's and Lack's reports are appended to Alexander Wetmore to S. W. Boggs, December 27, 1939, RU 7006, Box 99, Folder 4, AWP/SIA.

57. Larson, *Evolution's Workshop*, 175–76.

58. Alexander Wetmore to Harold J. Coolidge, July 7, 1943, RU 7006, Box 79, Folder 1, AWP/SIA.

59. Alexander Wetmore to Childs Frick, July 7, 1943, RU 7006, Box 79, Folder 1, AWP/SIA.

60. Dorsey, "Hawks and Doves," 2–3; Lane Simonian, *Defending the Land of the Jaguar: A History of Conservation in Mexico* (Austin: University of Texas Press, 1995), 78.

61. E. W. Nelson to T. Gilbert Pearson, February 21, 1920, Records of the Department of Agriculture, Bureau of the Biological Survey, RG 22, Box 40, U.S. National Archives, College Park, Md. (hereafter cited as BBS/NA).

62. The biographical sketch of Goldman that follows comes from William E. Cox, "Edward A. Goldman," in *Biographical Dictionary of American and Canadian Naturalists and Environmentalists*, ed. Keir B. Sterling et al. (Westport, Conn.: Greenwood Press, 1997), 312–14; Ludwig Caminita, Jr., "Naturalist," *American Wildlife* 29 (1940): 168–72; and Henry M. Reeves, "Edward Alphonso Goldman," in *Flyways: Pioneering Waterfowl Management in North America*, ed. A. S. Hawkins et al. (Washington, D.C.: Government Printing Office, 1984), 75–81. A convenient source of biographical information on Nelson is James R. Glenn, "Edward William Nelson," in *Biographical Dictionary of American and Canadian Naturalists and Environmentalists*, ed. Keir B. Sterling et al. (Westport, Conn.: Greenwood Press, 1997), 571–73.

63. Wetmore's pioneering studies are detailed in Henry M. Reeves, "Alexander Wetmore," in *Flyways: Pioneering Waterfowl Management in North America*, ed. A. S. Hawkins et al. (Washington, D.C.: Government Printing Office, 1984), 75–81. On the development of bird banding, see Frederick C. Lincoln, "Bird Banding," in *Fifty Years' Progress of American Ornithology, 1883–1933*, ed. Frank M. Chapman and T. S. Palmer (Lancaster, Pa..: American Ornithologists' Union, 1933), 65–87; and Mark V. Barrow, Jr., *A Passion for Birds: American Ornithology after Audubon* (Princeton, N.J.: Princeton University Press, 1998), 169–72.

64. On Lincoln's life and work, see John K. Terres, "Big Brother to the Waterfowl," *Audubon Magazine* 49 (1947): 150–58; and Ira N. Gabrielson, "Obituary," *The Auk* 79 (1962): 495–99.

65. See Frederick C. Lincoln, *The Waterfowl Flyways of North America*, U.S. Department of Agriculture, Circular No. 342 (Washington, D.C.: Government Printing Office, 1935); Frederick C. Lincoln, *The Migra-*

tion of North American Birds, U.S. Department of Agriculture, Circular No. 363 (Washington, D.C.: Government Printing Office, 1935); and A. S. Hawkins et al., eds., *Flyways: Pioneering Waterfowl Management in North America* (Washington, D.C.: Government Printing Office, 1984). See also Robert M. Wilson, "Directing the Flow: Migratory Waterfowl, Scale, and Mobility in Western North America," *Environmental History* 7, no. 2 (2002): 247–66.

66. Dorsey, "Hawks and Doves," 4–6.

67. Simonian, *Defending the Land of the Jaguar*, 85.

68. Dorsey, "Hawks and Doves," 6, also stresses the importance of the appointment of Josephus Daniels as ambassador to Mexico.

69. Cárdenas's main concession was minor: a willingness to negotiate bird protection and fisheries regulation in tandem, not as a single agreement. Dorsey, "Hawks and Doves," 7.

70. See E. A. Goldman to Henry Norweb, April 28, 1935, RG 22, Box 45, BBS/NA.

71. The list that follows is adapted from Dorsey, "Hawks and Doves," 8.

72. Rexford Tugwell to Cordell Hull, December 10, 1934, RG 22, Box 45, BBS/NA.

73. Henry Norweb to Cordell Hull, April 20, 1935, RG 22, Box 34, BBS/NA.

74. Ding Darling to Rexford Tugwell, June 14, 1935, RG 22, Box 45, BBS/NA, quoted in Dorsey, "Hawks and Doves," 9. See also Ding Darling to Rexford Tugwell, May 31, 1935, in that same collection.

75. Dorsey, "Hawks and Doves," 9.

76. Rosalie Edge, *The Migratory Bird Treaty with Mexico: Double-Crossing Conservationists and Migratory Birds* (New York: Emergency Conservation Committee, 1936), 1, claims it was Zinzer who made the announcement at the banquet on February 7, but there is no record of him doing so in the first *Proceedings of the North American Wildlife Conference* (1936); there is, however, report of an announcement by Darling (on p. 269).

77. A copy of the treaty is found in "Convention between the United States of America and Mexico for the Protection of Migratory Birds and Game Animals," *U. S. Statutes* 50, pt. 2 (1937): 1311–16.

78. "Convention between the United States of America and Mexico," 1311.

79. Rexford Tugwell to Cordell Hull, February 6, 1936, General Records of the Department of State, RG 59, File 711.129, National Archives, College Park, Md. (hereafter cited as GRDS/NA); quoted in Dorsey, "Hawks and Doves," 10.

80. The phrase comes from the subtitle of the pamphlet, Edge, *The Migratory Bird Treaty with Mexico*.

81. [Harold J. Coolidge?] to Julian Huxley, February 11 1935, Box 4, ACIWLP/WCS.

82. "Notes on Research Program of American Committee," 4, ms. in Harold J. Coolidge Papers, HUGFP 78.10, Box 26, Folder: ACIC Correspondence-1941, Harvard University Archives, Cambridge, Mass. (hereafter cited as HJCP/HUA). This is undated, but internal evidence points to a creation date of early 1936.

83. On the Pan American Institute, see Pan American Institute of Geography and History, *The Pan American Institute of Geography and History: Its Creation, Development and Current Program, 1929–1954* (Mexico: Pan American Institute of Geography and History, 1954); and Leopold F. Rodríguez, "Pan-American Institute of Geography and History," in *Latin American History and Culture*, ed. Barbara A. Tenenbaum (New York: Charles Scribner's Sons, 1996), 4:274; on the resolutions, see Wallace W. Atwood, "The Second General Assembly of the Pan American Institute of Geography," *Science* 82, no. 2135 (1935) 521–22; on Atwood, see Trent A. Mitchell, "Wallace Walter Atwood," *American National Biography* 1 (1999): 736–38.

84. "Notes on Research Program of American Committee," 5.

85. See, for example, "Minutes, American

Committee for International Wild Life Protection Seventh Annual Meeting," December 18, 1936, ms. in RU 7006, Box 79, Folder 3, AWP/SIA.

86. John C. Phillips to Alexander Wetmore, August 17, 1936, RU 7006, Box 78, Folder 2, AWP/SIA. The Pan American Union emerged out of a series of meetings between U.S. and Latin American officials, beginning with the First International Conference of American States, held in Washington from 1889 to 1890. The hope was to set up a body that would promote stability, cooperation, and solidarity in the hemisphere by addressing the arbitration of financial and territorial claims, the codification of international law, copyrights and patents, and other legal, commercial, and social issues of common concern between American nations. While agreements were made on each of these fronts, some Latin Americans worried about U.S. domination of the Pan American Union, which was so named in 1910. Formal meetings of the organization were held every year in a variety of Latin American capitals, but its permanent headquarters was in Washington, D.C., its activities were largely financed by the State Department, and its early years coincided with a period of increasing U.S intervention in the affairs of Central American and Caribbean nations. See Richard V. Salisbury, "Pan-Americanism," in *Encyclopedia of Latin American History and Culture*, ed. Barbara A. Tenenbaum (New York: Simon and Shuster Macmillan, 1996), 4:274–75; and L. S. Rowe, *The Pan American Union and Pan American Conferences* (Washington, D.C.: Pan American Union, 1940).

87. John C. Phillips to Alexander Wetmore, August 17, 1936, RU 7006, Box 78, Folder 2, AWP/SIA.

88. See "Notes on J. C. Phillips' Visit to Dr. Atwood," July 30, 1937, Box 1, ACIWLP/WCS.

89. John C. Phillips to Wallace Atwood, August 2, 1937, Box 1, ACIWLP/WCS.

90. Coolidge and the rest of the American Committee's Executive Committee ruffled feathers within the organization when they submitted an initial draft to the State Department without Wetmore's input. See Harold J. Coolidge to Alexander Wetmore, October 4, 1938, RU 7006, Box 78, Folder 2, AWP/SIA.

91. Harold J. Coolidge to Alexander Wetmore, October 6, 1938, RU 7006, Box 78, Folder 2, AWP/SIA.

92. A copy of this packet may be found in Box 7, ACIWLP/WCS.

93. See the copy in Box 7, Folder: Newsletter, 1939, ACIWLP/WCS; emphasis added.

94. See "Memorandum, Questions to be Discussed with Dr. Wetmore and Dr. Rowe on Washington Visit," February 21 [1939], Box 25, Folder: ACIC Correspondence, 1939, HJCP/HUA.

95. Harold J. Coolidge to W. Reid Blair, February 24, 1939, Box 25, Folder: ACIC Correspondence, 1939, HJCP/HUA.

96. National governments also appointed individual committees of experts to gather information for the official delegates of the Inter-American Committee of Experts.

97. Harold J. Coolidge to W. Reid Blair, February 14, 1939, Box 25, Folder: ACIC Correspondence, 1939, HJCP/HUA.

98. Harold J. Coolidge to Robert C. Murphy, March 29, 1939, Box 25, Folder: ACIC Correspondence, 1939, HJCP/HUA.

99. Robert Cushman Murphy, "S.O.S. for a Continent," *Natural History* 43 (1939): 150–51.

100. The article was published as Harold J. Coolidge, "Protección internacional de los recursos naturales," *Bulletin of the Pan American Union* 73, no. 7 (1939): 399–412. The quotes are from the English ms., Harold J. Coolidge, Jr., "International Nature Protection Article for the Bulletin of the Pan American Union," Box 2, Folder: Coolidge, H. J., 4. Photographs for Pan American Union International Nature Protection Article," ACIWLP/WCS.

101. Coolidge, "International Nature Protection," 1.

102. Coolidge, "International Nature Protection," 5.

103. Coolidge, "International Nature Protection," 5–9.

104. Coolidge, "International Nature Protection," 10.

105. Coolidge, "International Nature Protection," 14–15.

106. Coolidge, "International Nature Protection," 15.

107. Details are appended to Harold J. Coolidge to Alexander Wetmore, April 6, 1939, RU 7006, Box 78, Folder 3, AWP/SIA.

108. The conference and meeting of the Inter-American Committee of Experts actually took place in May 1940.

109. See Alexander Wetmore to Harold J. Coolidge, May 22, 1939, and "Notes on Special Meeting Called by Chairman of the Pan-American Committee," May 22, 1939, Box 25, Folder: ACIC Correspondence, 1939, HJCP/HUA.

110. "Report of the Pan American Committee Meeting," October 2, 1939, Box 25, ACIC Correspondence, 1939, HJCP/HUA, and Box 2, Folder: Harold J. Coolidge, June–December 1939, ACIWLP/WCS. Two days later, Coolidge met with José L. Colom, the chief of the division of agricultural cooperation of the Pan American Union, who had been charged with organizing the meeting of the Committee of Experts in May. After the meeting Coolidge reported with pride that the Pan American Committee's version of the convention "met with his entire approval."

111. In a letter urging him to reconsider, Coolidge credited him with "arousing the Pan-American interest which is now running so strong" within the American Committee. Moore answered that he hoped "the Pan-American Committee will be able to speed up activities, so something can be done for the Galápagos fauna before the forms, which are now in danger, shall go extinct." Robert Moore to Harold J. Coolidge, July 21, 1939, Box 25, Folder: ACIC Correspondence, 1939, HJCP/HUA.

112. Julian Huxley to Harold J. Coolidge, May 5, 1939, May 15, 1939, and June 29, 1939; Harold J. Coolidge to Julian Huxley, May 26, 1939 and July 13, 1939, Box 25, Folder: ACIC Correspondence, 1939, HJCP/HUA.

113. Harold J. Coolidge to Henry S. Maurice, October 13, 1939, Box 25, Folder: ACIC Correspondence, 1939, HJCP/HUA.

114. At about this time, Pearson became increasingly interested in Latin American conservation. See Thomas Gilbert Pearson, *Where the Game Laws Are Needed* (New York: U.S. Section of the International Committee for Bird Preservation, 1938), 2; John Baker to Harold J. Coolidge, February 14, 1939, Box 25, Folder: ACIC Correspondence, 1939, HJCP/HUA; Ruth Dauchy to Harold J. Coolidge, August 17, 1939, Box 25, Folder: ACIC Correspondence, 1939, HJCP/HUA; T. Gilbert Pearson, "Notes on the Games Laws of the Argentina Republic," ms. attached to Ruth Dauchy to Alexander Wetmore, February 1, 1940, RU 7006, Box 78, Folder: International Wild Life Protection, American Committee, Correspondence, 1939–42, AWP/SIA; T. Gilbert Pearson, "Some Notes on Game Protection in Brazil," Box 25, Folder: ACIC Correspondence, 1939, HJCP/HUA; Harold J. Coolidge to T. Gilbert Pearson, June 18, 1940, Box 25, Folder: ACIC Correspondence, 1940, HJCP/HUA.

115. See, for example, John C. Phillips to Tordis Graim, July 20, 1936, Box 3, ACIWLP/WCS; John C. Phillips to John Baker, July 4 [, 1936] and July 8, 1936, Box A-143, Folder: John C. Phillips, National Audubon Society Records, New York Public Library, New York (hereafter cited as NASR/NYPL).

116. See, for example, John Baker to Kermit Roosevelt, July 28, 1936, Box B-8, NASR/NYPL.

117. John Baker to Kermit Roosevelt, December 16, 1937, Box B-8, NASR/NYPL.

118. Harold J. Coolidge to Alexander Wetmore, January 26, 1940, Box 25, Folder: ACIC Correspondence, 1940, HJCP/HUA.

119. Harold J. Coolidge to Childs Frick, May 28, 1940, Box 25, Folder: ACIC Correspondence, 1940, HJCP/HUA.

120. A copy of the convention may be found in *Congressional Record,* vol. 87, pt. 3, 77th Congress, 1st sess., 3070–71.

121. Harold J. Coolidge, Jr., "A New Pan American Treaty," *Science* 92, no. 2394 (1940), on pp. 458 and 459.

122. Anonymous, *Convention on Nature Protection and Wild Life Preservation in the Western Hemisphere* (New York: American Committee for International Wild Life Protection, n.d.), on p. 12.

123. William G. Sheldon, "A Pan-American Treaty on Nature Protection," *American Wildlife* 30 (1941), on p. 41.

124. Earl of Onslow to Harold J. Coolidge, September 1, 1941, Box 25, Folder: ACIC Correspondence, 1941, HJCP/HUA.

125. Harold J. Coolidge to William Phillips, October 24, 1940, Box 25, Folder: ACIC Correspondence, 1940, HJCP/HUA.

126. Harold J. Coolidge to Alexander Wetmore, December 3, 1940; and Alexander Wetmore to Harold J. Coolidge, December 6, 1940, Box 25, Folder: ACIC Correspondence, 1941, HJCP/HUA.

127. Harold J. Coolidge to Lester Markel, January 22, 1941, Box 25, Folder: ACIC Correspondence, 1941, HJCP/HUA.

128. Harold J. Coolidge to Fairfield Osborn, October 24, 1940; Fairfield Osborn to Harold J. Coolidge, October 31, 1940, Box 25, Folder: ACIC Correspondence, 1941, HJCP/HUA.

129. Coolidge, "A New Pan American Treaty."

130. Biographical information on Vogt and his conservation activities may be found in Anonymous, "William Vogt," in *Current Biography 1953,* ed. Marjorie D. Candee (New York: H. W. Wilson, 1954), 638–40; Richard

Harmond, "William Vogt," *American National Biography* 22 (1999): 390–91; Stephen Fox, *The American Conservation Movement: John Muir and His Legacy* (Madison: University of Wisconsin Press, 1985), 197–98, 226, 307–10, 338; and Gregory Todd Cushman, "The Birth of the Environmental Technocrat in Peru: Marine Science, State-Led Development, and the Compañía Administradora del Guano, 1906–1964" (Ph.D. diss., University of Texas–Austin, 1999).

131. Bill Vogt to Harold J. Coolidge, June 22, 1940, Box 25, Folder: ACIC Correspondence, 1940, HJCP/HUA.

132. Harold J. Coolidge to Laurance Rockefeller, January 24, 1941, May 14, 1941, and July 29, 1941, Box 25, Folder: ACIC Correspondence, 1941, HJCP/HUA.

133. Goodspeed convinced the Conservation Committee of the New York Zoological Society to give $500 to the American Committee to pay for copies of the motion pictures and slides he planned to use to illustrate his lectures. "Minutes, Meeting of Executive Committee," November 18, 1941, Box 28, Folder: Minutes of Committee Meetings, HJCP/HUA; and Harold J. Coolidge to Childs Frick, November 17, 1941, Box 26, Folder: Colleagues and Friends: Childs Frick, HJCP/HUA.

134. "Minutes, American Committee for International Wild Life Protection," December 11, 1941, 5, attached to W. Reid Blair to Alexander Wetmore, January 5, 1942, RU 7006, Box 79, Folder 4, AWP/SIA.

135. "Minutes, American Committee for International Wild Life Protection," December 11, 1942, 4, Box 79, Folder 4, AWP/SIA.

136. William Vogt, *Road to Survival* (New York: W. Sloane Associates, 1948).

137. For a thorough analysis of these differing viewpoints and other relevant details about the negotiations that led to the agreement, see Keri E. Lewis, "Negotiating for Nature: Conservation Diplomacy and the Convention on Nature Protection and Wild-

life Preservation in the Western Hemisphere, 1929–1976" (Ph.D. diss., University of New Hampshire, 2007), which I learned about only after drafting this chapter.

138. Kathleen Rogers and James A. Moore, "Revitalizing the Convention on Nature Protection and Wild Life Preservation in the Western Hemisphere: Might Awakening a Visionary but 'Sleeping' Treaty Be the Key to Preserving Biodiversity and Threatened Natural Areas in the Americas?" *Harvard International Law Journal* 36 (1995): 465–508, on p. 466.

139. The arguments that follow are adapted from Dorsey, "Diplomats and Vertebrates," 16–18.

140. Lewis, "Negotiating Nature," chaps. 6–7, and conclusion, on p. 1.

CHAPTER 7

1. Francis B. Sumner, "The Need for a More Serious Effort to Rescue a Few Fragments of Vanishing Nature," *Scientific Monthly* 10, no. 3 (1920): 236–48, on pp. 237, 239, 240, and 241. On Sumner's view of the relationship between biology and society, see his earlier article: Francis B. Sumner, "Some Perils Which Confront Us as Scientists," *Scientific Monthly* 8, no. 3 (1919): 258–74.

2. Francis B. Sumner, "The Responsibility of the Biologist in the Matter of Preserving Natural Conditions," *Science* 54, no. 1385 (1921): 39–43.

3. Sumner, "Need for a More Serious Effort," 237–38.

4. Biographical information on Sumner may be found in Elizabeth Noble Shor, "Francis Bertody Sumner," *Dictionary of Scientific Biography*, 13:150–51; Carl L. Hubbs, "Francis Bertody Sumner," *Dictionary of American Biography, Supplement 3 (1941–1945)*, 752–53; Keir B. Sterling, "Francis Bertody Sumner," *American National Biography*, 21:141–42; and Francis B. Sumner, *The Life History of an*

American Naturalist (Lancaster, Pa..: Jacques Cattell Press, 1945).

5. Robert E. Kohler, *Landscapes and Labscapes: Exploring the Lab-Field Border in Biology* (Chicago: University of Chicago Press, 2002).

6. For a general historical discussion of biologists and their missionary zeal for applying their science, see Philip J. Pauly, *Biologists and the Promise of American Life: From Meriwether Lewis to Alfred Kinsey* (Princeton, N.J.: Princeton University Press, 2000). See also Mark V. Barrow, Jr., "Naturalists as Conservationists: American Scientists, Social Responsibility, and Political Activism before the Bomb," in *Science, History and Social Activism: A Tribute to Everett Mendelsohn*, ed. Garland C. Allen and Roy M. MacLeod (Dordrecht: Kluwer Academic Publishers, 2001), 217–33; John C. Merriam, "Some Responsibilities with Relation to Government," *Science* 80, no. 2087 (1934): 597–601; and John C. Merriam, "Science and Conservation," *Science* 79, no. 2057 (1934): 496–97.

7. See the discussion in the introduction.

8. The subject of how and why biological education and practice changed in the late nineteenth- and early twentieth-century America has long been a source of contention among historians of science. See, for example, Jane Maienschein, Ronald Rainger, and Keith Benson, "Introduction: Were American Morphologists in Revolt?" *Journal of the History of Biology* 14 (1981): 83–88; and the rest of the essays in the "Special Section on American Morphology at the Turn of the Century" appearing in that same issue; Ronald Rainger, Keith Benson, and Jane Maienschein, eds., *The American Development of Biology* (Philadelphia: University of Pennsylvania Press, 1988); and Kohler, *Landscapes and Labscapes*.

9. On the relationship between science, the establishment of graduate programs, and the development of the modern university in America, see Laurence R. Vesey, *The Emergence of the American University* (Chicago:

University of Chicago Press, 1965); Roger L. Geiger, *To Advance Knowledge: The Growth of American Research Universities, 1900–1940* (New York: Oxford University Press, 1986); and Robert E. Kohler, "The Ph.D. Machine: Building on the Collegiate Base," *Isis* 81 (1990): 638–62.

10. See the extensive discussion in Kohler, *Landscapes and Labscapes,* 23–40.

11. Charles C. Adams, "The New Natural History—Ecology," *American Museum Journal* 17 (1917): 491–94, on p. 492.

12. Convenient entry points into the large and growing literature on the history of ecology include Joe Cain, "Ecology," in *History of Modern Science,* ed. Brian S. Baigrie (New York: Charles Scribner's Sons, 2002), 3:44–68; Eugene Cittadino, "Ecology," in *The Oxford Companion to the History of Modern Science,* ed. J. L. Heilbron (Oxford: Oxford University Press, 2003), 229–32; and Joel B. Hagen, "Ecology," in *The History of Science in the United States: An Encyclopedia,* ed. Marc Rothenberg (New York: Garland, 2001), 168–69. Longer overviews include Robert P. McIntosh, *The Background of Ecology: Concept and Theory* (Cambridge: Cambridge University Press, 1985); Leslie Real and James Brown, eds., *Foundations of Ecology: Classic Papers with Commentaries* (Chicago: University of Chicago Press, 1991); Donald Worster, *Nature's Economy: A History of Ecological Ideas,* 2nd ed. (Cambridge: Cambridge University Press, 1994); Peder Anker, *Imperial Ecology: Environmental Order in the British Empire, 1895–1945* (Cambridge, Mass.: Harvard University Press, 2001); and Sharon E. Kingsland, *The Evolution of American Ecology, 1890–2000* (Baltimore: Johns Hopkins University Press, 2005).

13. Quoted in Worster, *Nature's Economy,* 192.

14. Charles Darwin, *On the Origin of Species* (Cambridge, Mass.: Harvard University Press, 1964 [1859]), 74.

15. Eugene Cittadino, "Ecology and the Professionalization of Botany in America, 1890–1905," *Studies in History of Biology* 4 (1980): 171–98.

16. Ronald C. Tobey, *Saving the Prairies: The Life Cycle of the Founding School of American Plant Ecology, 1895–1955* (Berkeley: University of California Press, 1981); and Richard A. Overfield, *Science with Practice: Charles E. Bessey and the Maturing of American Botany* (Ames: Iowa State University Press, 1993).

17. Clements laid out the basics of his methodological and theoretical framework for plant ecology in Frederic E. Clements, *Research Methods in Ecology* (Lincoln, Neb.: University Publishing, 1905); and Frederic E. Clements, *Plant Succession: An Analysis of the Development of Vegetation* (Washington, D.C.: Carnegie Institution of Washington, 1916). On his research and pervasive influence, see Tobey, *Saving the Prairies;* Joel B. Hagen, *An Entangled Bank: The Origins of Ecosystem Ecology* (New Brunswick, N.J.: Rutgers University Press, 1992); and Kohler, *Landscapes and Labscapes.*

18. See the series of papers in "Crossing the Borderlands: Biology at Chicago," a special issue of *Perspectives on Science* 1 (1993): 359–559; and Jane Maienschein, "Whitman at Chicago: Establishing a Chicago Style of Biology?" in *The American Development of Biology,* ed. Ronald Rainger, Keith Benson, and Jane Maienschein (Philadelphia: University of Pennsylvania Press, 1988), 151–83.

19. On Cowles and the critical importance of the Indiana Dunes to American ecology, see Eugene Cittadino, "A 'Marvelous Cosmopolitan Preserve': The Dunes, Chicago, and the Dynamic Ecology of Henry Cowles," *Perspectives on Science* 1 (1993): 520–59.

20. Worster, *Nature's Economy,* 207.

21. For a fascinating study of how U.S. Steel impacted the environment of Gary, see Andrew Hurley, *Environmental Inequalities: Class, Race, and Industrial Pollution in Gary, Indiana, 1945–1980* (Chapel Hill: University of North Carolina Press, 1995).

22. Charles C. Adams, *Guide to the Study of Animal Ecology* (New York: Macmillan, 1913). Sketches of Adams's work may be found in Ralph S. Palmer, "Resolution of Respect: Dr. Charles C. Adams (1873–1955)," *Bulletin of the Ecological Society of America* 37, no. 4 (1956): 103–5; and Paul B. Sears, "Charles C. Adams, Ecologist," *Science* 123, no. 3205 (1956): 974; on his scientific output, see H. D. Adams and T. S. Robinson, "An Ecological Bibliography of Charles Christopher Adams," *Occasional Papers of the Adams Center of Ecology* 2 (1961): 1–6.

23. Victor E. Shelford, *Animal Communities in Temperate America* (Chicago: University of Chicago Press, 1913); for a review of both books, see Anonymous, "Principles and Methods of Animal Ecology," *Journal of Ecology* 3 (1915): 56–64.

24. On Elton and his ecological legacy, see Peter Crowcroft, *Elton's Ecologists: A History of the Bureau of Animal Population* (Chicago: University of Chicago Press, 1991); David L. Cox, "Charles Elton and the Emergence of Modern Ecology" (Ph.D. diss., Washington University, 1979); Anker, *Imperial Ecology;* and McIntosh, *Background of Ecology*, 88–92.

25. Henry C. Cowles, "The Work of the Year 1903 in Ecology," *Science* 19, no. 493 (1904): 879–85, on p. 879.

26. On the formation and early history of the Ecological Society of America, see Victor E. Shelford, ed., "Handbook of the Ecological Society of America," *Bulletin of the Ecological Society of America* 1, no. 3 (1917): 1–57; Victor E. Shelford, "The Organization of the Ecological Society of America, 1914–19," *Ecology* 19, no. 1 (1938): 164–66; and Robert L. Burgess, "The Ecological Society of America: Historical Data and Some Preliminary Analyses," in *History of American Ecology*, ed. Frank N. Egerton (New York: Arno Press, 1977), 1–24. Cowles's call was issued as Henry C. Cowles, "A Proposed Ecological Society," *Science* 42, no. 1084 (1915): 496. On the history of the British Ecological Society, see John Sheail, *Seventy-Five Years in Ecology: The British Ecological Society* (Oxford: Blackwell Scientific, 1987).

27. Burgess, "The Ecological Society of America," 3–5.

28. Anonymous, "Proceedings: Business Meetings of the Ecological Society of America at Cleveland, Ohio, December 31, 1930 and January 1, 1931," *Ecology* 12, no. 2 (1931): 427–38, on p. 433.

29. The counts are based on searches in WorldCat and Dissertation Abstracts. During the period between 1903 and 1945, the University of Chicago led the way with sixty dissertations that included "ecology" in the title, Cornell came in second with fourteen dissertations, and Nebraska came in third with six.

30. Shelford, "The Organization of the Ecological Society of America." On Huntington and his ecological ideas, see Kingsland, *The Evolution of American Ecology*, 134–41.

31. M. M. Sherfy, "The National Park Service and the First World War," *Journal of Forest History* 22 (1978): 203–5.

32. Shelford and his work for the ESA preservation committee have been the subject of several historical studies, including Robert A. Croker, *Pioneer Ecologist: The Life and Work of Victor Ernest Shelford, 1877–1968* (Washington, D.C.: Smithsonian Institution Press, 1991); Sara F. Tjossem, "Preservation of Nature and Academic Respectability: Tensions in the Ecological Society of America, 1915–1979" (Ph.D. diss., Cornell University, 1994); and Geoffroy J. McQuilkin, "Saving the Living Laboratory: The Natural Areas Preservation Efforts of the Ecological Society of America" (senior honors thesis, Harvard University, 1991). I have relied on all three in my discussion of the committee's work. Other important biographical sketches include S. Charles Kendeigh, "Victor Ernest Shelford: Eminent Ecologist," *Bulletin of the Ecological Society of America* 49, no. 3 (1968): 97–100; and William C. Kimler, "Victor Ernest

Shelford," *Dictionary of Scientific Biography*, 18:811–13.

33. Quoted in Kendeigh, "Victor Ernest Shelford," 100.

34. Adams, *Guide to the Study of Animal Ecology*, 30, 25.

35. Shelford, *Animal Communities in Temperate America*, 10.

36. Willard G. Van Name, "Zoological Aims and Opportunities," *Science* 50, no. 1282 (1919), 81–84, on p. 82.

37. Committee on the Preservation of Natural Conditions, *Preservation of Natural Conditions* (Springfield, Ill.: Schneph and Barnes, 1922), 11.

38. Committee on the Preservation of Natural Conditions, *Preservation of Natural Conditions*, 9.

39. Committee on the Preservation of Natural Conditions, *Preservation of Natural Conditions*, 5.

40. Quoted in McQuilkin, "Saving the Living Laboratory," 26.

41. Anonymous, "Proceedings: Meetings of the Ecological Society of America at Toronto Meeting of December 28, 1921," *Ecology* 3, no. 2 (1922): 166–76, on p. 166.

42. Anonymous, "The Naturalist's Guide," *Ecology* 6 (1925): 469; and Victor E. Shelford, ed., *Naturalist's Guide to the Americas* (Baltimore: Williams and Wilkins, 1926).

43. Shelford, *Naturalist's Guide*, 14.

44. On the history of the National Park Service, see Alfred Runte, *National Parks: The American Experience*, 2nd ed. (Lincoln: University of Nebraska Press, 1987); and John Ise, *Our National Park Policy: A Critical History* (Baltimore: Resources for the Future by Johns Hopkins University Press, 1961). On science, wildlife management, and nature preservation in the National Parks, see Richard West Sellars, *Preserving Nature in the National Parks* (New Haven, Conn.: Yale University Press, 1997); R. Gerald Wright, *Wildlife Research and Management in the National Parks* (Urbana: University of Illinois Press, 1992);

James A. Pritchard, *Preserving Yellowstone's Natural Conditions* (Lincoln: University of Nebraska Press, 1999); and Alfred Runte, *Yosemite: The Embattled Wilderness* (Lincoln: University of Nebraska Press, 1990).

45. Forest Service management policies are examined in Paul W. Hirt, *A Conspiracy of Optimism: Management of the National Forests since World War Two* (Lincoln: University of Nebraska Press, 1994); David C. Clary, *Timber and the Forest Service* (Lawrence: University Press of Kansas, 1986); and Harold K. Steen, *The U.S. Forest Service: A History* (Seattle: University of Washington Press, 1976). On wildlife and ecosystem management in the National Forests, see Theodore Catton and Lisa Mighetto, *The Fish and Wildlife Job on the National Forests: A Century of Game and Fish Conservation, Habitat Protection, and Ecosystem Management* (Washington, D.C.: USDA Forest Service, 1998).

46. The story of the formation of Glacier Bay National Park is told in Theodore Catton, *Inhabited Wilderness: Indians, Eskimos, and National Parks in Alaska* (Albuquerque: University of New Mexico Press, 1997).

47. On Cooper's life and his studies at Glacier Bay, see Catton, *Inhabited Wilderness*, 19–26, and 31; and D. B. Lawrence, *Obituary: William Skinner Cooper* (St. Paul: Department of Botany, University of Minnesota, 1979).

48. Quoted in Catton, *Inhabited Wilderness*, 21.

49. The site became the basis for a series of his research publications, including William S. Cooper, "The Recent Ecological History of Glacier Bay, Alaska: I. The Interglacial Forests of Glacier Bay," *Ecology* 4, no. 2 (1923): 93–125; William S. Cooper, "The Recent Ecological History of Glacier Bay, Alaska: II. The Present Vegetation Cycle," *Ecology* 4, no. 3 (1923): 223–45; and William S. Cooper, "The Recent Ecological History of Glacier Bay, Alaska: III. Permanent Quadrats at Glacier Bay: An Initial Report upon a Long Period Study," *Ecology* 4, no. 4 (1923): 355–72.

50. Anonymous, "Proceedings: Business Meetings of the Ecological Society of America at Boston: Meeting of December 28, 1922," *Ecology* 4 (1923): 202–9, on p. 207.

51. Anonymous, "Proceedings: Business Meetings of the Ecological Society of America at Cincinnati," *Ecology* 5 (1924): 208–19, on p. 208

52. Anonymous, "Proceedings: Business Meetings of the Ecological Society of America at Kansas City, Missouri," *Ecology* 7 (1926): 235–46, on p. 238.

53. Victor E. Shelford, "Twenty-Five-Year Effort at Saving Nature for Scientific Purposes," *Science* 98, no. 2543 (1943): 280–81.

54. On the natural and environmental history of the Everglades, see David McCally, *The Everglades: An Environmental History* (Gainesville: University Press of Florida, 1999); Charlton W. Tebeau, *Man in the Everglades: 2,000 Years of Human History in the Everglades National Park* (Coral Gables, Fla.: University of Miami Press, 1968); and Michael Grunwald, *The Swamp: The Everglades, Florida and the Politics of Paradise* (New York: Simon and Shuster, 2006). On the drive to preserve part of the Everglades as a national park, see Runte, *National Parks,* 108–9, 128–40.

55. Quoted in Runte, *National Parks,* 131.

56. Anonymous, "Proceedings (1931)," 430.

57. Quoted in Runte, *National Parks,* 135–36.

58. Anonymous, "Proceedings (1931)," 430.

59. Barrington Moore, "Importance of Natural Conditions in National Parks," in *Hunting and Conservation: The Book of the Boone and Crockett Club,* ed. George Bird Grinnell and Charles Sheldon (New Haven, Conn.: Yale University Press, 1925), on pp. 346, 347.

60. Anonymous, "Proceedings (1926)," 246.

61. On Wright and his influence on the national parks, see Jerry Emory and Pamela Wright Lloyd, "George Melendez Wright, 1904–1936: A Voice on the Wing," *George Wright Forum* 17, no. 4 (2000): 14–45; Richard West Sellars, "The Significance of George Wright," *George Wright Forum* 17, no. 4 (2000): 46–50; Sellars, *Preserving Nature,* 86–101; and Craig L. Shafer, "Conservation Biology Trailblazers: George Wright, Ben Thompson, and Joseph Dixon," *Conservation Biology* 15, no. 2 (April 2001): 332–44.

62. Not surprisingly, their headquarters were at Berkeley.

63. George M. Wright, Joseph S. Dixon, and Ben H. Thompson, *A Preliminary Survey of Faunal Relations in National Parks* (Washington, D.C.: Government Printing Office, 1933). Wright and Thompson published a second study two years later: George M. Wright and Ben H. Thompson, *Wildlife Management in the National Parks* (Washington, D.C.: Government Printing Office, 1935). Quoted in Sellars, *Preserving Nature,* 96, 97.

64. Sellars, *Preserving Nature,* 99.

65. Sellars, *Preserving Nature,* 109. It's not clear how widely Cammerer's order was publicized because a year later the ESA Preservation Committee was still talking about the idea as though it had not yet been implemented. See Anonymous, "Proceedings: Business Meeting of the Ecological Society of America at New Orleans, Louisiana, December 29, 1931," *Ecology* 13, no. 2 (1932): 202–11, on pp. 202–3.

66. S. Charles Kendeigh, "Research Areas in the National Parks, January 1942," *Ecology* 23, no. 2 (1942): 236–38.

67. See Aldo Leopold, "The Wilderness and Its Place in Forest Recreational Policy," *Journal of Forestry* 19, no. 3 (1921): 718–19. The episode is discussed in Curt Meine, *Aldo Leopold: His Life and Work* (Madison: University of Wisconsin Press, 1988), 196–97, 200–201, 205, 224; and Paul S. Sutter, *Driven Wild: How the Fight against Automobiles Launched the Modern Wilderness Movement*

(Seattle: University of Washington Press, 2002), 71–72.

68. G. A. Pearson, "Preservation of Natural Areas in the National Forests," *Ecology* 3, no. 4 (1922): 284–87, on pp. 286, 284.

69. Anonymous, "Proceedings: Business Meetings of the Ecological Society of America at New York City, New York, December 28 and 29, 1928," *Ecology* 10, no. 2 (1929): 256–67, on p. 266.

70. The new policy was announced in Anonymous, "Forest Service Policy Covering Preservation of Natural Areas," *Ecology* 10, no. 4 (1929): 557–58.

71. These figures are from Kendeigh, "Research Areas in the National Parks, January 1942."

72. Victor E. Shelford, "Nature Sanctuaries," *Science* 75, no. 1949 (1932): 481.

73. Victor E. Shelford, "The Preservation of Natural Biotic Communities," *Ecology* 14, no. 2 (1933): 240–45, on p. 240.

74. Shelford, "Preservation of Natural Biotic Communities," 240–41.

75. Shelford, "Preservation of Natural Biotic Communities," 241–42.

76. Shelford, "Preservation of Natural Biotic Communities," 244.

77. The literature on attitudes and policies regarding predators, especially wolves, is vast and still growing. Some useful places to start include Thomas R. Dunlap, *Saving America's Wildlife* (Princeton, N.J.: Princeton University Press, 1988); Barry Holston Lopez, *Of Wolves and Men* (New York: Charles Scribner's Sons, 1978); and Jon T. Coleman, *Vicious: Wolves and Men in America* (New Haven, Conn.: Yale University Press, 2004).

78. Dunlap, *Saving America's Wildlife*, 5.

79. Quoted in Dunlap, *Saving America's Wildlife*, 16.

80. On the history of the Bureau of the Biological Survey, see Mark V. Barrow, Jr., *A Passion for Birds: American Ornithology after Audubon* (Princeton, N.J.: Princeton University Press, 1998), 59–61; Jenks Cameron,

Bureau of the Biological Survey: Its History, Activities and Organization (Baltimore: Johns Hopkins University Press, 1929); Keir B. Sterling, "Builders of the U.S. Biological Survey, 1885–1930," *Journal of Forest History* 33 (1989): 180–87; and Dunlap, *Saving America's Wildlife*.

81. Worster, *Nature's Economy*, 264.

82. Dunlap, *Saving America's Wildlife*, 49.

83. Dunlap, *Saving America's Wildlife*, 50

84. Quoted in Dunlap, *Saving America's Wildlife*, 54, 56.

85. Anonymous, "Proceedings (1931)," 431.

86. Shelford, "Preservation of Natural Biotic Communities," 244.

87. Anonymous, "Proceedings: Business Meetings of the Ecological Society of America at St. Louis, Missouri, December 31, 1935, and January 1 and 2, 1936," *Ecology* 17, no. 2 (1936): 305–21, on pp. 316 and 317.

88. Aldo Leopold, *A Sand County Almanac and Sketches Here and There* (New York: Oxford University Press, 1949), on pp. 224–25.

89. On Leopold's life and the development of his ideas, see Meine, *Aldo Leopold;* Susan L. Flader, *Thinking Like a Mountain: Aldo Leopold and the Evolution of an Ecological Attitude toward Deer, Wolves, and Forests* (Lincoln: University of Nebraska Press, 1978); J. Baird Callicott, ed., *Companion to a* Sand County Almanac: *Interpretive and Critical Essays* (Madison: University of Wisconsin Press, 1987); Richard L. Knight and Suzanne Riedel, eds., *Aldo Leopold and the Ecological Conscience* (New York: Oxford University Press, 2002); and Curt Meine and Richard L. Knight, eds., *The Essential Aldo Leopold: Quotations and Commentaries* (Madison: University of Wisconsin Press, 1999). Convenient sources for many of his published and unpublished essays include Susan L. Flader and J. Baird Callicott, eds., *The River of the Mother of God and Other Essays by Aldo Leopold* (Madison: University of Wisconsin Press, 1991); and J. Baird Callicott and Eric T. Freyfogle, eds., *For the*

Health of the Land: Previously Unpublished Essays and Other Writings (Washington, D.C.: Island Press, 1999).

90. Meine, *Aldo Leopold*, 293; Aldo Leopold, *Game Management* (New York: Charles Scribner's Sons, 1933).

91. See the series of historical articles in the fiftieth anniversary edition of the *Wildlife Society Bulletin* 15, no. 1 (Spring 1987): 1–152.

92. Quoted in Meine, *Aldo Leopold*, 295.

93. Quoted in Meine, *Aldo Leopold*, 284; see also the discussion in Anker, *Imperial Ecology*, 107–8.

94. See Thomas R. Dunlap, "That Kaibab Myth," *Journal of Forest History* 32, no. 2 (1988): 60–68; and Christian C. Young, *In the Absence of Predators: Conservation and Controversy on the Kaibab Plateau* (Lincoln: University of Nebraska Press, 2002).

95. Aldo Leopold, "The Conservation Ethic," *Journal of Forestry* 31, no. 6 (1933): 634–43; quoted in Flader and Callicott, *River of the Mother of God*, 182.

96. Flader and Callicott, *River of the Mother of God*, 190.

97. Harvey Broome, "Origins of the Wilderness Society," *Living Wilderness* 5, no. 5 (1940): 13–15; Stephen Fox, "We Want No Straddlers," *Wilderness* 48 (Winter 1984): 5–19; and especially, Sutter, *Driven Wild*.

98. Quoted in Meine, *Aldo Leopold*, 345.

99. Aldo Leopold, "A Biotic View of the Land," *Journal of Forestry* 37, no. 9 (1939): 727–30; for convenience, I have quoted from the reprint of this essay found in Flader and Callicott, *River of the Mother of God*, 266–73.

100. Flader and Callicott, *River of the Mother of God*, 266–67.

101. Flader and Callicott, *River of the Mother of God*, 268.

102. Flader and Callicott, *River of the Mother of God*, 268–70.

103. Flader and Callicott, *River of the Mother of God*, 272.

104. Aldo Leopold, "Wilderness as a Land Laboratory," *Living Wilderness* 6 (July 1941):

3; reproduced in Flader and Callicott, *River of the Mother of God*, 287–89.

105. For background on this essay, see Meine, *Aldo Leopold*, 458–59.

106. Leopold, *Sand Country Almanac*, 130.

107. The episode is related in Sellars, *Preserving Nature in the National Parks*, 111–12.

108. Sellars, *Preserving Nature in the National Parks*, 112; and Wright, *Wildlife Research and Management of the National Parks*, 18–19, argues that there is no evidence these areas were ever utilized as intended.

109. The rise and decline of the Wildlife Division's influence, see Sellars, *Preserving Nature in the National Parks*, 145–47; Wright, *Wildlife Research and Management*, 14–19.

110. Daniel B. Beard et al., *Fading Trails: The Story of Endangered American Wildlife* (New York: Macmillan, 1942), on p. 8. When members of the American Committee got wind of the project, they feared it might duplicate the inventory that Harper was completing. See Charles M. B. Cadwalader to Alexander Wetmore, June 18, 1940 and Alexander Wetmore to Charles M. B. Cadwalader, June 19, 1940, Box 26, Folder: ACIC Colleagues and Friends: Charles Cadwalader, Harold Jefferson Coolidge Papers, HUGFP 78.10, Harvard University Archives, Cambridge, Mass.

111. Beard et al., *Fading Trails*, 11, 19.

112. Beard et al., *Fading Trails*, ix.

113. An anonymous reviewer for *American Naturalist* highlighted the former shortcoming: Anonymous, "Review of Fading Trails," *American Naturalist* 76, no. 767 (1942): 618–19.

114. O. H. Robertson, "Review of American Wildlife," *Ecology* 24, no. 1 (1943): 132–33.

115. The first instance is mentioned briefly in Anonymous, "Proceedings (1929)," 259; Shelford himself requested the second review, as recorded in Anonymous, "Proceedings: Business Meetings of the Ecological Society of America at Pittsburgh, Pennsylvania, December 27 and 28, 1934," *Ecology* 16, no. 2 (1935):

266–77, on p. 275; the results of that review were reported in Anonymous, "Proceedings (1936)," 312. The events that led to the dissolution of the preservation committee are analyzed in much more detail in McQuilkin, "Saving the Living Laboratory," 49–65; Croker, *Pioneer Ecologist*, 138–46; and Tjossem, "Preservation of Nature and Academic Respectability," 49–64.

116. Shelford, "Twenty-Five-Year Effort at Saving Nature for Scientific Purposes," *Science* 98, no 2543 (1943): 280–81.

117. Croker, *Pioneer Ecologist*, 140.

118. Quoted in Croker, *Pioneer Ecologist*, 143.

119. See Paul Boyer, *By the Bomb's Early Light: American Thought and Culture at the Dawn of the Nuclear Age* (New York: Pantheon, 1985); and Allan Winkler, *Life under a Cloud: American Anxiety about the Atom* (New York: Oxford University Press, 1993).

120. The history of the Ecologists' Union and its transformation into the Nature Conservancy is recorded in Croker, *Pioneer Ecologist*, 144–46; and Ralph W. Dexter, "History of the Ecologists' Union: Spin-Off from the E.S.A. and Prototype of the Nature Conservancy," *Bulletin of the Ecological Society of America* 59, no. 3 (1978): 146–47. On the Nature Conservancy's British counterpart, founded by ecologists in 1949, see Stephen Bocking, "Conserving Nature and Building a Science: British Ecologists and the Origins of the Nature Conservancy," in *Science and Nature: Essays in the History of Environmental Sciences*, ed. Michael Shortland (Stanford in the Vale: British Society for the History of Science, 1993), 89–114; and Stephen Bocking, *Ecologists and Environmental Politics: A History of Contemporary Ecology* (New Haven, Conn.: Yale University Press, 1997). For a recent critique of the Nature Conservancy and its approach to conservation, see Timothy W. Luke, "The Nature Conservancy or the Nature Cemetery: Buying and Selling 'Perpetual Care' as Environmental Resistance," *Capital-*

ism, Nature, Socialism 6, no. June (1995): 1–20.

121. Quoted in McQuilkin, "Saving the Living Laboratory," 63–64.

CHAPTER 8

1. On the symbolism associated with the bald eagle and its incorporation into the national seal, see Richard S. Patterson and Richardson Dougall, *The Eagle and the Shield: A History of the Great Seal of the United States* (Washington: U.S. Department of State, 1978); and Elizabeth Atwood Lawrence, "Symbol of a Nation: The Bald Eagle in American Culture," *Journal of American Culture* 13 (1990): 63–69.

2. The act, subsequent amendments to it, and the litigation it engendered are briefly detailed in Michael J. Bean, *The Evolution of National Wildlife Law*, rev. ed. (New York: Praeger Publishers, 1983), 89–98; and Thomas A. Lund, *American Wildlife Law* (Berkeley: University of California Press, 1980), 135, 154, 155, 164.

3. Convenient entry points into this voluminous literature include Thomas R. Dunlap, *Saving America's Wildlife* (Princeton, N.J.: Princeton University Press, 1988); Jon T. Coleman, *Vicious: Wolves and Men in America* (New Haven, Conn.: Yale University Press, 2004); and Michael J. Robinson, *Predatory Bureaucracy: The Extermination of Wolves and the Transformation of the West* (Boulder: University Press of Colorado, 2005).

4. John C. Phillips, "Conservation of Birds and Mammals," in *Hunting and Conservation: The Book of the Boone and Crockett Club*, ed. George Bird Grinnell and Charles Sheldon (New Haven, Conn.: Yale University Press, 1925), 29–65, on p. 30.

5. Rosalie Edge, "Good Companions in Conservation: An Implacable Widow," 9a, Manuscripts, Special Collections, University Archives, University of Washington Libraries, Seattle.

6. Metaphors of eradication and their implications are discussed in Edmund Russell, *War and Nature: Fighting Humans and Insects with Chemicals from World War I to Silent Spring* (Cambridge: Cambridge University Press, 2001).

7. On the development of economic ornithology, see Matthew D. Evenden, "The Laborers of Nature: Economic Ornithology and the Role of Birds as Agents of Biological Control in North American Agriculture, ca. 1880–1930," *Forest and Conservation History* 39 (1995): 172–83; and T. S. Palmer, "A Review of Economic Ornithology in the United States," *Yearbook of the U. S. Department of Agriculture for 1899* (Washington, D.C.: Government Printing Office, 1900), 259–92. In declaring birds to be "good" or "bad," economic ornithologists and nature writers also frequently judged their habits by the standards of Victorian morality. See Peter Schmitt, *Back to Nature: The Arcadian Myth in Urban America* (Oxford: Oxford University Press, 1969), 36–38; Dunlap, *Saving America's Wildlife*, 15–16; and Evenden, "The Laborers of Nature," 175–77.

8. See chapter 5.

9. F. A. Lucas et al. to Members of the Committee on Bird Protection of the AOU, March 9, 1920; and A. K. Fisher to F. A. Lucas et al., March 15, 1920, A. K. Fisher Papers, Box 23, Library of Congress, Washington, D.C. The ongoing struggle for authority over game management in Alaska is discussed in Morgan Sherwood, *Big Game in Alaska: A History of Wildlife and People* (New Haven, Conn.: Yale University Press, 1981). Morgan reports (on p. 43) that until the 1930s, the Bureau of the Biological Survey encouraged individuals to raise foxes on a number of islands within its extensive Aleutian Island Reservation, an activity that became known as "fox farming."

10. Albert K. Fisher, *The Hawks and Owls of the United States in Their Relation to Agriculture* (Washington, D.C.: Government

Printing Office, 1893), on p. 3. Fisher's findings were also presented in summary form as "Hawks and Owls as Related to the Farmer," *Yearbook of the United States Department of Agriculture* (Washington, D.C.: Government Printing Office, 1894). Fisher's ideas about "good" and "bad" birds remained influential in conservation circles for many years. See, for example, [Louis Agassiz Fuertes], "American Birds of Prey—a Review of Their Value," *National Geographic* 38 (December 1920): 460–67; and J. P. H., "Good and Bad Hawks," *Bird-Lore* 26 (1924): 376–78.

11. See, for example, Anonymous, "The Government Exterminating Our National Bird," *Literary Digest* 65 (1920): 130; Anonymous, "A Renewed Attempt to Save the Eagle," *American Review of Reviews* 68 (December 1920): 659–60; William T. Hornaday, "Alaska Can Save the American Eagle: The Bird of Our National History Threatened with Extinction," *Natural History* 20 (1920): 117–19; Willard G. Van Name, "Threatened Extinction of the Bald Eagle," *Ecology* 2 (1921): 76–78; and William L. and Irene Finley, "A War against American Eagles," *Nature* 2 (1923): 261–70.

12. See the discussion in chapter 7.

13. On the duck crisis, see Theodore W. Cart, "'New Deal' for Wildlife: A Perspective on Federal Conservation Policy, 1933–40," *Pacific Northwest Quarterly* 63 (1972): 113–20; and Jared Orsi, "From Horicon to Hamburgers and Back Again: Ecology, Ideology, and Wildfowl Management, 1917–1935," *Environmental History Review* 18 (1994): 19–40.

14. *More Game Birds by Controlling Their Natural Enemies*, rev. ed. (New York: More Game Birds in America, 1936), 6. Similar sentiments are expressed in other publications from this organization: *More Waterfowl by Assisting Nature* (New York: More Game Birds in America, 1931); *Game Birds: How to Make Them Pay on Your Farm* (New York: More Game Birds in America, n.d.); *Small Refuges* for *Waterfowl* (New York: More Game Birds in America, 1933). In 1937, the More Game Birds in America Foundation merged with another organization, American Wild Fowlers, to form Ducks Unlimited. See Jon R. Tennyson, *A Singleness of Purpose: The Story of Ducks Unlimited* (Chicago: Ducks Unlimited, 1977).

15. On bird clubs, see Mark V. Barrow, Jr., *A Passion for Birds American Ornithology after Audubon* (Princeton, N.J.: Princeton University Press, 1998), 69, 163–64, 169, and 171; and the special edition of *Bird-Lore* 8 (September 1915): 347–71.

16. On the life and philosophy of Jack Miner, see his autobiographies, *Jack Miner and the Birds and Some Things I Know about Nature* (Chicago: Reilly and Lee, 1923); and *Jack Miner: His Life and Religion* (Kingsville, Ontario: Jack Miner Migratory Bird Foundation, 1969); see also Molly Clare Wilson, "Wild Goose Chase: The Communal Science of Waterfowl Migration Study in North America, 1880–1940," (senior honors thesis, Harvard University, 2006); and Harry McDougall, "Jack Miner's Bird Sanctuary," *Canadian Geographical Journal* 83 (1971): 102–8. Miner did make an exception for the red-tailed and the red-shouldered hawks, which he considered "too big and clumsy to be very destructive on our birds" (Miner, *Jack Miner and the Birds*, 28). The quote is from John B. Kennedy, "The Birds Aren't So Wild," *Collier's* 80 (December 3, 1927): 21 and 38, on p. 38.

17. T. Gilbert Pearson, *Conservative Conservation*, circular no. 9 (New York: National Association of Audubon Societies, 1924).

18. James P. Chapin, "In Memoriam: Waldron DeWitt Miller, 1879–1929," *The Auk* 44 (1932): 1–8.

19. Chapin, "In Memoriam," 7. See Miller's "Handwritten Notes on the Food of Hawks and Owls," ms. in the Department of Ornithology at the American Museum of Natural History, New York.

20. On Van Name, see Barrow, *Passion for Birds*, 148–50; Edge, "Good Companions

in Conservation: An Implacable Widow," 25–30; Willard G. Van Name to Robert Cushman Murphy, September 25, 1924, Box B-11, National Audubon Society Records, Manuscripts and Archives Division, New York Public Library, New York (hereafter cited as NASR/NYPL); and Stephen Fox, *The American Conservation Movement: John Muir and His Legacy* (Madison: University of Wisconsin Press, 1985), 174.

21. On the history of the Linnaean Society of New York, see Barrow, *Passion for Birds*, 193–95; Eugene Eisenmann, "Seventy-Five Years of the Linnaean Society of New York," *Proceedings of the Linnaean Society of New York*, nos. 63–65 (1954): 1–15; and *Reminiscences by Members Collected on the Occasion of the Centennial of the Linnaean Society of New York* (New York: Linnaean Society of New York, 1978).

22. The quotes are from copies of the Linnaean Society and DVOC resolutions that accompany T. Gilbert Pearson (hereafter cited as TGP) to Charles A. Urner, December 28, 1925, and Julian Potter to TGP, February 18, 1928, Box A-131, Folder: 1926, Hawks and Owls, NASR/NYPL.

23. Aretas Saunders to TGP, February 21, 1926, Box A-131, Folder: 1926, Hawks and Owls, NASR/NYPL; see also Aretas Saunders to Waldron D. Miller, February 19, 1926, Historic Correspondence File, Department of Ornithology, American Museum of Natural History, New York (hereafter cited as HCF/AMNH).

24. The quotes are from TGP to Myron Ackland, January 28, 1926, Box A-131, Folder: 1926, Hawks and Owls, NASR/NYPL. On Pearson's suspicions about Van Name, see Mabel Osgood Wright to TGP, February 18, 1926, and TGP to Mabel Osgood Wright, February 23, 1926, Box A-131, Folder: 1926, Hawks and Owls, NASR/NYPL. Much additional correspondence related to this issue is also found in this folder.

25. See the long series of letters from

Henry Carey to Waldron DeWitt Miller, HCF/AMNH. See also Henry Carey, "Hawk Extermination," *The Auk* 43 (1926): 275–76; and Ernest G. Holt, "Nature-Wasters and Sentimentalists," *The Auk* 43 (1926): 409–10.

26. Waldron DeWitt Miller et al., *Save These Birds!* (n.p., [1926]), from a copy in the reprint file of the Department of Ornithology, AMNH. See also George M. Sutton, "How Can the Bird-Lover Help to Save the Hawks and Owls?" *The Auk* 46 (1929): 190–95; and Witmer Stone, "The Hawk Question: Editorial and Correspondence," *The Auk* 47 (1930): 208–17.

27. TGP to A. F. Ganier, April 23, 1926, Box A-131, Folder: 1926, Hawks and Owls, NASR/NYPL.

28. Anonymous, "Destructive Birds Again," *Bird-Lore* 29 (1927): 83–85, quote on p. 84.

29. T. Gilbert Pearson, "Eagles and the Alaskan Bounty," *Bird-Lore* 30 (1928): 86–90, quote on p. 90. Pearson followed up by creating a packet of material that included twenty color pictures of Alaskan birds, descriptive text by his assistant Alden H. Hadley, and his own eight-page circular he entitled "The Value of Birds." Pearson's circular failed to mention the bald eagle or other birds of prey. He arranged for ten thousand copies of this packet to be sent to school children and teachers in Alaska. T. Gilbert Pearson, *The Value of Birds (A Letter to the Children of Alaska)*, circular no. 13 (New York: National Association of Audubon Societies, n.d.). See also "Bird Protection in Alaska," *Bird-Lore* 31 (1929): 160.

30. W. DeWitt Miller, Willard G. Van Name, and Davis Quinn, *A Crisis in Conservation: Serious Danger of Extinction of Many North American Birds* (New York, 1929). The pamphlet did not specifically name the National Association of Audubon Societies, but anyone familiar with wildlife conservation during the period would have been aware

that the organization was one of its principal targets.

31. More details on Edge's background and the specific events that led to the creation of the Emergency Conservation Committee are found in Edge, "Good Companions in Conservation: An Implacable Widow," quote on p. 6; and Peter Edge, "A Determined Lady," unpublished ms., in Manuscripts, Special Collections, University Archives, University of Washington Libraries, Seattle. For additional biographical information on Edge and background on the ECC, see Robert Lewis Taylor, "Profiles: Oh, Hawk of Mercy!" *New Yorker* 24 (April 17, 1948): 31–43; Laura Kathleen Sumner, "Rosalie Edge and the American Conservation Movement" (master's thesis, Oklahoma State University, 1993); Robin Epstein, "Rediscovering Rosalie Edge" (senior thesis, Duke University, 1985); Irving Brant, *Adventures in Conservation with Franklin D. Roosevelt* (Flagstaff, Ariz.: Northland Publishing, 1988); and Fox, *American Conservation Movement*, 173–82, and passim. See also Dyana Furmansky, *Rosalie Edge, Hawk of Mercy: The Activists Who Saved Nature from Conservationists* (Athens: University of Georgia Press, forthcoming), which went to press after I completed this chapter.

32. Willard G. Name apparently played a major role in authoring numerous Emergency Conservation Committee pamphlets, including many that were published under Rosalie Edge's name.

33. The disparaging title for the Bureau of the Biological Survey comes from another ECC pamphlet: *The United States Bureau of Destruction and Extermination: The Misnamed and Perverted "Biological Survey"* (New York: Emergency Conservation Committee, 1934). A complete run of ECC pamphlets may be found in the Rosalie Edge Papers, Conservation Collection, Denver Public Library, Denver, Colo.

34. Quinn's pamphlet was first issued as Davis Quinn, *"Framing" the Birds of Prey*

(New York, December 1929), quote on p. 1. The ECC later republished several revised editions of this pamphlet.

35. *The Bald Eagle, Our National Emblem: Danger of Its Extinction by the Alaska Bounty* (New York: Emergency Conservation Committee, April 1930), quote on p. 17.

36. Various (and often contradictory) accounts of the events that led up to the passage of the Norbeck-Andresen Migratory Bird Conservation Act may be found in William T. Hornaday, *Thirty Years War for Wild Life* (Stamford, Conn.: Permanent Wild Life Protection Fund, 1931), 235–44; Thomas Gilbert Pearson, *Adventures in Bird Protection* (New York: D. Appleton-Century, 1937), 289–303; and Cart, "'New Deal' for Wildlife." For an extensive behind-the-scenes view of Pearson's efforts on behalf of the bill, see Box A-143, NASR/NYPL. See especially, John Burnham to TGP, February 12, 1929; and TGP to George Bird Grinnell, January 29, 1929, in Folder: Misc. A–G of that box.

37. The address was published as T. Gilbert Pearson, *The Case of the Hawk*, circular no. 17 (New York: National Association of Audubon Societies, 1929), and under an identical title in *Bird-Lore* 32 (1930): 87–89. Box C-13, NASR/NYPL, contains a complete collection of Audubon circulars.

38. T. Gilbert Pearson, "A Bill to Protect the Bald Eagle," *Bird-Lore* 32 (1930): 86–87, on p. 86. The fate of the bill was closely followed in subsequent editorials: "The Bald Eagle Bill," *Bird-Lore* 32 (1930): 164–67; "The Bald Eagle Bill," *Bird-Lore* 32 (1930): 395; "The Eagle Bill," *Bird-Lore* 33 (1931): 159–60.

39. The quote is from TGP to Bayard Christy, January 18, 1930, Box A-143, Folder: Misc. A–G, NASR/NYPL; Pearson's pessimism about the fate of the bill is found in TGP to Reed Holloman, February 28, 1930, Box A-143, Folder: Misc. H–M, NASR/NYPL; Pearson's discussions with Norbeck about the bill are found in Box A-143, Folder: Misc. N–S, NASR/NYPL. This collection contains much additional correspondence related to the bald eagle bill campaign, including hundreds of copies of letters written to members of Congress in support of the legislation and their replies. Among the articles on bald eagle protection that appeared during this period are Arthur H. Fisher, "Our National Bird," *Nature* 32 (1929): 321–25; and Ben East, "He Needs Protection: The Bald Eagle, Emblem of American throughout the World, Is Preyed upon by the People He Represents," *American Forests and Forest Life* 36 (1930): 14–16.

40. "American Eagle Protection," Hearing before the Committee on Agriculture, House of Representatives, 71st Congress, 2nd Sess., January 31, 1930, Serial D (Washington: D.C.: Government Printing Office, 1930), quote on pp. 1, 11, and 14. The response of the House Committee on Agriculture and Forestry is found in Augustus H. Andresen to TGP, February 7, 1931, Box A-143, Folder: Misc., A–G, NASR/NYPL, and also reproduced in *Bird-Lore* 33 (1931): 159–69. The brief debate on the legislation in the Senate and the amended bill is found in Congressional Record, Senate, vol. 72, pt. 6, 71st Congress, 2nd Sess., April 7 1930, 6612–13.

41. The history of the organization can be followed through its periodical, *Hawk and Owl Society Bulletin* 1–5 (1932–35). See especially the account in issue no. 3 (March 1933): 2–5. The quotes are from the inside cover of that same issue. See also Warren F. Eaton to Frank M. Chapman, March 2, 1932, HCF/AMNH. On Eaton's life and work, see Charles A. Urner, "Warren Francis Eaton, 1900–1936," *Proceedings of the Linnaean Society of New York*, no. 47 (1935): 10–11, on p. 10; and "Reminiscences by Members," 2. Pearson also agreed to serve on the six-person organizing committee of the Hawk and Owl Society.

42. *Hawk and Owl Society Bulletin* 3 (March 1933): 5. Among the regular features in the organization's bulletin were reports of anti-predatory-bird campaigns, accounts of efforts to secure state protection for hawks and owls, summaries of existing laws, and bibliographies of publications on the subject. Two of the articles that the Hawk and Owl Society widely circulated include S. Prentiss Baldwin, S. Charles Kendeigh, and Roscoe W. Franks, "The Protection of Hawks and Owls in Ohio," *Ohio Journal of Science* 32 (1932): 403–24; and George E. Hix, *Birds of Prey for Boy Scouts* (New York: Published by the Author, n.d.).

43. This account of the origin of Hawk Mountain Sanctuary is largely drawn from Keith L. Bildstein and Robert A. Compton, "Mountaintop Science: The History of the Hawk Mountain Sanctuary," in *Contributions to the History of North American Ornithology*, vol. 2, ed. William E. Davis and Jerome Jackson (Cambridge, Mass.: Nuttall Ornithological Club, 2000), 153–81, quote on 157; and Maurice Broun, *Hawks Aloft: The Story of Hawk Mountain* (New York: Dodd, Mead, 1948). See also the long series of correspondence on the subject between Rosalie Edge and Willard G. Van Name, Box B-11, NASR/NYPL.

44. Pough's and Collins's published reports appeared as Richard H. Pough, "Wholesale Killing of Hawks in Pennsylvania," *Bird-Lore* 34 (1932): 429–40; and Henry H. Collins, Jr., "Hawk Slaughter at Drehersville," *Hawk and Owl Society Bulletin* 3 (1933): 10–18.

45. Pearson's offer to purchase Hawk Mountain is discussed in May 20, 1934, Rosalie Edge to Willard G. Van Name, Box B-11, Folder: Van Name, Willard G., NASR/NYPL.

46. Pearson's letter to the Montana Fish and Game Commission is found in *Bird-Lore* 35 (1933): 75–76, while his press release on the bald eagle is reproduced in *Bird-Lore* 35 (1933): 76–77.

47. Pearson's address was published as *Evils That Lurk in the Bounty System*, circular no. 22 (New York: National Association of Audubon Societies, 1933).

48. Pearson was granted the title president

emeritus. For more on the specific circumstances that lead to Baker's ascendancy and Pearson's decline, see Frank Graham, Jr., *The Audubon Ark: A History of the National Audubon Society* (New York: Alfred A. Knopf, 1990), 107–21; and Fox, *American Conservation Movement*, 173–82.

49. An announcement of the resolution came in *Bird-Lore* 36 (1934): 334–35.

50. The campaign was announced in Anonymous, "Campaign for Hawk and Owl Protection," *Bird-Lore* 36 (1934): 333–35, which also includes a copy of the Civilian Conservation Corps poster. See also John B. May, *The Hawks of North America: Their Field Identification and Feeding Habits* (New York: National Association of Audubon Societies, 1935); and John H. Baker, "Predators," typescript, Box C-48, Folder: Birds of Prey, NASR/NYPL.

51. Warren F. Eaton, "Feathered vs. Human Predators," *Bird-Lore* 37 (1935): 122–26, quote on p. 122. Errington's importance in promoting a more ecological perspective on predators is also discussed in Dunlap, *Saving America's Wildlife*, 73–74.

52. Eaton, "Feathered v. Human Predators," 123; emphasis added. For additional evidence of a new concern with ecology within the National Association of Audubon Societies during this period, see Baker, "Predators"; Paul Errington, review of John B. May, *The Hawks of North America*, *Bird-Lore* 37 (1935): 283; Leonard Wing, "Predation Is Not What It Seems," *Bird-Lore* 38 (1936): 401–5. On the close relationship between Errington and Leopold, see Curt Meine, *Aldo Leopold: His Life and Work* (Madison: University of Wisconsin Press, 1988), 274–75. On the diffusion of ecological ideas to the public, see Dunlap, *Saving America's Wildlife*, chap. 7.

53. A copy of the bill is found in U.S. Fish and Wildlife Service Records, RG 22, E252, Box 11, "Bills, Congressional—Conservation," National Archives, College Park, Md. The favorable report from the Senate Special

Committee on Conservation of Wildlife Resources is found in "Preservation of the American Eagle," 74th Congress, 1st Sess., Senate, Report no. 899. The ECC pamphlet was published anonymously, *Save the Bald Eagle: Shall We Allow Our National Emblem to Become Extinct?* (New York: Emergency Conservation Committee, 1935).

54. Richard H. Pough, *Enter Hawk—Exit Mouse* (New York: National Association of Audubon Societies, n.d.). Biographical sketches of Pough may be found in Graham, *The Audubon Ark*, 160–61; Anonymous, "Richard Pough to Direct Hawk and Owl Campaign," *Bird-Lore* 38 (1936): 127; and Richard Stroud, ed., *National Leaders of American Conservation* (Washington, D.C.: Smithsonian Institution Press, 1985), 311–12.

55. On the practice of falconry in the United States during this period, see R. L. Meredith, *American Falconry in the Twentieth Century* (Boise, Idaho: Archives of American Falconry, 1999), which was written in the 1930s; George G. Goodwin, "Winged Monarchs of the Air," *Natural History* 36 (1935): 51–61; Ellsworth Lumley, "Falconry," *Nature* 29 (1937): 299–300; Rosalie Edge, "The Falcons of Manhattan," *Nature* 28 (1936): 305; Lewis Wayne Walker, "Hounds of the Sky," *Popular Science Monthly* 132 (1938): 48–49; Edwin Teale, "Falcons: Feathered Hunters for Man," *Popular Science Monthly* 139 (1941): 70–74; Anonymous, "Hunting Hawks: Falconry, Age-Old Sport, Finds New Enthusiasm in U.S.," *Literary Digest* 123 (1937): 36–38. On the cultural history of falcons and falconry more generally, by far the best source is Helen Macdonald, *Falcon* (London: Reaktion Books, 2006).

56. Louis Agassiz Fuertes, "Falconry, the Sport of Kings," *National Geographic* 38 (1920): 429–60.

57. Alva G. Nye, Jr., "American Falconry," *American Falconry* 1 (1942): 6–8, on p. 6; and Walter R. Spofford, "Falconry and

Conservation," *Nature* 38 (1945): 257–61, 74–275, on p. 275.

58. Rosalie Edge, *The Duck Hawk and the Falconers* (New York: Emergency Conservation Committee, 1944), on pp. 3 and 6.

59. See, for example, Frank and John Craighead, "Adventures with Birds of Prey," *National Geographic* 72 (1932): 109–34; John and Frank Craighead, "Skyriders," *Nature* 30 (1937): 220–23; Frank and John Craighead, "Bad Boy: The Story of an Interesting Hawk Pet," *Nature* 26 (1935): 77–80; John and Frank Craighead, "In Quest of the Golden Eagle: Over the Lonely Mountain and Prairie Soars This Rare and Lordly Bird, but Three Youths from the East Catch Up with Him at Last," *National Geographic* 77 (1940): 692–710; Frank and John Craighead, *Hawks in the Hand: Adventures in Photography and Falconry* (Boston: Houghton Mifflin, 1939). The Craighead brothers were later important in developing radio-tracking and satellite biotelemetry techniques in wildlife research and in promoting the Wild and Scenic Rivers Act; see Gregg Mitman, "When Nature Is the Zoo: Vision and Power in the Art and Science of Natural History," *Osiris* 11 (1996): 117–43.

60. On the status of state legislation in 1939, see John Baker, "The Director Reports to You," *Bird-Lore* 41 (1939): 165–67.

61. On the Bald Eagle Protection Act, see "Protection of the Eagle," House of Representatives, 76th Congress, 3rd Sess., Report no. 2104; "Preserving from Extinction the American Eagle, Emblem of Sovereignty of the United States of America," Senate, 76th Congress, 3rd Sess., Report no. 1589. On the role of Maud Phillips in authoring the bill, finding sponsors for it, and testifying before Congress, see "Backs Bill to Protect Eagle," *New York Times*, March 12, 1940, 25; "Protecting the Eagle," *Newsweek* 15 (June 10, 1940): 37; "Maud Phillips Dies; Founder of Blue Cross," *Springfield* (Mass.) *Daily News*, July 30, 1951, 1; "Maud Gillett Phillips," *Springfield* (Mass.) *Daily News*, July 31, 1951,

10; Stan Berchulski, "Bald Eagles—Saved by a Local Woman—Are in Danger of Extinction Again," *Springfield* (Mass.) *Daily News,* May 17, 1958. Alaska's exemption from the Bald Eagle Protection law did not come until 1959; by this time, more than one hundred thousand eagles had fallen victim to the bounty.

62. Carl T. Keller to Ellison D. Smith, May 10, 1939, Records of the U.S. Senate, RG 46, Box 45, 76th Congress, Papers Relating to Specific Bills and Resolutions, Folder: S.1494, National Archives, Washington, D.C., which also includes a copy of the bill and numerous letters written in support of the Bald Eagle Protection Act.

63. On raptor-watching at Hawk Mountain, see the references cited in n. 43 above and Edwin Way Teale, "Hawkways," *Natural History* 54 (1945): 207–11.

64. On the elusiveness of wolves, see Barry Holston Lopez, *Of Wolves and Men* (New York: Charles Scribner's Sons, 1978), 65–66.

65. The imagery associated with these creatures is more complex that this abbreviated discussion suggests; both groups have been framed in negative and positive terms. See, for example, Lopez, *Of Wolves and Men;* Alexander Wetmore, "The Eagle, King of Birds, and His Kin," *National Geographic* 64 (1933): 43–88; Beryl Rowland, *Birds with Human Souls: A Guide to Bird Symbolism* (Knoxville: University of Tennessee Press, 1978); and Beryl Rowland, *Animals with Human Faces: A Guide to Animal Symbolism* (Knoxville: University of Tennessee Press, 1973).

66. On the growth and decline in hunting in postwar America, see U.S. Department of Interior, Fish and Wildlife Service, and U.S. Department of Commerce, Bureau of the Census, *1996 National Survey of Fishing, Hunting, and Wildlife-Associated Recreation* (Washington, D.C., 1997), appendix B; and Daniel J. Herman, *Hunting and the American Imagination* (Washington, D.C.: Smithsonian Institution Press, 2001).

67. On the growth of birdwatching in the United States, see Barrow, *Passion for Birds;* Joseph Kastner, *A World of Watchers* (New York: Alfred A. Knopf, 1986); Felton Gibbons and Deborah Strom, *Neighbors to the Birds: A History of Birdwatching in America* (New York: W. W. Norton, 1988); and Scott Weidensaul, *Of a Feather: A Brief History of American Birding* (Orlando, Fla.: Harcourt, 2007).

68. Amendments to the Bald Eagle Protection Act are discussed in Bean, *Evolution of National Wildlife Law,* 89–93. The figures on state predatory bird legislation are from R. C. Clement, *Status of Raptor Protection in the United States and Canada* (New York: National Audubon Society, 1966). The progress toward protective legislation can be followed in earlier articles as well: Richard Stuart Phillips, "A Fair Deal for Our Birds of Prey," *Audubon Magazine* 51 (1949): 377–81, 392–97; and Kenneth D. Morrison, "Bird Protection Laws Show Progress: Trend Is toward Protected Status for All Hawks and Owls," *Audubon Magazine* 57 (September 1955): 222–25.

CHAPTER 9

1. The story of the heath hen has recently been told in masterful fashion in Christopher Cokinos, *Hope Is the Thing with Feathers: A Personal Chronicle of Vanished Birds* (New York: Jeremey P. Tarcher/Putnam, 2000); and Alfred O. Gross, *The Heath Hen, Memoirs of the Boston Society of Natural History* vol. 6, no. 4 (Boston: Boston Society of Natural History, 1928); I have relied heavily on both in the account that follows.

2. Burgess is an important and neglected figure in the American nature study movement. For a summary of his life and work, see Janet Spaeth, "Thorton Waldo Burgess," in *American National Biography,* ed. John A. Garraty and Mark C. Carnes (New York: Oxford University Press, 1999), 3:943–44; see also Thorton W. Burgess, *Now I Remember: Au-*

tobiography of an Amateur Naturalist (Boston: Little, Brown, 1960).

3. The episode is described in Cokinos, *Hope Is the Thing with Feathers*, 173–74, 177–78.

4. Thorton W. Burgess to Waldron DeWitt Miller, July 10, 1929, Historic Correspondence File, Department of Ornithology, American Museum of Natural History, New York.

5. Both quotes are from Cokinos, *Hope Is the Thing with Feathers*, 180.

6. On the transformation of American natural history and the development of biology, see the introduction.

7. On the history of American wildlife management, see Thomas R. Dunlap, *Saving America's Wildlife* (Princeton, N.J.: Princeton University Press, 1988); Curt Meine, *Aldo Leopold: His Life and Work* (Madison: University of Wisconsin Press, 1988); Kurkpatrick Dorsey, *The Dawn of Conservation Diplomacy: U.S.-Canadian Wildlife Protection Treaties in the Progressive Era* (Seattle: University of Washington Press, 1998); Aldo Leopold, *Game Management* (New York: Charles Scribner's Sons, 1933); the series of articles in the special fiftieth anniversary edition of *Wildlife Society Bulletin* 15, no. 1 (1987); and Keir Sterling, "Zoological Research, Wildlife Management, and the Federal Government," 19–65, and Bonnie Christensen, "From Divine Nature to Umbrella Species: The Development of Wildlife Sciences in the United States," 209–29, in *Forest and Wildlife Science in America*, ed. Harold K. Steen (Durham, N.C.: Forest History Society, 1999).

8. Gross, *The Heath Hen*, 522.

9. Gross, *The Heath Hen*, 542–43, is the main source of the description that follows.

10. According to Gross, *The Heath Hen*, 543, one prominent naturalist learned to imitate the call by blowing across the top of an empty bottle.

11. Gross, *The Heath Hen*, 546–50.

12. Gross, *The Heath Hen*, 493.

13. Gross, *The Heath Hen*, 496.

14. Catesby failed to locate the heath hen during his visit to America and instead drew it after his return to England using live specimens kept at a nobleman's estate in Cheswick. Alan Feduccia, ed., *Catesby's Birds of Colonial America* (Chapel Hill: University of North Carolina Press, 1985), 37–38.

15. The information on legislation and the decline in range is from Gross, *The Heath Hen*, 497–500.

16. William Brewster, "The Heath Hen of Massachusetts," *The Auk* 2 (1885): 80–84, on pp. 81 and 84.

17. William Brewster, "The Heath Hen," *Forest and Stream* 35 (1890): 188.

18. Hoyle's activities are mentioned briefly in Gross, *The Heath Hen*, 505. Hoyle eventually published a short popular article on the species: Charles E. Hoyle, "Heath Hen (Hethen)," *American Ornithology* 1 (1901): 197–201.

19. C. E. Hoyle to William Brewster, May 8, 1895, William Brewster Papers, Archives of the Ernst Mayr Library, Museum of Comparative Zoology, Harvard University, Cambridge, Mass. (hereafter cited as WBP/AEML-HU).

20. C. E. Hoyle to William Brewster, February 17, 1898, WBP/AEML-HU.

21. Gross, *The Heath Hen*, 505–6.

22. William Dutcher, "Report of the President," *Bird-Lore* 7 (1905): 329.

23. Gross, *The Heath Hen*, 506.

24. The story of the reservation is recounted in Gross, *The Heath Hen*, 507–11.

25. While naturalists and game officials were clearly interested in the species during this period, little information about its plight filtered down to the public. Some exceptions include George W. Field, "A Sketch of a Bird Now on the Verge of Extinction," *Bird-Lore* 9 (1907): 249–55; Katherine B. Tippetts, "A Heath Hen Quest," *Bird-Lore* 11 (1909): 244–48; T. Gilbert Pearson, "Last Days of the Heath Hen," *Bird-Lore* 25 (1923): 223–24; and Herbert K. Job, "The Imperiled Heath Hen," *Bird-Lore* 25 (1923): 363, 365.

26. Gross, *The Heath Hen*, 511.

27. Cokinos, *Hope Is the Thing with Feathers*, 143.

28. Details on the ill-fated experiment are found in Gross, *The Heath Hen*, 512–13.

29. Gross, *The Heath Hen*, 512.

30. Gross, *The Heath Hen*, 515.

31. Gross, *The Heath Hen*, 516. On the Nuttall Ornithological Club, see William E. Davis, Jr., *History of the Nuttall Ornithological Club, 1873–1986* (Cambridge, Mass.: Nuttall Ornithological Club, 1987); on the Brookline Bird Club, see Charles B. Floyd, "The Brookline Bird Club," *Bird-Lore* 17 (1915): 358–62; and on the Federation of New England Bird Clubs, see Francis H. Allen, *The Federation of the Bird Clubs of New England: A Record of Its First Ten Years* (Boston: Published by the Federation, 1934).

32. A short biographical sketch of Gross is found in Raymond A. Paynter, Jr., "In Memoriam: Alfred Otto Gross," *The Auk* 88 (1971): 520–27.

33. On Forbes and his legacy, see Robert A. Croker, *Stephen Forbes and the Rise of American Ecology* (Washington, D.C.: Smithsonian Institution Press, 2001); and Daniel W. Schneider, "Local Knowledge, Environmental Politics, and the Founding of Ecology in the United States: Stephen Forbes and 'The Lake as a Microcosm' (1887)," *Isis* 91 (2000): 681–705.

34. See Mark V. Barrow, Jr., "Gentlemanly Specialists in the Age of Professionalization: The First Century of Ornithology at Harvard's Museum of Comparative Zoology," in *Contributions to the History of North American Ornithology*, ed. William E. Davis, Jr., and Jerome A. Jackson (Cambridge, Mass.: Nuttall Ornithological Club, 1995), 55–94.

35. Gross, *The Heath Hen*, 531–32.

36. Gross, *The Heath Hen*, 517.

37. Gross, *The Heath Hen*, 536.

38. Gross, *The Heath Hen*, 519.

39. Gross, *The Heath Hen*, 491.

40. Gross, *The Heath Hen*, 522.

41. Gross, *The Heath Hen*, 496.

42. The discussion that follows is found in Gross, *The Heath Hen*, 522–31, quoted on p. 526.

43. Gross, *The Heath Hen*, 529–31.

44. Gross, *The Heath Hen*, 551.

45. Alfred O. Gross, "The Last Heath Hen," *Scientific Monthly* 32, no. 4 (1931): 382–84, on p. 382.

46. Gross, "The Last Heath Hen," *Bird-Lore* 31 (1929): 252–54, on p. 252.

47. Cokinos, *Hope Is the Thing with Feathers*, 178.

48. Gross, "The Last Heath Hen" (1931), 384.

49. Cokinos, *Hope Is the Thing with Feathers*, 182–84; Henry B. Hough, *The Heath Hen's Journey to Extinction, 1792–1933* (Edgartown, Mass.: Dukes County Historical Society, 1933), 22.

50. Hough, *The Heath Hen's Journey to Extinction*, 29.

51. The story of Tanner's study is told in Cokinos, *Hope Is the Thing with Feathers*, 61–117. On the fate of the ivory-billed woodpecker from an environmental history perspective, see Mikko Saikku, "'Home in the Big Forest': Decline of the Ivory-Billed Woodpecker and Its Habitat in the United States," in *Encountering the Past: Essays in Environmental History*, ed. Mikko Saikku and Timo Myllyntaus (Athens: Ohio University Press, 2001), 94–140. Though my emphasis is different, I have relied on both of these publications to reconstruct the story that follows. Several excellent books, which appeared after I had drafted this chapter, provide additional details: Jerome A. Jackson, *In Search of the Ivory-Billed Woodpecker* (Washington, D.C.: Smithsonian Books, 2004); Phillip Hoose, *The Race to Save the Lord God Bird* (New York: Farrar, Straus and Giroux, 2004); and Tim Gallagher, *The Grail Bird: Hot on the Trail of the Ivory-Billed Woodpecker* (Boston: Houghton Mifflin, 2005).

52. On the history of American wildlife

management, see the works cited in n. 7 above.

53. On the elevation of the doctoral degree as the standard entry-level degree in science, see Robert Kohler, "The Ph.D. Machine: Building on the Collegiate Base," *Isis* 81 (1990): 638–62. On the transformation of American natural history and the development of biology, see Ronald Rainger, Keith R. Benson, and Jane Maienschein, *The American Development of Biology* (Philadelphia: University of Pennsylvania Press, 1988); Keith R. Benson, Jane Maienschein, and Ronald Rainger, *The Expansion of American Biology* (New Brunswick, N.J.: Rutgers University Press, 1991); and Philip J. Pauly, *Biologists and the Promise of American Life* (Princeton, N.J.: Princeton University Press, 2000). For an insider's account of the program at Cornell, see Albert H. Wright, "Biology at Cornell University," *Bios* 24 (1953): 122–45; on Berkeley, see Richard M. Eakin, "History of Biology at the University of California, Berkeley," *Bios* 27 (1956): 67–80; and Richard M. Eakin, *History of Zoology at Berkeley* (Berkeley: University of California, 1988); on Michigan, see A. Franklin Shull, "The Department of Zoology," in *The University of Michigan: An Encyclopedic Survey*, ed. Wilfred B. Shaw (Ann Arbor: University of Michigan Press, 1951), 2:738–50.

54. Catesby's engraving of the ivorybill and the accompanying text can be found in a convenient modern edition of his *Natural History of Carolina, Florida and the Bahama Islands* (1731–43): Alan Feduccia, ed., *Catesby's Birds of Colonial America* (Chapel Hill: University of North Carolina Press, 1985), following pp. xvi and 88–89.

55. Excerpts from the text of Audubon and Wilson are collected in accessible reprint editions: Scott Russell Sanders, ed., *Audubon Reader: The Best Writings of John James Audubon* (Bloomington: Indiana University Press, 1986); and Alexander Wilson, *Wilson's American Ornithology*, reprint ed.

(New York: Arno and the New York Times, 1970).

56. On the role of Chapman and Brewster in the Audubon movement, see Frank Graham, *The Audubon Ark: A History of the National Audubon Society* (New York: Alfred Knopf, 1990); and Mark V. Barrow, Jr., *A Passion for Birds: American Ornithology after Audubon* (Princeton, N.J.: Princeton University Press, 1998).

57. See Elizabeth Austin, ed., *Frank M. Chapman in Florida: His Journal and Letters* (Gainesville: University Press of Florida, 1967); the correspondence between Chapman and Brewster from this period in the WBP/AEML-HU and William Brewster and Frank M. Chapman, "Notes on the Birds of the Lower Suwanee River," *The Auk* 8 (1891): 136–37.

58. Paul Hahn, *Where Is That Vanished Bird? An Index of Known Specimens of Extinct North American Species* (Toronto: Royal Ontario Museum, University of Toronto, 1963), lists the locations of over four hundred specimens. Jerome A. Jackson, "The Ivory-Billed Woodpecker," in *Rare and Endangered Biota of Florida*, vol 5., *Birds*, ed. James A. Rodgers, Jr., Herbert W. Kale, and Henry T. Smith (Gainesville: University Press of Florida, 1996), reports more than twenty additional specimens.

59. Reported in Cokinos, *Hope Is the Thing with Feathers*, 67. Biographical information on Allen may be found in Olin S. Pettingill, Jr., "In Memoriam: Arthur A. Allen," *The Auk* 85 (1968): 192–202; Edwin Way Teale, "Arthur A. Allen: Ten Thousand Bird Students Have Learned from Him," *Audubon Magazine* 45 (1943): 84–89; and Richard B. Fisher, "Ambassador of Birdlife," *Audubon Magazine* 67 (1965): 26–31.

60. The history of the Cornell program is chronicled in Arthur A. Allen, "Ornithological Education in America," in *Fifty Years' Progress of American Ornithology, 1883–1933*, ed. Frank M. Chapman and T. S. Palmer

(Lancaster, Pa.: American Ornithologists' Union, 1933), 215–29; Arthur A. Allen, "Cornell's Laboratory of Ornithology," *Living Bird* 1 (1962): 7–36; and Gregory S. Butcher and Kevin McGowan, "History of Ornithology at Cornell University," in *Contributions to the History of North American Ornithology,* ed. William E. Davis, Jr., and Jerome A. Jackson (Cambridge, Mass.: Nuttall Ornithological Club, 1995), 223–45.

61. On recording birds at the Cornell program, see Butcher and McGowan, "Ornithology at Cornell," 233–34; Peter Paul Kellogg, "Bird-Sound Studies at Cornell," *Living Bird* 1 (1962): 37–48; and J. Boswell and D. Couzens, "Fifty Years of Bird Sound Publication in North America: 1931–1981," *American Birds* 36 (1982): 924–43.

62. See the extensive correspondence between Brand and Allen related to the planning of the expedition in Box 52, Arthur A. Allen Papers, Kroch Library, Cornell University, Ithaca, N.Y. (hereafter cited as AAAP/Cornell)

63. Quoted from Allen's account of the expedition: Arthur A. Allen, "Hunting with a Microphone the Voices of Vanishing Birds," *National Geographic* 71 (1937): 696–723, on p. 699. See also Arthur A. Allen and Peter Paul Kellogg, "Recent Observations of the Ivory-Billed Woodpecker," *The Auk* 54 (1937): 164–84.

64. George Miksch Sutton, *Birds in the Wilderness: Adventures of an Ornithologist* (New York: Macmillan, 1936), on p. 195.

65. See the glowing report in Anonymous, "Audubon Association Holds Annual Meeting," *Bird-Lore* 37 (1935): 429–35, on p. 431.

66. Waldron DeWitt Miller, Willard G. Van Name, and Davis Quinn, *A Crisis in Conservation: Serious Danger of Extinction of Many North American Birds* (New York: n.p., 1929). For the circumstances leading to this publication, see Barrow, *Passion for Birds,* 146–50; and chapter 8 above.

67. On reforms in the National Associa-

tion of Audubon Societies during this period, see Stephen Fox, *The American Conservation Movement: John Muir and His Legacy* (Madison: University of Wisconsin Press, 1985), 173–82; Graham, *The Audubon Ark,* 108–44.

68. John Baker to Guy Emerson, November 27, 1940, Box B-4, Folder: Guy Emerson, National Audubon Society Records, Manuscripts and Archives Division, New York Public Library, New York (hereafter cited as NASR/NYPL). Leopold was invited on to the board of the National Association of Audubon Societies in 1936.

69. Aldo Leopold to John Baker, September 30, 1935, Box B-5, Folder: Ideas File, NASR/NYPL.

70. See, for example, the copy of Leopold's "Proposal for a Conservation Inventory of Threatened Species," in Box B-9, Folder: Speeches by Others, NAS/NYPL. Other copies may be found in Box 58, AAAP/Cornell; and Folder: Leopold, Correspondence Files, Museum of Vertebrate Zoology, University of California, Berkeley (hereafter cited as Correspondence Files, MVZ). Leopold published a version of this paper as Aldo Leopold, "Threatened Species: A Proposal to the Wildlife Conference for an Inventory of the Needs of Near-Extinct Birds and Animals," *American Forests* 42 (1936): 116–19. This paper has also been reprinted in Aldo Leopold, *The River of the Mother of God and Other Essays by Aldo Leopold,* ed. Susan L. Flader and J. Baird Callicott (Madison: University of Wisconsin Press, 1991), 230–34.

71. Leopold, "Proposal," 1.

72. Leopold, "Proposal," 2.

73. Leopold, "Proposal," 2–3.

74. Arthur A. Allen to Aldo Leopold, April 21, 1936, Box 58, AAAP/Cornell.

75. Aldo Leopold to Arthur A. Allen, April 28, 1936, Box 58, AAAP/Cornell. See also Arthur A. Allen to Aldo Leopold, May 4, 1936; Aldo Leopold to Arthur Allen, May 13, 1936; and Arthur A. Allen to Aldo Leopold, May 25, 1936, for additional discussion. Allen was hoping that the Technical Committee of recently created American Wildlife Institute, a committee that Leopold headed, might provide funds for the fellowship, but Baker rejected this idea.

76. Much of this speech is reproduced in Anonymous, "The Audubon Fellowship Plan," *Bird-Lore* 38 (1936): 444–46. Additional information on the fellowship plan may be found in Anonymous, "Action on Threatened Species," *Bird-Lore* 39 (1937): 20–24. See also Richard Pough, "An Inventory of Threatened and Vanishing Species," *Transactions of the Second North American Wildlife Conference* (1937): 599–604.

77. John Baker to Arthur A. Allen, November 18, 1936, Box 57, AAAP/Cornell.

78. See the long series of exchanges between Allen and Baker regarding the fellowship, November and December 1936, Box 57, AAAP/Cornell.

79. Arthur A. Allen to John Baker, December 17, 1936, Box 57, AAAP/Cornell.

80. This phase of the study is discussed in his final report: James T. Tanner, *The Ivory-Billed Woodpecker*, Research Report no. 1 (New York: National Audubon Society, 1942), 20–29.

81. Tanner, *Ivory-Billed Woodpecker*, 20 and 99.

82. Tanner, *Ivory-Billed Woodpecker*, 99.

83. Tanner, *Ivory-Billed Woodpecker*, 100.

84. Tanner, *Ivory-Billed Woodpecker*, 100.

85. Tanner, *Ivory-Billed Woodpecker*, 94.

86. Tanner, *Ivory-Billed Woodpecker*, 94–97.

87. Tanner's conclusions were first published in James T. Tanner, "The Life History and Ecology of the Ivory-Bill" (Ph.D. diss., Cornell University, 1940). He also published a popular account of his studies: James T. Tanner, "Three Years with the Ivory-Billed Woodpecker: America's Rarest Bird," *Audubon Magazine* 43 (1941): 5–14.

88. Tanner, *Ivory-Billed Woodpecker*, ii.

89. Tanner, *Ivory-Billed Woodpecker*, iii.

90. In addition to the history in Tanner's report (37–39), see also Richard H. Pough, "Report to the Executive Director, National Audubon Society, on the Present Condition of the Tensas River Forests of Madison Parish, Louisiana, and the Status of the Ivory-billed Woodpecker in this Area as of January, 1944," Box B-9, Folder: Singer Tract, 1944, NASR/NYPL.

91. Tanner, *Ivory-Billed Woodpecker*, 37.

92. Tanner, *Ivory-Billed Woodpecker*, 90.

93. See John H. Baker, "Statement with Regard to Establishment of Wildlife Refuge in Louisiana in an Effort, among Other Things, to Preserve America's Rarest Bird," ms. of speech delivered at Convention of Outdoor Writers Association of America, February 22, 1944, Box B-9, Folder: Singer Tract, 1944, NASR/NYPL; and Cokinos, *Hope Is the Thing with Feathers*, 101.

94. See the estimate in V. H. Sonderegger, "Inspection Report on the Singer Reserve in Madison Parish, Louisiana, March 31, 1933," Box B-9, Folder: Singer Tract, 1936–43, NASR/NYPL.

95. From a copy of the petition in Box B-9, Folder: Singer Tract, 1936–43, NASR/NYPL.

96. "For the Confidential Information of Directors of National Audubon Society, December 15, 1943," Box B-9, Folder: Singer Tract, 1936–43, NASR/NYPL.

97. Reported in Don Eckleberry, *Discovery: Great Moments in the Lives of Outstanding Naturalists* (Philadelphia and New York: J. B. Lippincott, 1961), 195–207.

98. The species was officially listed as endangered on March 11, 1967.

99. There were several reliable sightings of a ivorybills in Cuba during the mid-1980s. See, for example, Lester L. Short and Jennifer F. M. Horne, "'I Saw It!' in the Rugged Mountains of Eastern Cuba, Scientists Spot the Ivory-Billed Woodpecker," *International Wildlife* 17 (March–April 1987): 22–24; and Lester L. Short and Jennifer F. M. Horne,

"The Ivorybill Still Lives," *Natural History* 95 (July 1986): 26–28. Unfortunately, recent efforts to locate the species in Cuba have failed. See Martjan Lammertink, "No More Hope for the Ivory-Billed Woodpecker," *Cotinga* 3 (February 1995), available at http://www.neotropicalbirdclub.org/feature/ivory.html. A series of credible recent claims of ivorybill sightings in Louisiana (1999), Arkansas (beginning in 2004), and Florida (beginning in 2006) have electrified the ornithological and birding community, raising hopes that the species might still be hanging on by a slender thread. See Gallagher, *The Grail Bird*; Geoffrey E. Hill et al., "Evidence Suggesting that Ivory-billed Woodpeckers (*Campephilus principalis*) Exist in Florida," *Avian Conservation and Ecology* 1, no. 3, article 2, available at http://www.ace-eco.org/vol1/iss3/art2/; and Geoffrey E. Hill, *Ivorybill Hunters: The Search for Proof in a Flooded Wilderness* (New York: Oxford University Press, 2007).

100. The biological and historical information on the condor that follows comes from a variety of sources, including David Darlington, *In Condor Country* (Boston: Houghton Mifflin, 1987); Ian McMillan, *Man and the California Condor* (New York: E. P. Dutton, 1968); Dick Smith and Robert Easton, *California Condor: Vanishing America; a Study of an Ancient and Symbolic Giant of the Sky* (Charlotte, N.C.: McNally and Loftin, 1964); Noel Snyder and Helen Snyder, *The California Condor: A Saga of Natural History and Conservation* (San Diego: Academic Press, 2001); Sanford R. Wilbur, *The California Condor, 1966–76: A Look at Its Past and Future,* North American Fauna, No. 72 (Washington, D.C.: U.S. Department of Interior, Fish and Wildlife Service, 1978); John Nielsen, *Condor: To the Brink and Back—the Life and Times of One Giant Bird* (New York: HarperCollins, 2006); and Peter Alagona, "Biography of a 'Feathered Pig': The California Condor Conservation Controversy," *Journal of the History of Biology* 37, no. 3 (2004): 557–83. A convenient

recent summary of current research on the California condor is Noel Snyder and Helen Snyder, "California Condor (*Gymnogyps Californianus*)," in *The Birds of North America,* ed. A. Poole and F. Gill (Philadelphia: Birds of North America, Inc., 2002), no. 610.

101. Snyder and Snyder, *The California Condor,* 10.

102. See the discussion in Snyder and Snyder, *The California Condor,* 10–13.

103. That is, many condor scholars accept the basic notion of the "Pleistocene overkill" hypothesis first formulated in the mid-1960s by the paleontologist Paul Martin. He and his followers argue that when humans first arrived in North American about twelve thousand years ago, they quickly killed off numerous large mammals as they moved across the continent and into South America. What remains unclear is why the condor could not have survived on the ample supply of large mammals that remained in the eastern United States after the Pleistocene extinctions. A recent sympathetic review of Martin's ideas is found in Shepard Krech III, *The Ecological Indian: Myth and History* (New York: W. W. Norton, 1999), 29–43.

104. D. D. Simons, "Interactions between California Condors and Humans in Prehistoric Far Western North America," in *Vulture Biology and Management,* ed. Sanford R. Wilbur and Jerome A. Jackson (Berkeley: University of California Press, 1983): 470–94, argues against the idea that Native American captures played a significant role in reducing the population of California condors, while Snyder and Snyder, *The California Condor,* 43–45, argue that "ceremonial practices may have had major impacts."

105. The most thorough summary of the early accounts of the condor is Harry Harris, "The Annals of *Gymnogyps* to 1900," *The Condor* 43, no. 1 (1941): 3–55.

106. Joseph Grinnell, "Archibald Menzies, First Collector of California Birds," *The Condor* 34, no. 6 (1932): 243–52.

107. Harris, "The Annals of *Gymnogyps* to 1900," 13–17. While Lewis and Clark failed to bring an entire specimen of the large bird back East, they did deposit a skull and primary wing feather in Peale's Museum.

108. Harris, "The Annals of *Gymnogyps* to 1900," 27–29.

109. James G. Cooper, "A Doomed Bird," *Zoe* 1 (1890): 248–49. On Cooper's life and career, see Eugene V. Coan, *James Graham Cooper, Pioneer Western Naturalist* (Moscow: University Press of Idaho, 1981).

110. Wilbur, *The California Condor*, 21. Wilbur (72–82) includes a list of all known California condor casualties between 1792 and 1976, including many that were collected.

111. Henry R. Taylor, "Open Letter to W. F. Webb," *Nidiologist* 2, no. 7 (1895): 100.

112. Truesdale's collecting activities are extensively documented in McMillan, *Man and the California Condor*, 30–49.

113. Wilbur, *The California Condor*, 20.

114. Snyder and Snyder, *The California Condor*, 56.

115. Michael L. Smith, *Pacific Visions: California Scientists and the Environment, 1850–1915* (New Haven, Conn.: Yale University Press, 1987), details how California scientists became engaged in conservation initiatives during this period, but curiously omits any discussion of the Cooper Ornithological Club. The early history of this organization is recounted in Henry B. Kaeding, "Retrospective," *The Condor* 10, no. 6 (1908): 215–18; Harold C. Bryant, "The Cooper Club Member and Scientific Work," *The Condor* 16, no. 3 (1914), 101–7; and Henry Swarth, *History of the Cooper Ornithological Club* (1929?).

116. Quoted in Bryant, "The Cooper Club Member and Scientific Work," 102.

117. William L. Finley, "Life History of the California Condor: Part II—Historical Data and Range of the Condor," *The Condor* 10, no. 1 (1908): 5–10, on pp. 5–6.

118. William T. Hornaday, *Our Vanishing Wildlife: Its Extermination and Preservation* (New York: New York Zoological Society, 1913), 23–24.

119. John C. Phillips, "An Attempt to List the Extinct and Vanishing Birds of the Western Hemisphere, with Some Notes on Recent Status, Location of Specimens, Etc.," in *Verhandlungen des VI. Internationalen Ornithologen-Kongresses in Kopenhagen 1926*, ed. F. Steinbacher (Berlin: Dornblüth, 1929), 506.

120. Miller et al., *A Crisis in Conservation*, 2.

121. Alexander Wetmore, "The Eagle, King of Birds, and His Kin," *National Geographic* 64, no. 1 (1933): 43–95.

122. The account that follows relies on Ray Ford, "Saving the Condor: Robert E. Easton's Fight to Create the Sisquoc Condor Sanctuary," *Noticias—Santa Barbara Historical Society* 32, no. 4 (1986): 75–83, who unfortunately does not indicate locations for the letters he cites.

123. Ernest I. Dyer, "Meeting the Condor on Its Own Ground," *The Condor* 37 (1935): 5–11, on p. 11.

124. See John H. Baker to Joseph Grinnell, February 6, 1935; Joseph Grinnell to John H. Baker, March 5, 1935; Correspondence Files, MVZ.

125. John H. Baker to Joseph Grinnell, March 16, 1935, Correspondence Files, MVZ.

126. Joseph Grinnell to John H. Baker, April 8, 1935, Correspondence Files, MVZ.

127. There is an account of his visit in Cyril S. Robinson, "Notes on the California Condor Collected on Los Padres National Forest" (copy of mimeographed report in author's possession [1939?]), 4–5.

128. These concerns are detailed in Ford, "Saving the Condor."

129. Baker's discussion with Silcox is referenced in John H. Baker to Joseph Grinnell, March 12, 1935; while his discussion with the Cooper Ornithological Club is mentioned in John H. Baker to Joseph Grinnell, April 30, 1936; both in Correspondence Files, MVZ.

130. Cyril S. Robinson, "A Report on

Study of Life Habits of the California Condor—May 1936" (copy of mimeographed report in author's possession).

131. Reported in Theodore Catton and Lisa Mighetto, *The Fish and Wildlife Job on the National Forests: A Century of Game and Fish Conservation, Habitat Protection, and Ecosystem Management* (Washington, D.C.: U.S. Department of Agriculture, Forest Service, 1998), 173.

132. On the early activities and mission of the Forest Service, see David A. Clary, *Timber and the Forest Service* (Lawrence: University Press of Kansas, 1986); Harold K. Steen, *The U.S. Forest Service: A History* (Seattle: University of Washington Press, 1976). On wildlife management within the agency, see Catton and Mighetto, *The Fish and Wildlife Service Job;* and Dennis Roth, "A History of Wildlife Management in the Forest Service" (Forest History Society Library, Durham, N.C., 1989).

133. See Christian C. Young, *In the Absence of Predators: Conservation and Controversy on the Kaibab Plateau* (Lincoln: University of Nebraska Press, 2002); and Thomas R. Dunlap, "That Kaibab Myth," *Journal of Forest History* 32, no. 2 (1988): 60–68.

134. Forest Service, U.S. Department of Agriculture, *Code of Federal Regulations of Forest Service* (Washington, D.C.: Government Printing Office, 1940), on p. 49.

135. A version of Baker's address to the National Association of Audubon Societies may be found in Baker, "The Audubon Research Fellowship Plan."

136. John H. Baker to Joseph Grinnell, October 22, 1936, Correspondence Files, MVZ.

137. The biographical information on Grinnell that follows is from Hilda Wood Grinnell, "Joseph Grinnell: 1877–1939," *The Condor* 42 (1940): 3–34; Jean M. Linsdale, "In Memoriam: Joseph Grinnell," *The Auk* 59 (1942): 269–85; and William E. Ritter, "Joseph Grinnell," *Science* 90, no. 2326 (1939):

75–76, while the information on the Museum of Vertebrate Zoology and its relationship to Berkeley is found in Susan Leigh Star and James R. Griesemer, "Institutional Ecology, 'Translations' and Boundary Objects: Amateurs and Professionals in Berkeley's Museum of Vertebrate Zoology, 1907–39," *Social Studies of Science* 19 (1989): 387–420; Eakin, "History of Zoology at the University of California, Berkeley"; Eakin, *History of Zoology at Berkeley;* Ned K. Johnson, "Ornithology at the Museum of Vertebrate Zoology," in *Contributions to the History of North American Ornithology,* ed. William E. Davis, Jr., and Jerome A. Jackson (Cambridge, Mass.: Nuttall Ornithological Club, 1995), 183–221; and Barbara R. Stein, *On Her Own Terms: Annie Montague Alexander and the Rise of Science in the American West* (Berkeley: University of California Press, 2001).

138. Joseph Grinnell, "Why We Need Birds and Animals," *Scientific Monthly* 41, no. 6 (1935): 553–56, on p. 556.

139. See "Comments of Alden Miller on Baker Letter of Nov. 5, 1936," Folder: John H. Baker, 1934–38, Correspondence Files, MVZ.

140. John H. Baker to Joseph Grinnell, November 27, 1936, Folder: John H. Baker, 1934–38, Correspondence Files, MVZ.

141. A copy of the circular he used is found in J. R. Pemberton, "To My Audience," circular letter, Folder: California Condor II, Box C-49, NASR/NYPL. See also the *Save the Condor* leaflet found in this folder.

142. The biographical information that follows comes primarily from an interview of Koford completed in 1979 and partially transcribed in David Phillips and Hugh Nash, eds., *The Condor Question: Captive or Forever Free?* (San Francisco: Friends of the Earth, 1981), 67–97; and Anonymous, "Carl Buckingham Koford," in *American Men and Women of Science: The Physical and Biological Sciences* (New York: Bowker, 1986), 2716.

143. From Alden Miller's preface to Carl B. Koford, *The California Condor,* Research

Report, no. 4 (New York: National Audubon Society, 1953), vii.

144. Quoted in Darlington, *In Condor Country*, 86–87.

145. Phillips and Nash, *The Condor Question*, 85; Joseph Grinnell to "To Whom It May Concern," February 27, 1939, Correspondence Files, MVZ.

146. Carl Koford to Gentlemen [Joseph Grinnell and Alden Miller], March 7, 1939, Correspondence Files, MVZ.

147. Koford, *The California Condor*, 1, which is also the source for the figures cited in the previous sentence.

148. Miller's early involvement in the project is documented in the Koford correspondence found in the MVZ.

149. John H. Baker to Alden H. Miller, July 10, 1941, Correspondence Files, MVZ.

150. Alden H. Miller to John H. Baker, July 15, 1941, Correspondence Files, MVZ.

151. See, for example, John H. Baker to Alden H. Miller, May 12, 1942, Correspondence Files, MVZ.

152. John H. Baker to Alden H. Miller, April 21, 1941, April 26, 1941, August 16, 1941, Correspondence Files, MVZ.

153. John H. Baker to C. S. Robinson, December 17, 1941; C. S. Robinson to John H. Baker, April 29, 1942; John H. Baker to Alden H. Miller, May 12, 1942; Correspondence Files, MVZ.

154. See Carl B. Koford to John H. Baker, January 14, 1947, and September 29, 1947, Correspondence Files, MVZ. Numerous additional documents related to the expansion of the Sespe Sanctuary may be found in these files.

155. Koford, *The California Condor*, 137.

156. H. H. Sheldon, "What Price Condor?" *Field and Stream* 45 (September 1939): 22–23, 61.

157. John H. Baker to Alden H. Miller, May 12, 1942, Correspondence Files, MVZ.

158. Tanner, *The Ivory-Billed Woodpecker;* Robert Porter Allen, *The Roseate Spoonbill,*

Research Report no. 2 (New York: National Audubon Society, 1942); and Robert Porter Allen, *The Whooping Crane*, Research Report, no. 3 (New York: National Audubon Society, 1952). The desert mountain sheep fellowship failed to produce a published report.

159. Phillips and Nash, *The Condor Question*, 68.

160. The continued delays are detailed in Koford's letters from this period, Correspondence Files, MVZ.

161. Koford, *The California Condor*, v.

162. Koford, *The California Condor*, vii–viii.

163. See the map in Koford, *The California Condor*, 10.

164. Koford, *The California Condor*, 17, 19. Curiously, in his dissertation he had hedged his bets a bit here: "If my estimate of 60 condors is in error it is too conservative. If there are 100 condors, so much the better for the survival of the species." Quoted in Snyder and Snyder, *The California Condor*, 63.

165. Koford, *The California Condor*, 23.

166. This point is made emphatically in Snyder and Snyder, *The California Condor*, 64–65, who argue that Koford overstated the sensitivity of the species in order to secure additional protection for it.

167. Koford, *The California Condor*, 129. The list that follows comes from pp. 129–35 of the report.

168. The discussion of proposals to save the species comes from Koford, *The California Condor*, 135–38.

169. See, for example, Alden H. Miller, "The Case against Trapping California Condors," *Audubon Magazine* 55, no. 6 (1953): 261–62.

170. Koford, *The California Condor*, 135.

171. Koford, *The California Condor*, 136.

172. Koford, *The California Condor*, 137–38.

173. Harvey I. Fisher, "Review of Carl B. Koford, the California Condor," *The Auk* 71 (1954): 91–93; Joe T. Marshall, Jr., "Review of Carl B. Koford, the California Condor," *Wilson Bulletin* 66 (1954): 75–76. In their recent study *The California Condor*, Snyder and Snyder are a bit more ambivalent. On the one hand, they praise Koford's research as "awesome," but they also point out he was ultimately limited in what he could accomplish by the "general state of ecological knowledge at the time" and the "difficulties inherent in condor studies" (33). Koford was, for example, unable to state with certainty whether the decline of the condor was due to reproductive failure or to excess mortality, information that was crucial in devising effective responses to the species' decline.

174. Alden H. Miller, Ian McMillan, and Eben McMillan, *The Current Status and Welfare of the California Condor*, Research Report, no. 6 (New York: National Audubon Society, 1965).

175. Snyder and Snyder, *California Condor*, 61–63.

176. Snyder and Snyder, *California Condor*, 71.

CHAPTER 10

1. The discussion about Erickson and his early ideas about captive breeding of whooping cranes that follows is from Jan DeBlieu, *Meant to Be Wild: The Struggle to Save Endangered Species through Captive Breeding* (Golden, Colo.: Fulcrum Publishing, 1991), 108–12; and Alston Chase, *In a Dark Wood: The Fight over Forests and the Rising Tyranny of Ecology* (Boston: Houghton Mifflin, 1995), 81–83.

2. See Anonymous, *Ray Charles Erickson*, http://www.pwrc.usgs.gov/resshow/perry/bios/EricksonRay.htm.

3. Ray C. Erickson, "Breeding Habits of the Canvasback, *Nyroca valisineria* (Wilson) on the Malheur National Wildlife Refuge, Oregon" (master's thesis, Iowa State College, 1942); Ray C. Erickson, "Life History and Ecology of the Canvasback, *Nyroca valisineria*

(Wilson), in Southeastern Oregon" (Ph.D. diss., Iowa State College, 1948). On the environmental history of the Malheur Refuge, see Nancy Langston, *Where Land and Water Meet: A Western Landscape Transformed* (Seattle: University of Washington Press, 2003).

4. The story of the whooping crane that follows comes from Robert Porter Allen, *The Whooping Crane* (New York: National Audubon Society, 1952); Robin W. Doughty, *Return of the Whooping Crane* (Austin: University of Texas Press, 1989); Thomas R. Dunlap, "Organization and Wildlife Preservation: The Case of the Whooping Crane in North America," *Social Studies of Science* 21 (1991): 192–221; James C. Lewis, "Whooping Crane (*Grus americana*)," in *The Birds of North America*, ed. A Poole and F. Gill (Washington, D.C.: American Ornithologists' Union, 1995), no. 153; and Faith McNulty, *The Whooping Crane: The Bird That Defies Extinction* (New York: E. P. Dutton, 1966).

5. Allen, *Whooping Crane*, 75–78, concludes that there were about 1,300 or so whooping cranes by 1870. More recent estimates suggest that only about 500–700 whooping cranes were alive by then. See Lewis, "Whooping Crane," 16.

6. See McNulty, *Whooping Crane*, 40–50, for an account of the founding and early history of Aransas.

7. McNulty, *Whooping Crane*, 46–47. A convenient compilation of whooping crane population data is found in Ray C. Erickson and S. R. Derrickson, "The Whooping Crane," in *Crane Research around the World*, ed. James C. Lewis and Hiroyuki Masatomi (Baraboo, Wisc.: International Crane Foundation, 1981), 107.

8. Allen's research is detailed in Allen, *Whooping Crane*, and in his memoir, Robert Porter Allen, *On the Trail of Vanishing Birds* (New York: McGraw-Hill, 1957).

9. A Canadian forestry official spotted a pair of whooping cranes in June 1954, while on a helicopter flight to observe a fire that had broken out in a remote, relatively unexplored section of Wood Buffalo National Park, in northeast Alberta. On Lynch's proposal, see Dunlap, "Organization and Wildlife Preservation," 206–7.

10. Their efforts and arguments are chronicled in Dunlap, "Organization and Wildlife Preservation," 204–5; and McNulty, *Whooping Crane,* 112–13, 137–38, and 151.

11. McNulty, *Whooping Crane,* 151. Apparently, the Whooping Crane Conservation Association was not formally constituted until 1961. See Lorne Scott, "The Whooping Crane Conservation Association," in *Proceedings of the International Crane Workshop, 3–6 September 1965, International Crane Foundation, Baraboo, Wisconsin,* ed. James C. Lewis (Stillwater: Oklahoma State Printing and Publishing, 1976), 223–24.

12. See, for example, Dennis Chitty, "The Technology of Wildlife Management," *Journal of Animal Ecology* 19, no. 1 (1950): 78–80.

13. Quoted in McNulty, *Whooping Crane,* 136. The arguments echoed those that Koford, Miller, and the Audubon Society offered in opposition to captive breeding of the California condor.

14. Erickson and Derrickson, "The Whooping Crane," 113; McNulty, *Whooping Crane,* 178.

15. The history of the Patuxent facility is told in Matthew C. Perry, *The Evolution of Patuxent as a Research Refuge,* http://www.pwrc.usgs.gov/history/perryhist.htm. Erickson's vision for the facility and the larger program of which it forms a part is detailed in Ray C. Erickson, "A Federal Research Program for Endangered Wildlife," *Transactions of the North American Wildlife and Natural Resources Conference* 33 (1968): 418–33.

16. DeBlieu, *Meant to Be Wild,* 111. The quote comes from "Udall Warns Battle to Save Endangered Species Being Lost," U.S. Fish and Wildlife Service Press Release, January 4, 1966, Department of Interior, Box 12,

Lyndon Baines Johnson Library, Austin, Tex. (hereafter cited as LBJ Library).

17. Donald Worster, *Nature's Economy: A History of Ecological Ideas,* 2nd ed. (Cambridge: Cambridge University Press, 1994), 342.

18. See, for example, Paul Boyer, *By the Bomb's Early Light: American Thought and Culture at the Dawn of the Atomic Age* (New York: Pantheon Books, 1985); Margot A. Henriksen, *Dr. Strangelove's America: Society and Culture in the Atomic Age* (Berkeley: University of California Press, 1997); and Allen Winkler, *Life under a Cloud: American Anxiety about the Atom* (New York: Oxford University Press, 1993).

19. Boyer, *By the Bomb's Early Light,* 182–95; Robert J. Lifton and Greg Mitchell, *Hiroshima in America: Fifty Years of Denial* (New York: G. P. Putnam's Sons, 1995), 32–35.

20. Quoted in Boyer, *By the Bomb's Early Light,* 8. See the series of similar responses on pp. 243–87.

21. On the creation and early years of the publication, see also Boyer, *By the Bomb's Early Light,* 63–64, and especially, Alice Kimball Smith, *A Peril and a Hope: The Scientists' Movement in America, 1945–1946* (Chicago: University of Chicago Press, 1965). On the Doomsday Clock and its movements over the years, see Mike Moore, *Midnight Never Came: The History of the Doomsday Clock,* http://www.thebulletin.org/clock/nd95moore1.html.

22. Joel B. Hagen, *An Entangled Bank: The Origins of Ecosystem Ecology* (New Brunswick, N.J.: Rutgers University Press, 1992), 107–12.

23. Besides Hagen, see also Betty Jean Craige, *Eugene Odom: Ecosystem Ecologist and Environmentalist* (Athens: University of Georgia Press, 2001). Beyond funding university-based research, the AEC also established ecological programs at Oak Ridge and other national laboratories. See Stephen Bocking, *Ecologists and Environmental Politics: A History of Contemporary Ecology* (New Haven, Conn.: Yale University Press, 1997).

24. Aldo Leopold's *Sand County Almanac*, published posthumously in 1949, also offered accessible discussions of food chains and biological communities, but it failed to attract a wide readership until the 1960s. Other accessible books that presented basic ecological concepts from this period include John H. Storer, *The Web of Life: A First Book of Ecology* (New York: Devin-Adair, 1953); Lorus J. Milne and Margery J. G. Milne, *The Balance of Nature* (New York: Knopf, 1960); and Marston Bates, *The Forest and the Sea: A Look at the Economy of Nature and the Ecology of Man* (New York: Random House, 1960). A convenient source on the growing public fear about fallout from atmospheric testing, see Winkler, *Life under a Cloud*, 84–108.

25. On the growing concern about strontium 90, see Ralph Lutts, "Chemical Fallout: Rachel Carson's 'Silent Spring,' Radioactive Fallout, and the Environmental Movement," *Environmental Review* 9, no. 3 (1985): 214–25; on Barry Commoner's role in expanding knowledge and concern about fallout, see Michael Egan, *Barry Commoner and the Science of Survival: The Remaking of American Environmentalism* (Cambridge, Mass.: MIT Press, 2008).

26. Steve Ryfle, *Japan's Favorite Mon-Star (the Unauthorized Biography of "the Big G")* (Toronto: ECW Press, 1999); Jerome F. Shapiro, *Atomic Bomb Cinema* (New York: Routledge, 2002); and William M. Tsutsui, "Looking Straight at Them! Understanding the Big Bug Movies of the 1950s," *Environmental History* 12 (April 2007): 237–53.

27. Rachel Carson, *Silent Spring* (Boston: Houghton Mifflin, 1962). On the origins and reception of this landmark book, see Thomas R. Dunlap, *DDT: Scientists, Citizens, and Public Policy* (Princeton, N.J.: Princeton University Press, 1981); Frank Graham, Jr., *Since Silent Spring* (Boston: Houghton

Mifflin, 1970); and Linda Lear, *Rachel Carson: Witness for Nature* (New York: Henry Holt, 1997).

28. Carson, *Silent Spring*, 118–19, 122; Charles E. Broley, "The Bald Eagle in Florida," *Atlantic Naturalist* 12 (1957): 230–31; Charles E. Broley, "The Plight of the American Eagle," *Audubon Magazine* 60 (1958): 162–63; and Jonathan M. Gerrard, *Charles Broley: An Extraordinary Naturalist* (Headingley, Manitoba: White Horse Plains Publisher, 1983).

29. Carson, *Silent Spring*, 119–20.

30. On Carson's use of concern about strontium 90 to further her argument about the dangers of synthetic pesticides, see Lutts, "Chemical Fallout."

31. Carson, *Silent Spring*, v.

32. On the creation and history of the IUCN, see two insider sources—Martin Holdgate, *The Green Web: A Union for World Conservation* (London: Earthscan Publications, 1999); and the official IUCN Web site at http://www.iucn.org/50/background.html—and two more disinterested accounts: Robert Boardman, *International Organization and the Conservation of Nature* (Bloomington: Indiana University Press, 1981); and John McCormick, *Reclaiming Paradise: The Global Environmental Movement* (Bloomington: Indiana University Press, 1989).

33. Quoted in Holdgate, *Green Web*, 37.

34. By 1955, the fledgling organization had representatives from 197 organizations and agencies of forty-five nations, five international organizations, and eight governments.

35. Harold J. Coolidge, "An Outline of the Origins and Growth of the IUCN Survival Service Commission," *Transactions of the North American Wildlife and Natural Resources Conference* 33 (1968): 407–17, on p. 409. See also Harold J. Coolidge, "The Survival Service of the IUPN: Brief History," November 15, 1954, Box 6, Folder: International Relations 1954:International Unions: Protection of Nature: Survival Service, Harold Jefferson

Coolidge Papers, Administrative Papers of the IUCN and Other Conservation Organizations, ca. 1941–69, HUGFP 78.14, Harvard University Archives, Cambridge, Mass. (hereafter cited HJCAP/HUA).

36. Coolidge, "Outline," 410.

37. Even before the Fountainebleau meeting, he had been pushing for the creation of a "world information office" devoted to the problem of "vanishing species."

38. Holdgate, *Green Web,* especially p. 107, documents Coolidge's essential role in the IUCN during its first two decades.

39. See, Harold J. Coolidge, "The Survival Service of the IUPN."

40. International Union for the Conservation of Nature, *Les Fossiles de demain: Treize mammifères menacés d'extinction* (Paris: Société d'édition d'enseignement supérieur, 1954).

41. "Tomorrow's Fossils," *Life* 38, no. 11 (March 14, 1955): 91–98. On the plans for an English edition, see Report of the Survival Service Committee, International Union for the Protection of Nature, Fourth General Assembly, August 25–September 3, 1954, Box 6, Folder: Folder: International Relations 1954: International Unions: Protection of Nature: Survival Service, HJCAP/HUA.

42. See Jean-Marie Vrydagh to J. S. Greenway, August 3, 1950; Harold J. Coolidge to Jean-Paul Harroy, July 10, 1950; and Minutes of the International Union for the Protection of Nature, Second Session of the General Assembly, First Technical Meeting, October 19, 1950, Box 6, Folder: International Relations 1950: International Unions: Protection of Nature, HJCAP/HUA. See also Report of the Committee on Survival Service [1951], Box 6, Folder: International Relations 1951: International Relations: Protection of Nature: Survival Service, HJCAP/HUA. The ICPB proved slow to get started on the work. See Phyllis Baclay-Smith to Harold J. Coolidge, May 10, 1955, Box 6, Folder: International Relations, 1955: International Unions: Protec-

tion of Nature: Survival Service: General, HJCAP/HUA.

43. In his first formal report to the IUPN, presented in 1950, Vrydagh complained that "threatened species" had generally failed to become "the subject of ecological study" because researchers preferred working with more abundant and readily accessible organisms. Minutes of the International Union for the Protection of Nature, Second Session of the General Assembly, First Technical Meeting, October 19, 1950, Box 6, Folder: Folder: International Relations 1950: International Unions: Protection of Nature: Survival Service, HJCAP/HUA.

44. Harold J. Coolidge to Jean-Paul Harroy, November 16, 1954, Box 6, Folder: International Relations 1954: International Unions: Protection of Nature: Survival Service; Russell M. Arundel to Harold J. Coolidge, October 10, 1955; and Harold J. Coolidge to Russell M. Arundel, October 14, 1955, Box 6, Folder: International Relations: International Unions: Protection of Nature: Survival Service: Middle East and Asia Survey, HJCAP/HUA.

45. See the copy of Talbot's curriculum vitae in Box 6, Folder: International Relations: International Unions: Protection of Nature: Survival Service: Middle East and Asia Survey, HJCAP/HUA.

46. Lee Merriam Talbot, *A Look at Threatened Species: A Report on Some Animals of the Middle East and Southern Asia Which Are Threatened with Extermination* (London: Fauna Preservation Society for the International Union for the Preservation of Nature, 1960). On the spin-off projects suggested by Talbot's initial trip, see, for example, Lee Talbot to Fairfield Osborn [ca. 1955] and accompanying documents, Box 6, Folder: International Relations: International Unions: Protection of Nature: Survival Service: Middle East and Asia Survey, HJCAP/HUA.

47. The information on the Arabian oryx that follows is largely from Robert Silverberg,

The Auk, the Dodo, and the Oryx: Vanished and Vanishing Creatures (New York: Thomas Y. Crowell, 1967), 211–14.

48. Report of the Survival Service Committee, IUPN, Fourth General Assembly, August 25–September 3, 1954, Box 6, Folder: International Relations 1954: International Unions: Protection of Nature: Survival Service, HJCAP/HUA.

49. C. L. Boyle, "Survival Service Memorandum no. 2," August 14, 1959, Box 8, Folder: Government Board: Office International Relations 1959: International Unions: IUC: Survival Service Commission, HJCAP/HUA.

50. C. L. Boyle, "Survival Service Memorandum no. 3," January 23, 1960, Box 8, Folder: Government Board: Office International Relations 1960: International Unions: IUC: Survival Service Commission, HJCAP/HUA.

51. C. L. Boyle, "Considerations Affecting the Capture of Endangered Species with a View to Forming Breeding Herds in Captivity," attached to C. L. Boyle, Survival Service Memorandum no. 11, October 22, 1962, Box 9, Folder: Government Board: Office International Relations 1962: International Unions: IUC: Survival Service Commission, HJCAP/HUA.

52. Richard Fitter, "Conservation by Captive Breeding: A General Survey," *Oryx* 9 (1967–68): 87–96, on p. 88. Fitter's paper was delivered at a follow-up meeting at the San Diego Zoo in 1966. See also William G. Conway, "The Opportunity for Zoos to Save Vanishing Species," *Oryx* 9 (1967–68): 154–60; Caroline Jarvis, "The Value of Zoos for Science and Conservation," *Oryx* 9 (1967–68): 127–36; and Peter Scott, "The Role of Zoos in Wildlife Conservation," *Oryx* 9 (1967–68): 82–86.

53. C. L. Boyle, "Survival Service Memorandum no. 7," April 28, 1961, and "Operation Oryx," Box 9, Folder: Government Board: Office International Relations 1960:

International Unions: IUC: Survival Service Commission, HJCAP/HUA.

54. See the papers in Alexandra Dixon and David Jones, eds., *Conservation and Biology of Desert Antelopes* (London: Christopher Helm, 1988); and Catherine Tsagarakis, *History and Distribution*, http://www.arabian-oryx.com/history.htm.

55. The full story of the establishment of the Darwin Biological Research Station is told in Edward J. Larson, *Evolution's Workshop: God and Science in on the Galapagos Islands* (New York: Basic Books, 2001), 175–87; see also Box 6, Folder: Government Board: International Relations 1958: International Unions: IUC: Galapagos Research and Conservation Station, HJCAP/HUA.

56. See the discussion in chapter 6 above.

57. Larson, *Evolution's Workshop*, 187.

58. On the early history of the World Wildlife Fund, see Holdgate, *Green Web*, 65, 79–87; and Raymond Bonner, *At the Hand of Man: Peril and Hope for Africa's Wildlife* (New York: Alfred A. Knopf, 1993), 61–80.

59. Quoted in Bonner, *At the Hand of Man*, 61.

60. Holdgate, *Green Web*, 85.

61. Peter Scott, "Rhino Campaign," March 25, 1962, appended to C. L. Boyle, "Survival Service Memorandum no. 10," September 11, 1962, Box 10, Folder: Government Board: Office International Relations 1962: International Unions: IUC: Survival Service Commission, HJCAP/HUA. See also C. L. Boyle, "The Future of the Survival Service Commission," April 2, 1963, Box 12, Folder: Administration: International Relations: International Unions: IUC: Survival Service Commission, HJCAP/HUA.

62. Peter Scott, "Reorganization of the Survival Service Commission of the IUCN," October 1, 1963, Box 12, Folder: Administration: International Relations: International Unions: IUC: Survival Service Commission, HJCAP/HUA. He also proposed eight subcommittees to serve as liaisons with other

IUCN commissions and outside conservation organizations or to deal with the special problems of island species. By 1964, only ten of the specialist groups had been created.

63. Peter Scott, "Draft Proposal for Collation and Distribution of Information Necessary for Allocation of World Wildlife Fund Grants," December 1961, Box 9, Folder: Government Board: Office of International Relations 1961: International Unions: IUC: Survival Service Commission, HJCAP/HUA.

64. Holdgate, *Green Web,* 91.

65. On discussion about possible publication of the Red Book, see Minutes of the Survival Service Commission, June 22 and 23, 1964; and Minutes of Survival Service Commission, November 29 and 30, 1964, Box 15, Folder: Administration: International Relations 1964: International Unions: IUCN: Survival Service Commission; Minutes of Survival Service Commission, February 17 and 18, 1965; and Minutes of the Survival Service Commission, September 28 and 29, 1965, Box 18, Folder: Administration: International Relations 1965: International Unions: IUCN: Survival Service Commission, HJCAP/HUA.

66. Noel Simon, *Mammalia: A Compilation,* Red Data Book, Vol. 1 (Lausanne: IUCN, 1966); Jack Vincent, *Aves,* Red Data Book, Vol. 2 (Morges, Switzerland: IUCN, Species Survival Commission, 1966); James Fisher, Noel Simon, and Jack Vincent, *Wildlife in Danger* (New York: Viking Press, 1969).

67. René E. Honneger, *Amphibia and Reptilia,* Red Data Book, Vol. 3 (Morges, Switzerland: IUCN, Survival Service Commission, 1968); Robert Rush Miller, *Pisces: Freshwater Fishes,* Red Data Book, Vol. 4 (Morges, Switzerland: IUCN, Species Survival Commission, 1969); and Ronald Melville, *Angiospermae: A Compilation,* Red Data Book, Vol. 5 (Morges, Switzerland: IUCN, Species Survival Commission, 1970).

68. For copies of the speech Udall delivered in Nairobi and other papers associated with this trip, see Box 111, Stewart Udall Pa-

pers, AZ 372, University of Arizona Library, Special Collections, Tucson.

69. Thomas G. Smith, "John Kennedy, Steward Udall, and the New Frontier Conservation," *Pacific Historical Review* 64 (1995): 329–62.

70. On Udall's background, transformation from old-school conservationist, and role in the Johnson administration, see Barbara Laverne Blythe LeUnes, "The Conservation Philosophy of Stewart L. Udall, 1961–1968" (Ph.D. diss., Texas A&M University, 1977); Martin V. Melosi, "Lyndon Johnson and Environmental Policy," in *The Johnson Years,* vol. 2, *Vietnam, the Environment, and Science,* ed. Robert A. Divine (Lawrence: University Press of Kansas, 1987), 113–49; Anonymous, "LBJ Gives Udall Wider Role," *Business Week* (April 3, 1965): 98–100, 105, 107; and Transcript, Stewart L. Udall Oral History Interview I, April 18, 1969; II, May 16, 1969; III, July 29, 1969; IV, October 31, 1969; and V, December 16, 1969, LBJ Library.

71. Stewart L. Udall, "The Legacy of Rachel Carson," *Saturday Review* 47 (1964): 23, 59; and Stewart L. Udall, *The Quiet Crisis* (New York: Holt, Rinehart and Winston, 1963).

72. Udall, *Quiet Crisis,* 190.

73. Udall, *Quiet Crisis,* 196.

74. Clarence F. Pautzke, *Annual Report of the Commission of the Fish and Wildlife Service to the Secretary of the Interior* (Washington, D. C.: Department of the Interior, 1961).

75. A. Starker Leopold, "Predator and Rodent Control," *Transactions of the North American Wildlife and Natural Resources Conference* 29 (1964): 27–47.

76. Ira N. Gabrielson, "What Can We Do about Our Rare and Vanishing Species?" *Scientific American* 158 (January 1938): 5–8. On Gabrielson's life and legacy, see Henry M. Reeves and David B. Marshall, "In Memoriam: Ira Noel Gabrielson," *The Auk* 102 (1985): 412–36.

77. On the trumpeter swan, see Winston E. Banko, *The Trumpeter Swan, North American Fauna*, No. 63 (Washington, D.C.: U.S. Fish and Wildlife Service, 1960). On the agency's failure to act to protect the ivory-billed woodpecker, see chapter 9 above.

78. Daniel B. Beard et al., *Fading Trails: The Story of Endangered American Wildlife* (New York: Macmillan, 1942). See also Hartley H. T. Jackson, "Conserving Endangered Wildlife Species," *Annual Report of the Smithsonian Institution for 1945* (Washington, D.C.: Government Printing Office, 1946): 247–71.

79. Anonymous, "XIII Conference of the International Council for Bird Preservation, New York City, June 11–16, 1962," *Bulletin of the International Council for Bird Preservation* 9 (1963): 33–46; and Anonymous, "120 Species of Birds Facing Extinction," *Washington Post*, June 19, 1962, A1. See also the paper that U.S. Fish and Wildlife staff member John Aldrich delivered at the meeting: "Endangered Species of Birds in the United States," *Bulletin of the International Council for Bird Preservation* 9 (1963): 80–86.

80. Holdgate, *Green Web*, 90; Alexander B. Adams, ed., *First World Conference on National Parks: Proceedings of a Conference: Seattle, Washington, June 30–July 7, 1962* (Washington, D.C.: National Park Service, U.S. Department of Interior, 1964).

81. Anonymous, "Natural History Chapter," *Living Wilderness*, no. 83 (1963): 24. Several months later, Udall published a mildly toned-down version of this basic message in Stewart L. Udall, "To Save Wildlife and Aid Us, Too," *New York Times Magazine*, September 15, 1963, 44–47. See also Stewart L. Udall, "What Price Resources for the Good Life?" *Transactions of the North American Wildlife and Natural Resources Conference* 29 (1964): 52–57.

82. Quoted in Charles C. Mann and Mark L. Plummer, *Noah's Choice: The Future of Endangered Species* (New York: Alfred A. Knopf, 1995), 152.

83. Quoted in Mann and Plummer, *Noah's Choice*, 153.

84. On CREWS, see the letter from Daniel H. Janzen, Special Assistant for Endangered Species, November 6, 1964, appended to copy of Committee on Rare and Endangered Wildlife Species, *Rare and Endangered Fish and Wildlife of the United States* (Washington, D.C.: U.S. Department of Interior, 1964), found in Natural Resources Library, Department of Interior, Washington, D.C. See also Fish and Wildlife Service News Releases, "Interior Department Steps Up Fight to Save Near-Extinct Wildlife," July 6, 1964, and "Interior Department Seeks Information on Endangered Wildlife," November 27, 1964, Department of Interior Papers, Box 17, LBJ Library.

85. That decision alarmed some members of the Species Survival Commission, who feared that the American committee would steal its thunder by publishing before it did. See Minutes of the Science Survival Commission, November 26, 1964, Box 15, Folder: Administration: International Relations 1964: IUCN: Science Survival Commission, HJCAP/HUA.

86. William H. Blair, "U.S. Studying Way to Save Wildlife," *New York Times*, July 7, 1964, 16; Martha Cole, "60 U.S. Wildlife Species Face Extinction Peril," *Washington Post*, July 7, 1964, A7.

87. DeBleiu, *Meant to Be Wild*, 110–11.

88. On Mundt's central role in pushing for the new center, see Lyndon Baynes Johnson to Karl E. Mundt, September 14, 1965, and series of memos and press releases appended to that letter, White House Central File, Box 9, Folder NR2 Fish-Wildlife, LBJ Library. On the center's organization and activities, especially as they related to the whooping crane, see Ray C. Erickson, "New Federal Research Station," *Modern Game Breeding* 2, no. 11 (1966): 21–23, 39–42; Ray C. Erickson, "A Federal Research Program for Endangered Wildlife"; Ray C. Erickson, "Captive Breeding of Whooping Cranes at the Patuxent Wildlife Research Center," in *Breeding Endan-*

gered Species in Captivity, ed. R. D. Martin (London: Academic Press, 1975), 99–114; and Ray C. Erickson, "Whooping Crane Studies at the Patuxent Wildlife Research Center," in Proceedings of the International Crane Workshop, 3–6 September 1965, International Crane Foundation, Baraboo, Wisconsin, ed. James C. Lewis (Stillwater: Oklahoma State University Publishing and Printing, 1976), 166–67.

89. On the history of the 1966 Endangered Species Act, see Mann and Plummer, Noah's Choice, 153–54, 159; Shannon Petersen, "Congress and Charismatic Megafauna: A Legislative History of the Endangered Species Act," Environmental Law 29 (1999): 463–91; Shannon Petersen, Acting for Endangered Species: The Statutory Ark (Lawrence: University Press of Kansas, 2002), 23–28; Steven Lewis Yaffee, Prohibitive Policy: Implementing the Endangered Species Act (Cambridge, Mass.: MIT Press, 1982), 39–42; and Kathryn A. Kohm, "The Act's History and Framework," in Balancing on the Brink of Extinction: The Endangered Species Act and Lessons for the Future, ed. Kathryn A. Kohm (Washington, D.C.: Island Press, 1991), 12–13. See also Department of Interior News Release, "Interior Department Requests Legislation to Protect Endangered Species of Wildlife," June 15, 1965, Department of Interior, Box 12, LBJ Library; the Congressional hearings held on the legislation—U.S. House of Representatives. Subcommittee on Fisheries and Wildlife Conservation of the Committee on Merchant Marine and Fisheries, Endangered Species, 89th Congress, 1st Sess., July 15, 1965; and U.S. Senate, Merchant Marine and Fisheries Subcommittee of the Committee on Commerce, Conservation, Protection, and Propagation of Endangered Species of Fish and Wildlife, August 12, 1965—and the committee reports—U.S. House of Representatives, Committee on Merchant Marine and Fisheries, Report to Accompany H.R. 9424, 89th Congress, 1st Sess., Report no. 1186; U. S. Senate, Committee on Commerce, Re-

port to Accompany H.R. 9424, 89th Congress, 2d Sess., Report no. 1463; and U.S. House of Representatives, Committee of Conference, Conference Report to Accompany H.R. 9424, 89th Congress, 2d Sess., Report no. 2205.

90. Quoted from the letter accompanying the House version, reproduced in Stewart L. Udall to John W. MCormack, June 5, 1965, U.S. House of Representatives, Endangered Species, 123.

91. On the specific terms of the act, see Petersen, Acting for Endangered Species, 23–24; Yaffee, Prohibitive Policy, 40–41, 166–71; and Michael J. Bean, The Evolution of National Wildlife Law, rev. ed. (New York: Praeger, 1981), 319–21, 329.

92. See, for example, Fish and Wildlife Service Press Release, "Interior Department Seeks to Strengthen Wildlife Law Enforcement," April 28, 1966, Department of Interior, Box 17, LBJ Library.

93. Some convenient entry points into the vast literature on attitudes toward mammalian predators in America, see Thomas R. Dunlap, Saving America's Wildlife (Princeton, N.J.: Princeton University Press, 1988); Barry H. Lopez, Of Wolves and Men (New York: Charles Scribner's Sons, 1978); and Jon T. Coleman, Vicious: Wolves and Men in America (New Haven, Conn.: Yale University Press, 2004).

94. On the wide range of souvenirs featuring the species before 1930, see Larry Roberts, Florida's Golden Age of Souvenirs (Gainesville: University Press of Florida, 2001).

95. Hugh M. Smith, "Report on the Fisheries of the South Atlantic States," Bulletin of the United States Fish Commission 11 (1893): 271–356, on p. 344.

96. The first scientific monograph to treat the species was Albert M. Reese, The Alligator and Its Allies (New York: G. P. Putnam's Sons, 1915). See also Edward A. McIlhenny, The Alligator's Life History (Boston: Christopher Publishing House, 1935). The species received limited protection in Louisiana beginning in 1926 and in Florida beginning in 1943.

97. The classic treatment on the impact of air conditioning on southern life is Raymond Arsenault, "The End of the Long Hot Summer: The Air Conditioner and Southern Culture," *Journal of Southern History* 50 (1984): 597–628. On DDT spraying in Florida, see Gordon Patterson, *The Mosquito Wars: A History of Mosquito Control in Florida* (Gainesville: University Press of Florida, 2004). On the environmental history of the Sunshine State, see Mark Derr, *Some Kind of Paradise: A Chronicle of Man and Land in Florida* (New York: William R. Morrow, 1989); and Jack E. Davis and Raymond Arsenault, eds., *Paradise Lost? The Environmental History of Florida* (Gainesville: University Press of Florida, 2005).

98. On Carr and his conservation activities, see Frederick Rowe Davis, *The Man Who Saved Sea Turtles: Archie Carr and the Origins of Conservation Biology* (New York: Oxford University Press, 2007).

99. Archie Carr, "Alligators: Dragons in Distress," *National Geographic* 131, no. 1 (1967): 133–38, on pp. 133–34.

100. Carr, "Dragons in Distress," 134.

101. Carr, "Dragons in Distress," 148. A year after the publication of this article, Carr joined with several fellow gator enthusiasts to create a loose-knit organization, the American Alligator Council, dedicated to "restoring the alligator to a balanced population as an observable part of the natural environment." The activities of the organization are chronicled in the Archie Carr Papers, Department of Special and Area Studies Collections, George A. Smathers Library, University of Florida, Gainesville.

102. On the development of wildlife television shows and their impact, see Gregg Mitman, *Reel Nature: America's Romance with Wildlife on Film* (Cambridge, Mass.: Harvard University Press, 1999); and Cynthia Chris, *Watching Wildlife* (Minneapolis: University of Minnesota Press, 2006).

103. See Mitman, *Reel Nature;* Chris, *Watching Wildlife;* and Derek Bousé, *Wildlife*

Films (Philadelphia: University of Pennsylvania Press, 2000).

104. Roy Pinney, *Vanishing Wildlife* (New York: Dodd, Mead, 1963); Vinzenz Ziswiler, *Extinct and Vanishing Animals: A Biology of Extinction and Survival* (New York: Springer-Verlag, 1967); and Silverberg, *The Auk, the Dodo, and the Oryx.*

105. Fisher et al., *Wildlife in Danger,* 7. See a similar claim in Department of the Interior News Release, "Udall Asks Wildlife Advisors to Study How to Complete National Refuge System," April 25, 1966, Department of Interior, Box 12, LBJ Library.

106. The discussion that follows is drawn from Yaffee, *Prohibitive Policy* and the various committee hearings and reports leading up to the passage of the Endangered Species Conservation Act of 1969, including U.S. House of Representatives, Hearings before the Subcommittee on Fisheries and Wildlife Conservation of the Committee on Merchant Marine and Fisheries, Hearings, Endangered Species, 90th Congress, 1st Sess., October 4, 1967, Serial No. 90-12 (Washington, D.C.: Government Printing Office, 1967); U.S. House of Representatives, Hearings before the Subcommittee on Fisheries and Wildlife Conservation of the Committee on Merchant Marine and Fisheries, Hearings, Bills to Prevent the Importation of Endangered Species . . . , 91st Congress, 1st Sess., February 19 and 20, 1969, Serial No. 91-2 (Washington, D.C.: Government Printing Office, 1969); U.S. Senate, Subcommittee on Merchant Marine and Fisheries of the Committee on Commerce, Hearings, To Prevent the Importation of Endangered Species . . . , 90th Congress, 2nd Sess., July 28, 1968, Serial No. 90-77 (Washington, D.C.: Government Printing Office, 1968); U.S. Senate, Subcommittee on Merchant Marine and Fisheries of the Committee on Commerce, Hearings, Endangered Species, 91st Congress, 1st Sess., Serial No. 91-10 (Washington, D.C.: Government Printing Office, 1969).

107. On the specific terms of the Endangered Species Conservation Act of 1969, see Yaffee, *Prohibitive Policy*, 42–47, 166–71; Bean, *Evolution of National Wildlife Law*, 321–24; Petersen, *Acting for Endangered Species*, 25–28; and Kohm, "The Act's History," 13–15.

108. That legislation, which committed the federal government to a broad range of environmental goals, established the president's Council on Environmental Quality and mandated Environmental Impact Statements for all federal projects. On Nixon's environmental legacy, see J. Brooks Flippen, *Nixon and the Environment* (Albuquerque: University of New Mexico Press, 2000); James Rathlesberger, ed., *Nixon and the Environment: The Politics of Devastation* (New York: Village Voice Books, 1972); and John C. Whitaker, *Striking a Balance: Environment and Natural Resources Policy in the Nixon-Ford Years* (Washington, D.C.: American Enterprise Institute for Public Policy Research, 1976). On the growth of environmentalism during this period, see Samuel P. Hays, *Beauty, Health, and Permanence: Environmental Politics in the United States, 1955–1985* (Cambridge: Cambridge University Press, 1987); Hal K. Rothman, *The Greening of a Nation? Environmentalism in the United States since 1945* (Fort Worth, Tex.: Harcourt Brace College Publishers, 1998); and Philip Shabecoff, *A Fierce Green Fire: The American Environmental Movement* (New York: Hill and Wang, 1993).

109. Quoted in Flippen, *Nixon and the Environment*, 51.

110. Riley E. Dunlap, "Trends in Public Opinion toward Environmental Issues: 1965–1990," in *American Environmentalism: The U.S. Environmental Movement, 1970–1990*, ed. Riley E. Dunlap and Angela G. Mertig (Washington, D.C.: Taylor and Francis, 1992), 89–116.

111. Peter R. Janssen, "The Age of Ecology," in *Ecotactics: The Sierra Club Handbook for Environmental Activists*, ed. John G. Mitchell (New York: Pocket Books, 1970), 53–62, on p. 53.

112. It featured either the letters *e* and *o* (standing for environment and organism) superimposed over one another or the Greek letter *theta*, which the editors associated with the word *thanatos* (or death). See http://atlasgeo .span.ch/fotw/flags/us_eco.html#ecol.

113. On the earlier bills, see Yaffee, *Prohibitive Policy*, 49–56. The quote from Nixon's 1972 address is found in Petersen, *Acting for Endangered Species*, 27.

114. The spotted cats discussed in this section are known to science as *Panthera tigris* (tiger), *Panthera pardus* (leopard), *Acinonyx jubatus* (cheetah), *Panthera onca* (jaguar), *Leoparda pardalis* (ocelot), *Leopardus wiedii* (margay), and *Leopardus tigrinis* (oncilla).

115. On the growth in demand for spotted cat furs, see Greta Nilsson et al., *Facts about Furs* (Washington, D.C.: Animal Welfare Institute, 1980); Elizabeth Ewing, *Fur in Dress* (London: B. T. Batsford, 1981); and Anna Municchi, *Ladies in Furs, 1940–1990* (Modena, Italy: Zanfi Editori, 1993). The latter publication argues that initial interest in these furs began earlier, when furriers trying to revive a declining industry, began promoting the use of nontraditional furs, like spotted cats.

116. Kristin Nowell and Peter Jackson, eds., *Wild Cats: Status Survey and Conservation Action Plan*, http://lynx.uio.no/catfolk/public82.htm.

117. Examples include Anonymous, "Fun Furs," *Defenders of Wildlife News* 42 (1967): 279–82; Norman Myers, "Spotted Cats in Trouble," *International Wildlife* 1 (1971): 4–6; Jack O'Conner, "The Big Cats in Trouble," *Outdoor Life* 146, no. 4 (1970): 63–67, 167–71; and Clive Gammon, "The Sad Tale of the Tiger," *Sports Illustrated* 33 (1970): 42–43.

118. The protests, boycott, newspaper ad, and response from furriers are discussed in Margaret Crimmins, "Is Furor over Furs Only Fashionable," *Washington Post*, June 28, 1970, E10–E11.

119. Anonymous, "Furs, Fashion, and Conservation," *Vogue* 156, no. 3 (September 1970): 144. See additional coverage of the issue in Anonymous, "The Fashion for Fakes," *Vogue* 156, no. 1 (July 1970): 94–103; Anonymous, "Furs for the Woman with a Conscience," *McCall's* 98 (October 1970): 96–101; and Anonymous, "The Fur Flies in Defense of the Great Cats," *Life* 68 (February 27, 1970): 68–70.

120. F. Wayne King, "Adventures in the Skin Trade," *Natural History* 80, no. 5 (1971): 8–16.

121. This had been one of the concessions granted to the fur and leather industry when it complained about the impact of the bill that became the Endangered Species Act of 1969.

122. Crimmins, "Furor over Furs."

123. Norman Myers, "The Spotted Cats and the Fur Trade," in *The World's Cats*, vol. 1, *Ecology and Conservation*, ed. Randall L. Eaton (Winston, Ore.: World Wildlife Safari, 1973), 276–309.

124. Randall L. Eaton, ed., *The World's Cats*, vol. 1, *Ecology and Conservation* (Winston, Ore.: World Wildlife Safari, 1973), 328.

125. Eaton, *The World's Cats*, 1:343.

126. The episode is recounted in Lewis G. Regenstein, *The Politics of Extinction: The Shocking Story of the World's Endangered Wildlife* (New York: Macmillan Publishing, 1975), 141–42. See also Lewis G. Regenstein, "Can Spotted Cats Survive Despite the Fur Industry?" *Washington Post*, August 9, 1971, A30; and his master's thesis, Lewis G. Regenstein, "A History of the Endangered Species Act of 1973 and an Analysis of Its History, Strengths and Weaknesses, Administration, and Probable Future Effectiveness" (M.A. thesis, Emory University, 1975), 33–39.

127. An edited version of that report was published as John L. Paradiso, *Status Report on Cats (Felidae) of the World, 1971*, Special Scientific Report: Wildlife no. 157 (Washington, D.C.: U.S. Department of the Interior, Fish and Wildlife Service, 1972).

128. Quoted in Regenstein, "History of the Endangered Species Act," 34.

129. Jack Anderson, "Hardship Safaris," *Washington Post*, September 16, 1972, D15.

130. On the exhibition of dead whales and their skeletons, see Richard Ellis, *Men and Whales* (New York: Alfred A. Knopf, 1991), 371–85. On the exhibition of live marine mammals at zoos and aquariums, see Vernon Kisling, ed., *Zoo and Aquarium History: Ancient Animal Collections to Zoological Gardens* (Boca Raton, Fla.: CRC Press, 2001); and Randall R. Reeves and James G. Mead, "Marine Mammals in Captivity," in *Conservation and Management of Marine Mammals*, ed. John R. Twiss, Jr., and Randall R. Reeves (Washington, D.C.: Smithsonian Institution Press, 1999), 412–36.

131. Marineland served as a site for public entertainment, an underwater motion picture studio, and a research facility for the Atlantic bottle-nosed dolphin. Mitman, *Reel Nature*, 159–77.

132. See Susan G. Davis, *Spectacular Nature: Corporate Culture and the Sea World Experience* (Berkeley: University of California Press, 1997).

133. David M. Lavigne, Victor B. Scheffer, and Stephen R. Kellert, "The Evolution of American Attitudes toward Marine Mammals," in *Conservation and Management of Marine Mammals*, ed. John R. Twiss, Jr., and Randall R. Reeves (Washington, D.C.: Smithsonian Institution Press, 1999), 10–47, on p. 17; see also Erich Hoyt, "Whale Watching," in *Encyclopedia of Marine Mammals*, ed. William F. Perrin, Bernd Würsig, and J. G. M. Thewissen (San Diego: Academic Press, 2002), 1305–10.

134. Bemi DeBus, *A Brief History of the Founding of the American Cetacean Society*, http://www.acsonline.org/aboutus/history/.

135. Ellis, *Men and Whales*, 460.

136. Scott McVay, "The Last of the Great Whales," *Scientific American* 215, no. 2 (1966): 13–21.

137. Lavigne et al., "Attitudes toward

Marine Mammals," 26; Ellis, *Whales and Men,* 436–37.

138. On the role of Flipper in transforming the dolphin into a glamour species, see Mitman, *Reel Nature,* 178–79.

139. On Lilly and his legacy, see Mitman, *Reel Nature,* 173–77; and Andrew C. Revkin, "John C. Lilly Dies at 86," *New York Times,* October 7, 2001.

140. The story of the controversy is well told in Michael L. Gosliner, "The Tuna-Dolphin Controversy," in *Conservation and Management of Marine Mammals,* ed. John R. Twiss, Jr., and Randall R. Reeves (Washington, D.C.: Smithsonian Institution Press, 1999), 120–55.

141. William F. Perrin, "The Porpoise and the Tuna," *Sea Frontiers* 14 (1968), 166–74; and William F. Perrin, "Using Porpoises to Catch Tuna," *World Fishing* 18 (1969): 42–45.

142. Barry Kent MacKay, *Seal Slaughtering in the Northwest Atlantic: It's Not Over and It's Worse than Ever* (1997), http://www.api4animals.org/491.htm; David M. Lavigne and K. M. Kovacs, *Harps and Hoods: Ice-Breeding Seals of the Northwest Atlantic* (Waterloo: University of Waterloo Press, 1988); and Douglass Pimlott, "The Whitecoat in Peril," *Audubon* 69, no. 5 (1967): 76–81.

143. Quoted in MacKay, "Seal Slaughtering."

144. Lavigne et al., "Attitudes toward Marine Mammals," 24.

145. K. Radway Allen, *Conservation and Management of Whales* (Seattle: Washington Sea Grant, 1980), 11–12. On the history of whaling, see also Ellis, *Men and Whales.*

146. On the regulation of whaling, see Allen, *Conservation and Management of Whales;* Kurk Dorsey, "The International Context of Whale Conservation on the High Seas, 1931–1965" (paper presented at the Annual Meeting of the American Society for Environmental History, Denver, Colo., 2002); and Patricia Birnie, ed., *International Regulation of Whaling: From Conservation of Whaling*

to *Conservation of Whales and Regulation of Whale Watching,* 2 vols. (New York: Ocean Publications, 1985).

147. Allen, *Conservation and Management of Whales,* 24.

148. Dorsey, "The International Context"; and Frank C. Whitmore, "Remington Kellogg: October 5, 1892–May 8, 1969," *Biographical Memoirs of the National Academy of Sciences* 46 (1975): 159–89.

149. The episode is discussed in Mann and Plummer, *Noah's Choice,* 155–56; and Regenstein, *Politics of Extinction,* 67–68.

150. The taxonomy of the many marine mammals bills introduced comes from Gosliner, "Tuna-Dolphin Controversy," 121–22.

151. On the final terms and subsequent amendments of the Marine Mammal Protection Act (1972), see Bean, *Evolution of National Wildlife Law,* 281–317; Donald C. Baur, Michael J. Bean, and Michael L. Gosliner, "The Laws Governing Marine Mammal Conservation in the United States," in *Conservation and Management of Marine Mammals,* ed. John R. Twiss, Jr., and Henry M. Reeves (Washington, D.C.: Smithsonian Institution Press, 1999), 48–86; Lavonne R. Dye, "The Marine Mammal Protection Act: Maintaining the Commitment to Marine Mammal Conservation," *Case Western Law Review* 43 (1993): 1410–48; and Susan C. Alker, "The Marine Mammal Protection Act: Refocusing the Approach to Conservation," *UCLA Law Review* 44 (1996): 527–77.

152. Quoted in Alker, "Marine Mammal Protection Act," 534.

153. Quoted in Bean, *Evolution of National Wildlife Law,* 292.

154. Any authorized taking had to be done using humane methods that did not inflict unnecessary pain on the animals being harvested. The bill also called for additional research on species and their habitats and directed the State Department and other federal agencies to negotiate protective agreements and treaties. And it placed oversight of the administration of the act under an independent Marine Mammal Commission.

155. William G. Conway, "What's 'Endangered'? New Legislation Fails to Come to Grips with a Difficult Problem," *Animal Kingdom* 73 (1970): 2–9.

156. For a general assessment of CITES, see John Hutton and Barnabas Dickson, *Endangered Species, Threatened Convention: The Past, Present and Future of CITES* (London: Earthscan Publications Ltd., 2000).

157. On the history of CITES, see Boardman, *International Organization*, 88–95; see also Russell E. Train, *Politics, Pollution, and Pandas: An Environmental Memoir* (Washington, D.C.: Island Press, 2003), 143–45; and Holdgate, *Green Web*, 114–15.

158. These are detailed in Boardman, *International Organization*, 89–90; and Robert Gillette, "Endangered Species: Diplomacy Tries Building an Ark," *Science* 179, no. 4075 (1973): 777, 779–80.

159. Robert Gillette, "Endangered Species: Moving toward a Cease-Fire," *Science* 179, no. 4078 (1973): 1107–9.

160. Morris Kaplan, "An Illegal Wild-Fur Ring Broken Up Here by U.S.," *New York Times*, February 22, 1973, 78.

161. On the specific terms of CITES, see Bean, *Evolution of National Wildlife Law,* 324–29; and Gillette, "Endangered Species."

162. Bean, *Evolution of National Wildlife Law,* 325.

163. The specific actions leading to passage of the Endangered Species Act of 1973 are detailed in Mann and Plummer, *Noah's Choice,* 156–63; Yaffee, *Prohibitive Policy,* 47–57; Petersen, *Acting for Endangered Species,* 27–35; and Chase, *In a Dark Wood,* 89–93. On the legislative history of the act, see Congressional Research Service, *A Legislative History of the Endangered Species Act of 1973, as Amended in 1976, 1977, 1978, 1979, and 1980* (Washington, D.C.: U.S. Government Printing Office, 1982).

164. Quoted in Petersen, *Acting for Endangered Species,* 27.

165. Peterson, *Acting for Endangered Species,* 28–29.

166. Peterson, *Acting for Endangered Species,* 29.

167. Richard M. Nixon, "Statement on Signing of the Endangered Species Act of 1973, December 28, 1973," *Public Papers of the Presidents of the United States: Richard Nixon* (Washington, D.C.: Government Printing Office, 1975), 1027–28.

168. A convenient summary of the bill's major provisions may be found in Yaffee, *Prohibitive Policy,* 56–57, 166–71.

169. The initial version of that report was completed in 1975, and a revised version was published as Edward S. Ayensu and Robert A. DeFilipps, *Endangered and Threatened Plants of the United States* (Washington, D.C.: Smithsonian Institution Press and World Wildlife Fund, 1978).

170. U.S. Supreme Court, *TVA v. Hill,* 437 U.S. 153 (1978).

171. Erickson and Derrickson, "The Whooping Crane," 114.

172. Lewis, *Whooping Crane,* 19.

173. Recent developments are covered in Matthiessen, *Birds of Heaven: Travels with Cranes* (New York: North Point Press, 2001), 274–300.

174. Matthiessen, *Birds of Heaven,* 295, 296.

CONCLUSION

1. Charles Darwin, *On the Origin of Species by Means of Natural Selection, or the Preservation of Favoured Races in the Struggle for Life* (London: John Murray, 1859), 317, 322.

2. See, for example, Holly Doremus, "Science and Controversy," and Mary Ruckelshaus and Donna Darm, "Science and Implementation," in *The Endangered Species Act at Thirty: Conserving Biodiversity in Human-Dominated Landscapes,* ed. J. Michael Scott, David D. Goble, and Frank W. Davis (Washington, D.C.: Island Press, 2006), 2:91–103 and 2:104–26.

3. On the snail darter controversy, see Kenneth Murchison, *The Snail Darter Case: TVA versus the Endangered Species Act* (Lawrence: University Press of Kansas, 2007); and Shannon Petersen, *Acting for Endangered Species: The Statutory Ark* (Lawrence: University Press of Kansas, 2002), 39–77.

4. Etnier's discovery brought to 140 the total number of North American species of fish commonly known as darters.

5. U.S. Supreme Court, *TVA v. Hill*, 437 U.S. 153 (1978).

6. Murchison, *Snail Darter Case*, 110.

7. On the California condor, see chapter 9; Carl B. Koford, *The California Condor* (New York: National Audubon Society, 1953); Noel Snyder and Helen Snyder, *The California Condor: A Saga of Natural History and Conservation* (San Diego: Academic Press, 2000); and John Nielson, *Condor: The Brink and Back—the Life and Times of One Giant Bird* (New York: HarperCollins, 2006).

8. Snyder and Snyder, *California Condor*, 135.

9. "Milestones in California Condor Conservation," Conservation and Research for Endangered Species, Zoological Society of San Diego, http://cres.sandiegozoo.org/projects/sp_condors_milestones.html.

10. Steven Lewis Yaffee, *The Wisdom of the Spotted Owl: Policy Lessons for a New Century* (Washington, D.C.: Island Press, 1994), 14–15. See also William Dietrich, *The Final Forest: The Battle for the Last Great Trees of the Pacific Northwest* (New York: Penguin, 1993); and Petersen, *Acting for Endangered Species*, 81–118.

11. Yaffee, *Wisdom of the Spotted Owl*, 99.

12. On the listing record of the Fish and Wildlife Service, see J. Michael Scott, Dale D. Goble, Leona K. Svancara, and Anna Pigorna, "By the Numbers," in *The Endangered Species Act at Thirty: Renewing the Conservation Promise*, ed. Dale D. Goble, J. Michael Scott, and Frank W. Davis (Washington, D.C.: Island Press, 2006), 16–35. On Reagan's environ-

mental record, see Jeffrey K. Stine, "Natural Resources and Environmental Policy," in *The Reagan Presidency: Pragmatic Conservatism and Its Legacies,* ed. W. Elliot Brownlee and Hugh Davis Graham (Lawrence: University Press of Kansas, 2003), 233–56.

13. Yaffee, *Wisdom of the Spotted Owl*, 116.

14. Yaffee, *Wisdom of the Spotted Owl*, 132; Petersen, *Acting for Endangered Species,* 98–99. After several months of public comment on the proposal, the Fish and Wildlife Service reduced the amount of critical habitat to 8.2 million acres.

15. John Cushman, "Judge Approves Plan for Logging in Forests Where Rare Owls Live," *New York Times*, December 22, 1994, A1, B12; Brian Tokar, "Between the Loggers and the Owls: The Clinton Northwest Forest Plan," *Ecologist* 23, no. 4 (July/August 2004): 149–53; and Petersen, *Acting for Endangered Species,* 111–12.

16. Norman Myers, *The Sinking Ark: A New Look at the Problem of Vanishing Species* (Oxford: Pergamon Press, 1979), 5.

17. Paul Ehrlich and Anne Ehrlich, *Extinction: The Causes and Consequences of the Disappearance of Species* (New York: Random House, 1981), xiv.

18. These figures are based on a search on WorldCat using "extinction" as the title and limiting returns only to those books that are listed in five or more libraries, which helps avoid duplicate counts based on minor variations in cataloging data.

19. The National Academy of Sciences is an honorific society of distinguished scholars that has been offering scientific advice to the federal government and the public since its founding in 1863.

20. On the conference and the origins of the term biodiversity, see David Takacs, *The Idea of Biodiversity: Philosophies of Paradise* (Baltimore: Johns Hopkins University Press, 1996), 34–40.

21. Takacs, *Idea of Biodiversity*, 38.

22. Wilson edited a volume based on the

forum: E. O. Wilson, ed., *Biodiversity* (Washington, D.C.: National Academy Press, 1988).

23. Joan Hamilton, "Babbitt's New Life List: Interior Secretary Bruce Babbitt Establishes National Biological Survey," *Sierra* 78, no. 6 (November/December 1993): 52–54.

24. James Conaway, "Babbitt in the Woods: The Clinton Environmental Revolution That Wasn't," *Harper's* 287, no. 1723 (December 1993): 52–60, on p. 54; Jeffrey P. Cohn, "The National Biological Survey," *BioScience* 43, no. 8 (September 1993): 521–22; Laura Tangley, "A National Biological Survey," *BioScience* 35, no. 11 (December 1985): 686–90.

25. The symposium resulted in an edited volume: Ke Chung Kim and Lloyd Knutson, eds., *Foundations for a National Biological Survey* (Lawrence, Kans.: Association of Systematics Collections, 1986). Systematic biology is scientific study of the earth's biodiversity and the relationships among living things through time.

26. See Committee on the Formation of the National Biological Survey, *A Biological Survey for the Nation* (Washington, D.C.: National Academy Press, 1993).

27. Frederic H. Wagner, "Whatever Happened to the National Biological Survey," *BioScience* 49, no. 3 (March 1999): 219–21. Conservatives in Congress also imposed a moratorium on listing of new species under the Endangered Species Act until September 1996. Petersen, *Acting for Endangered Species*, 113–14.

28. On the history and content of this report, see "Systematics Agenda 2000: Integrating Biological Diversity and Societal Needs," *Systematic Botany* 16, no. 4 (1991): 758–61; Systematics Agenda 2000, *Systematics Agenda 2000: Charting the Biosphere* (New York: Society of Systematics Biologists, American Society of Plant Taxonomists, Willi Hennig Society, and Association of Systematics Collections, 1994); Michael F. Claridge, "Introducing Systematics Agenda 2000,"

Biodiversity and Conservation 4 (1994): 451–54; W. Hardy Eshbaugh, "Systematics Agenda 2000: An Historical Perspective," *Biodiversity and Conservation* 4 (1994): 455–62; and the rest of the articles in that special issue.

29. Jane Lubchenco et al., "The Sustainable Biosphere Initiative: An Ecological Research Agenda," *Ecology* 72, no. 3 (1991): 371–412.

30. Claridge, "Introducing Systematics Agenda 2000," 451.

31. See, for example, James E. Rodman and Jeannine H. Cody, "The Taxonomic Impediment Overcome: NSF's Partnerships for Enhancing Expertise in Taxonomy (PEET) as a Model," *Systematic Biology* 52, no. 3 (June 2003): 428–35; and Stephen Blackmore, "Biodiversity Update: Progress in Taxonomy," *Science* 298, no. 5592 (11 October 2002): 365.

32. On the history of the Convention on Biological Diversity, see Martin Holdgate, *The Green Web: A Union for World Conservation* (London: Earthscan Publications, 1999), 213–16; see also William K. Stevens, "Talks Seek to Prevent Huge Loss of Species," *New York Times*, March 3, 1992, C4; and Stas Burgiel, "Convention on Biological Diversity: A Progress Report," SciDev.Net: Science and Development Network, http://www.scidev.net/dossiers/index.cfm?fuseaction=policybrief&dossier=11&policy=45.

33. The full text of the Convention on Biological Diversity is available at http://www.cbd.int/convention/convention/shtml. The three broad principles mentioned here are adapted from David E. Pitt, "A Biological Treaty to Save Species Becomes Law," *New York Times*, January 2, 1994, 4.

34. Keith Schneider, "White House Snubs U.S. Envoy's Plea to Sign Rio Treaty," *New York Times*, June 5, 1992, A1, A6.

35. "Biodiversity Pact on the Ropes," *New York Times*, September 26, 1994, A16.

36. Andrew C. Revkin, "Biologists Sought a Treaty; Now They Fault It," *New York Times*, May 7, 2002, F1.

37. The full history of conservation biology has yet to be written but a convenient starting place is Curt Meine, Michael Soulé, and Reed Noss, "'A Mission-Oriented Discipline': The Growth of Conservation Biology," *Conservation Biology* 20, no. 3 (2006): 631–51; see also Timothy J. Farnham, *Saving Nature's Legacy: The Origins of the Idea of Biological Diversity* (New Haven, Conn.: Yale University Press, 2007).

38. Ann Gibbons, "Conservation Biology in the Fast Lane," *Science* 255, no. 5040 (January 3, 1992): 20.

39. Michael Soulé and Bruce A. Wilcox, eds., *Conservation Biology: An Evolutionary-Ecological Perspective* (Sunderland, Mass.: Sinauer Associates, 1980), 1.

40. Thomas Lovejoy, "Foreword," in Soulé and Wilcox, eds., *Conservation Biology,* ix.

41. Michael E. Soulé, "History of the Society for Conservation Biology: How and Why We Got Here," *Conservation Biology* 1, no. 1 (May 1987): 4–5.

42. Michael E. Soulé, "What Is Conservation Biology?" *BioScience* 35, no. 11 (December 1985): 727–34.

43. See, for example, James G. Teer, "Conservation Biology—a Book Review," *Wildlife Society Bulletin* 17, no. 3 (1989): 337–89.

44. The figures and quote are from Gibbons, "Conservation Biology in the Fast Lane," 20–22.

45. Reed F. Noss, "The Naturalists Are Dying Off," *Conservation Biology* 10, no. 1 (February 1996): 1–3.

46. Noss's editorial was one of a series of publications decrying the decline of natural history at the time. See, for example, David S. Wilcove and Thomas Eisner, "The Impending Extinction of Natural History," *Chronicle Review* 47, no. 3 (September 15, 2000): B-24; and David J. Schmidly, "What It Means to Be a Naturalist and the Future of Natural History at American Universities," *Journal of Mammalogy* 86, no. 3 (June 2005): 449–56.

47. Wilson has also been a leading advocate of embracing the term "naturalist" to encompass those who study biodiversity and has even named his own autobiography *Naturalist.*

48. Edward O. Wilson, "On the Future of Conservation Biology," *Conservation Biology* 14, no. 1 (February 2000): 1–3. Richard Louv, *Last Child in the Woods: Saving Our Children from Nature-Deficit Disorder* (Chapel Hill, N.C.: Algonquin Books, 2005), expands the argument about the importance of continued contact with the natural world from naturalists to all humans.

SELECT BIBLIOGRAPHY

Adams, William M. *Against Extinction: The Story of Conservation.* London: Earthscan, 2004.

Alagona, Peter. "Biography of a 'Feathered Pig': The California Condor Conservation Controversy." *Journal of the History of Biology* 37, no. 3 (2004): 557–83.

Allard, Dean C. *Spencer Fullerton Baird and the United States Fish Commission.* New York: Arno Press, 1978.

Allen, Arthur A. "Ornithological Education in America." In *Fifty Years' Progress of American Ornithology, 1883–1933,* edited by Frank M. Chapman and T. S. Palmer, 215–29. Lancaster, Penn.: American Ornithologists' Union, 1933.

———. "Hunting with a Microphone the Voices of Vanishing Birds." *National Geographic* 71 (1937): 696–723.

Allen, Arthur A., and Peter Paul Kellogg. "Recent Observations of the Ivory-Billed Woodpecker." *The Auk* 54 (1937): 164–84.

Allen, David Elliston. *The Naturalist in Britain: A Social History.* 2nd ed. Princeton, N.J.: Princeton University Press, 1994.

Allen, Garland C. *Life Science in the Twentieth Century.* Cambridge: Cambridge University Press, 1978.

———. "Naturalists and Experimentalists: The Genotype and the Phenotype." *Studies in History of Biology* 3 (1979): 179–209.

Allen, Glover M. *Extinct and Vanishing Mammals of the Western Hemisphere, with the Marine Species of All the Oceans.* Special Publication no. 11. New York: American Committee for International Wild Life Protection, 1942.

Allen, J. A. *The American Bisons, Living and Extinct. Memoirs of the Museum of Comparative Zoology* 4, no. 10. Cambridge, Mass.: Welch, Bigelow, 1876.

———. "The Extirpation of the Large Indigenous Mammals of the United States." *The Penn Monthly* 7 (1876): 794–806.

———. "The North American Bison and Its Extermination." *The Penn Monthly* 7 (1876): 214–24.

———. "On the Decrease of Birds." *The Penn Monthly* 7 (1876): 931–44.

———. *Autobiographical Notes and a Bibliography of the Scientific Publications of Joel Asaph Allen.* New York: American Museum of Natural History, 1916.

Allen, K. Radway. *Conservation and Management of Whales.* Seattle: Washington Sea Grant, 1980.

Allen, Robert Porter. *The Roseate Spoonbill.* Research Report no. 2. New York: National Audubon Society, 1942.

————. *The Whooping Crane*. Research Report no. 3. New York: National Audubon Society, 1952.

————. *On the Trail of Vanishing Birds*. New York: McGraw-Hill Book, 1957.

Anderson, Atholl. *Prodigious Birds: Moas and Moa-Hunting in Prehistoric New Zealand*. Cambridge: Cambridge University Press, 1989.

Andrei, Mary Anne. *Nature's Mirror: The Taxidermists Who Shaped America's Natural History Museums and Saved Our Endangered Species*. Chicago: University of Chicago Press, forthcoming.

Anker, Peder. *Imperial Ecology: Environmental Order in the British Empire, 1895–1945*. Cambridge, Mass.: Harvard University Press, 2001.

Appel, Toby A. "Science, Popular Culture and Profit: Peale's Philadelphia Museum." *Journal of the Society for the Bibliography of Natural History* 9 (1980): 619–34.

Arsenault, Raymond. "The End of the Long Hot Summer: The Air Conditioner and Southern Culture." *Journal of Southern History* 50 (1984): 597–628.

Atwood, Elizabeth Lawrence. "Symbol of a Nation: The Bald Eagle in American Culture." *Journal of American Culture* 13 (1990): 63–69.

Atwood, Wallace W. *The Protection of Nature in the Americas*. Mexico City: Antigua imprenta de E. Murguia, 1940.

Audubon, Maria R., ed. *Audubon and His Journals*. 2 vols. New York: Charles Scribner's Sons, 1897.

Ayensu, Edward S., and Robert A. DeFilipps. *Endangered and Threatened Plants of the United States*. Washington, D.C.: Smithsonian Institution Press and World Wildlife Fund, 1978.

Banko, Winston E. *The Trumpeter Swan*. North American Fauna, no. 63. Washington, D.C.: U.S. Fish and Wildlife Service, 1960.

Baratay, Eric, and Elisabeth Hardouin-Fugier. *Zoo: A History of Zoological Gardens in the West*. London: Reaktion Books, 2002.

Barrow, Mark V., Jr. "Gentlemanly Specialists in the Age of Professionalization: The First Century of Ornithology at Harvard's Museum of Comparative Zoology." In *Contributions to the History of North American Ornithology*, edited by William E. Davis, Jr., and Jerome A. Jackson, 55–94. Cambridge, Mass.: Nuttall Ornithological Club, 1995.

————. *A Passion for Birds: American Ornithology after Audubon*. Princeton, N.J.: Princeton University Press, 1998.

————. "The Specimen Dealer: Entrepreneurial Natural History in America's Gilded Age." *Journal of the History of Biology* 33 (2000): 493–534.

————. "Naturalists as Conservationists: American Scientists, Social Responsibility, and Political Activism before the Bomb." In *Science, History and Social Activism: A Tribute to Everett Mendelsohn*, edited by Garland C. Allen and Roy M. MacLeod, 217–33. Dordrecht: Kluwer Academic Publishers, 2001.

Barsness, Larry. *Heads, Hides, and Horns: The Compleat Buffalo Book*. Fort Worth: Texas Christian University Press, 1985.

Bates, Marston. *The Nature of Natural History*. New York: Charles Scribner's Sons, 1950.

Bean, Michael J. *The Evolution of National Wildlife Law*. Rev. ed. New York: Praeger, 1981.

Beard, Daniel B., Frederick C. Lincoln, Victor C. Cahalane, Hartley H. T. Jackson, and Ben H. Thompson. *Fading Trails: The Story of Endangered American Wildlife*. New York: Macmillan, 1942.

Bedini, Silvio A. *Thomas Jefferson and American Vertebrate Paleontology*. Virginia Division of

Mineral Resources Publication 61. Charlottesville, Va.: Department of Mines, Minerals and Energy, 1985.

———. *Thomas Jefferson: Statesman of Science.* New York: Macmillan, 1990.

Beebe, William. *Galápagos: World's End.* New York: G. P. Putnam's Sons, 1924.

Beers, Diane L. *For the Prevention of Cruelty: The History and Legacy of Animal Rights Activism in the United States.* Athens: Ohio University Press, 2006.

Bell, Whitfield J., Jr. "A Box of Old Bones: A Note on the Identification of the Mastodon, 1766–1806." *Proceedings of the American Philosophical Society* 93, no. 2 (1949): 169–77.

Benedict, R. C. "Game Laws for the Conservation of Wild Plants." *Science* 58, no. 1490 (1923): 39–41.

Benson, Keith. "From Museum Research to Laboratory Biology: The Transformation of Natural History into Academic Biology." In *The American Development of Biology,* edited by Ronald Rainger, Keith Benson, and Jane Maienschein, 49–83. Philadelphia: University of Pennsylvania Press, 1988.

Benson, Keith R., Jane Maienschein, and Ronald Rainger, eds. *The Expansion of American Biology.* New Brunswick, N.J.: Rutgers University Press, 1991.

Berry, Andrew, ed. *Infinite Tropics: An Alfred Russel Wallace Anthology.* London: Verso, 2002.

Bildstein, Keith L., and Robert A. Compton. "Mountaintop Science: The History of Hawk Mountain Sanctuary." In *Contributions to the History of North American Ornithology,* edited by William E. Davis and Jerome A. Jackson, 2:153–81. Cambridge: Nuttall Ornithological Club, 2000.

Birnie, Patricia, ed. *International Regulation of Whaling: From Conservation of Whaling to Conservation of Whales and Regulation of Whale Watching.* 2 vols. New York: Ocean Publications, 1985.

Blockstein, David E. "Passenger Pigeon: *Ectopistes migratorius.*" In *The Birds of North America,* edited by A. Poole and F. Gill, no. 611. Philadelphia: Birds of North America, 2002.

Blunt, Wilfred. *The Compleat Naturalist: A Life of Linnaeus.* New York: Viking, 1971.

Boardman, Robert. *International Organization and the Conservation of Nature.* Bloomington: Indiana University Press, 1981.

Bocking, Stephen. "Conserving Nature and Building a Science: British Ecologists and the Origins of the Nature Conservancy." In *Science and Nature: Essays in the History of Environmental Sciences,* edited by Michael Shortland, 89–114. Stanford in the Vale: British Society for the History of Science, 1993.

———. *Ecologists and Environmental Politics: A History of Contemporary Ecology.* New Haven, Conn.: Yale University Press, 1997.

Bodry-Sanders, Penelope. *Carl Akeley: Africa's Collector, Africa's Savior.* New York: Paragon House, 1991.

Bogue, Margaret Beattie. "To Save the Fish: Canada, the United States, the Great Lakes, and the Joint Commission of 1892." *Journal of American History* 79, no. 4 (1993): 1429–54.

Bonner, Raymond. *At the Hand of Man: Peril and Hope for Africa's Wildlife.* New York: Alfred A. Knopf, 1993.

Bousé, Derek. *Wildlife Films.* Philadelphia: University of Pennsylvania Press, 2000.

Bowler, Peter J. *Fossils and Progress: Paleontology and the Idea of Progressive Evolution in the Nineteenth Century.* New York: Science History Publications, 1976.

———. *Evolution: The History of an Idea.* Rev. ed. Berkeley: University of California Press, 1989.

————. *The Norton History of the Environmental Sciences.* 1st American ed. New York: W. W. Norton, 1993.

Boyd, Julian. "The Megalonyx, the Megatherium, and Thomas Jefferson's Memory Lapse." *Proceedings of the American Philosophical Society* 102, no. 5 (1958).

Boyer, Paul. *By the Bomb's Early Light: American Thought and Culture at the Dawn of the Atomic Age.* New York: Pantheon Books, 1985.

Brechin, Gray. "Conserving the Race: Natural Aristocracies, Eugenics, and the U.S. Conservation Movement." *Antipode* 28 (1996): 229–45.

Brewster, William. "The Heath Hen of Massachusetts." *The Auk* 2 (1885): 80–84.

————. "The Present Status of the Wild Pigeon (*Ectopistes migratorius*) as a Bird of the United States, with Some Notes on Its Habits." *The Auk* 6 (1889): 285–91.

————. "The Heath Hen." *Forest and Stream* 35 (1890): 188.

Bridges, William. *Gathering of Animals: An Unconventional History of the New York Zoological Society.* New York: Harper and Row, 1974.

Brigham, David R. *Public Culture in the Early Republic: Peale's Museum and Its Audience.* Washington, D.C.: Smithsonian Institution Press, 1995.

Brockway, Lucile. *Science and Colonial Expansion: The Role of the British Royal Botanic Gardens.* Reprint ed. New Haven, Conn.: Yale University Press, 2002.

Broun, Maurice. *Hawks Aloft: The Story of Hawk Mountain.* New York: Dodd, Mead, 1948.

Brouwer, G. A. *The Organisation of Nature Protection in the Various Countries.* Special Publication no. 9. New York: American Committee for International Wild Life Protection, 1938.

Browne, Janet. *The Secular Ark: Studies in the History of Biogeography.* New Haven, Conn.: Yale University Press, 1983.

————. *Charles Darwin: Voyaging.* New York: Alfred A. Knopf, 1995.

————. *Charles Darwin: The Power of Place.* New York: Alfred A. Knopf, 2002.

————. "Natural History." In *The Oxford Companion to the History of Modern Science,* edited by J. L. Heilbron, 556–63. Oxford: Oxford University Press, 2003.

Buick, T. Lindsay. *The Mystery of the Moa: New Zealand's Avian Giant.* New Plymouth, New Zealand: Thomas Avery and Sons, 1931.

————. *The Discovery of Dinornis.* New Plymouth, New Zealand: Thomas Avery and Sons, 1936.

————. *The Moa-Hunters of New Zealand.* New Plymouth, New Zealand: Thomas Avery and Sons, 1937.

Bull, Bartle. *Safari: A Chronicle of Adventure.* New York: Viking, 1988.

Burgess, Robert L. "The Ecological Society of America: Historical Data and Some Preliminary Analyses." In *History of American Ecology,* edited by Frank N. Egerton, 1–24. New York: Arno Press, 1977.

Burgess, Thorton W. *Now I Remember: Autobiography of an Amateur Naturalist.* Boston: Little, Brown, 1960.

Burkhardt, Richard W. *Spirit of System: Lamarck and Evolutionary Biology.* Cambridge, Mass.: Harvard University Press, 1977.

————. *Patterns of Behavior: Konrad Lorenz, Niko Tinbergen, and the Founding of Ethology.* Chicago: University of Chicago Press, 2005.

Busch, Briton. *The War against the Seals: A History of the North American Seal Fishery.* Montreal: McGill-Queen's University Press, 1985.

Butcher, Gregory S., and Kevin McGowan. "History of Ornithology at Cornell University." In

Contributions to the History of North American Ornithology, edited by William E. Davis, Jr., and Jerome A. Jackson, 223–45. Cambridge, Mass.: Nuttall Ornithological Club, 1995.

Bynum, William F. "The Great Chain of Being after Forty Years: An Appraisal." *History of Science* 13 (1975): 1–28.

Cain, Joe. "Ecology." In *History of Modern Science*, edited by Brian S. Baigrie, 3:44–68. New York: Charles Scribner's Sons, 2002.

Callicott, J. Baird, ed. *Companion to* A Sand County Almanac. Madison: University of Wisconsin Press, 1987.

Callicott, J. Baird, and Eric T. Freyfogle, eds. *For the Health of the Land: Previously Unpublished Essays and Other Writings*. Washington, D.C.: Island Press, 1999.

Camerini, Jane R., ed. *The Alfred Russel Wallace Reader*. Baltimore: Johns Hopkins University Press, 2002.

Cameron, Jenks. *Bureau of the Biological Survey: Its History, Activities and Organization*. Baltimore: Johns Hopkins University Press, 1929.

Carr, Archie. "Alligators: Dragons in Distress." *National Geographic* 131, no. 1 (1967): 133–48.

Carson, Rachel. *Silent Spring*. Boston: Houghton Mifflin, 1962.

Cart, Theodore Whaley. "'New Deal' for Wildlife: A Perspective on Federal Conservation Policy, 1933–40." *Pacific Northwest Quarterly* 63 (1972): 113–20.

———. "The Lacey Act: America's First Nationwide Wildlife Statute." *Forest History* 17 (October 1973): 4–13.

Catlin, George. *North American Indians: Being Letters and Notes on Their Manners, Customs, and Conditions*. 2 vols. London: George Catlin, 1880.

Catton, Theodore. *Inhabited Wilderness: Indians, Eskimos, and National Parks in Alaska*. Albuquerque: University of New Mexico Press, 1997.

Catton, Theodore, and Lisa Mighetto. *The Fish and Wildlife Job on the National Forests: A Century of Game and Fish Conservation, Habitat Protection, and Ecosystem Management*. Washington, D.C.: U.S. Department of Agriculture, Forest Service, 1998.

Chase, Alston. *In a Dark Wood: The Fight over Forests and the Rising Tyranny of Ecology*. Boston: Houghton Mifflin, 1995.

Chris, Cynthia. *Watching Wildlife*. Minneapolis: University of Minnesota Press, 2006.

Christensen, Bonnie. "From Divine Nature to Umbrella Species: The Development of Wildlife Science in the United States." In *Forest and Wildlife Science in America*, edited by Harold K. Steen, 209–29. Durham, N.C.: Forest History Society, 1999.

Cittadino, Eugene. "Ecology and the Professionalization of Botany in America, 1890–1905." *Studies in History of Biology* 4 (1980): 171–98.

———. "A 'Marvelous Cosmopolitan Preserve': The Dunes, Chicago, and the Dynamic Ecology of Henry Cowles." *Perspectives on Science* 1 (1993): 520–59.

———. "Ecology." In *The Oxford Companion to the History of Modern Science*, edited by J. L. Heilbron, 229–32. Oxford: Oxford University Press, 2003.

Clary, David A. *Timber and the Forest Service*. Lawrence: University Press of Kansas, 1986.

Cohen, Claudine. *The Fate of the Mammoths: Fossils, Myth, and History*. Chicago: University of Chicago Press, 2002.

Cokinos, Christopher. *Hope Is the Thing with Feathers: A Personal Chronicle of Vanished Birds*. New York: Jeremy P. Tarcher/Putnam, 2000.

Coleman, Jon T. *Vicious: Wolves and Men in America*. New Haven, Conn.: Yale University Press, 2004.

Coleman, William. *Georges Cuvier, Zoologist: A Study in the History of Evolution Theory.* Cambridge, Mass.: Harvard University Press, 1964.

Committee on the Formation of the National Biological Survey. *A Biological Survey for the Nation.* Washington, D.C.: National Academy Press, 1993.

Committee on Rare and Endangered Wildlife Species. *Rare and Endangered Fish and Wildlife of the United States.* Washington, D.C.: U.S. Department of Interior, 1964.

Congressional Research Service. *A Legislative History of the Endangered Species Act of 1973, as Amended in 1976, 1977, 1978, 1979, and 1980.* Washington, D.C.: U.S. Government Printing Office, 1982.

Conway, James. "Babbitt in the Woods: The Clinton Environmental Revolution That Wasn't." *Harper's Magazine* 287, no. 1723 (December 1993): 52–60.

Conway, William G. "The Opportunity for Zoos to Save Vanishing Species." *Oryx* 9 (1967–68): 154–60.

———. "Zoos: Their Changing Role." *Science* 163, no. 3862 (1969): 48–52.

———. "What's 'Endangered'? New Legislation Fails to Come to Grips with a Difficult Problem." *Animal Kingdom* 73 (October 1970): 2–9.

Coolidge, Harold J., Jr. "Wanted—One Giant Gorilla." *The Sportsman* 4, no. 5 (1928): 41–44, 98.

———. "Protección internacional de los recursos naturales." *Bulletin of the Pan American Union* 73, no. 7 (1939): 399–412.

———. "Notes on Conservation in the Americas." *Chronica Botanica* 7, no. 4 (1942): 155–63.

———. "An Outline of the Origins and Growth of the IUCN Survival Service Commission." *Transactions of the North American Wildlife and Natural Resources Conference* 33 (1968): 407–17.

Craige, Betty Jean. *Eugene Odom: Ecosystem Ecologist and Environmentalist.* Athens: University of Georgia Press, 2001.

Craighead, Frank, and John Craighead. *Hawks in the Hand: Adventures in Photography and Falconry.* Boston: Houghton Mifflin, 1939.

Croker, Robert A. *Pioneer Ecologist: The Life and Work of Victor Ernest Shelford, 1877–1968.* Washington, D.C.: Smithsonian Institution Press, 1991.

———. *Stephen Forbes and the Rise of American Ecology.* Washington, D.C.: Smithsonian Institution Press, 2001.

Cronon, William. *Changes in the Land: Indians, Colonists, and the Ecology of New England.* New York: Hill and Wang, 1983.

———. "The Trouble with Nature; or, Getting Back to the Wrong Nature." *Environmental History* 1 (1996): 7–28.

Crosby, Alfred W. *Ecological Imperialism: The Biological Expansion of Europe, 900–1900.* Cambridge: Cambridge University Press, 1986.

Cumbler, John C. "The Early Making of Environmental Consciousness: Fish, Fish Commissions, and the Connecticut River." *Environmental History Review* 15 (1991): 73–91.

Cushman, Gregory Todd. "The Birth of the Environmental Technocrat in Peru: Marine Science, State-Led Development, and the Compañía Administradora del Guano, 1906–1964." Ph.D. dissertation, University of Texas–Austin, 1999.

Cutright, Paul Russell. *Lewis and Clark, Pioneering Naturalists.* Urbana: University of Illinois Press, 1969.

Cuvier, Georges. "Mémoire sur les espèces d'éléphans tant vivantes que fossiles." *Magasin encyclopédique* 3 (1796): 440–45.

————. "Memoir upon Living and Fossil Elephants." *Philosophical Magazine* 26 (1806): 158–69, 204–11, 302–13.

————. "Additional Memoir upon Living and Fossil Elephants." *Philosophical Magazine* 28:258–64, 359–66; 29:52–65, 244–54; 30:15–25 (1807–8).

Czech, Brian, and Paul R. Krausman. *The Endangered Species Act: History, Conservation, Biology, and Public Policy.* Baltimore: Johns Hopkins University Press, 2001.

Danz, Harold P. *Of Bison and Man.* Niwot: University of Colorado Press, 1997.

Darlington, David. *In Condor Country.* Boston: Houghton Mifflin, 1987.

Darwin, Charles. *Journal of Researches into the Geology and Natural History of the Various Countries Visited by the HMS Beagle, under the Command of Captain Fitzroy, R.N., from 1832 to 1836.* London: Henry Colburn, 1839.

————. *On the Origin of Species by Means of Natural Selection, or the Preservation of Favoured Races in the Struggle for Life.* London: John Murray, 1859.

Dary, David A. *The Buffalo Book: The Full Saga of the American Animal.* Chicago: Sage Books, 1974.

Davis, Frederick Rowe. *The Man Who Saved Sea Turtles: Archie Carr and the Origins of Conservation Biology.* New York: Oxford University Press, 2007.

Davis, Jack E., and Raymond Arsenault, eds. *Paradise Lost? The Environmental History of Florida.* Gainesville: University Press of Florida, 2005.

Davis, Susan G. *Spectacular Nature: Corporate Culture and the Sea World Experience.* Berkeley: University of California Press, 1997.

DeBlieu, Jan. *Meant to Be Wild: The Struggle to Save Endangered Species through Captive Breeding.* Golden, Colo.: Fulcrum Publishing, 1991.

Dehler, Gregory J. "An American Crusader: William Temple Hornaday and Wildlife Protection in America, 1840–1940." Ph.D. dissertation, Lehigh University, 2001.

Delacour, Jean. *The Living Air: Memoirs of an Ornithologist.* London: Country Life Limited, 1966.

Derr, Mark. *Some Kind of Paradise: A Chronicle of Man and Land in Florida.* New York: William R. Morrow, 1989.

Desmond, Adrian, and James Moore. *Darwin: The Life of a Tormented Evolutionist.* New York: Warner Books, 1991.

Dexter, Ralph W. "History of the Ecologists' Union: Spin-Off from the E.S.A. and Prototype of the Nature Conservancy." *Bulletin of the Ecological Society of America* 59, no. 3 (1978): 146–47.

Dietrich, William. *The Final Forest: The Battle for the Last Great Trees of the Pacific Northwest.* New York: Penguin, 1993.

Dixon, Alexandra, and David Jones, eds. *Conservation and Biology of Desert Antelopes.* London: Christopher Helm, 1988.

Dolph, James Andrew. "Bringing Wildlife to Millions: William Temple Hornaday, the Early Years: 1854–1896." Ph.D. dissertation, University of Massachusetts, 1975.

Doremus, Holly. "Science and Controversy." In *The Endangered Species Act at Thirty: Conserving Biodiversity in Human-Dominated Landscapes,* edited by J. Michael Scott, Dale D. Goble, and Frank W. Davis, 91–103. Washington, D.C.: Island Press, 2006.

Dorman, Robert L. *A Word for Nature: Four Pioneering Environmental Advocates, 1845–1913.* Chapel Hill: University of North Carolina Press, 1998.

Dorsey, Kurkpatrick. "Putting a Ceiling on Sealing: Conservation and Cooperation in the International Arena, 1909–1911." *Environmental History Review* 15, no. 3 (1991): 27–45.

———. "Scientists, Citizens, and Statesmen: U.S.-Canadian Wildlife Protection Treaties in the Progressive Era." *Diplomatic History* 19, no. 3 (1995): 407–29.

———. *The Dawn of Conservation Diplomacy: U.S.-Canadian Wildlife Protection Treaties in the Progressive Era*. Seattle: University of Washington Press, 1998.

Doughty, Robin W. *Feather Fashions and Bird Preservation: A Study in Nature Protection*. Berkeley: University of California Press, 1975.

———. *Return of the Whooping Crane*. Austin: University of Texas Press, 1989.

Drake, Victoria Croninshield. "The Pioneering International Vision of Harold Jefferson Coolidge: American Conservationist." B.A. thesis, Harvard University, 1983.

Drayton, Richard H. *Nature's Government: Science, Imperial Britain, and the "Improvement" of the World*. New Haven, Conn.: Yale University Press, 2000.

Dunlap, Riley E. "Trends in Public Opinion toward Environmental Issues: 1965–1990." In *American Environmentalism: The U.S. Environmental Movement, 1970–1990*, edited by Riley E. Dunlap and Angela G. Mertig, 89–116. Washington, D.C.: Taylor and Francis, 1992.

Dunlap, Thomas R. *DDT: Scientists, Citizens, and Public Policy*. Princeton, N.J.: Princeton University Press, 1981.

———. "That Kaibab Myth." *Journal of Forest History* 32, no. 2 (1988): 60–68.

———. *Saving America's Wildlife*. Princeton, N.J.: Princeton University Press, 1988.

———. "Organization and Wildlife Preservation: The Case of the Whooping Crane in North America." *Social Studies of Science* 21 (1991): 192–221.

———. *Nature and the English Diaspora: Environment and History in the United States, Canada, Australia, and New Zealand*. Cambridge: Cambridge University Press, 1999.

Dutcher, William. "History of the Audubon Movement." *Bird-Lore* 7 (1905): 45–57.

Eakin, Richard. "History of Zoology at the University of California, Berkeley." *Bios* 27 (1956): 67–90.

———. *History of Zoology at Berkeley*. Berkeley: Department of Zoology, University of California, 1988.

Eaton, Randall L., ed. *The World's Cats*. Vol. 1, *Ecology and Conservation*. Winston, Ore.: World Wildlife Safari, 1973.

———, ed. *The World's Cats*. Vol. 2, *Biology, Behavior and Management of Reproduction*. Seattle: Feline Research Group, Woodland Park Zoo, 1974.

Edge, Rosalie. *The Migratory Bird Treaty with Mexico: Double-Crossing Conservationists and Migratory Birds*. New York: Emergency Conservation Committee, 1936.

Egan, Michael. *Barry Commoner and the Science of Survival: The Remaking of American Environmentalism*. Cambridge, Mass.: MIT Press, 2008.

Egerton, Frank N. "Changing Concepts of the Balance of Nature." *Quarterly Review of Biology* 48 (1973): 322–50.

Ehrlich, Paul, and Anne Ehrlich. *Extinction: The Causes and Consequences of the Disappearance of Species*. New York: Random House, 1981.

Ellis, Richard. *Men and Whales*. New York: Alfred A. Knopf, 1991.

Elton, Charles. *Animal Ecology*. London: Sidgwick and Jackson, 1927.

Erickson, Ray C. "A Federal Research Program for Endangered Wildlife." *Transactions of the North American Wildlife and Natural Resources Conference* 33 (1968): 418–33.

———. "Captive Breeding of Whooping Cranes at the Patuxent Wildlife Research Center." In *Breeding Endangered Species in Captivity*, edited by R. D. Martin, 99–114. London: Academic Press, 1975.

Erickson, Ray C., and Scott R. Derrickson. "The Whooping Crane." In *Crane Research around the World*, edited by James C. Lewis and Hiroyuki Masatomi, 104–34. Baraboo, Wisc.: International Crane Foundation, 1981.

Evenden, Matthew D. "The Laborers of Nature: Economic Ornithology and the Role of Birds as Agents of Biological Pest Control in North American Agriculture, ca. 1880–1930." *Forest and Conservation History* 39 (1995): 172–83.

Ewan, Joseph, and Nesta Dunn Ewan. *Benjamin Smith Barton, Naturalist and Physician in Jeffersonian America*. Edited by Victoria C. Hollowell, Eillen P. Duggan and Marshall Crosby. St. Louis: Missouri Botanical Garden Press, 2007.

Farber, Paul Lawrence. *The Emergence of Ornithology as a Scientific Discipline, 1760–1850*. Dordrecht: D. Reidel, 1982.

———. *Finding Order in Nature: The Naturalist Tradition from Linnaeus to E. O. Wilson*. Baltimore: Johns Hopkins University Press, 2000.

Farnham, Timothy J. *Saving Nature's Legacy: The Origins of the Idea of Biological Diversity*. New Haven, Conn.: Yale University Press, 2007.

Feduccia, Alan, ed. *Catesby's Birds of Colonial America*. Chapel Hill: University of North Carolina Press, 1985.

Findlen, Paula. *Possessing Nature: Museums, Collecting, and Scientific Culture in Early Modern Italy*. Berkeley: University of California Press, 1994.

Fisher, Albert K. *The Hawks and Owls of the United States in Their Relation to Agriculture*. Washington, D.C.: Government Printing Office, 1893.

Fisher, James, Noel Simon, and Jack Vincent. *Wildlife in Danger*. New York: Viking Press, 1969.

Fitter, Richard, and Peter Scott. *The Penitent Butchers: The Fauna Preservation Society, 1903–1978*. London: Fauna Preservation Society, 1978.

Flader, Susan L. *Thinking Like a Mountain: Aldo Leopold and the Evolution of an Ecological Attitude toward Deer, Wolves, and Forests*. Lincoln: University of Nebraska Press, 1978.

Flader, Susan L., and J. Baird Callicott, eds. *The River of the Mother of God and Other Essays by Aldo Leopold*. Madison: University of Wisconsin Press, 1991.

Fleming, Donald. "Roots of the New Conservation Movement." *Perspectives in American History* 6 (1972): 7–91.

Fleming, John. *The Philosophy of Zoology; or a General View of the Structure, Functions, and Classification of Animals*. 2 vols. Edinburgh: Archibald Constable, 1822.

———. "Remarks Illustrative of the Influence of Society on the Distribution of British Animals." *Edinburgh Philosophical Journal* 11 (1824): 287–305.

Flippen, J. Brooks. *Nixon and the Environment*. Albuquerque: University of New Mexico Press, 2000.

Ford, Alice. *John James Audubon: A Biography*. New York: Abbeville Press, 1988.

Ford, Ray. "Saving the Condor: Robert E. Easton's Fight to Create the Sisquoc Condor Sanctuary." *Noticias—Santa Barbara Historical Society* 32, no. 4 (1986): 75–83.

Fox, Stephen. *The American Conservation Movement: John Muir and His Legacy*. Madison: University of Wisconsin Press, 1985.

Frängsmyr, Tore, ed. *Linnaeus, the Man and His Work*. Berkeley: University of California Press, 1983.

Freethy, Ron. *Auks: An Ornithologist's Guide*. New York: Facts on File, 1987.

Fuller, Errol. *The Great Auk*. New York: Harry N. Abrams, 1999.

———. *Extinct Birds*. Rev. ed. Ithaca: Cornell University Press, 2001.

————. *Dodo: A Brief History*. New York: Universe Publishing, 2002.

Furmansky, Dyana. *Rosalie Edge, Hawk of Mercy: The Activist Who Saved Nature from Conservationists*. Athens: University of Georgia Press, forthcoming.

Gallagher, Tim. *The Grail Bird: Hot on the Trail of the Ivory-Billed Woodpecker*. Boston: Houghton Mifflin, 2005.

Garretson, Martin S. *The American Bison: The Story of Its Extermination as a Wild Species and Its Restoration under Federal Protection*. New York: New York Zoological Society, 1938.

Gaskell, Jeremy. *Who Killed the Great Auk?* Oxford: Oxford University Press, 2000.

Gaston, Anthony, and Ian L. Jones. *The Auks: Alcidae*. Oxford: Oxford University Press, 1998.

Geiger, Roger L. *To Advance Knowledge: The Growth of American Research Universities, 1900–1940*. New York: Oxford University Press, 1986.

Gerbi, Antonello. *The Dispute of the New World: The History of a Polemic, 1750–1900*. Translated by Jeremy Moyle. Pittsburgh: University of Pittsburgh Press, 1973.

Gerrard, Jonathan M. *Charles Broley: An Extraordinary Naturalist*. Headingley, Manitoba: White Horse Plains Publisher, 1983.

Gerstner, Patsy A. "Vertebrate Paleontology, an Early Nineteenth-Century Transatlantic Science." *Journal of the History of Biology* 3 (1970): 137–48.

Gibbons, Ann. "Conservation Biology in the Fast Lane." *Science* 255, no. 5040 (January 3, 1992): 20.

Gibbons, Felton, and Deborah Strom. *Neighbors to the Birds: A History of Birdwatching in America*. New York: W. W. Norton, 1988.

Gillespie, Neal C. "Natural History, Natural Theology, and Social Order: John Ray and the 'Newtonian Ideology.'" *Journal of the History of Biology* 20 (1987): 1–49.

Glacken, Clarence J. *Traces on the Rhodian Shore: Nature and Culture in Western Thought from Ancient Times to the End of the Eighteenth Century*. Berkeley: University of California Press, 1967.

Glasgow, Vaughn L. *A Social History of the American Alligator: The Earth Trembles with His Thunder*. New York: St. Martins, 1991.

Goble, Dale D., J. Michael Scott, and Frank W. Davis, eds. *The Endangered Species Act: Renewing the Conservation Promise*. Washington, D.C.: Island Press, 2006.

Golley, Frank Benjamin. *A History of the Ecosystem Concept in Ecology: More than the Sum of Its Parts*. New Haven, Conn.: Yale University Press, 1993.

Goodwin, Harry A. "Concern for Endangered Species in the United States." *International Association of Game, Fish and Conservation Commissioners* 57 (1967): 56–61.

————. "Status of Endangered Species." *Transactions of the North American Wildlife and Natural Resources Conference* 36 (1971): 331–42.

Gorges, Raymond. *Ernest Harold Baynes: Naturalist and Crusader*. Boston: Houghton Mifflin, 1928.

Gosliner, Michael L. "The Tuna-Dolphin Controversy." In *Conservation and Management of Marine Mammals*, edited by John R. Twiss, Jr., and Randall R. Reeves, 120–55. Washington, D.C.: Smithsonian Institution Press, 1999.

Gould, Carol Grant. *The Remarkable Life of William Beebe: Explorer and Naturalist*. Washington, D.C.: Island Press, 2004.

Gould, Stephen Jay. "The Dodo in the Caucus Race." *Natural History* 105, no. 11 (1996): 22–33.

Graham, Frank, Jr. *Since Silent Spring*. Boston: Houghton Mifflin, 1970.

————. *The Audubon Ark: A History of the National Audubon Society*. New York: Alfred A. Knopf, 1990.

Grayson, Donald K. "Vicissitudes and Overkill: The Development of Explanations of the Pleistocene Extinctions." In *Advances in Archaeological Method and Theory*, edited by Michael B. Schiffer, 357–403. New York: Academic Press, 1980.

———. "Nineteenth-Century Explanations of Pleistocene Extinctions: A Review and Analysis." In *Quaternary Extinctions: A Prehistoric Revolution*, edited by Paul Martin and Richard G. Klein, 5–39. Tucson: University of Arizona Press, 1984.

Greene, John C. *The Death of Adam: Evolution and Its Impact on Western Thought*. Ames: Iowa State University Press, 1959.

———. *American Science in the Age of Jefferson*. Ames: Iowa State University Press, 1984.

Greenway, James C. *Extinct and Vanishing Birds of the World*. Special Publication no. 13. New York: American Committee for International Wild Life Protection, 1958.

Grinnell, Joseph. "Conserve the Collector." *Science* 41 (1915): 229–32.

———. "Why We Need Birds and Animals." *Scientific Monthly* 41, no. 6 (1935): 553–56.

Gross, Alfred O. *The Heath Hen*. Memoirs of the Boston Society of Natural History 6, no. 4. Boston: Boston Society of Natural History, 1928.

Grove, Richard H. *Green Imperialism: Colonial Expansion, Tropical Island Edens and the Origins of Environmentalism, 1600–1860*. Cambridge: Cambridge University Press, 1995.

Haber, Francis C. *The Age of the World: Moses to Darwin*. Baltimore: Johns Hopkins University Press, 1959.

Hachisuka, Masauji. *The Dodo and Kindred Birds, or the Extinct Birds of the Mascarene Islands*. London: H. F. and G. Witherby, 1953.

Hagen, Joel B. "Organism and Environment: Frederic Clements's Vision of a Unified Plant Ecology." In *The American Development of Biology*, edited by Ronald Rainger, Keith Benson, and Jane Maienschein, 257–80. Philadelphia: University of Pennsylvania Press, 1988.

———. *An Entangled Bank: The Origins of Ecosystem Ecology*. New Brunswick, N.J.: Rutgers University Press, 1992.

———. "Ecology." In *The History of Science in the United States: An Encyclopedia*, edited by Marc Rothenberg, 168–69. New York: Garland, 2001.

Hahn, Paul. *Where Is That Vanished Bird? An Index to the Known Specimens of Extinct North American Species*. Toronto: Royal Ontario Museum, University of Toronto, 1963.

Hanson, Elizabeth. *Animal Attractions: Nature on Display in American Zoos*. Princeton, N.J.: Princeton University Press, 2002.

Haraway, Donna. *Primate Visions: Gender, Race, and Nature in the World of Modern Science*. New York: Routledge, 1989.

Harlan, Richard. *Fauna Americana: Being a Description of the Mammiferous Animals Inhabiting North America*. Philadelphia: Anthony Finley, 1825.

Harper, Francis. *Extinct and Vanishing Mammals of the Old World*. Special Publication no. 12. New York: American Committee for International Wild Life Protection, 1945.

Harris, Harry. "The Annals of *Gymnogyps* to 1900." *The Condor* 43, no. 1 (1941): 3–55.

Hawkins, A. S., R. C. Hanson, H. K. Nelson, and H. M. Reeves, eds. *Flyways: Pioneering Waterfowl Management in North America*. Washington, D.C.: Government Printing Office, 1984.

Hayden, Sherman Strong. *The International Protection of Wild Life: An Examination of Treaties and Other Agreements for the Preservation of Birds and Mammals*. New York: Columbia University Press, 1942.

Hays, Samuel P. *Beauty, Health, and Permanence: Environmental Politics in the United States, 1955–1985*. Cambridge: Cambridge University Press, 1987.

Hedeen, Stanley. *Big Bone Lick: The Cradle of American Paleontology.* Lexington: University Press of Kentucky, 2008.

Henriksen, Margot A. *Dr. Strangelove's America: Society and Culture in the Atomic Age.* Berkeley: University of California Press, 1997.

Herman, Daniel Justin. *Hunting and the American Imagination.* Washington, D.C.: Smithsonian Institution Press, 2001.

Herrick, Francis Hobart. *Audubon the Naturalist: A History of His Life and Time.* New York: D. Appleton-Century, 1938.

Hill, Geoffrey. *Ivorybill Hunters: The Search for Proof in a Flooded Wilderness.* New York: Oxford University Press, 2007.

Hindle, Brooke. *The Pursuit of Science in Revolutionary America, 1735–1789.* Chapel Hill: University of North Carolina Press, 1965.

Hirt, Paul W. *A Conspiracy of Optimism: Management of the National Forests since World War Two.* Lincoln: University of Nebraska Press, 1994.

Hoage, R. J., and William A. Deiss, eds. *New Worlds, New Animals: From Menagerie to Zoological Park in the Nineteenth Century.* Baltimore: Johns Hopkins University Press, 1996.

Holdgate, Martin. *The Green Web: A Union for World Conservation.* London: Earthscan, 1999.

Hone, Elisabeth. *African Game Protection: An Outline of Existing Game Reserves and National Parks of Africa with Notes on Certain Species of Big Game Nearing Extinction, or Needing Additional Protection.* Special Publication, vol. 1, no. 3. Cambridge, Mass.: American Committee for International Wild Life Protection, 1933.

Honneger, René E. *Amphibia and Reptilia.* Red Data Book 3. Morges, Switzerland: IUCN, Survival Service Commission, 1968.

Hoose, Phillip. *The Race to Save the Lord God Bird.* New York: Farrar, Straus and Giroux, 2004.

Hornaday, William Temple. "The Passing of the Buffalo." *The Cosmopolitan* 4 (1887): 85–98, 231–43.

———. *Our Vanishing Wildlife: Its Extermination and Preservation.* New York: New York Zoological Society, 1913.

———. *Thirty Years War for Wild Life.* Stamford, Conn.: Permanent Wild Life Protection Fund, 1931.

———. *The Extermination of the American Bison.* Washington, D.C.: Smithsonian Institution Press, 2002. Originally published 1889.

Horowitz, Helen L. "The National Zoological Park: 'City of Refuge' or Zoo?" *Records of the Columbia Historical Society* 49 (1973–74): 405–29.

———. "Animal and Man in the New York Zoological Park." *New York History* 56 (1975): 426–55.

Horrocks, Thomas. "Thomas Jefferson and the Great Claw." *Virginia Cavalcade* 35, no. 2 (1985): 70–79.

Hough, Henry B. *The Heath Hen's Journey to Extinction, 1792–1933.* Edgartown, Mass.: Dukes County Historical Society, 1933.

Hurley, Andrew. *Environmental Inequalities: Class, Race, and Industrial Pollution in Gary, Indian, 1945–1980.* Chapel Hill: University of North Carolina Press, 1995.

Hutton, John, and Barnabas Dickson. *Endangered Species, Threatened Convention: The Past, Present and Future of CITES.* London: Earthscan Publications Ltd., 2000.

Ilerbaig, Juan. "Allied Sciences and Fundamental Problems: C. C. Adams and the Search for Method in Early American Ecology." *Journal of the History of Biology* 32 (1999): 439–63.

Imperato, Pascal James, and Eleanor M. Imperato. *They Married Adventure: The Wandering Lives of Martin and Osa Johnson.* New Brunswick, N.J.: Rutgers University Press, 1992.

Impey, Oliver, and Arthur MacGregor. *The Origins of Museums: Cabinets of Curiosities in Sixteenth- and Seventeenth-Century Europe.* Oxford: Oxford University Press, 1985.

International Union for the Conservation of Nature. *Les Fossiles de demain: Treize mammifères menacés d'extinction.* Paris: Société d'édition d'enseignement supérieur, 1954.

Irmscher, Christoph. *The Poetics of Natural History: From John Bartram to William James.* New Brunswick, N.J.: Rutgers University Press, 1999.

Ise, John. *Our National Park Policy: A Critical History.* Baltimore: Resources for the Future by Johns Hopkins University Press, 1961.

Isenberg, Andrew C. *The Destruction of the Bison: An Environmental History, 1750–1920.* Cambridge: Cambridge University Press, 2000.

Jackson, Donald, ed. *Letters of the Lewis and Clark Expedition, with Related Documents, 1783–1854.* 2nd ed. Urbana: University of Illinois Press, 1978.

Jackson, Hartley H. T. "Conserving Endangered Wildlife Species." *Annual Report of the Board of Regents of the Smithsonian Institution . . . for . . . 1945* (1946): 247–71.

Jackson, Jerome A. *In Search of the Ivory-Billed Woodpecker.* Washington, D.C.: Smithsonian Books, 2004.

Jacoby, Karl. *Crimes against Nature: Squatters, Poachers, Thieves, and the Hidden History of American Conservation.* Berkeley: University of California Press, 2001.

Jahn, Melvin E., and Daniel J. Woolf. *The Lying Stones of Dr. Johann Bartholomew Adam Beringer, Being His Lithographiæ Wirceburgensis.* Berkeley: University of California Press, 1963.

Jansen, A. J. "An Analysis of 'Balance in Nature' as an Ecological Concept." *Acta Biotheoretica* 21 (1972): 86–114.

Jardine, Nicholas, James A. Secord, and Emma C. Spary, eds. *Cultures of Natural History.* Cambridge: Cambridge University Press, 1996.

Jefferson, Thomas. "A Memoir on the Discovery of Certain Bones of a Clawed Kind in the Western Parts of Virginia." *Transactions of the American Philosophical Society* 4 (1799): 246–60.

———. *Notes on the State of Virginia.* Edited by William Peden. Chapel Hill: University of North Carolina Press, 1955. Originally published 1787.

Johnson, Ned K. "Ornithology at the Museum of Vertebrate Zoology." In *Contributions to the History of North American Ornithology,* edited by William E. Davis, Jr., and Jerome A. Jackson, 183–221. Cambridge, Mass.: Nuttall Ornithological Club, 1995.

Johnson, Osa. *I Married Adventure: The Lives and Adventures of Martin and Osa Johnson.* New York: William Morrow, 1989.

Judd, Richard W. *Common Lands, Common People: The Origins of Conservation in New England.* Cambridge, Mass.: Harvard University Press, 1997.

Kastner, Joseph. *A World of Watchers: An Informal History of the American Passion for Birds—from Its Scientific Beginnings to the Great Birding Boom of Today.* New York: Alfred A. Knopf, 1986.

Kellert, Stephen R. "Social and Perceptual Factors in the Preservation of Animal Species." In *The Preservation of Species: The Value of Biological Diversity,* edited by Bryan G. Norton, 50–73. Princeton, N.J.: Princeton University Press, 1986.

Keynes, Richard Darwin. *Charles Darwin's Beagle Diary.* Cambridge: Cambridge University Press, 1988.

————. *Fossils, Finches and Fuegians: Darwin's Adventures and Discoveries on the* Beagle. Oxford: Oxford University Press, 2003.

Kim, Chung Kem, and Lloyd Knutson, eds. *Foundations for a National Biological Survey.* Lawrence, Kans.: Association of Systematics Collections, 1986.

King, F. Wayne. "Adventures in the Skin Trade." *Natural History* 80, no. 5 (1971): 8–16.

Kingsland, Sharon. *The Evolution of American Ecology, 1890–2000.* Baltimore: Johns Hopkins University Press, 2005.

Kinsey, Darin. "'Seeding the Water as the Earth': The Center and Periphery of a Western Aquaculture Revolution." *Environmental History* 11 (2006): 527–66.

Kisling, Vernon N., Jr., ed. *Zoo and Aquarium History: Ancient Animal Collections to Zoological Gardens.* Boca Raton, Fla.: CRC Press, 2001.

Kitchener, Andrew C. "On the External Appearance of the Dodo, *Raphus cucullatus* (L., 1758)." *Archives of Natural History* 20, no. 2 (1993): 279–301.

Knight, Richard L., and Suzanne Riedel, eds. *Aldo Leopold and the Ecological Conscience.* New York: Oxford University Press, 2002.

Koerner, Lisbet. *Linnaeus: Nature and Nation.* Cambridge, Mass.: Harvard University Press, 1999.

Koford, Carl B. *The California Condor.* Research Report no. 4. New York: National Audubon Society, 1953.

Kohler, Robert E. "The Ph.D. Machine: Building on the Collegiate Base." *Isis* 81 (1990): 638–62.

————. *Landscapes and Labscapes: Exploring the Lab-Field Border in Biology.* Chicago: University of Chicago Press, 2002.

————. *All Creatures: Naturalists, Collectors, and Biodiversity, 1850–1950.* Princeton, N.J.: Princeton University Press, 2006.

Kohlstedt, Sally G. "Henry A. Ward: The Merchant Naturalist and American Museum Development." *Journal of the Society for the Bibliography of Natural History* 9 (1980): 647–61.

Kohm, Kathryn A., ed. *Balancing on the Brink of Extinction: The Endangered Species Act and Lessons for the Future.* Washington, D.C.: Island Press, 1991.

Kohn, David, ed. *The Darwinian Heritage.* Princeton, N.J.: Princeton University Press, 1985.

Krech, Shepard, III. *The Ecological Indian: Myth and History.* New York: W. W. Norton, 1999.

Kuklick, Henrika, and Robert Kohler, eds. *Science in the Field.* Osiris 11. Chicago: University of Chicago Press, 1996.

Langston, Nancy. *Where Land and Water Meet: A Western Landscape Transformed.* Seattle: University of Washington Press, 2003.

Larson, Edward J. *Evolution's Workshop: God and Science in the Galápagos Islands.* New York: Basic Books, 2001.

Larson, James L. *Reason and Experience: The Representation of Natural Order in the Work of Carl von Linné.* Berkeley: University of California Press, 1971.

Lavigne, David M. *Harps and Hoods: Ice-Breeding Seals of the Northwest Atlantic.* Waterloo: University of Waterloo Press, 1988.

Lavigne, David M., Victor B. Scheffer, and Stephen R. Kellert. "The Evolution of American Attitudes toward Marine Mammals." In *Conservation and Management of Marine Mammals,* edited by John R. Twiss, Jr., and Randall R. Reeves, 10–47. Washington, D.C.: Smithsonian Institution Press, 1999.

Lear, Linda. *Rachel Carson: Witness for Nature.* New York: Henry Holt, 1997.

Leopold, Aldo. "The Wilderness and Its Place in Forest Recreational Policy." *Journal of Forestry* 19, no. 3 (1921): 718–19.

———. "The Conservation Ethic." *Journal of Forestry* 31, no. 6 (1933): 634–43.

———. *Game Management.* New York: Charles Scribner's Sons, 1933.

———. "Threatened Species: A Proposal to the Wildlife Conference for an Inventory of the Needs of Near-Extinct Birds and Animals." *American Forests* 42 (1936): 116–19.

———. "A Biotic View of the Land." *Journal of Forestry* 37, no. 9 (1939): 727–30.

———. *A Sand County Almanac and Sketches Here and There.* New York: Oxford University Press, 1949.

———. *The River of the Mother of God and Other Essays by Aldo Leopold.* Edited by Susan L. Flader and J. Baird Callicott. Madison: University of Wisconsin Press, 1991.

LeUnes, Barbara Laverne Blythe. "The Conservation Philosophy of Stewart L. Udall, 1961–1968." Ph.D. dissertation, Texas A&M University, 1977.

Lewis, James C. "Whooping Crane (*Grus americana*)." In *The Birds of North America*, edited by A Poole and F. Gill, no. 153. Washington, D.C.: American Ornithologists' Union, 1995.

Lewis, Keri E. "Negotiating for Nature: Conservation Diplomacy and the Convention on Nature Protection and Wildlife Preservation in the Western Hemisphere, 1929–1976." Ph.D. dissertation, University of New Hampshire, 2007.

Lifton, Robert J., and Greg Mitchell. *Hiroshima in America: Fifty Years of Denial.* New York: G. P. Putnam's Sons, 1995.

Lincoln, Frederick C. *The Waterfowl Flyways of North America,* U.S. Department of Agriculture, Circular no. 342. Washington, D.C.: Government Printing Office, 1935.

Lindroth, Sten. "The Two Faces of Linnaeus." In *Linnaeus: The Man and His Work*, edited by Tore Frängsmyr, 1–62. Berkeley: University of California Press, 1983.

Linnaeus, Carl. "On the Police of Nature." In *Select Dissertations from the Amœnitates Academicœ*, edited by F. J. Brand. London: G. Robinson and J. Robson, 1781.

———. "The Oeconomy of Nature." In *Miscellaneous Tracts Relating to Natural History, Husbandry, and Physick*, edited by Benjamin Stillingfleet, 39–129. London: J. Dodsley, Baker and Leigh, and T. Payne, 1791.

Loo, Tina. *States of Nature: Conserving Canada's Wildlife in the Twentieth Century.* Vancouver: UBC Press, 2006.

Lopez, Barry Holston. *Of Wolves and Men.* New York: Charles Scribner's Sons, 1978.

Louv, Richard. *Last Child in the Woods: Saving Our Children from Nature-Deficit Disorder.* Chapel Hill, N.C.: Algonquin Books, 2005.

Lovejoy, Arthur O. *The Great Chain of Being: A Study in the History of an Idea.* Cambridge, Mass.: Harvard University Press, 1966.

Lowenthal, David. *George Perkins Marsh: Prophet of Conservation.* Seattle: University of Washington Press, 2000.

Luke, Timothy W. "The Nature Conservancy or the Nature Cemetery: Buying and Selling 'Perpetual Care' as Environmental Resistance." *Capitalism, Nature, Socialism* 6 (June 1995): 1–20.

Lund, Thomas A. *American Wildlife Law.* Berkeley: University of California Press, 1980.

Lutts, Ralph. "Chemical Fallout: Rachel Carson's 'Silent Spring,' Radioactive Fallout, and the Environmental Movement." *Environmental Review* 9, no. 3 (1985).

Lyell, Charles. *Principles of Geology, Being an Attempt to Explain Former Changes of the Earth's Surface by Reference to Causes Now in Operation.* Vol. 2. London: John Murray, 1832.

Macdonald, Helen. *Falcon*. London: Reaktion Books, 2006.

MacKenzie, John M. *The Empire of Nature: Hunting, Conservation, and British Imperialism*. Manchester: Manchester University Press, 1988.

Maienschein, Jane, Ronald Rainger, and Keith Benson. "Introduction: Were American Morphologists in Revolt?" *Journal of the History of Biology* 14 (1981): 83–88.

Mann, Charles C., and Mark L. Plummer. *Noah's Choice: The Future of Endangered Species*. New York: Alfred A. Knopf, 1995.

Mark, Stephen R. *Preserving the Living Past: John C. Merriam's Legacy in the State and National Parks*. Berkeley: University of California Press, 2005.

Marsh, George Perkins. *Man and Nature: Or, Physical Geography as Modified by Human Action*. Edited by David Lowenthal. Seattle: University of Washington Press, 2003. Originally published 1864.

Mathewson, Worth. *William L. Finley: Pioneer Wildlife Photographer*. Corvallis: Oregon State University, 1986.

Matijasic, Thomas D. "Science, Religion, and the Fossils at Big Bone Lick." *Journal of the History of Biology* 20 (1987): 413–21.

Matthiessen, Peter. *Birds of Heaven: Travels with Cranes*. New York: North Point Press, 2001.

Mayor, Adrienne. *The First Fossil Hunters: Paleontology in Greek and Roman Times*. Princeton, N.J.: Princeton University Press, 2000.

————. *Fossil Legends of the First Americans*. Princeton, N.J.: Princeton University Press, 2005.

McClung, Robert M. *Lost Wild America: The Story of Extinct and Vanishing Wildlife*. New York: William Morrow, 1969.

McCormick, John. *Reclaiming Paradise: The Global Environmental Movement*. Bloomington: Indiana University Press, 1989.

McEvoy, Arthur F. *The Fisherman's Problem: Ecology and Law in the California Fisheries, 1850–1980*. Cambridge: Cambridge University Press, 1986.

McIntosh, Robert P. *The Background of Ecology: Concept and Theory*. Cambridge: Cambridge University Press, 1985.

McIntyre, Rick, ed. *War against the Wolf: America's Campaign to Exterminate the Wolf*. Stillwater, Minn.: Voyageur Press, 1995.

McKinley, Daniel. "The Last Days of the Carolina Parrakeet: Life in the Zoos." *Avicultural Magazine* 83 (1977): 42–42.

————. *The Carolina Parakeet in Florida*. Edited by John William Hardy. Special Publication No. 2. Gainesville: Florida Ornithological Society, 1985.

McMillan, Ian. *Man and the California Condor*. New York: E. P. Dutton, 1968.

McNulty, Faith. *The Whooping Crane: The Bird That Defies Extinction*. New York: E. P. Dutton, 1966.

McQuilkin, Geoffroy J. "Saving the Living Laboratory: The Natural Areas Preservation Efforts of the Ecological Society of America." Senior honors thesis, Harvard University, 1991.

McVay, Scott. "The Last of the Great Whales." *Scientific American* 215, no. 2 (1966): 13–21.

Mearns, Barbara, and Richard Mearns. *The Bird Collectors*. San Diego: Academic Press, 1998.

Meine, Curt. *Aldo Leopold: His Life and Work*. Madison: University of Wisconsin Press, 1988.

Meine, Curt, and Richard L. Knight, eds. *The Essential Aldo Leopold: Quotations and Commentaries*. Madison: University of Wisconsin Press, 1999.

Meine, Curt, Michael Soulé, and Reed Noss. "'A Mission-Oriented Discipline': The Growth of Conservation Biology." *Conservation Biology* 20, no. 3 (2006): 631–51.

Melosi, Martin V. "Lyndon Johnson and Environmental Policy." In *The Johnson Years*. Vol. 2, *Vietnam, the Environment, and Science*, edited by Robert A. Divine, 113–49. Lawrence: University Press of Kansas, 1987.

Melville, Ronald. *Angiospermae: A Compilation*. Red Data Book 5. Morges, Switzerland: IUCN, Species Survival Commission, 1970.

———. "Plant Conservation and the Red Book." *Biological Conservation* 2, no. 3 (1970): 185–88.

Meredith, R. L. *American Falconry in the Twentieth Century*. Boise, Idaho: Archives of American Falconry, 1999.

Mighetto, Lisa. *Wild Animals and American Environmental Ethics*. Tucson: University of Arizona Press, 1991.

Miller, Lillian B. "Charles Willson Peale as History Painter: The Exhumation of the Mastodon." *American Art Journal* 13 (1981): 47–68.

Miller, Lillian B., Sidney Hart, and David C. Ward, eds. *The Selected Papers of Charles Willson Peale and His Family*. 5 vols. New Haven, Conn.: Yale University Press, 1983–2000.

Miller, Robert Rush. *Pisces: Freshwater Fishes*. Red Data Book 4. Morges, Switzerland: IUCN, Species Survival Commission, 1969.

Miller, Waldron DeWitt, Willard G. Van Name, and Davis Quinn. *A Crisis in Conservation: Serious Danger of Extinction of Many North American Birds*. New York: n.p., 1929.

Mitchell, John H. "A Man Called Bird." *Audubon Magazine* 89 (March 1987): 81–104.

Mitman, Gregg. "When Nature Is the Zoo: Vision and Power in the Art and Science of Natural History." *Osiris* 11 (1996): 117–43.

———. *Reel Nature: America's Romance with Wildlife on Film*. Cambridge, Mass.: Harvard University Press, 1999.

Montevecchi, William A., and David A. Kirk. "Great Auk (*Pinguinus impennis*)." In *The Birds of North America*, edited by A. Poole and F. Gill, no. 260. Washington, D.C.: American Ornithologists' Union, 1996.

Moore, Barrington. "Importance of Natural Conditions in National Parks." In *Hunting and Conservation: The Book of the Boone and Crockett Club*, edited by George Bird Grinnell and Charles Sheldon, 340–55. New Haven, Conn.: Yale University Press, 1925.

Moss, Stephen. *A Bird in the Bush: A Social History of Birdwatching*. London: Aurum Press, 2004.

Muir, John. *A Thousand-Mile Walk to the Gulf*. Boston: Houghton Mifflin, 1916.

Municchi, Anna. *Ladies in Furs, 1940–1990*. Modena, Italy: Zanfi Editori, 1993.

Murchison, Kenneth. *The Snail Darter Case: TVA versus the Endangered Species Act*. Lawrence: University Press of Kansas, 2007.

Murphy, Robert Cushman. "S.O.S. for a Continent." *Natural History* 43 (1939): 150–51.

Myers, Norman. *The Sinking Ark: A New Look at the Problem of Vanishing Species*. Oxford: Pergamon Press, 1979.

———. *A Wealth of Wild Species: Storehouse for Human Welfare*. Boulder: Westview Press, 1982.

Nash, Roderick Frazier. *Wilderness and the American Mind*. 3rd ed. New Haven, Conn.: Yale University Press, 1982.

———. *The Rights of Nature: A History of Environmental Ethics*. Madison: University of Wisconsin Press, 1989.

Neumann, Thomas W. "Human-Wildlife Competition and the Passenger Pigeon: Population Growth from System Destabilization." *Human Ecology* 13, no. 4 (1985): 389–410.

Nielson, John. *Condor: To the Brink and Back—the Life and Times of One Giant Bird*. New York: Harper/Collins, 2006.

Nilsson, Greta, et al. *Facts about Furs*. Washington, D.C.: Animal Welfare Institute, 1980.

Noss, Reed. "The Naturalists Are Dying Off." *Conservation Biology* 10, no. 1 (February 1996): 1–3.

Novak, Barbara. *Nature and Culture: American Landscape Painting, 1825–1875*. New York: Oxford University Press, 1980.

Nye, David E. *America as Second Creation: Technology and Narratives of New Beginnings*. Cambridge, Mass.: MIT Press, 2003.

Nyhart, Lynn. "Natural History and the 'New' Biology." In *Cultures of Natural History*, edited by N. Jardine, J. Secord, and E. C. Spary, 426–43. Cambridge: Cambridge University Press, 1996.

Orsi, Jared. "From Horicon to Hamburgers and Back Again: Ecology, Ideology, and Wildfowl Management, 1917–1935." *Environmental History Review* 18 (1994): 19–40.

Osborn, Henry Fairfield. *Preservation of Wild Animals of North America*. Washington, D.C.: Boone and Crockett Club, 1904.

Osborne, Michael A. *Nature, the Exotic, and the Science of French Colonialism*. Bloomington: Indiana University Press, 1994.

Ospovat, Dov. *The Development of Darwin's Theory: Natural History, Natural Theology, and Natural Selection, 1838–1859*. Cambridge: Cambridge University Press, 1981.

Ovenell, R. F. "The Tradescant Dodo." *Archives of Natural History* 19, no. 2 (1992): 145–52.

Overfield, Richard A. *Science with Practice: Charles E. Bessey and the Maturing of American Botany*. Ames: Iowa State University Press, 1993.

Owen, Richard. "On the Bone of an Unknown Struthius Bird from New Zealand." *Proceedings of the Zoological Society of London* 7 (1839): 169–71.

———. *Memoir on the Dodo* (Didus ineptus, *Linn.*). London: Taylor and Francis, 1866.

———. *Memoirs on the Extinct Wingless Birds of New Zealand*. 2 vols. London: John Van Voorst, 1879.

Palmer, T. S. "A Review of Economic Ornithology in the United States." In *Yearbook of the U.S. Department of Agriculture for 1899*, 259–92. Washington, D.C.: Government Printing Office, 1900.

Palmer, William D. "Endangered Species Protection: A History of Congressional Action." *Environmental Affairs* 4 (1975): 255–93.

Paradiso, John L. *Status Report on Cats (Felidae) of the World, 1971*. Special Scientific Report: Wildlife no. 157. Washington, D.C.: U.S. Department of the Interior, Fish and Wildlife Service, 1972.

Pauly, Philip J. "The Appearance of Academic Biology." *Journal of the History of Biology* 17 (1984): 369–97.

———. *Biologists and the Promise of American Life: From Meriwether Lewis to Alfred Kinsey*. Princeton, N.J.: Princeton University Press, 2000.

Peale, Rembrandt. *Historical Disquisition on the Mammoth, or, Great American Incognitum, an Extinct, Immense, Carnivorous Animal Whose Fossil Remains Have Been Found in North America*. London: E. Lawrence, 1803.

Pearson, Thomas Gilbert. *Adventures in Bird Protection: An Autobiography*. New York: D. Appleton-Century, 1937.

Perrin, William F. "The Porpoise and the Tuna." *Sea Frontiers* 14 (1968): 166–74.

————. "Using Porpoises to Catch Tuna." *World Fishing* 18 (1969): 42–45.

Peteresen, Shannon. "Congress and Charismatic Megafauna: A Legislative History of the Endangered Species Act." *Environmental Law* 29 (1999): 463–91.

————. *Acting for Endangered Species: The Statutory Ark.* Lawrence: University Press of Kansas, 2002.

Phillips, David, and Hugh Nash, eds. *The Condor Question: Captive or Forever Free?* San Francisco: Friends of the Earth, 1981.

Phillips, John C. "An Attempt to List the Extinct and Vanishing Birds of the Western Hemisphere, with Some Notes on Recent Status, Location of Specimens, Etc." In *Verhandlungen des VI Internationalen Ornithologen-Kongresses in Kopenhagen 1926,* edited by F. Steinbacher, 503–34. Berlin: Dornblüth, 1929.

————. *Migratory Bird Protection in North America.* Special Publication, vol. 1, no. 4. Cambridge, Mass.: American Committee for International Wild Life Protection, 1934.

————. "The Work of the American Committee for International Wild Life Protection." *Proceedings, North American Wildlife Conference, February 3–7, 1936* (1936): 51–55.

Phillips, John C., and Harold J. Coolidge, Jr. *The First Five Years: The American Committee for International Wild Life Protection.* Cambridge, Mass.: American Committee for International Wild Life Protection, 1934.

Phillips, John C., and Frederick C. Lincoln. *American Waterfowl: Their Present Situation and Outlook for Their Future.* Boston: Houghton Mifflin, 1930.

Pinney, Roy. *Vanishing Wildlife.* New York: Dodd, Mead, 1963.

Pinto-Correia, Clara. *Return of the Crazy Bird: The Sad, Strange Tale of the Dodo.* New York: Copernicus Books, 2003.

Porter, Charlotte M. *The Eagle's Nest: Natural History and American Ideas, 1812–1842.* University: University of Alabama Press, 1986.

Pough, Richard. "An Inventory of Threatened and Vanishing Species." *Transactions of the Second North American Wildlife Conference* (1937): 599–604.

Prendergast, David K., and William M. Adams. "Colonial Wildlife Conservation and the Origins of the Society for the Preservation of Wild Fauna of the Empire (1903–1914)." *Oryx* 37 (April 2003): 251–60.

Price, Jennifer. *Flight Maps: Adventures with Nature in Modern America.* New York: Basic Books, 1999.

Pritchard, James A. *Preserving Yellowstone's Natural Conditions: Science and the Perception of Nature.* Lincoln: University of Nebraska Press, 1999.

Punke, Michael. *Last Stand: George Bird Grinnell, the Battle to Save the Buffalo, and the Birth of the New West.* New York: Smithsonian Books/Collins, 2007.

Pyle, Robert Michael. "The Rise and Fall of Natural History." *Orion* 17 (Autumn 2001): 17–23.

Quammen, David. *The Song of the Dodo: Island Biogeography in an Age of Extinctions.* New York: Scribner, 1996.

Quinn, Davis. *"Framing" the Birds of Prey.* New York: n.p., 1929.

Raby, Peter. *Alfred Russel Wallace: A Life.* Princeton, N.J.: Princeton University Press, 2001.

Rainger, Ronald. *An Agenda for Antiquity: Henry Fairfield Osborn and Vertebrate Paleontology at the American Museum of Natural History, 1890–1935.* Tuscaloosa: University of Alabama Press, 1991.

Rainger, Ronald, Keith R. Benson, and Jane Maienschein, eds. *The American Development of Biology.* Philadelphia: University of Pennsylvania Press, 1988.

Rathlesberger, James, ed. *Nixon and the Environment: The Politics of Devastation*. New York: Village Voice Books, 1972.

Raven, C. E. *John Ray, Naturalist: His Life and Works*. Cambridge: Cambridge University Press, 1950.

Ray, John. *Observations Topographical, Moral and Physiological, Made in a Journey through Part of the Low Countries, Germany, Italy, and France*. London: J. Martyn, 1673.

Reeves, Randall R., and James G. Mead. "Marine Mammals in Captivity." In *Conservation and Management of Marine Mammals*, edited by John R. Twiss, Jr., and Randall R. Reeves, 412–36. Washington, D.C.: Smithsonian Institution Press, 1999.

Regal, Brian. *Henry Fairfield Osborn: Race and the Search for the Origins of Man*. Aldershot, England: Ashgate, 2002.

Regenstein, Lewis G. "A History of the Endangered Species Act of 1973 and an Analysis of Its History, Strengths and Weaknesses, Administration, and Probable Future Effectiveness." M.A. thesis, Emory University, 1975.

———. *The Politics of Extinction: The Shocking Story of the World's Endangered Wildlife*. New York: Macmillan, 1975.

Rehbock, Philip F. *The Philosophical Naturalists: Themes in Early Nineteenth-Century British Biology*. Madison: University of Wisconsin Press, 1983.

———. "John Fleming (1785–1857) and the Economy of Nature." In *From Linnaeus to Darwin: Commentaries on the History of Biology and Geology*, edited by Alywne Wheeler and James H. Price, 129–40. London: Society for the History of Natural History, 1985.

Reiger, John F. *American Sportsmen and the Origins of Conservation*. New York: Winchester Press, 1975.

Repcheck, Jack. *The Man Who Found Time: James Hutton and the Discovery of Earth's Antiquity*. New York: Perseus, 2003.

Rhodes, Richard. *John James Audubon: The Making of an American*. New York: Alfred A. Knopf, 2004.

Rice, Gary. "Trailing a Celebrity: Press Coverage of Theodore Roosevelt's African Safari, 1909–1910." *Theodore Roosevelt Association Journal* 21 (1996): 4–16.

Rice, Howard C., Jr. "Jefferson's Gift of Fossils to the Museum of Natural History in Paris." *Proceedings of the American Philosophical Society* 95, no. 6 (1951): 597–627.

Rice, Tony. *Voyages of Discovery: Three Centuries of Natural History Exploration*. New York: Clarkson N. Potter, 1999.

Richardson, Edgar P., Brooke Hindle, and Lillian B. Miller. *Charles Willson Peale and His World*. New York: Harry N. Abrams, 1982.

Richardson, Robert D. *Henry Thoreau: A Life of the Mind*. Berkeley: University of California Press, 1986.

Ricklefs, Robert E. *Report of the Advisory Panel on the California Condor*. Audubon Conservation Report no. 6. New York: National Audubon Society, 1978.

Rigal, Laura. "Peale's Mammoth." In *American Iconology: New Approaches to Nineteenth-Century Art and Literature*, edited by David C. Miller, 18–38. New Haven, Conn.: Yale University Press, 1993.

Robinson, Michael J. *Predatory Bureaucracy: The Extermination of Wolves and the Transformation of the West*. Boulder: University Press of Colorado, 2005.

Roger, Jacques. *Buffon: A Life in Natural History*. Translated by Sara Lucille Bonnefoi. Edited by L. Pearce Williams. Ithaca, N.Y.: Cornell University Press, 1997.

Rogers, Kathleen, and James A. Moore. "Revitalizing the Convention on Nature Protection and Wild Life Preservation in the Western Hemisphere: Might Awakening a Visionary but 'Sleeping' Treaty Be the Key to Preserving Biodiversity and Threatened Natural Areas in the Americas?" *Harvard International Law Journal* 36 (1995): 465–508.

Rolfe, W. D. "William Hunter (1718–1783) on Irish 'Elk' and Stubb's Moose." *Archives of Natural History* 11, no. 2 (1983): 263–90.

———. "William and John Hunter: Breaking the Chain of Being." In *William Hunter and the Eighteenth-Century Medical World*, edited by W. F. Bynum and Roy Porter, 297–319. Cambridge: Cambridge University Press, 1985.

Roosevelt, Theodore. *African Game Trails: An Account of the African Wanderings of an American Hunter-Naturalist*. New York: Charles Scribner's Sons, 1920.

Rothfels, Nigel. *Savages and Beasts: The Birth of the Modern Zoo*. Baltimore: Johns Hopkins University Press, 2002.

Rothman, Hal K. *The Greening of a Nation? Environmentalism in the United States since 1945*. Fort Worth, Tex.: Harcourt Brace College Publishers, 1998.

Rothschild, Lionel Walter. *Extinct Birds: An Attempt to Unite in One Volume a Short Account of Those Birds Which Have Become Extinct in Historical Times—That Is, within the Last Six or Seven Hundred Years: To Which Are Added a Few Which Still Exist, but Are on the Verge of Extinction*. London: Hutchinson, 1907.

Rothschild, Miriam. *Dear Lord Rothschild: Birds, Butterflies and History*. Philadelphia: Balaban Publishers, 1983.

Ruckelshaus, Mary, and Donna Darm. "Science and Implementation." In *The Endangered Species Act at Thirty: Conserving Biodiversity in Human-Dominated Landscapes*, edited by J. Michael Scott, Dale D. Goble, and Frank W. Davis, 104–26. Washington, D.C.: Island Press, 2006.

Rudwick, Martin J. S. *The Meaning of Fossils: Episodes in the History of Paleontology*. 2nd ed. Chicago: University of Chicago Press, 1985.

———. *Scenes from Deep Time: Early Pictorial Representations of the Prehistoric World*. Chicago: University of Chicago Press, 1992.

———. *Georges Cuvier, Fossil Bones, and Geological Catastrophes: New Translations and Interpretations of the Primary Texts*. Chicago: University of Chicago Press, 1997.

———. *Bursting the Limits of Time: The Reconstruction of Geohistory in the Age of Revolution*. Chicago: University of Chicago Press, 2005.

Runte, Alfred. *National Parks: The American Experience*. 2nd ed. Lincoln: University of Nebraska Press, 1987.

———. *Yosemite: The Embattled Wilderness*. Lincoln: University of Nebraska Press, 1990.

Rupke, Nicolaas. "The Study of Fossils in the Romantic Philosophy of History and Nature." *History of Science* 21 (1983): 389–413.

———. *Richard Owen: Victorian Naturalist*. New Haven, Conn.: Yale University Press, 1994.

Russell, Edmund. *War and Nature: Fighting Humans and Insects with Chemicals from World War I to Silent Spring*. Cambridge: Cambridge University Press, 2001.

Ryfle, Steve. *Japan's Favorite Mon-Star (the Unauthorized Biography of "the Big G")*. Toronto: ECW Press, 1999.

Saikku, Mikko. "The Extinction of the Carolina Parakeet." *Environmental History Review* 14 (1990): 1–18.

———. "'Home in the Big Forest': Decline of the Ivory-Billed Woodpecker and Its Habitat in the United States." In *Encountering the Past: Essays in Environmental History*, edited by Mikko Saikku and Timo Myllyntaus, 94–140. Athens: Ohio University Press, 2001.

Sale, Kirkpartick. *The Green Revolution: The American Environmental Movement, 1962–1992*. New York: Hill and Wang, 1993.

Sanders, Scott Russell, ed. *Audubon Reader: The Best Writings of John James Audubon*. Bloomington: Indiana University Press, 1986.

Schiebinger, Londa. *Plants and Empire: Colonial Bioprospecting in the Atlantic World*. Cambridge, Mass.: Harvard University Press, 2004.

Schmidly, David J. "What It Means to Be a Naturalist and the Future of Natural History at American Universities." *Journal of Mammalogy* 86, no. 3 (June 2005): 449–56.

Schmitt, Peter. *Back to Nature: The Arcadian Myth in Urban America*. New York: Oxford University Press, 1969.

Schneider, Daniel W. "Local Knowledge, Environmental Politics, and the Founding of Ecology in the United States: Stephen Forbes and 'The Lake as a Microcosm' (1887)." *Isis* 91 (2000): 681–705.

Schofield, Robert. "The Science Education of an Enlightenment Entrepreneur: Charles Willson Peale and His Philadelphia Museum, 1784–1827." *American Studies* 30 (1989): 21–40.

Schorger, A. W. *The Passenger Pigeon: Its Natural History and Extinction*. Madison: University of Wisconsin Press, 1955.

Schoultz, Lars. *Beneath the United States: A History of U.S. Policy toward Latin America*. Cambridge, Mass.: Harvard University Press, 1998.

Schreiner, Keith M, and C. E. Ruhr. "Progress in Saving Endangered Species." *Transactions of the North American Wildlife and Natural Resources Conference* 39 (1974): 127–35.

Schrepfer, Susan R. *The Fight to Save the Redwoods: A History of Environmental Reform, 1917–1978*. Madison: University of Wisconsin Press, 1983.

Schroeder, M. A., and L. A. Robb. "Greater Prairie-Chicken: *Tympanuchus cupido*." In *The Birds of North America*, edited by A. Poole, P. Stettenheim, and F. Gill, no. 36. Washington, D.C.: American Ornithologists' Union, 1993.

Scott, J. Michael, Dale D. Goble, Leona K. Svancara, and Anna Pigorna. "By the Numbers." In *The Endangered Species Act at Thirty: Renewing the Conservation Promise*, edited by Dale D. Goble, J. Michael Scott, and Frank W. Davis. Washington, D.C.: Island Press, 2006.

Sellars, Richard. *Preserving Nature in the National Parks: A History*. New Haven, Conn.: Yale University Press, 1997.

Sellers, Charles Coleman. "Unearthing the Mastodon: Peale's Greatest Triumph." *American Heritage* 30, no. 5 (1979): 16–23.

————. *Mr. Peale's Museum: Charles Willson Peale and the First Popular Science Museum of Natural Science and Art*. New York: W. W. Norton, 1980.

Semonin, Paul. "'Nature's Nation': Natural History as Nationalism in the New Republic." *Northwest Review* 30 (1992): 6–41.

————. *American Monster: How the Nation's First Prehistoric Creature Became a Symbol of National Identity*. New York: New York University Press, 2000.

Shabecoff, Philip. *A Fierce Green Fire: The American Environmental Movement*. New York: Hill and Wang, 1993.

Shafer, Craig L. "Conservation Biology Trailblazers: George Wright, Ben Thompson, and Joseph Dixon." *Conservation Biology* 15, no. 2 (April 2001): 332–44.

Shapiro, Jerome F. *Atomic Bomb Cinema*. New York: Routledge, 2002.

Sheail, John. *Seventy-Five Years in Ecology: The British Ecological Society.* Oxford: Blackwell Scientific, 1987.

Sheets-Pyenson, Susan. *Cathedrals of Science: The Development of Colonial Natural History Museums during the Late Nineteenth Century.* Kingston: McGill-Queen's University Press, 1988.

Shelford, Victor E. *Animal Communities in Temperate America.* Chicago: University of Chicago Press, 1913.

———. "The Preservation of Natural Conditions." *Science* 52, no. 1317 (1920): 316–17.

———, ed. *Naturalist's Guide to the Americas.* Baltimore: Williams and Wilkins, 1926.

———. "The Preservation of Natural Biotic Communities." *Ecology* 14, no. 2 (1933): 240–45.

———. "Nature Sanctuaries: A Means of Saving Natural Biotic Communities." *Science* 77, no. 1994 (1933): 281–82.

———. "Twenty-Five-Year Effort at Saving Nature for Scientific Purposes." *Science* 98, no. 2543 (1943): 280–81.

Shell, Hannah Rose. "Last of the Wild Buffalo." *Smithsonian Magazine* 30, no. 11 (February 2000): 26–30.

Shermer, Michael. *In Darwin's Shadow: The Life and Science of Alfred Russel Wallace: A Biographical Study on the Psychology of History.* Oxford: Oxford University Press, 2002.

Sherwood, Morgan. *Big Game in Alaska: A History of Wildlife and People.* New Haven, Conn.: Yale University Press, 1981.

Silverberg, Robert. *The Auk, the Dodo, and the Oryx: Vanished and Vanishing Creatures.* New York: Thomas Y. Crowell, 1967.

Simon, Noel. *Mammalia: A Compilation.* Red Data Book 1. Lausanne: IUCN, 1966.

Simonian, Lane. *Defending the Land of the Jaguar: A History of Conservation in Mexico.* Austin: University of Texas Press, 1995.

Simons, D. D. "Interactions between California Condors and Humans in Prehistoric Far Western North America." In *Vulture Biology and Management,* edited by Sanford R. Wilbur and Jerome A. Jackson, 470–94. Berkeley: University of California Press, 1983.

Simpson, George Gaylord. "The Beginnings of Vertebrate Paleontology in North America." *Proceedings of the American Philosophical Society* 86 (1942): 130–88.

———. "The Discovery of Fossil Vertebrates in North America." *Journal of Paleontology* 17, no. 1 (1943): 26–38.

———. "Extinction." *Proceedings of the American Philosophical Society* 129 (1985): 407–15.

Simpson, George Gaylord, and H. Tobien. "The Rediscovery of Peale's Mastodon." *Proceedings of the American Philosophical Society* 98, no. 4 (1954): 279–81.

Slevin, Joseph Richard. "The Galápagos Islands: A History of Their Exploration." *Occasional Papers of the California Academy of Sciences* 25 (1949): 1–150.

Slotkin, Richard. "Nostalgia and Progress: Theodore Roosevelt's Myth of the Frontier." *American Quarterly* 33 (1981): 608–35.

Smith, Alice Kimball. *A Peril and a Hope: The Scientists' Movement in America, 1945–1946.* Chicago: University of Chicago Press, 1965.

Smith, Charles H., ed. *Alfred Russel Wallace: An Anthology of His Shorter Writings.* Oxford: Oxford University Press, 1991.

Smith, Dick, and Robert Easton. *California Condor: Vanishing America; a Study of an Ancient and Symbolic Giant of the Sky.* Charlotte, N.C.: McNally and Loftin, 1964.

Smith, Michael L. *Pacific Visions: California Scientists and the Environment, 1850–1915.* New Haven, Conn.: Yale University Press, 1987.

Smith, Peter H. *Talons of the Eagle: Dynamics of U.S.-Latin American Relations.* New York: Oxford University Press, 1996.

Smith, Thomas G. "John Kennedy, Steward Udall, and the New Frontier Conservation." *Pacific Historical Review* 64 (1995): 329–62.

Smits, David. "The Frontier Army and the Destruction of the Buffalo, 1865–1883." *Western Historical Quarterly* 25 (1994): 312–38.

Snyder, Noel, and Helen Snyder. *The California Condor: A Saga of Natural History and Conservation.* San Diego: Academic Press, 2001.

———. "California Condor (*Gymnogyps californianus*)." In *The Birds of North America,* edited by A. Poole and F. Gill, no. 610. Philadelphia: Birds of North America, 2002.

Snyder, Noel F. R. *The Carolina Parakeet: Glimpses of a Vanished Bird.* Princeton, N.J.: Princeton University Press, 2004.

Soulé, Michael E. "What Is Conservation Biology?" *BioScience* 35, no. 11 (December 1985): 727–34.

———. "History of the Society for Conservation Biology: How and Why We Got Here." *Conservation Biology* 1, no. 1 (May 1987): 4–5.

Soulé, Michael E., and Bruce A. Wilcox, eds. *Conservation Biology: An Evolutionary-Ecological Perspective.* Sunderland, Mass.: Sinauer Associates, 1980.

Souder, William. *Under a Wild Sky: John James Audubon and the Making of the Birds of America.* New York: North Point Press, 2004.

Spence, Mark. *Dispossessing the Wilderness: Indian Removal and the Making of National Parks.* New York: Oxford University Press, 1999.

Star, Susan Leigh, and James R. Griesemer. "Institutional Ecology, 'Translations' and Boundary Objects: Amateurs and Professionals in Berkeley's Museum of Vertebrate Zoology, 1907–39." *Social Studies of Science* 19 (1989): 387–420.

Steen, Harold K. *The U.S. Forest Service: A History.* Seattle: University of Washington Press, 1976.

———, ed. *Forest and Wildlife Science in America.* Durham, N.C.: Forest History Society, 1999.

Stein, Barbara R. *On Her Own Terms: Annie Montague Alexander and the Rise of Science in the American West.* Berkeley: University of California Press, 2001.

Steinberg, Ted. *Nature Incorporated: Industrialization and the Waters of New England.* Cambridge: Cambridge University Press, 1991.

———. *Down to Earth: Nature's Role in American History.* New York: Oxford University Press, 2002.

Sterling, Keir B. "Builders of the U.S. Biological Survey, 1885–1930." *Journal of Forest History* 33 (1989): 180–87.

———. "Zoological Research, Wildlife Management, and the Federal Government." In *Forest and Wildlife Science in America,* edited by Harold K. Steen, 19–65. Durham, N.C.: Forest History Society, 1999.

Stern, Minna Alexandra. *Eugenic Nation: The Faults and Frontiers of Better Breeding in Modern America.* Berkeley: University of California Press, 2005.

Stine, Jeffery K. "Natural Resources and Environmental Policy." In *The Reagan Presidency: Pragmatic Conservatism and Its Legacies,* edited by W. Elliot Brownlee and Hugh Davis Graham, 233–56. Lawrence: University Press of Kansas, 2003.

Stott, R. Jeffrey. "The Historical Origins of the Zoological Park in American Thought." *Environmental Review* 5 (1981): 52–65.

Strickland, Hugh E., and Alexander G. Melville. *The Dodo and Its Kindred; or the History, Af-*

finities, and Osteology of the Dodo, Solitaire, and Other Extinct Birds of the Islands Mauritius, Rodriguez, and Bourbon. London: Reeve, Benham, and Reeve, 1848.

Stroud, Richard, ed. National Leaders of American Conservation. Washington, D.C.: Smithsonian Institution Press, 1985.

Sulloway, Frank J. "Darwin's Conversion: The Beagle Voyage and Its Aftermath." Journal of the History of Biology 15 (1982): 325–96.

———. "Darwin and His Finches: The Evolution of a Legend." Journal of the History of Biology 15 (1982): 1–53.

———. "Darwin's Early Intellectual Development: An Overview of the Beagle Voyage (1831–1836)." In The Darwinian Heritage, edited by David Kohn, 121–54. Princeton, N.J.: Princeton University Press, 1985.

Sumner, Francis B. "Some Perils Which Confront Us as Scientists." Scientific Monthly 8, no. 3 (1919): 258–74.

———. "The Need for a More Serious Effort to Rescue a Few Fragments of Vanishing Nature." Scientific Monthly 10, no. 3 (1920): 236–48.

———. "The Responsibility of the Biologist in the Matter of Preserving Natural Conditions." Science 54, no. 1385 (1921): 39–43.

———. The Life History of an American Naturalist. Lancaster, Penn.: Jacques Cattell Press, 1945.

Sutter, Paul S. Driven Wild: How the Fight against Automobiles Launched the Modern Wilderness Movement. Seattle: University of Washington Press, 2002.

Sutton, George Miksch. Birds in the Wilderness: Adventures of an Ornithologist. New York: Macmillan, 1936.

Systematics Agenda 2000. Systematics Agenda 2000: Charting the Biosphere. New York: Society of Systematic Biologists, American Society of Plant Taxonomists, Willi Hennig Society, and Association of Systematics Collections, 1994.

Takacs, David. The Idea of Biodiversity: Philosophies of Paradise. Baltimore: Johns Hopkins University Press, 1996.

Talbot, Lee Merriam. A Look at Threatened Species: A Report on Some Animals of the Middle East and Southern Asia Which Are Threatened with Extermination. London: Fauna Preservation Society for the International Union for the Conservation of Nature, 1960.

Tanner, James T. "Three Years with the Ivory-Billed Woodpecker." Audubon Magazine 43 (1941): 5–14.

———. The Ivory-Billed Woodpecker. Research Report no. 1. New York: National Audubon Society, 1942.

Taylor, Joseph. Making Salmon: An Environmental History. Seattle: University of Washington Press, 1999.

Teale, Edwin Way. "Arthur A. Allen: Ten Thousand Bird Students Have Learned from Him." Audubon Magazine 45 (1943): 84–89.

Tennyson, Jon R. A Singleness of Purpose: The Story of Ducks Unlimited. Chicago: Ducks Unlimited, 1977.

Terres, John K. "Smithsonian 'Bird Man': A Biographical Sketch of Alexander Wetmore." Audubon Magazine 50 (1948): 161–67.

Thoreau, Henry David. The Journal of Henry David Thoreau. Edited by Bradford Torrey and Francis H. Allen. Boston: Houghton Mifflin, 1949.

Tjossem, Sara F. "Preservation of Nature and Academic Respectability: Tensions in the Ecological Society of America, 1915–1979." Ph.D. dissertation, Cornell University, 1994.

Tober, James A. *Who Owns the Wildlife? The Political Economy of Conservation in Nineteenth-Century America.* Westport, Conn.: Greenwood Press, 1981.

Tobey, Ronald C. *Saving the Prairies: The Life Cycle of the Founding School of American Plant Ecology, 1895–1955.* Berkeley: University of California Press, 1981.

Townsend, Charles Haskins. "Impending Extinction of the Galápagos Tortoises." *Zoological Society Bulletin* 27, no. 2 (1924): 55–56.

———. "The Galápagos Tortoises in Their Relation to the Whaling Industry: A Study of Old Logbooks." *Zoologica* 4, no. 3 (1925): 55–135.

———. "The Whaler and the Tortoise." *Scientific Monthly* 21, no. 2 (1925): 166–72.

———. "Old Times with the Birds: Autobiographical." *The Condor* 29, no. 5 (1927): 224–32.

———. "The Galápagos Islands Revisited." *Bulletin of the New York Zoological Society* 31 (1928): 148–69.

———. "Giant Tortoises: Nearing Extinction in the Galápagos, They Are Being Propagated in the United States." *Scientific American* 144 (1931): 42–44.

Train, Russell E. *Politics, Pollution, and Pandas: An Environmental Memoir.* Washington, D.C.: Island Press, 2003.

Trefethen, James B. *An American Crusade for Wildlife.* New York: Winchester Press and the Boone and Crockett Club, 1975.

Trumblack, Stephen C., ed. *So Great a Vision: The Conservation Writings of George Perkins Marsh.* Hanover, N.H.: Middlebury College Press, 2001.

Tsutsui, William M. "Looking Straight at Them! Understanding the Big Bug Movies of the 1950s." *Environmental History* 12 (2007): 237–53.

Turner, James. *Reckoning with the Beast: Animals, Pain, and Humanity in the Victorian Mind.* Baltimore: Johns Hopkins University Press, 1980.

Twiss, John R., Jr., and Randall R. Reeves, eds. *Conservation and Management of Marine Mammals.* Washington, D.C.: Smithsonian Institution Press, 1999.

Udall, Stewart L. *The Quiet Crisis.* New York: Holt, Rinehart and Winston, 1963.

U.S. Fish and Wildlife Service. *Protecting Our Endangered Birds.* Circular No. 61. Washington, D.C.: Government Printing Office, 1959.

———. *Survival or Surrender for Our Endangered Wildlife.* Washington, D.C.: Government Printing Office, 1965.

Van Name, Willard G. "Zoological Aims and Opportunities." *Science* 50, no. 1282 (1919): 81–84.

Van Wissen, Ben, ed. *Dodo: Raphus cucullatus* [Didus ineptus]. Amsterdam: ISP/Zoölogisch Museum-Universiteit van Amsterdam, 1995.

Vesey, Laurence R. *The Emergence of the American University.* Chicago: University of Chicago Press, 1965.

Vincent, Jack. *Aves.* Red Data Book 2. Morges, Switzerland: IUCN, Species Survival Commission, 1966.

Vogt, William. *Road to Survival.* New York: W. Sloane Associates, 1948.

Von Hagen, Victor Wolfgang. *Ecuador the Unknown: Two and a Half Years' Travels in the Republic of Ecuador and the Galápagos Islands.* London: Jarrods Publishers, 1939.

———. *Ecuador and the Galápagos Islands.* Norman: University of Oklahoma Press, 1949.

Wallace, Alfred Russel. *Island Life, or the Phenomena and Causes of Insular Faunas and Floras, Including a Revision and Attempted Solution to the Problem of Geological Climates.* 3rd ed. London: Macmillan, 1911.

Walls, Laura Dassow. *Seeing New Worlds: Henry David Thoreau and Nineteenth-Century Natural Science*. Madison: University of Wisconsin Press, 1995.

Warren, Louis S. *The Hunter's Game: Poachers and Conservationists in Twentieth-Century America*. New Haven, Conn.: Yale University Press, 1997.

Weart, Spencer. *Nuclear Fear: A History of Images*. Cambridge, Mass.: Harvard University Press, 1988.

Weidensaul, Scott. *Of a Feather: A Brief History of American Birding*. Orlando, Fla.: Harcourt, 2007.

Weiner, Jerome. *The Beak of the Finch: A Story of Evolution in Our Time*. New York: Alfred A. Knopf, 1994.

Welch, Margaret. *The Book of Nature: Natural History in the United States, 1825–1875*. Boston: Northeastern University Press, 1998.

Welker, Robert Henry. *Natural Man: The Life of William Beebe*. Bloomington: Indiana University Press, 1975.

Whitaker, John C. *Striking a Balance: Environment and Natural Resources Policy in the Nixon-Ford Years*. Washington, D.C.: American Enterprise Institute for Public Policy Research, 1976.

Wilbur, Sanford R. *The California Condor, 1966–76: A Look at Its Past and Future*. North American Fauna, no. 72. Washington, D.C.: U.S. Department of Interior, Fish and Wildlife Service, 1978.

Wilcove, David S., and Thomas Eisner. "The Impending Extinction of Natural History." *Chronicle Review*, September 15 2000, B24.

Wilson, Alexander. *Wilson's American Ornithology*. Reprint ed. New York: Arno and the New York Times, 1970.

Wilson, Edward O., ed. *Biodiversity*. Washington, D.C.: National Academy Press, 1988.

———. *Naturalist*. Washington, D.C.: Island Press, 1994.

———. "On the Future of Conservation Biology." *Conservation Biology* 14, no. 1 (February 2000): 1–3.

Wilson, Molly Clare. "Wild Goose Chase: The Communal Science of Waterfowl Migration Study in North America, 1880–1950." Senior honors thesis, Harvard University, 2006.

Winkler, Allan M. *Life under a Cloud: American Anxiety about the Atom*. New York: Oxford University Press, 1993.

Wonders, Karen. *Habitat Dioramas: Illusions of Wilderness in Museums of Natural History*. Figura Nova Series 25. Uppsala: Acta Universitatis Upsaliensis, 1993.

Wood, Casey A. "Lessons in Aviculture from English Aviaries." *The Condor* 28 (1926): 1–30.

Wood, William. *New England's Prospect*. Edited by Alden Vaughan. Amherst: University of Massachusetts Press, 1977. Originally published 1634.

Worster, Donald. *Nature's Economy: A History of Ecological Ideas*. 2nd ed. Cambridge: Cambridge University Press, 1994.

Worthy, Trevor H., and Richard N. Holdaway. *The Lost World of the Moa: Prehistoric Life of New Zealand*. Bloomington: Indiana University Press, 2002.

Wright, Albert H. "Biology at Cornell University." *Bios* 24 (1953): 122–45.

Wright, George M., Joseph S. Dixon, and Ben H. Thompson. *A Preliminary Survey of Faunal Relations in National Parks*. Washington, D.C.: Government Printing Office, 1933.

Wright, George M., and Ben H. Thompson. *Wildlife Management in the National Parks*. Washington, D.C.: Government Printing Office, 1935.

Wright, R. Gerald. *Wildlife Research and Management in the National Parks.* Urbana: University of Illinois Press, 1992.

Yaffee, Steven Lewis. *Prohibitive Policy: Implementing the Endangered Species Act.* Cambridge, Mass.: MIT Press, 1982.

———. *The Wisdom of the Spotted Owl: Policy Lessons for a New Century.* Washington, D.C.: Island Press, 1994.

Young, Christian C. *In the Absence of Predators: Conservation and Controversy on the Kaibab Plateau.* Lincoln: University of Nebraska Press, 2002.

Ziswiler, Vinzenz. *Extinct and Vanishing Animals: A Biology of Extinction and Survival.* New York: Springer-Verlag, 1967.

INDEX

Page references followed by f refer to the figures.

American Committee for International Wild
   Life Protection (*continued*)
   Islands conservation, 172–83, 193–94;
   initially focuses on Africa, 148–52, 167;
   and International Committee for Bird
   Preservation, 149, 194; and New York
   Zoological Society, 162; and Pan American
   Convention, 188–93, 194–97, 347; reliance
   on science, 166; resolution on collect-
   ing rare species, 154–56; turns to Latin
   America, 167
American Fish Culturists' Association, 92
American Fisheries Society, 92
American Game Protective and Propagation
   Association, 132, 242
American Humane Association, 101
American identity: shaped by nature, 9; as
   argument for bison protection, 120
American incognitum: American Philo-
   sophical Society seeks bones of, 35; Cuvier
   names, 41–42; Hunter on, 29; Jefferson's
   account of, 17, 18; as refutation of Buffon,
   17; Turner on, 32; Peale on, 38–39. *See also*
   mammoth; mastodon
American Museum of Natural History:
   African Hall, 145–46; Allen joins, 80; hosts
   American Ornithologists' Union (AOU)
   bird protection committee, 101; naturalists
   affiliated with, petition AOU, 238; praised
   for involvement in wildlife conservation,
   132; Whitney South Seas Expedition,
   152–54
American Ornithologists' Union (AOU): bird
   protection committee, 101–3, 104, 105;
   model bird law, 102–3, 104, 105; and northern spotted owl, 351; petitioned by naturalists of American Museum of Natural
   History to oppose Alaskan eagle bounty,
   238; raptor resolution, 245; resolution on
   collecting rare species, 156
American Ornithology (Wilson), 88
American Philosophical Society, 30–32; home
   to Peale's Museum, 34; issues call for intact
   skeleton of American incognita, 35
American Society for the Prevention of Cruelty to Animals, 83

American Society of Mammalogists: and con-
   troversy over predator control, 223, 239;
   Harper reports to, 161
American Society of Taxidermists, 110
Amory, Cleveland, 330
Andean condor, 297
Andrews Bald, 230–31
anglers, 90–91
animistic view of nature, 85
AOU. *See* American Ornithologists' Union
Arabian oryx, 313–15
Aransas Migratory Wildlife Refuge, 303–4,
   319, 341
argument from design, 20
artificial propagation, 90–91. *See also* captive
   breeding
Ashmolean Museum, 54
Asian lion, 313
Asian rhinoceros, 313
Asiatic Primate Expedition, 156
Association of Systematics Collections, 354
Astor, Victor, 174–75
Astor Expedition, 174–75
atomic bomb, 306–7
Atomic Energy Commission. *See* Division of
   Biology and Medicine (Atomic Energy
   Commission)
"Attempt to List the Extinct and Vanishing
   Birds of the Western Hemisphere, An"
   (Phillips), 159
Attwater's prairie chicken, 319
Atwood, Wallace W., 188, 189
Audubon, John James, 64f; and California
   condor, 285; and Carolina parakeet,
   127f, 128; Cuvier praises, 88; decries
   bird slaughter, 64; depicts great auk, 62f;
   on ivorybill, 275; nationalistic impulses
   inform, 88; on passenger pigeon, 97, 98;
   travels to Labrador, 63–65
Audubon movement: Chapman's role in,
   11; early call for, 83; early history of,
   100–104; revival of, 104–6; expansion of,
   107; focuses on nongame birds, 141. *See
   also* Audubon Society; National Audubon
   Society
Audubon Society: early activities, 100;

founded by Grinnell, 103–4. *See also*
National Audubon Society

aurochs, 53, 82–83

Autobiography of a Bird-Lover (Chapman), 10

automobile, concern about impact of, 202,
219, 228

aviculture: and Carolina parakeet, 128–29;
Peter Scott and, 315–16; and rare birds,
305, 321; and rare parrots, 129–30; and
whooping crane, 304, 306. *See also* captive
breeding

Babbitt, Bruce, 353–54

Bachman, John, 88

Bailey, Vernon, 223

Baird, Spencer F., 92, 109, 114

Baker, John: attempts to merge International
Committee for Bird Preservation (ICBP)
and American Committee, 194; attempts
to protect Singer Tract, 282–83; and
California condor, 288–92, 294, 295–96;
corresponds with Leopold, 278; creates
graduate fellowship program, 263, 279–80,
290–91, 292; on ecological perspectives in
conservation, 233; reforms National Association of Audubon Societies, 253–55, 278

balance of nature, 21–23; birds maintain,
102, 143; Blumenbach rejects, 40; humans
upset, 101, 191, 241, 254; Jefferson invokes,
31; Leopold on, 228; Lyell on, 45; in national parks, 217; predators maintain, 224,
254; predator control disrupts, 271

bald eagle, 235f; Alaska bounty on, 238–39,
243, 246, 248, 249, 258; National Association of Audubon Societies funds study of,
254; number of states protecting, 257, 259.
See also Bald Eagle Protection Act (1940)

Bald Eagle Protection Act (1940), 234–35,
236, 319; amended to include Alaska, 259;
passes, 257–58; Pearson campaigns for,
249–50; second attempt at, 255

Baldwin, S. Prentiss, 156

Bangs, Outram, 158

Barbour, Thomas: advises the American
Committee, 148; begins catalog of rare
species, 159; encourages Museum of Comparative Zoology (MCZ) to collect rare
species, 158; and heath hen, 268; reports on
Harper report, 162

Bard, Fred, 304

Barro Colorado Island, 177

Barton, Benjamin Smith, 42–43

Bartsch, Paul 129

Bates, Henry Walter, 73

Bavin, Clark, 339

Baynes, Ernest Harold, 118–20, 119f

Baysinger, Earl, 339

Beard, Daniel B., 231–32

Beebe, William: compares nature to work
of art, 7; on Galápagos Islands, 172;
leads campaign to locate passenger pigeon,
124

Belgium Congo: Akeley travels to, 146;
Coolidge travels to, 135–37; Johnson captures live gorilla in, 154

Bell, W. B., 223

Bendire, Charles, 128

Bergh, Henry, 83

Berringer, Johann Bartholomew Adam,
25–26

Bessey, Charles, 206

Bible, authority of, 23

Big Bone Lick, Kentucky, 17, 27, 33, 39

biodiversity, term coined, 353

Biological Resources Division of U.S. Geological Survey, 354

biologists, and nature preservation, 201–2,
211, 212. *See also* natural areas; naturalists

biology: emergence of, 203; evolution helps
synthesize, 76; laboratory revolution in,
205; threatens to eclipse natural history,
205; training in, 204, 205. *See also* conservation biology; ecology; natural history

bioprospecting, 3, 355

biotic pyramid, 208, 228–29

"Biotic View of the Land, A" (Leopold),
228–29

Bird-Lore, 107

bird of paradise, 73

bird protection movement. *See* Audubon
movement

bird sanctuaries, fail to protect raptors, 241

on, 220–21. *See also* Committee on the Preservation of Natural Conditions; Hawk Mountain Sanctuary; national forests; national parks; Sespe Condor Sanctuary; Sisquoc Condor Sanctuary; wilderness; wildlife refuges

natural history: broadens scope, 262, 274; global species inventory as focus of, 48; graduate training in, 5, 205–6, 262, 274; history of, 3–6; importance of field experience in, 357–58; marginalized, 5; methods of, used to craft endangered species lists, 166; museums, 3; and nationalism, 9, 86–87; "new natural history," 204; as patriotic duty, 33; popular interest in, 3, 4–5; popularization of, 4–5; reform of, 205–6; specialization in, 4; taxonomy as focus of, x, 4, 5, 109, 205; Wilson calls for more emphasis on, 358. *See also* biology; conservation biology; ecology; systematics; taxonomy

Natural History of Carolina, Florida and the Bahama Islands (Catesby), 28, 264

Naturalist's Guide to the Americas (Shelford, et al.), 212–13

naturalists: criticized for lack of commitment to wildlife conservation, 132, 211; early role in wildlife conservation, 101, 133, 346; emotional attachment to nature, 10, 133, 359; and endangered species, 11; engagement with extinction, 3, 5, 6, 13, 299, 341, 358–59; focus on game species, 133; focus on North American species, 134; and heath hen, 262; importance of field experience in, 358–59; initially indifferent to extinction, 6, 45, 54, 346; plea for engagement of, in preservation, 201–2; and raptor protection, 236, 238; reject migratory bird treaty with Latin American nations, 170–71; resist wildlife laws, 11; seek rare and endangered species, 11, 158; and spotted cat protection, 329. *See also names of individual naturalists*

natural theology, 20–21

nature, contact with, x, 4, 121, 130, 206, 443n68. *See also* strenuous life

nature, emotional bond with: birdwatching promotes, 236, 243, 258; naturalists experience, 10, 133, 358; public experiences, 100

Nature Conservancy, 233

nature faking, 150

nature lovers: call for asserting rights of, 245; as part of Audubon coalition, 105; and raptors, 236, 244; and wildlife conservation, 262

nature protection: aesthetic arguments, 7, 87, 102, 121, 147, 192, 202, 211, 212, 219, 236 (*see also* aesthetics); cultural arguments, 9, 120, 147; ecological arguments, 8, 102, 202–3, 204, 210–11, 212, 213, 228, 233, 347; economic arguments, 7, 102, 191–92; ethical arguments, 9–10, 147, 211, 225, 242, 291–92, 310, 320; evolutionary arguments, 8, 76; inventory of endangered landscapes as first step in, 212–13; scientific arguments, 9, 202–3, 212. *See also* nationalism; nostalgia

nature sanctuaries. *See* Committee on the Preservation of Natural Conditions; Hawk Mountain Sanctuary; national forests; national parks; natural areas; Sespe Condor Sanctuary; Sisquoc Condor Sanctuary; wilderness; wildlife refuges

Neck, Jacob Cornelius van, 52

Nelson, E. W.: as Goldman's mentor, 184–85; on Latin American bird treaties, 171, 183–84

nene, 316f, 319, 321

Newton, Alfred: interest in dodo, 56; studies great auk, 65–66

Newton, Edward, 56

New York Audubon Society, 107

New York Zoological Society: American Committee relocates to, 194; Cooper donates California condor to, 286; creation of, 113–14, 115–16, 117; Hornaday praises, 132; recruits Hornaday, 113, 115–16; role in saving bison, 11, 188, 119–20, 122, 123; sponsors Galápagos expeditions, 174–75

New Zealand, 58–61

Nicholson, Max, 315–16

Nixon, Richard M., 2, 328, 335, 339–40

noble savage, 86

Randolph, Jennings, 255
raptors: calls for eradication of, 237; 240f; campaign to garner support for, 235, 242–46; compared with mammalian predators, 258–59; ECC defends, 247–48; efforts to control, 239–42, 240f; feeding habits studied, 238–39; Hawk and Owl Society seeks support for, 250–51; killed during seasonal migrations, 252f; Leopold opposes control of, 255; Migratory Bird Treaty fails to protect, 238; National Association of Audubon Societies campaigns for, 253–55, 258; Pearson devotes more attention to, 248–50; symbolism associated with, 258–59. *See also* bald eagle
Raven, Peter, 353
Ray, John: argues number of species constant, 23; on balance of nature, 22; on fossils, 25; and natural theology, 20
Red Rock Lakes National Waterfowl Refuge, 319
red wolf, 329, 342
Redington, Paul, 223
Reed, Charles K., 124
Reed, Chester, 124
Reed, Nathaniel, 338
relict species, 76, 295, 296
Reunion solitary, 51, 373n18
Ridgway, Robert, 129
rights of nature, 10
right whale, 334
Ripley, S. Dillon, 304, 315, 320, 321
Robinson, Cyril, 289, 295
Rockefeller, Laurance, 197
Roderiguez solitary, 51, 373n18
Rogers, Kathleen, 199–20
romanticism: and bison protection, 120; Catlin influenced by, 94; diffusion of, 346; and predator control, 236; rise of, 86–87
Roosevelt, Franklin D., 167, 282, 303
Roosevelt, Kermit, 145f; on African safari, 144; and American Committee, 148; and Astor Expedition, 174
Roosevelt, Theodore, 145f; advocates strenuous life, 88, 116; African expedition, 144; creates wildlife refuges, 106, 119, 123;

founds Boone and Crockett Club, 116; and Hornaday, 112; as president of American Bison Society, 120
roseate spoonbill, 295
Rosen, Walter G., 353
Rothschild, Miriam, 158
Rothschild, Walter: collects Galápagos Islands species, 172; interest in rare species, 158
Rousseau, Jean-Jacques, 86
Rowe, L. S., 189, 190
Royal Society of London, 27
Rudwick, Martin, 24, 25, 41
Rule, John, 57
Russell, C. P., 193

sage grouse, 280
Sand Country Almanac, A (Leopold), 225, 318
sandhill crane, 301, 320, 321, 322
San Diego Zoo, 297, 350
Sandy (bison calf), 110, 111f, 114
Santa Barbara National Forest. *See* Los Padres National Forest
Saunders, Aretas A., 244
Savery, John, 54
Savery, Roland, 54, 56
Save These Birds! (Miller et al.), 245–46
Schorger, A. W., 96
Schultz, Adolph, 156
Schweitzer, Albert, 310
Scott, Peter, 315–17, 316f, 321
Scripps Institution of Oceanographic Research, 204
sea otter, 196
Sea World, 331
sei whale, 334
Semonin, Paul, 33
Sennett, George B., 101
Sergeant, David, 333
Sespe Condor Sanctuary, 294–95, 298, 351
Seton, Ernest Thompson, 245
Shelbourne, Lord, 28
Sheldon, H. H., 295
Sheldon, William G., 193, 196
Shelford, Victor, 210f, 359; background, 209–10; and beginnings of animal ecology, 207–8; and Committee on the Preserva-

tion of Natural Conditions, 204–5, 209, 212–14, 215, 216–21, 223–25, 232–33; decries specialization, 220; first president of Ecological Society of America, 208; helps found Ecologists' Union, 233; organizes nature sanctuary conference, 220; on science and conservation policy, 211

Sheridan, Philip: discourages protection for bison, 95; visits Hornaday, 111–12

Sherman, William T., 95

Shields, G. O., 105

Shreve, Forrest, 212

Shufeldt, R. W., 127

Silcox, Ferdinand, 289, 290

Silent Spring (Carson), 308–10, 318

Simba, King of the Beasts, 146

Simon, Noel, 317

Simpson, George Gaylord, 28

Singer Tract, 277f; Allen visits, 276–77; failure to protect, 300, 319; National Audubon Society seeks preservation of, 282–83; Tanner studies ivorybill at, 280–81

Sisquoc Condor Sanctuary, 288–90, 298

Smithsonian Institution, 340, 353. *See also* U.S. National Museum; U.S. National Zoo

Smits, David, 95

snail darter, 342, 349

snow leopard, 330

Snyder, Helen, 284, 300

Snyder, Noel, 284, 300

Society for Conservation Biology, 357

Society for the Preservation of Wild Fauna of the Empire (SPWFE), 156; activities of, 147–48; American Committee intends to support, 149; and Galápagos Islands conservation, 193–94

Society for the Protection of Native Plants, 12

Society of American Foresters, 228

Soulé, Michael, 10–11, 356, 357, 359

species, intrinsic value of, 10

species inventories: focus of natural history, x, 4, 6, 48, 80; and nationalism, 33, 87; renewed call for, 354–55, 358; Strickland urges speed in creating, 9, 54. *See also* endangered species inventories

species protection. *See* nature protection

species senescence, 69

sperm whale, 334

spinner dolphins, 332–33

sportsmen: decline in numbers of, 259; and heath hen, 262; Hornaday praises, 131; promote game laws, 82; and raptor control, 239–41; and wildlife conservation, 89, 105. *See also* Boone and Crockett Club; Society for the Preservation of Wild Fauna of the Empire (SPWFE)

spotted cats, 328–31, 340f

spotted dolphins, 332–33

spotted owl. *See* northern spotted owl

SPWFE. *See* Society for the Preservation of Wild Fauna of the Empire (SPWFE)

St. Louis Committee for Nuclear Information, 308

Steno, Nicolas, 25

Stockholm Conference, 337

Stoddard, Herbert, 274

strenuous life, 88, 90, 116, 147

Strickland, Hugh, 53–56

Strong, Robert, 150

strontium, 90, 308, 310

Stuart, John, 29–30

Sumatran rhinoceros, 313–14

Sumner, Francis B., 9, 201–3, 203f

superabundance of North American wildlife, 84, 85f, 231; challenges to, 93, 237

Survival Service Commission: and Arabian oryx, 313–15; and CITES, 337; commissions Talbot to survey wildlife in Middle East and Asia, 312–13; Coolidge's central role in, 311–12; crafts captive breeding guidelines, 313–14; creation of, 311; and ICBP, 312; publishes *Les Fossiles de demain*, 312; Scott leads, 316–17

Sustainable Biosphere Initiative, 355

Sutter, Paul, 228

Sutton, George M., 277–78

Swarth, Harry S., 175–76

Synopsis of Quadrupeds (Pennant), 29

Syrian wild ass, 313

systematics: concern about declining status of, 354; Ray's knowledge of, 20; Wetmore's contributions to, 170; Wilson's call for

Wilderness Society, 228

Wildflower Preservation Society, 12

Wild Kingdom, 325

wildlife: Allen charts decline of in North
America, 80–82; compared to work of art,
87, 236; myth of inexhaustibility of, 84;
sportsmen as protectors of, 89–90. *See
also* wildlife conservation; wildlife films;
wildlife legislation; wildlife management;
wildlife refuges; wildlife treaties; *names of
individual species*

wildlife conservation: as Anglo Saxon duty,
149, 167; charismatic species as focus
of, 134; debate over scope and goals of,
262–63; and ecology, 202, 205, 230–33, 255,
347; game species as focus of, 133, 231, 279;
individual species as focus of, 202, 220,
232, 347; and nativism, 117, 131, 143; and
paternalism, 149, 166–67, 188; and racism,
131, 166–67; and science, 254, 263, 283, 348;
sentiment as basis for, 160. *See also* wildlife
legislation; wildlife management; wildlife
refuges; wildlife treaties

wildlife films, 325; *Flipper,* 332; *Ingagi,* 150;
nature faking in, 150; *Simba, King of the
Beasts,* 146

wildlife legislation: Allen calls for, 82; for
American alligator, 324; Bald Eagle
Protection Act, 234–35, 236, 249–50,
255, 257–58, 259; based on common law
tradition, 13; for California condor, 286;
during colonial period, 86; for heath hen,
265, 266; Lacey Act, 105, 141, 324, 326;
Land and Water Conservation Fund Act,
320; Marine Mammal Protection Act,
335–36; Mason Act, 329; Migratory Bird
Act, 132; for nongame birds, 102–3, 104,
105–6; Norbeck-Andresen Migratory
Bird Conservation Bill, 248; for passenger
pigeon, 100; restricts scientific collecting,
133; sportsmen promote, 82, 90; Texas
considers protecting bison, 95; Vandegrift
Tariff, 178; in Yellowstone, 116. *See also*
American Ornithologists' Union (AOU):
model bird law; Endangered Species Act

(1973); Endangered Species Conservation
Act (1969); Endangered Species Preservation Act (1966)

wildlife management: calls for basing on science, 254, 274, 347–48; debate over scope
and goals of, 262–63; ecological perspective in, 8, 227; game species as focus of,
274; Leopold helps establish field of,
226–27; period of professionalization, 274.
See also wildlife conservation; wildlife legislation; wildlife refuges; wildlife treaties

Wild Life Protective Association, 132

wildlife refuges: Allen calls for, 82; Aransas
Migratory Wildlife Refuge, 303–4, 319,
341; first federal, 106; Grays Lake National
Wildlife Refuge, 344; Harper calls for, 161;
Monte Vista National Wildlife Refuge,
305; National Bison Range, 123; National
Key Deer Refuge, 319; National Wildlife
Refuge System, 322; Pelican Island, 106;
Red Rock Lakes National Waterfowl Ref-
uge, 319; Sespe Condor Sanctuary, 294–95,
298; Sisquoc Condor Sanctuary, 288–90,
298; Wichita National Game Reserve,
119–20, 121–22; Wind Cave National
Park, 123. *See also* natural areas

Wildlife Society, 226–27, 274

wildlife treaties: CITES, 336–38; Convention
for the Protection of Migratory Birds and
Game Mammals, 183–84, 187–88, 347;
Convention on Biological Diversity,
355; covering fish, 140; International Whaling Convention, 334; London Convention
(1900), 151; London Convention (1933),
150–53; Migratory Bird Treaty, 141–42,
193, 238; North Pacific Fur Seal Treaty,
141; Pan American Convention, 171,
188–97, 198–200, 347; U.S. Senate calls for,
with Latin American countries, 168

Williams, William, 59–60

Wilson, Alexander: as example of nationalism, 88; on ivorybill, 275; on passenger
pigeon, 97

Wilson, Edward O., 353, 358

Wind Cave National Park, 123